SCIENCE and JUDGMENT
in Risk Assessment

Committee on Risk Assessment of Hazardous Air Pollutants

Board on Environmental Studies and Toxicology

Commission on Life Sciences

National Research Council

NATIONAL ACADEMY PRESS
Washington, D.C.   1994

NATIONAL ACADEMY PRESS • 2101 Constitution Ave., N.W. • Washington, D.C. 20418

NOTICE: The project that is the subject of this report was approved by the Governing Board of the National Research Council, whose members are drawn from the councils of the National Academy of Sciences, the National Academy of Engineering, and the Institute of Medicine. The members of the committee responsible for the report were chosen for their special competencies and with regard for appropriate balance.

This report has been reviewed by a group other than the authors according to procedures approved by a Report Review Committee consisting of members of the National Academy of Sciences, the National Academy of Engineering, and the Institute of Medicine.

The project was supported by the U.S. Environmental Protection Agency under contract CR818293-01-0.

Library of Congress Cataloging-in-Publication Data

Science and judgment in risk assessment / Committee on Risk Assessment
    of Hazardous Air Pollutants, Board on Enviornmental Studies and
    Toxicology, Commission on Life Sciences,, National Research Council.
        p. cm
    Includes bibliographal references and index.
    ISBN 0-309-04894-X
    1. Air—Pollution—Toxicology—United States—Statistical methods.
    2. Health risk assessment—Statistical methods. I. National
Research Council (U.S.). Committee on Risk Assessment of Hazardous
Air Pollutants.
RA576.S365   1994
363.73'92'0973—dc20   94-17475
                                                                CIP

Printed in the United States of America

COMMITTEE ON RISK ASSESSMENT
OF HAZARDOUS AIR POLLUTANTS

KURT J. ISSELBACHER *(Chairman)*, Massachusetts General Hospital, Charlestown, Mass.

ARTHUR C. UPTON *(Vice-Chairman)*, New York University Medical Center (retired), N.Y.

JOHN C. BAILAR, McGill University School of Medicine, Montreal, Canada

KENNETH B. BISCHOFF, University of Delaware, Newark, Del.

KENNETH T. BOGEN, Lawrence Livermore National Laboratory, Livermore, Calif.

JOHN I. BRAUMAN, Stanford University, Calif.

DAVID D. DONIGER, Natural Resources Defense Council, Washington, D.C.*

JOHN DOULL, University of Kansas Medical Center, Kansas City, Kan.

ADAM M. FINKEL, Resources for the Future, Washington, D.C.

CURTIS C. HARRIS, National Cancer Institute, Bethesda, Md.

PHILIP K. HOPKE, Clarkson University, Potsdam, N.Y.

SHEILA S. JASANOFF, Cornell University, Ithaca, N.Y.

ROGER O. MCCLELLAN, Chemical Industry Institute of Toxicology, Research Triangle Park, N.C.

LINCOLN E. MOSES, Stanford University, Calif.

D. WARNER NORTH, Decision Focus, Inc., Mountain View, Calif.

CRAIG N. OREN, Rutgers University School of Law, Camden, N.J.

REBECCA T. PARKIN, Beccam Services, Plainsboro, N.J.

EDO D. PELLIZZARI, Research Triangle Institute, N.C.

JOSEPH V. RODRICKS, Environ Corporation, Arlington, Va.

ARMISTEAD G. RUSSELL, Carnegie-Mellon University, Pittsburgh, Penn.

JAMES N. SEIBER, University of Nevada, Reno, Nev.

STEVEN N. SPAW, Law Environmental Incorporated, Austin, Tex.

JOHN D. SPENGLER, Harvard University, Boston, Mass.

BAILUS WALKER, University of Oklahoma, Oklahoma City, Okla.

HANSPETER WITSCHI, University of California, Davis, Calif.

Staff

RICHARD D. THOMAS, Program Director

DEBORAH D. STINE, Study Director

MARVIN A. SCHNEIDERMAN, Senior Staff Scientist

GAIL CHARNELY, Senior Staff Officer

KATHLEEN STRATTON, Senior Staff Officer

RUTH E. CROSSGROVE, Information Specialist

ANNE M. SPRAGUE, Information Specialist

RUTH P. DANOFF, Project Assistant

SHELLEY A. NURSE, Senior Project Assistant

CATHERINE M. KUBIK, Senior Project Assistant

---

*Left committee in May 1993 upon becoming Deputy Director of the White House Office of Environmental Quality

v

The National Academy of Sciences is a private, non-profit, self-perpetuating society of distinguished scholars engaged in scientific and engineering research, dedicated to the furtherance of science and technology and to their use for the general welfare. Upon the authority of the charter granted to it by the Congress in 1863, the Academy has a mandate that requires it to advise the federal government on scientific and technical matters. Dr. Bruce M. Alberts is president of the National Academy of Sciences.

The National Academy of Engineering was established in 1964, under the charter of the National Academy of Sciences, as a parallel organization of outstanding engineers. It is autonomous in its administration and in the selection of its members, sharing with the National Academy of Sciences the responsibility for advising the federal government. The National Academy of Engineering also sponsors engineering programs aimed at meeting national needs, encourages education and research, and recognizes the superior achievements of engineers. Dr. Robert M. White is president of the National Academy of Engineering.

The Institute of Medicine was established in 1970 by the National Academy of Sciences to secure the services of eminent members of appropriate professions in the examination of policy matters pertaining to the health of the public. The Institute acts under the responsibility given to the National Academy of Sciences by its congressional charter to be an adviser to the federal government and, upon its own initiative, to identify issues of medical care, research, and education. Dr. Kenneth I. Shine is president of the Institute of Medicine.

The National Research Council was organized by the National Academy of Sciences in 1916 to associate the broad community of science and technology with the Academy's purposes of furthering knowledge and advising the federal government. Functioning in accordance with general policies determined by the Academy, the Council has become the principal operating agency of both the National Academy of Sciences and the National Academy of Engineering in providing services to the government, the public, and the scientific and engineering communities. The Council is administered jointly by both Academies and the Institute of Medicine. Dr. Bruce Alberts and Dr. Robert M. White are chairman and vice chairman, respectively, of the National Research Council.

Preface

In the Clean Air Act Amendments of 1990, Congress directed the administrator of the Environmental Protection Agency (EPA) to engage the National Academy of Sciences (NAS) in a review of the methods that EPA uses to estimates toxicological risk. The resulting charge to the National Research Council (NRC) can be summarized in a short set of questions:

1. Given that quantitative risk assessment is essential for EPA's implementation of the Clean Air Act, is EPA conducting risk assessments in the best possible manner?

2. Has EPA developed mechanisms for keeping its risk-assessment procedures current in the face of new developments in science?

3. Are adequate risk-related data being collected to permit EPA to carry out its mandates?

4. What, if anything, should be done to improve EPA's development and use of risk assessments?

To meet the congressional mandate, and in response to the request from the administrator of EPA, the National Research Council established the Committee on Risk Assessment of Hazardous Air Pollutants under the the Board on Environmental Studies and Toxicology. The committee consisted of 25 members with expertise in medicine, epidemiology, chemistry, chemical engineering, environmental health, law, pharmacology and toxicology, risk assessment, risk management, occupational health, statistics, air monitoring, and public health. It included academics, industry scientists, public advocates, and state and local public-health officials.

The first meeting of the committee was held on October 31, 1991. In the first several meetings, presentations were made to the committee by committee members and by individuals or representatives of groups with special concerns in the development and use of risk assessment. Among the latter were presenters on behalf of the American Industrial Health Council, the Chemical Manufacturers Association, the American Petroleum Institute, the American Iron and Steel Institute, the American Chemical Society, such official public-health groups as the Texas Air Control Board and the State and Territorial Air Pollution Program Administrators, and such public-interest groups as the Natural Resources Defense Council and the Environmental Defense Fund. Presentations were also made by the representative of a paint manufacturer and by a senior member of an environmental consulting company. The committee also was greatly aided by the previous reports and workshops of the NRC's Committee on Risk Assessment Methodology.

Early in the course of its deliberations the committee developed a set of issues for consideration and reply by EPA's Office of Air and Radiation and its Office of Research and Development. EPA's responses were presented to the committee during the committee's meetings in late March 1992.

James Powell, of the U.S. Senate staff, described to the committee both the legislative history of the Clean Air Act Amendments and the concerns of senators in the evolution of EPA's development of regulations. Greg Wetstone, of the U.S. House of Representatives staff, spoke to the committee about the need for accurate risk assessments and exposure measures. Henry Habicht, Michael Shapiro, Robert Kellum, and William Farland of EPA discussed where EPA was in risk assessment and how it got there. Their briefings enabled the committee to get off to a quick start in its work.

The committee was substantially helped in its activities by strong support from the NRC and BEST staff: Richard D. Thomas, the program director; Deborah D. Stine, the study director; Marvin A. Schneiderman, senior staff scientist; Norman Grossblatt, editor; Anne M. Sprague, information specialist; Ruth E. Crossgrove, information specialist; Ruth P. Danoff, project assistant; and Shelley A. Nurse and Catherine M. Kubik, senior project assistants.

Finally, we must express our thanks and appreciation to the hard-working members of the committee, who struggled through long meetings, read mountains of documents, listened with interest and concern to many presentations, and then prepared what we consider to be a thoughtful, comprehensive, and balanced report.

Kurt Isselbacher, M.D.
Chairman

Arthur Upton, M.D.
Vice Chairman

Contents

SCIENCE
and
JUDGMENT
in Risk Assessment

Executive Summary

In recent decades, the public has become increasingly aware of seemingly innumerable reports of health threats from the environment. Myriad announcements about pesticides in food, pollutants in the air, chemical contaminants in drinking water, and hazardous-waste sites have created public concern about the chemical products and byproducts of modern industrial society. Alongside that concern is public skepticism about the reliability of scientific predictions concerning possible threats to human health. The skepticism has arisen in part because scientists disagree. But it is also apparent that many people want to understand the methods for assessing how much their exposures to chemicals threaten their health and well-being.

Many environmental issues that have risen to public prominence involve carcinogens—substances that can contribute to the development of cancer. Sometimes the decision that a substance is a carcinogen is based on evidence from workers exposed to high concentrations in the workplace, but more often it is based on evidence obtained in animals exposed to high concentrations in the laboratory. When such substances are found to occur in the general environment (even in much lower concentrations), efforts are made to determine the exposed population's risk of developing cancer, so that rational decisions can be made about the need for reducing exposure. However, scientists do not have and will not soon have reliable ways to measure carcinogenic risks to humans when exposures are small. In the absence of an ability to measure risk directly, they can offer only indirect and somewhat uncertain estimates.

Responses to these threats, often reflected in legislation and regulations, have led to reduced exposures to many pollutants. In recent years, however, concerns have arisen that the threats posed by some regulated substances might have been overstated and, conversely, that some unregulated substances might pose greater threats than originally believed. Questions have also been raised about the economic costs of controlling or eliminating emissions of chemicals that might pose extremely small risks. Debates about reducing risks and controlling costs have been fed by the lack of universal agreement among scientists about which methods are best for assessing risk to humans.

Epidemiological studies—typically, comparisons of disease rates between exposed and unexposed populations—are not sufficiently precise to find that a substance poses a carcinogenic risk to humans except when the risk is very high or involves an unusual form of cancer. For this reason, animal studies generally provide the best means of assessing potential risks to humans. However, laboratory animals are usually exposed to toxicants at concentrations much higher than those experienced by humans in the general population. It is not usually known how similar the toxic responses in the test animals are to those in humans, and scientists do not have indisputable ways to measure or predict cancer risks associated with small exposures, such as those typically experienced by most people in the general environment.

Some hypotheses about carcinogens are qualitative. For example, biological data might suggest that any exposure to a carcinogen poses some health risk. Although some scientists disagree with that view or believe that it is not applicable to every carcinogen, its adoption provides at least a provisional answer to a vexing scientific question, namely whether people exposed to low concentrations of substances that are known to be carcinogenic at high concentrations are at *some* risk of cancer associated with the exposure. The view has dominated policy-making since the 1950s but is not always consistent with new scientific knowledge on the biological mechanisms of chemically induced cancer.

Beginning in the 1960s, toxicologists developed quantitative methods to estimate the risks associated with small exposures to carcinogens. If it were reliable, quantitative risk assessment could improve the ability of decision-makers and to some extent the public to discriminate between important and trivial threats and improve their ability to set priorities, evaluate tradeoffs among pollutants, and allocate public resources accordingly. In short, it could improve regulatory decisions that affect public health and the nation's economy.

During the 1970s and 1980s, methods of risk assessment continued to evolve, as did the underlying science. It became increasingly apparent that the process of carcinogenesis was complex, involving multiple steps and pathways. The concept that all cancer-causing chemicals act through mechanisms similar to those operative for radiation was challenged. Some chemicals were shown to alter DNA directly and hence to mimic radiation. But evidence developed that other chemicals cause cancer without directly altering or damaging DNA, for example, through hormonal pathways, by serving as mitogenic stimuli, or by causing excess cell death with compensatory cell proliferation. Biologically based and pharmacokinetic models were introduced in some cases to describe exposure-response relationships more accurately. During the same period, substantial advances were made in modeling the dispersion of airborne materials from sources to receptors and in conducting exposure assessments. Furthermore, important advances have been made in the last 10 years in understanding the basic biology of chemical toxicity. All these advances are beginning to have a major impact on the estimation of risks associated with hazardous air pollutants.

REGULATION OF HAZARDOUS AIR POLLUTANTS

Before the enactment of the Clean Air Act Amendments of 1990 (1990 Amendments), Section 112 of the Clean Air Act required that the Environmental Protection Agency (EPA) set emission standards for hazardous air pollutants "to protect the public health with an ample margin of safety." In 1987, the District of Columbia Circuit Court of Appeals, in *Natural Resources Defense Council v. EPA* (824 F.2d 1146) interpreted this language to mean that EPA must first determine the emissions level that is safe—one that represents an acceptable degree of risk—and then add a margin of safety in light of the uncertainties in scientific knowledge about the pollutant in question. The agency was permitted to consider technological feasibility in the second step but not in the first.

In response, EPA decided that it would base its regulatory decisions largely on quantitative risk assessment. The agency adopted a general policy that a lifetime cancer risk of one in 10,000 for the most exposed person might constitute acceptable risk and that the margin of safety should reduce the risk for the greatest possible number of persons to an individual lifetime risk no higher than one in 1 million (10^{-6}).

The 1990 Amendments rewrote Section 112 to place risk assessment in a key role but one secondary to technology-based regulation. As altered, Section 112 defines a list of substances as hazardous air pollutants, subject to addition or deletion by EPA. Sources that emit hazardous air pollutants will be regulated in two stages. In the first, technology-based emissions limits will be imposed. Each major source of hazardous air pollutants must meet an emission standard, to be issued by EPA, based on using the maximum achievable control technology (MACT). Smaller sources, known as area sources, must meet emissions standards based on using generally available control technology.

In the second stage, EPA must set "residual-risk standards that protect public health with an ample margin of safety if it concludes that the technology-based standards have not done so." The establishment of a residual-risk standard is required if the MACT emission standard leaves a lifetime cancer risk for the most exposed person of greater than one in a million. In actually setting the standard, though, EPA is free to continue to use its present policy of accepting higher risks. Quantitative risk assessment techniques will be relevant to this second stage of regulation, as well as to various decisions required in the first stage.

CHARGE TO THE STUDY COMMITTEE

Section 112(o) of the Act (quoted in full in Appendix M) directs the EPA to arrange for the National Academy of Sciences to:

• Review the methods used by EPA to determine the carcinogenic risk associated with exposure to hazardous air pollutants from sources subject to Section 112;

• Include in its review evaluations of the methods used for estimating the carcinogenic potency of hazardous air pollutants and for estimating human exposures to these air pollutants;

• Evaluate, to the extent practicable, risk-assessment methods for noncancer health effects for which safe thresholds might not exist.

The Academy's report must be considered by EPA in revising its present risk assessment guidelines.

CURRENT RISK-ASSESSMENT PRACTICES

Methods for estimating risk to humans exposed to toxicants have evolved steadily over the last few decades. Not until 1983, however, was the process codified in a formal way. In that year, the National Research Council released *Risk Assessment in the Federal Government: Managing the Process.* This publication, now known also as the Red Book, provided many of the definitions used throughout the environmental-health risk-assessment community today. The Red Book served as the basis for the general description of risk assessment used by the present committee.

Risk assessment entails the evaluation of information on the hazardous properties of substances, on the extent of human exposure to them, and on the characterization of the resulting risk. Risk assessment is not a single, fixed method of analysis. Rather, it is a systematic approach to organizing and analyzing scientific knowledge and information for potentially hazardous activities or for substances that might pose risks under specified conditions.

In brief, according to the Red Book, risk assessment can be divided into four steps: hazard identification, dose-response assessment, exposure assessment, and risk characterization.

• *Hazard identification* involves the determination of whether exposure to an agent can cause an increased incidence of an adverse health effect, such as cancer or birth defects, and characterization of the nature and strength of the evidence of causation.

• *Dose-response assessment* is the characterization of the relationship between exposure or dose and the incidence and severity of the adverse health effect. It includes consideration of factors that influence dose-response relationships, such as intensity and pattern of exposure and age and lifestyle variables that could affect susceptibility. It can also involve extrapolation of high-dose responses to low-dose responses and from animal responses to human responses.

• *Exposure assessment* is the determination of the intensity, frequency, and duration of actual or hypothetical exposures of humans to the agent in question. In general, concentrations of the substance can be estimated at various points from its source through the environment. An important component of exposure assessment is emission characterization, i.e., determination of the magnitude and properties of the emissions that result in exposures. This is usually accomplished by measuring and analyzing emissions, but that is not always possible. Therefore, modeling is often used instead to establish the relationship between emissions and environmental concentrations of the substance. Inputs to such a model should include data on residence and activities of the exposed population.

• *Risk characterization* combines the assessments of exposure and response under various exposure conditions to estimate the probability of specific harm to an exposed individual or population. To the extent feasible, this characterization should include the distribution of risk in the population. When the distribution of risk is known, it is possible to estimate the risk to individuals who are most exposed to the substance in question.

Closely related to risk assessment is risk management, the process by which the results of risk assessment are integrated with other information—such as political, social, economic, and engineering considerations—to arrive at decisions about the need and methods for risk reduction. The authors of the Red Book advocated a clear conceptual distinction between risk assessment and risk management, noting, for instance, that maintaining the distinction between the two would help to prevent the tailoring of risk assessments to the political feasibility of regulating the substance in question. But they also recognized that the choice of risk-assessment techniques could not be isolated from society's risk-management goals. The result should be a process that supports the risk-management decisions required by the Clean Air Act and that provides appropriate incentives for further research to reduce important uncertainties on the extent of health risks.

In 1986, EPA issued risk-assessment guidelines that were generally consistent with the Red Book recommendations. The guidelines deal with assessing risks of carcinogenicity, mutagenicity, developmental toxicity, and effects of chemical mixtures. They include default options, which are essentially policy judgments of how to accommodate uncertainties. They include various assumptions that are needed for assessing exposure and risk, such as scaling factors to be used for converting test responses in rodents to estimated responses in humans.

As risk-assessment methods have evolved and been applied with increasing frequency in federal and state regulation of hazardous substances, regulated industries, environmental organizations, and academicians have leveled a broad array of criticisms regarding the processes used by EPA. The concerns have included

• The lack of scientific data quantitatively relating chemical exposure to health risks.

• The divergence of opinion within the scientific community on the merits of the underlying scientific evidence.

• The lack of conformity among reported research results needed for risk characterization—e.g., the use of different methods for describing laboratory findings, which makes it difficult to compare the data from different laboratories and apply them in risk characterizations.

• The uncertainty of results produced by theoretical modeling, which is used in the absence of measurements.

• In response to its mandates, EPA has traditionally adopted risk assessments that for the most part incorporate conservative default options (i.e., those that are more likely to overstate than to understate human risk).

• As scientific knowledge increases, the science policy choices made by the agency and Congress should have less impact on regulatory decision-making. Better data and increased understanding of biological mechanisms should enable risk assessments that are less dependent on conservative default assumptions and more accurate as predictions of human risk.

STRATEGIES FOR RISK ASSESSMENT

The committee observed that several common themes cut across the various stages of risk assessment and arise in criticisms of each individual step. These themes are as follows:

• *Default options.* Is there a set of clear and consistent principles for modifying and departing from default options?

• *Data needs.* Is enough information available to EPA to generate risk assessments that are protective of public health and are scientifically plausible?

• *Validation.* Has the EPA made a sufficient case that its methods and models for carrying out risk assessments are consistent with current scientific information available?

• *Uncertainty.* Has EPA taken sufficient account of the need to consider, describe, and make decisions in light of the inevitable uncertainty in risk assessment?

• *Variability.* Has EPA sufficiently considered the extensive variation among individuals in their exposures to toxic substances and in their susceptibilities to cancer and other health effects?

• *Aggregation.* Is EPA appropriately addressing the possibility of interactions among pollutants in their effects on human health, and addressing the consideration of multiple exposure pathways and multiple adverse health effects?

By addressing each of those themes in each step in the risk-assessment process, EPA can improve the accuracy, precision, comprehensibility, and utility of the entire risk-assessment process in regulatory decision making.

Flexibility and the Use of Default Options

EPA's risk-assessment guidelines contain a number of "default options." These options are used in the absence of convincing scientific knowledge on which of several competing models and theories is correct. The options are not rules that bind the agency; rather, they constitute guidelines from which the agency may depart when evaluating the risks posed by a specific substance. For the most part, the defaults are conservative (i.e., they represent a choice that, although scientifically plausible given existing uncertainty, is more likely to result in overestimating than underestimating human risk).

EPA has acted reasonably in electing to formulate guidelines. EPA should have principles for choosing default options and for judging when and how to depart from them. Without such principles, the purposes of the default options could be undercut. The committee has identified a number of criteria that it believes ought to be taken into account in formulating such principles: protecting the public health, ensuring scientific validity, minimizing serious errors in estimating risks, maximizing incentives for research, creating an orderly and predictable process, and fostering openness and trustworthiness. There might be additional relevant criteria.

The choice of such principles goes beyond science and inevitably involves policy choices on how to balance such criteria. After extensive discussion, the committee found that it could not reach consensus on what the principles should be or on whether it was appropriate for this committee to recommend principles. Thus, the committee decided not to do so. Appendix N contains papers by several committee members containing varied perspectives on the appropriate choice of principles. Appendix N-1 advocates the principle of "plausible conservatism" and N-2 advocates the principle of the maximum use of scientific information in selection of default options. These papers do not purport to represent the views of all committee members.

The committee did agree, though, that EPA often does not clearly articulate in its risk-assessment guidelines that a specific assumption is a default option and that EPA does not fully explain in its guidelines the basis for each default option. Moreover, EPA has not stated all the default options in the risk-assessment process or acknowledged where defaults do not exist.

EPA's practice appears to be to allow departure from a default option in a specific case when it ascertains that there is a consensus among knowledgeable scientists that the available scientific evidence justifies departure from the default option. The agency relies on its Scientific Advisory Board and other expert bodies to determine when such a consensus exists. But EPA has not articulated criteria for allowing departures.

Recommendations

• EPA should continue to regard the use of default options as a reasonable way to deal with uncertainty about underlying mechanisms in selecting methods and models for use in risk assessment.

• EPA should explicitly identify each use of a default option in risk assessments.

• EPA should clearly state the scientific and policy basis for each default option.

• The agency should consider attempting to give greater formality to its criteria for a departure from default options, in order to give greater guidance to the public and to lessen the possibility of ad hoc, undocumented departures from default options that would undercut the scientific credibility of the agency's risk assessments. At the same time, the agency should be aware of the undesirability of having its guidelines evolve into inflexible rules.

• EPA should continue to use the Science Advisory Board and other expert bodies. In particular, the agency should continue to make the greatest possible use of peer review, workshops, and other devices to ensure broad peer and scientific participation to guarantee that its risk-assessment decisions will have access to the best science available through a process that allows full public discussion and peer participation by the scientific community.

Validation: Methods and Models

Some methods and models used in emission characterization, exposure assessment, hazard identification, and dose-response assessment are specified as default options. Others are sometimes used as alternatives to the default options. The predictive accuracy and uncertainty of these methods and models for risk assessment are not always clearly understood or clearly explained.

A threshold model (i.e., one that assumes that exposures below some level will not cause health effects) is generally accepted for reproductive and developmental toxicants, but it is not known how accurately it predicts human risk. The fact that current evidence on some toxicants, most notably lead, does not clearly reveal a safe threshold has raised concern that the threshold model might reflect the limits of scientific knowledge, rather than the limits of safety.

EPA has worked with outside groups to design studies to refine emission estimates. However, it does not have guidelines for the use of emission estimates in risk assessment, nor does it adequately evaluate the uncertainty in the estimates.

EPA has relied on Gaussian-plume models to estimate the concentrations of hazardous pollutants to which people are exposed. These representations of airborne transport processes are approximations. EPA focuses primarily on stationary outdoor emission sources of hazardous air pollutants. It does not have a specific statutory mandate to consider all sources of hazardous air pollutants, but this should not deter the agency from assessing indoor sources to provide perspective in considering risks from outdoor sources.

EPA uses the Human-Exposure Model (HEM) to evaluate exposures from stationary sources. It estimates exposures and risk for both individuals and populations. For individuals, it has traditionally used a technique to determine what is called the maximally exposed individual (MEI) by estimating the highest exposure concentration that might be found among the broad distribution of possible exposures. Estimation of the maximum exposure is based on a variety of conservative assumptions, e.g., that the MEI lives directly downwind from the pollution source for his or her entire 70-year lifetime and remains outdoors the entire time. Traditionally, only exposure by inhalation is considered. Recently, in accordance with recommendations of the agency's Science Advisory Board, EPA has begun to replace the MEI estimate with two others: the high-end exposure estimate (HEEE) and the theoretical upper-bound exposure (TUBE).

In dose-response assessment, EPA has traditionally treated almost all chemical carcinogens as inducing cancer in a similar manner, mimicking radiation. It assumes that a linearized multistage model can be used to extrapolate from epidemiological observations (e.g., occupational studies) or experimental observations at high doses in laboratory animals down to the low doses usually experienced by humans in the general population.

Recommendations

• EPA should more rigorously establish the predictive accuracy and uncertainty of its methods and models and the quality of data used in risk assessment.

• EPA should develop guidelines for the amount and quality of emission information required for particular risk assessments and for estimating and reporting uncertainty in emission estimates, e.g., the predictive accuracy and uncertainty associated with each use of the HEM for exposure assessment.

• EPA should evaluate the Gaussian-plume models under realistic conditions of acceptable distances (based on population characteristics) to the site boundaries, complex terrain, poor plant dispersion characteristics, and the presence of other structures in the vicinity. Furthermore, EPA should consider incorporating such state-of-the-art techniques as stochastic-dispersion models.

• EPA should use a specific conservative mathematical technique to estimate the highest exposure likely to be encountered by an individual in the exposure group of interest.

• EPA should use bounding estimates for screening assessments to determine whether further levels of analysis are necessary. For further analyses, the committee supports EPA's development of distributions of exposures based on actual measurements, results from modeling, or both.

• EPA should continue to explore and, when scientifically appropriate, incorporate pharmacokinetic models of the link between exposure and biologically effective dose (i.e., dose reaching the target tissue).

• EPA should continue to use the linearized multistage model as a default option but should develop criteria for determining when information is sufficient to use an alternative extrapolation model.

• EPA should develop biologically based quantitative methods for assessing the incidence and likelihood of noncancer effects in human populations resulting from chemical exposure. These methods should incorporate information on mechanisms of action and differences in susceptibility among populations and individuals that could affect risk.

• EPA should continue to use as one of its risk-characterization metrics, upper-bound potency estimates of the probability of developing cancer due to lifetime exposure. Whenever possible, this metric should be supplemented with other descriptions of cancer potency that might more adequately reflect the uncertainty associated with the estimates.

Priority-Setting and Data Needs

EPA does not have the exposure and toxicity data needed to establish the health risks associated with all 189 chemicals identified as hazardous air pollutants in the 1990 Amendments. Furthermore, EPA has not defined how it will determine the types, quantities, and quality of data that are needed to assess the risks posed by facilities that emit any of those 189 chemicals or how it will determine when site-specific emission and exposure data are needed.

Recommendations

• EPA should compile an inventory of the chemical, toxicological, clinical, and epidemiological literature on each of the 189 chemicals identified in the 1990 Amendments.

• EPA should screen the 189 chemicals to establish priorities according to procedures described by the committee for assessing health risks, identify data gaps, and develop incentives to expedite the generation of data by other government agencies (e.g., the National Toxicology Program, the Agency for Toxic Substances and Disease Registry, and state agencies), industry, and academe.

• In addition to stationary sources of hazardous air pollutants, EPA should consider mobile and indoor sources; the latter might be even more important than outdoor sources. The agency should also explicitly consider all direct and indirect routes of exposure, such as ingestion and dermal absorption.

• EPA should develop a two-part scheme for classifying evidence on carcinogenicity that would incorporate both a simple classification and a narrative evaluation. At a minimum, both parts should include the strength (quality) of the evidence, the relevance of the animal model and results to humans, and the relevance of the experimental exposures (route, dose, timing, and duration) to those likely to be encountered by humans.

Variability

Many types of variability enter into the risk-assessment process: variability within individuals, among individuals, and among populations. Types of variability include nature and intensity of exposure and susceptibility to toxic insult related to age, lifestyle, genetic background, sex, ethnicity, and other factors.

Interindividual variability is not generally considered in EPA's cancer risk assessments. The agency's consideration of variability has been limited largely to noncarcinogenic effects, such as asthmatic responses to sulfur dioxide exposure. Analyses of such variability usually form the basis of decisions about whether to protect both the general population and sensitive individuals.

Recommendations

• Federal agencies should sponsor molecular, epidemiological, and other types of research to examine the causes and extent of interindividual variability in susceptibility to cancer and the possible correlations between susceptibility and such covariates as age, race, ethnicity, and sex. Results should be used to refine estimates of risks to individuals and the general population.

• EPA should adopt a default assumption for differences in susceptibility among humans in estimating individual risks.

• EPA should increase its efforts to validate or improve the default assumption that humans on average have the same susceptibility as humans in epidemiological studies, the most sensitive animals tested, or both.

• EPA's guidelines should clearly state a default assumption of nonthreshold, low-dose linearity for genetic effects on which adequate data might exist (e.g., data on chromosomal aberrations or dominant or X-linked mutations) so that a reasonable quantitative estimate of genetic risk to the first and later generations can be made for environmental chemical exposure.

• The distinction between uncertainty and individual variability should be maintained rigorously in each component of risk assessment.

• EPA should assess risks to infants and children whenever it appears that their risks might be greater than those of adults.

Uncertainty

There are numerous gaps in scientific knowledge regarding hazardous air pollutants. Hence, there are many uncertainties in risk assessment. When the uncertainty concerns the magnitude of a quantity that can be measured or inferred from assumptions, such as exposure, the uncertainty can be quantified. Other uncertainties pertain to the models being used. These stem from a lack of knowledge needed to determine which scientific theory is correct for a given chemical and population at risk and thus which assumptions should be used to derive estimates. Such uncertainties cannot be quantified on the basis of data.

The upperbound point estimate of risk typically computed by EPA does not convey the degree of uncertainty in the estimate. Thus, decision-makers do not know the extent of conservatism, if any, that is provided in the risk estimate.

Formal uncertainty analysis can help to inform EPA and the public about the extent of conservatism that is embedded in the default assumptions. Uncertainty analysis is especially useful in identifying where additional research is likely to resolve major uncertainties.

Uncertainty analysis should be an iterative process, moving from the identification of generic uncertainties to more refined analyses for chemical-specific or industrial plant-specific uncertainties. The additional resources needed to conduct the more specific analyses can be justified when the health or economic impacts of the regulatory decision are large and when further research is likely to change the decision.

Recommendations

• EPA should conduct formal uncertainty analyses, which can show where additional research might resolve major uncertainties and where it might not.

• EPA should consider in its risk assessments the limits of scientific knowledge, the remaining uncertainties, and the desire to identify errors of either overestimation or underestimation.

• EPA should develop guidelines for quantifying and communicating uncertainty (e.g., for models and data sets) as it occurs into each step in the risk-assessment process.

• Despite the advantages of developing consistent risk assessments between agencies by using common assumptions (e.g., replacing surface area with body weight to the 0.75 power), EPA should indicate other methods, if any, that might be more accurate.

• When ranking risks, EPA should consider the uncertainties in each estimate, rather than ranking solely on the basis of point estimate value. Risk managers should not be given only a single number or range of numbers. Rather, they should be given risk characterizations that are as robust (i.e., complete and accurate) as can be feasibly developed.

Aggregation

Typically, people at risk are exposed to a mixture of chemicals, each of which might be associated with an increased probability of one or more health effects. In such cases, data are often available on only one of the adverse effects (e.g., cancer) associated with each chemical. At issue is how best to characterize and estimate the potential aggregate risk posed by exposure to a mixture of toxic chemicals. Furthermore, emitted substances might be carried to and deposited on other media, such as water and soil, and cause people to be exposed via routes other than inhalation, e.g., by dermal absorption or ingestion. EPA has not yet indicated whether it will consider multiple exposure routes for regulation under the 1990 Amendments, although it has done so in other regulatory contexts, e.g., under Superfund.

EPA adds the risks related to each chemical in a mixture in developing its risk estimate. This is generally considered appropriate when the only risk characterization needed is a point estimate for use in screening. When a more comprehensive uncertainty characterization is desired, EPA should adopt the following recommendations.

Recommendations

• EPA should consider using appropriate statistical (e.g., Monte Carlo) procedures to aggregate cancer risks from exposure to multiple compounds.

• In the analysis of animal bioassay data on the occurrence of multiple tumor types, the cancer potencies should be estimated for each relevant tumor type that is related to exposure, and the individual potencies should be summed for those tumors.

• Quantitative uncertainty characterizations conducted by EPA should appropriately reflect the difference between uncertainty and interindividual variability.

Communicating Risk

Certain expressions of probability are subjective, whether qualitative (e.g., that a threshold might exist) or quantitative (e.g., that there is a 90% probability that a threshold exists). Although quantitative probabilities could be useful in conveying the judgments of individual scientists to risk managers and to the public, the process of assessing probabilities is difficult. Because substantial disagreement and misunderstanding concerning the reliability of single numbers or even a range of numbers can occur, the basis for the numbers should be set forth clearly and in detail.

Recommendation

• Risk managers should be given characterizations of risk that are both qualitative and quantitative, i.e., both descriptive and mathematical.

An Iterative Approach

Resources and data are not sufficient to perform a full-scale risk assessment on each of the 189 chemicals listed as hazardous air pollutants in the 1990 Amendments, and in many cases no such assessment is needed. After MACT is applied, it is likely that some of the chemicals will pose only de minimis risk (a risk of adverse health effects of one in a million or less). For these reasons, the committee believes that EPA should undertake an iterative approach to risk assessment. An iterative approach would start with relatively inexpensive screening techniques—such as a simple, conservative transport model—and then for chemicals suspected of exceeding de minimis risk move on to more resource-intensive levels of data-gathering, model construction, and model application. To guard against serious underestimations of risk, screening techniques must err on the side of caution when there is uncertainty about model assumptions or parameter values.

Recommendations

• EPA should develop the ability to conduct iterative risk assessments that would allow improvements to be made in the estimates until (1) the risk is below the applicable decision-making level, (2) further improvements in the scientific knowledge would not significantly change the risk estimate, or (3) EPA, the emission source, or the public determines that the stakes are not high enough to warrant further analysis. Iterative risk assessments would also identify needs for further research and thus provide incentives for regulated parties to undertake research without the need for costly, case-by-case evaluations of each individual chemical. Iteration can improve the scientific basis of risk-assessment decisions while responding to risk-management concerns about such matters as the level of protection and resource constraints.

OVERALL CONCLUSIONS AND RECOMMENDATIONS

The committee's findings are dominated by four central themes:

• Because of limitations on time, resources, scientific knowledge, and available data, EPA should generally retain its conservative, default-based approach to risk assessment for screening analysis in standard-setting; however, several corrective actions are needed to make this approach more effective.

• EPA should develop and use an iterative approach to risk assessment. This will lead to an improved understanding of the relationship between risk assessment and risk management and an appropriate blending of the two.

• The iterative approach proposed by the committee allows for improvements in the default-based approach by improving both models and the data used in analysis. For this approach to work properly, however, EPA needs to provide justification for its current defaults and establish a procedure that permits departures from the default options.

- When EPA reports estimates of risk to decision-makers and the public, it should present not only point estimates of risk, but also the sources and magnitudes of uncertainty associated with these estimates.

Risk assessment is a set of tools, not an end in itself. The limited resources available should be spent to generate information that helps risk managers to choose the best possible course of action among the available options.

1

Introduction

In recent decades, there have been seemingly innumerable reports of health threats from the environment. Myriad announcements about pesticides in food, pollutants in the air, chemical contaminants in drinking water, and hazardous-waste sites have created public concern about the chemical products and byproducts of modern industrial society. Alongside that concern exists skepticism about many of the possible threats to human health. The skepticism has arisen in part because scientists disagree. But it is also apparent that most people want to understand whether and how much their exposures to chemicals threaten their health and well-being.

Many environmental issues that have risen to public prominence involve carcinogens—substances that can contribute to the development of cancer. Sometimes the decision that a substance is a carcinogen is based on evidence from workers exposed to high concentrations in the workplace, but more often it is based on evidence obtained in animals exposed to high concentrations in the laboratory. When such substances are found to occur in the general environment (even in much lower concentrations), efforts are made to determine the exposed population's risk of developing cancer, so that rational decisions can be made about the need for reducing exposure. However, scientists do not have and will not soon have reliable ways to measure carcinogenic risks when exposures are small. In the absence of an ability to measure risk directly, they can offer only indirect and somewhat uncertain estimates.

Some hypotheses about carcinogens are qualitative. For example, biological data suggests that any exposure to a carcinogen may pose some health risk. Although some scientists disagree with that view or believe that it is not applicable to every carcinogen, its adoption provides at least a provisional answer to a vexing scientific question, namely whether people exposed to low concentrations of substances that are known to be carcinogenic at high concentrations are at *some* risk of cancer associated with the exposure. That view has been prominent since the 1950s and has guided much decision-making. For example, the "Delaney clause" of the Food Additive Amendments of 1958 stipulated that no additive that was found to be carcinogenic could be allowed in the food supply, on the grounds that it was not possible to specify a safe human exposure to such an agent. The policies that have flowed from regulations like the Delaney clause involve, where possible, absolute prohibition of exposures to carcinogens, but more commonly, reductions of exposures to the "lowest technically feasible level."

A qualitative response to the question of carcinogenic risk is still viewed by many scientists to be the best that can now be offered, even in the face of impressive scientific advances in understanding chemical carcinogenesis. Nonetheless, it is increasingly recognized that division of the binary division of the world of chemicals into carcinogens and non-carcinogens is overly simplistic and does not provide an adequate basis for regulatory decision-making. Beginning in the 1960s and coming to full force in the 1970s, some scientists have attempted to offer more useful, quantitative information about the risks of low exposures to carcinogens. Quantitative risk assessment is attractive because, at least ideally, it allows decision-makers and the public to discriminate between important and trivial threats (thus going beyond qualitative findings that there is some risk, however small).

The results of risk assessments are important in influencing important regulatory decisions that affect both the nation's economy and public health. They influence decision-makers as they attempt to balance the view that emission of hazardous air pollutants should be minimized or even eliminated, versus the view that meeting stringent control standards might cause other problems unacceptable to society. Accurate risk assessments are also needed to determine whether public health protection is adequate.

CHARGE TO THE COMMITTEE

The charge to the committee comes from Section 112(o) of the Clean Air Act, as added by the Clean Air Act Amendments of 1990, which requires EPA to enter into a contract with the National Research Council (NRC). NRC created the Committee on Risk Assessment of Hazardous Air Pollutants in the Board on Environmental Studies and Toxicology. Its charge is summarized as follows:

1. Review the risk assessment methods used by EPA (Environmental Protection Agency).

2. Evaluate methods used for estimating the carcinogenic potencies of hazardous air pollutants.

3. Evaluate methods used for estimating human exposures to hazardous air pollutants.

4. To the extent practicable, evaluate risk-assessment methods for noncancer health effects for which safe thresholds might not exist.

5. Indicate revisions needed in EPA's risk-assessment guidelines.

The specific congressional language is provided in Appendix M. Section 112(o) requires that if EPA decides not to comply with all of the report's recommendations and the Science Advisory Board's views of the report, it must provide a detailed explanation in the *Federal Register* of the reasons that any of the recommendations in the report are not implemented.

In its charge to EPA, Congress assigned NRC the task of evaluating whether EPA's risk-assessment methods express in a scientifically supportable way the risks posed by a substance. We therefore ask whether EPA's methods are consistent with current scientific knowledge. We also ask whether EPA's methods give policy-makers and the public the information they need to make judgments about risk management. Such methods should be logical and consistent and should, in particular, reveal the inevitable uncertainties in the underlying science.

We make no judgment regarding the appropriate risk-management decision, e.g., the extent to which society should control hazardous air pollutants. Such decisions ultimately hinge on nonscientific issues; for instance, the extent of risk from hazardous air pollutants that society is willing to accept in return for other benefits. Such issues involve not only science or science-policy judgments, but also matters of value on which scientists cannot purport to have any special insight. Such issues are therefore ultimately the province of policy-makers and the public.

It was precisely for this reason, we believe, that Congress specified in the Clean Air Act Amendments of 1990 that this committee is to undertake an investigation of EPA's risk-assessment methods, rather than of the validity of EPA's regulatory decisions. We have therefore refrained from addressing such risk management issues. We do, however, note that risk assessment and risk management are integrally related. As we explain later, Congress has generally directed EPA to be protective of health ("conservative" in the lexicon of public health) in its risk-management decisions. It is therefore essential for us to appraise whether EPA's risk-assessment methods are capable of supporting a policy of protective public-health regulation.

In addition, in its charge to EPA, Congress indicated that noncancer effects should be addressed to the extent feasible, but time constraints reduced the committee's ability to focus fully on this issue.

Section 303 of the 1990 Amendments created the Risk Assessment and Management Commission, part of whose charge is to examine risk-management policy issues. Specific subjects that the commission is to address are

- The report of the NRC committee.
- The use and limitations of risk assessment in establishing emission or effluent standards, ambient standards, exposure standards, acceptable concentrations, tolerances, or the environmental criteria for hazardous substances that present a risk of carcinogenic or other chronic health effects and the suitability of risk assessment for such purposes.
- The most appropriate methods for measuring and describing cancer risks or risks of other chronic health effects associated with exposure to hazardous substances.
- Methods to reflect uncertainties in measurement and estimation techniques, the existence of synergistic or antagonistic effects among hazardous substances, the accuracy of extrapolating animal-exposure data to human health risks, and the existence of unquantified direct or indirect effects on human health in risk-assessment studies.
- Risk-management policy issues, including the use of lifetime cancer risks to the people most exposed, the incidence of cancer, the cost and technical feasibility of exposure-reduction measures, and the use of site-specific actual exposure information in setting emission standards and other limitations applicable to sources of exposure to hazardous substances.
- The degree to which it is possible or desirable to develop a consistent risk-assessment method, or a consistent standard of acceptable risk among various federal programs.

Besides the Academy's report and the activities of the commission, both EPA and the Surgeon General are to evaluate the methods for evaluating health risks, the significance of residual risks, uncertainties associated with this analysis, and recommend legislative changes.

As a result, the committee highlights here some important and controversial subjects in risk assessment and management that it felt were beyond its charge.

1. *The use of a specific individual lifetime cancer risk number (e.g., 10^{-4} or 10^{-6}) as a target for risk regulations.* The committee notes that Congress has set a standard for considering regulatory decisions. We note that such a number should be tied to a method and that uncertainty will always surround such estimates.

2. *The use of comparative risk analysis for the allocation of resources to minimize health impacts.* Congress decides how much of the country's economic and social resources should be spent on reducing threats to public health and how to allocate resources among the many threats present in our daily lives.

3. *The relative risk associated with synthetic or industrial byproducts ver-*

sus natural chemicals. A recent study (Gold et al., 1992) contends that natural chemicals make up the vast bulk of chemicals to which humans are exposed, that natural chemicals are not much different from synthetic chemicals in their toxicology, and that about half the natural chemicals tested in chronic studies in rats and mice are carcinogens. The implication is that humans are likely to be exposed to a large background of rodent carcinogens as defined by high-dose testing. Some believe that this has implications for the amount of resources currently devoted to the study and control of synthetic chemicals. However, other studies (e.g., Perera and Bofetta, 1988) question the scientific underpinnings of these conclusions. The issue of the degree to which natural versus synthetic chemicals should be regulated is a policy issue that we cannot address. The scientific aspects of the issue will be discussed in a forthcoming NRC report on the relative risks of natural carcinogens. It is important to note that the present study focuses on airborne hazardous air pollutants and that, although some natural carcinogens are in food and water, there is little evidence of their widespread presence in air.

4. *The setting of relative policy priorities regarding the regulation of all sources of hazardous air pollutants.* The focus of Section 112 is on stationary sources of hazardous air pollutants; therefore, it was not within the charge of this committee to conduct an analysis of all sources of hazardous air pollutants and recommend which ones should be regulated and which should not. Congress already determined the extent to which it wanted to do that in the 1990 Amendments. Therefore, although the committee points out later in the report the potential impact of indoor versus outdoor pollutants, it is beyond our charge to go further and say whether, when, and how to take action on nonstationary and indoor sources of hazardous air pollutants.

5. *The uncertainty in engineering and economic assumptions.* There is, of course, uncertainty in the engineering and economic assumptions leading to EPA's estimates of the impact on industry of a regulation mandating specified magnitudes of risk. However, the committee was asked only to address EPA's implementation of risk assessment relative to public health, not the economic consequences of such regulation.

6. *The extent to which chemicals should be on or off the list of chemicals in the 1990 Amendments.* Although this report discusses how to set priorities for the collection and analysis of chemicals on the list, it is a policy judgment as to whether these chemicals, once ranked, should be included on such a list. (That does not imply that outside review of the list is not appropriate.)

7. *The presentation of uncertainty in the context of background risk.* Although this committee does discuss the issue of presentation of uncertainty, it was beyond its charge to indicate the extent to which it was appropriate to place the 1990 Amendments or other legislation in the context of all societal risk. Risk communication is complicated and involves such issues as involuntary versus voluntary risks, costs, benefits, and values, both individual and societal.

CONCEPTUAL FRAMEWORK OF THE REPORT

This report is aimed at a multidisciplinary audience with different levels of technical understanding. In discussing the many controversial aspects of risk assessment, the committee decided to address three categories of issues:

- Background of risk assessment and current practice at EPA. We organize this section (Chapters 2-5) via the old Red Book four-step paradigm.
- Specific concerns in risk assessment, such as the use of defaults and extrapolations. For example, is EPA justified in assuming, in the absence of contrary evidence, that the linearized multistage model should be used in determining the dose-response relationship for carcinogens?
- Cross-cutting issues that affect all parts of risk assessment. For example, how should uncertainty be handled? How should the accuracy of a model be evaluated?
- Implementation issues related to Section 112 of the 1990 Amendments. For example, how should EPA accommodate the tension between the goals of providing stability in its process and staying abreast of changing scientific knowledge?

The report addresses each type of issue. Our categorization of the issues reflects the analytical framework used by the committee and influences the structure of its recommendations. Although that might lead to some repetition, the committee feels that a degree of repetition is desirable because of the need to address audiences with different levels of knowledge.

The committee attempted to address the specific issues that arise from the uses of risk assessment under Section 112 of the Clean Air Act, which mandates the regulation of hazardous air pollutants. As amended in 1990, Section 112 de-emphasizes risk assessment in the initial phase of regulation, in which EPA is to establish "technology-based" standards for categories of sources that emit hazardous air pollutants. Risk assessment's main role will be in the second phase of regulation, in which EPA must determine whether residual risk (the risk presented by the emissions remaining after compliance with technology-based standards) should be further reduced. Risk assessment will also be used in several other ways (e.g., to determine whether an entire source category may be exempted from technology-based standards on the grounds that no source in the category creates more than a one-in-a-million lifetime risk of cancer for the most exposed person).

The appendixes to the report include EPA's responses to questions from the committee and some important EPA documents not readily available. Risk assessment is an ever-changing process, and these documents illustrate its status within EPA during the time when the committee is making its recommendations.

Two documents were also prepared by some committee members to reflect the committee's inability to reach consensus on how EPA should choose and refine its "default options" for conducting risk assessments when basic scientific mechanisms are unknown. One view espouses a principle of "plausible conservatism," while the other advocates "making full use of science."

PART

I

Current Approaches to Risk Assessment

The first part of the report examines the background and current practices of risk assessment consistent with the paradigm first codified in the 1983 NRC report *Risk Assessment in the Federal Government: Managing the Process*, often known as the Red Book (See Figure I-1). Chapter 2 of this report discusses the historical, social, and regulatory contexts of quantitative risk assessment. Chapters 3, 4, and 5 describe the Environmental Protection Agency's approach in applying the Red Book paradigm for risk assessment. As shown in Figure I-2, assessing human-health risks associated with a pollutant requires analysis of three elements: the *source* of the pollutant, the transport of the pollutant into the *environment* (air, water, land, and food), and the intake of the pollutant by *people* who might suffer adverse health effects either soon after exposure or later. Scientists and engineers take four basic interrelated steps to evaluate the potential health impact on people who are exposed to a hazardous air pollutant: emission characterization, exposure assessment, toxicity assessment, and risk characterization. In emission characterization, the chemical's identity and the magnitude of its emissions are determined. Exposure assessment includes how the pollutant moves from a source through the environment (transport) until it is converted to other substances (fate) or comes into contact with humans. In assessment of toxicity, the specific forms of toxicity that can be caused by the pollutant and the conditions under which these forms of toxicity might appear in exposed humans are evaluated. In risk characterization, the results of the analysis are described. These steps are described in detail in Chapters 3, 4, and 5.

The increase in the sophistication of the field of risk assessment since the Red Book requires risk assessors to have the ability to recognize and address fully such cross-cutting issues as uncertainty, variability, and aggregation, in addition to having a more overarching view of the practice of risk assessment. Therefore, the committee supplements the Red Book paradigm with a second approach—one that is less fragmented (and hence more holistic), less linear and more interactive, and, most important, one organized not according to discipline or function, but according to the recurring conceptual issues that cut across all the stages of risk assessment. These cross-cutting issues are described in Part II of this report.

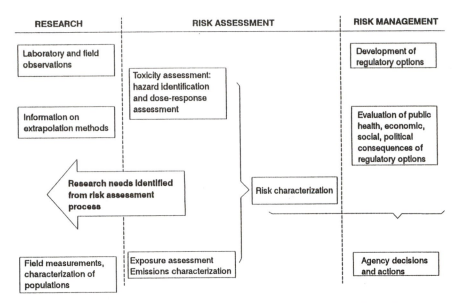

FIGURE I-1 NAS/NRC risk assessment/management paradigm. SOURCE: Adapted from NRC, 1983a.

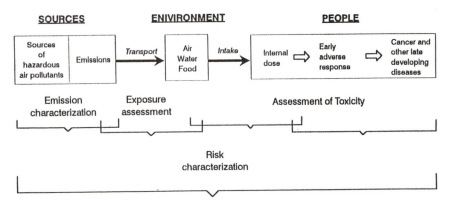

FIGURE I-2 Relationships in assessing human health risks of exposure to hazardous air pollutants. SOURCE: Adapted from NRC, 1983a.

2

Risk Assessment and Its Social and Regulatory Contexts

This chapter provides an overview of the origins and uses of quantitative risk assessment and the problems associated with it. Historical perspective is offered to aid understanding of how a method infused with so much uncertainty has still come to be seen by many as useful. Some attention is devoted to the important questions of how risk assessment has been used in decision-making and whether its use has improved decisions. The issues of public acceptance of the method and the degree to which decisions based on it are seen to provide adequate protection of the public health are also addressed. This chapter lists the major criticisms of risk assessment and the ways in which its results have been used, thus providing the justification for the selection of issues discussed in the succeeding chapters.

GENERAL CONCEPTS

This section briefly discusses some basic definitions and concepts concerning human-health risk assessment, its content, and its relationships to research and to decision-making. The definitions and concepts were first systematically formulated by a National Research Council committee in a report issued in 1983, *Risk Assessment in the Federal Government: Managing the Process*. The Red Book had a major influence on the practice of risk assessment and will be discussed extensively in this section of the report.

What is Risk Assessment?

Human-health risk assessment entails the evaluation of scientific information on the hazardous properties of environmental agents and on the extent of human exposure to those agents. The product of the evaluation is a statement regarding the probability that populations so exposed will be harmed, and to what degree. The probability may be expressed quantitatively or in relatively qualitative ways. There are other types of risk assessment that use similar processes but are outside the scope of this report, e.g., the risk assessment of the relative safety of a bridge.

Chemical hazards come in many forms. Some substances are radioactive, some explosive, some highly flammable. The particular hazard of concern here is chemical toxicity, including but not limited to carcinogenicity. Risk assessments can be carried out for any form of chemical toxicity. Risk assessment can be qualitative or quantitative. Many of the issues covered in this report concern quantitative expressions of risk.

How Is Risk Assessment Conducted?

The 1983 NRC report described a four-step analytic process for human-health risk assessment. A substance leaves a source (e.g., an industrial facility), moves through an environmental medium (e.g., the air), and results in an exposure (people breathe the air containing the chemical). The exposure creates a dose in the exposed people (the amount of the chemical entering the body, which may be expressed in any of several ways), and the magnitude, duration, and timing of the dose determine the extent to which the toxic properties of the chemical are realized in exposed people (the risk). This model is captured in the following analytic steps:

Step 1: Hazard Identification entails identification of the contaminants that are suspected to pose health hazards, quantification of the concentrations at which they are present in the environment, a description of the specific forms of toxicity (neurotoxicity, carcinogenicity, etc.) that can be caused by the contaminants of concern, and an evaluation of the conditions under which these forms of toxicity might be expressed in exposed humans. Information for this step is typically derived from environmental monitoring data and from epidemiologic and animal studies and other types of experimental work. This step is common to qualitative and quantitative risk assessment.

Step 2: Dose-Response Assessment entails a further evaluation of the conditions under which the toxic properties of a chemical might be manifested in exposed people, with particular emphasis on the quantitative relation between the dose and the toxic response. The development of this relationship may involve the use of mathematical models. This step may include an assessment of variations in response, for example, differences in susceptibility between young and old people.

Step 3: Exposure Assessment involves specifying the population that might be exposed to the agent of concern, identifying the routes through which exposure can occur, and estimating the magnitude, duration, and timing of the doses that people might receive as a result of their exposure.

Step 4: Risk Characterization involves integration of information from the first three steps to develop a qualitative or quantitative estimate of the likelihood that any of the hazards associated with the agent of concern will be realized in exposed people. This is the step in which risk-assessment results are expressed. Risk characterization should also include a full discussion of the uncertainties associated with the estimates of risk.

Not every risk assessment encompasses all four steps. Risk assessment sometimes consists only of a hazard assessment designed to evaluate the potential of a substance to cause human health effects. Regulators sometimes take the additional step of ranking the potency of a number of chemicals—what is known as hazard ranking. Sometimes potency information is combined with exposure data to produce a risk ranking. These techniques all use some, but not all, of the four steps of the quantitative risk-assessment process.

Much of this report is devoted to the technical contents of the four steps of the process, because therein lie the issues that affect the reliability, utility, and credibility of risk-assessment outcomes. One important feature of those steps, however, needs to be emphasized here.

The 1983 NRC committee recognized that completion of the four steps rests on many judgments for which a scientific consensus has not been established. Risk assessors might be faced with several scientifically plausible approaches (e.g., choosing the most reliable dose-response model for extrapolation beyond the range of observable effects) with no definitive basis for distinguishing among them. The earlier committee pointed out that selection of a particular approach under such circumstances involves what it called a *science-policy* choice. Science-policy choices are distinct from the policy choices associated with ultimate decision-making, as will be seen below. The science-policy choices that regulatory agencies make in carrying out risk assessments have considerable influence on the results and are the focus of much that follows in this report.

What is the Relationship Between Risk Assessment and Research?

Although the conduct of a risk assessment involves research of a kind, it is primarily a process of gathering and evaluating extant data and imposing science-policy choices. Risk assessment draws on research in epidemiology, toxicology, statistics, pathology, molecular biology, biochemistry, analytical chemistry, exposure modeling, dosimetry, and other disciplines; to the extent that it attempts to capture and take into account uncertainties, it also draws on the research efforts of decision analysts.

Risk assessment, at least in theory, can influence research directions. Because, at its best, risk assessment provides a highly organized profile of the current state of knowledge of particular issues and systematically elucidates scientific uncertainties, it can provide valuable guidance to research scientists regarding the types of data that can most effectively improve understanding. Little effort seems to have been made to use risk assessments in this way, although the Office of Technology Assessment has recently completed a study that describes the role of risk assessment in guiding research (OTA, 1993).

What is the Relationship Between Risk Assessment and Regulatory Decision-Making?

Risk management is the term used to describe the process by which risk-assessment results are integrated with other information to make decisions about the need for, method of, and extent of risk reduction. Policy considerations derived largely from statutory requirements dictate the extent to which risk information is used in decision-making and the extent to which other factors—such as technical feasibility, cost, and offsetting benefits—play a role.

Some statutes seem not to permit risk-assessment results to play a substantial role; they stress reductions of exposure to the "lowest technically feasible level" and usually require the best available technology. Proponents of such technology-based approaches often argue that they facilitate more rapid regulatory action and are especially suitable for making large and relatively inexpensive "first-cut" emission reductions. Proponents of quantitative risk assessment argue that such approaches are blind to the possibility that the risks remaining after application of such technology might still be unreasonably large or, in other situations, that they have been pushed to unnecessarily low values. As amended in 1990, Section 112 of the Clean Air Act gives quantitative risk-assessment results a secondary but still important role relative to technology-based controls.

What Is a Default Option?

EPA's guidelines set forth "default options." These are generic approaches, based on general scientific knowledge and policy judgment, that are applied to various elements of the risk assessment process when specific scientific information is not available. For instance, ambient doses of contaminants in humans are generally far lower than the doses that produce tumors in animals in controlled studies. The guidelines advise that, in assessing the magnitude of cancer risk to humans from low doses of a chemical based on the results of a high-dose experiment, "in the absence of adequate information to the contrary, the linearized multistage procedure will be employed" (EPA, 1986a, 1987a); that is, cancer risk in humans exposed to low doses will be estimated mathematically by using high-dose data and a curve-fitting procedure to extrapolate to low doses. Departure from the guideline is allowed if there is "adequate evidence" that the mechanism through which the substance is carcinogenic is more consistent with a different model; for instance, that there is a threshold below which a substance will not cause a risk. Thus, the guideline amounts to a "default" that guides a decision-maker in the absence of evidence to the contrary; in effect, it assigns the burden of persuasion to those wishing to show that the linearized multistage procedure should not be used. Similar guidelines cover such important issues as the calculation of effective dose, the consideration of benign tumors, and the procedure for scaling animal-test results to estimates of potency in humans. In the absence of information on some critical point in a risk assessment, default procedures seem essential. The question, then, is not whether to use defaults, but which defaults are most appropriate for a specific task and when it is appropriate to use an alternative to a default.

HISTORICAL ROOTS

It is helpful to provide a brief historical perspective on the origins and evolution of risk assessment, so that some of the reasons that led to the use of the technique can be seen. The review is divided into two main parts, with an intervening section devoted to the NRC study of 1983 that was so influential in the developments of the last decade.

Early Efforts to Establish Safe Limits of Exposure to Toxic Substances

About 50 years ago, toxicologists began to study the problem of establishing limits on exposures to hazardous substances that would protect human health. The early efforts began in the 1940s in connection with concerns about occupational exposures to chemicals and about residues of pesticides in foods. Toxicologists were guided by the principle that all substances could become harmful under some conditions of exposure—when the so-called threshold dose was exceeded—but that human health could be protected as long as those exposure conditions were avoided. Threshold doses were recognized to vary widely among chemicals, but as long as human exposures were limited to subthreshold doses, no injury to health would be expected. The threshold hypothesis thus involved rejection of the simplistic view that the world is divided into toxic and nontoxic substances and acceptance of the principle that, for all chemicals, there were ranges of exposure that were toxic and ranges that were not. The threshold hypothesis was based on both empirical observations and basic concepts of biology—that every organism, including the human, has the capacity to adapt to or otherwise tolerate some exposure to any substance and that the harmful effects of a substance would become manifest only when exposure exceeded that capacity. Even at that early stage, there were questions about whether carcinogens always had thresholds, but otherwise the threshold concept became widely accepted.

Although there was widespread acceptance of the threshold hypothesis (except among scientists working in genetics and in chemical carcinogenesis) (NRC, 1986), it was not apparent how the threshold dose was to be estimated for a large and diverse human population whose members have different thresholds of susceptibility. Experts in occupational health tended to rely heavily on observations of short-term toxicity in highly exposed workers and established acceptable exposure limits (the most prominent of which were the so-called threshold limit values, TLVs, first published by the American Conference of Governmental Industrial Hygienists in the 1950s) that were below the exposures that produced observable toxic effects. In the early 1950s, two Food and Drug Administration (FDA) scientists, O.G. Fitzhugh and A. Lehman, proposed a procedure for setting acceptable limits, which became known as acceptable daily intakes (ADIs), for dietary pesticide residues and food additives. Their procedure was based on the threshold hypothesis and first involved identification of a chemical's no-observed-effect level (NOEL) from the set of chronic animal-toxicity data in which the animals responded to the lowest dose tested—the "most sensitive" indication of the chemical's toxicity. Several response levels are characterized by acronyms. The first is the "no-observed-effect-level," NOEL. Earlier this was called the no-*observable*-effect level. *Observable* was changed to *observed* to be more in keeping with actual data ("observed"), rather than a rather vague potential "observable," which might be related to the size and sensitivity of the experiment. What is not observable in a small experiment might be easily observed in a large experiment. The word *adverse* was added to NOEL, making it NOAEL and making it clearer that *adverse* effects were of concern. The LOEL and LOAEL have a similar genesis and currently refer to the "lowest-observed-adverse-effect level"—the lowest dose at which an adverse effect was seen.

Fitzhugh and Lehman cited data suggesting that "average" human sensitivities might be up to 10 times those of laboratory animals and that some members of a large and diverse human population might be up to 10 times more sensitive than the "average" person. Thus came into use the safety factor of 100. The experimental NOEL was divided by 100 to arrive at a chemical-specific ADI. If human exposure was limited to daily amounts less than the ADI, then no toxicity was to be expected. In fact, Fitzhugh and Lehman, and later other authors and expert groups, including the World Health Organization, did not claim that an ADI arrived at in this fashion was risk-free, but only that it carried "reasonable certainty of no harm." No attempt was made to estimate the probability of harm. A variation of the safety-factor approach, often called margin of safety, is the estimate of the ratio of the NOEL to actual exposures. A judgment is made as to whether that ratio is acceptable. This margin-of-safety approach seems to be most common for substances already in general use, and in practice is often associated with lower ratios of NOEL to exposure than those based on safety factors.

The use of safety factors to establish ADIs was also recommended by various NRC committees (NRC, 1970, 1977, 1986) and adopted by the Joint Food and Agriculture Organization and World Health Organization expert committees on food additives (FAO/WHO, 1982) and pesticide residues (FAO/WHO, 1965).

Although it has since been modified in several minor ways, the basic procedure for setting limits on human exposures to chemicals in air, water, and food persists to this day. The threshold hypothesis has been criticized as inadequate to account for some toxic effects, and it has not been accepted by regulators as applicable to carcinogens, but it remains a cornerstone of other regulatory and public-health risk assessments. Section 112 of EPA's authority for regulating toxic air pollutants envisions a safety-factor approach for some kinds of risk assessment.

The Problem of Carcinogens

Not only is cancer a much-feared set of diseases, but public and scientific concerns about cancer-inducing chemicals in the environment have centered on the possibility that such substances might act through nonthreshold mechanisms; that is, that exposure to even one molecule of a carcinogen is associated with a small but non-zero increased risk of tumor induction. This possibility served as the basis for modern dose-response models, which were developed initially from observations of radiation-induced cancer. These models came into wide use and were promoted by the National Research Council's series of reports entitled *Biological Effects of Ionizing Radiation* and later incorporated into the regulatory decision-making of the Nuclear Regulatory Commission. Perhaps the earliest legislative acknowledgment of the possibility that chemical carcinogens might act in the same way came in the form of the "Delaney clause" of the Food Additive Amendments of 1958. Following the suggestions set forth by several FDA and National Cancer Institute (NCI) officials, Congress stipulated that no additive that concentrates in food during processing or is added to food during or after processing may be allowed in the food supply if it is found to be carcinogenic in animals. The basis for the Delaney clause was that it is not possible to specify a safe human exposure to a carcinogen in the same sense that a safe intake of a substance acting through threshold mechanisms could be identified.

Through the 1960s and into the early 1970s, toxicologists avoided the problem of identifying "acceptable" intakes of carcinogens. Where it was possible, regulators simply prohibited introduction of carcinogens into commerce. But where banning was difficult or even infeasible— for example, for environmental contaminants that were byproducts of manufacturing and energy production— choosing a maximal permissible human exposure, and acceptance of some risk. Limits were sometimes based on some concept of technical feasibility. The problem with such a criterion for setting limits was that it provided little confidence that human health was being adequately protected or, conversely, that risks were not being forced to unnecessarily low levels. In many cases, carcinogenic pollutants were simply ignored (NRC, 1983a).

Those approaches to the problem of regulatory exposure to environmental carcinogens became problematic in the face of two trends. First, government and industrial testing for carcinogenicity began to increase rapidly during the late 1960s; during the 1970s, regulators had to begin to deal with large numbers of newly identified carcinogens that were found among the many commercial products introduced after World War II. Second, analytic chemists became able to identify carcinogens in the environment at lower and lower concentrations. It became clear during the early to middle 1970s that a systematic approach to regulating carcinogens was needed.

Several authors had published methods for quantifying low-dose risks associated with chemical carcinogen exposure in the 1960s and 1970s, and regulatory agencies—FDA and EPA in particular—began adopting some of the methods in the middle 1970s. EPA, for example, estimated low-dose risks associated with several carcinogenic pesticides and relied in part on its assessments in actions to cancel or limit their registrations. FDA began using low-dose risk estimation to deal with so-called indirect food additives and some food contaminants that proved to be carcinogenic. The Occupational Safety and Health Administration (OSHA) at first rejected the use of risk quantification as it mounted a major effort during the late 1970s to regulate occupational carcinogens, because it believed that the statute under which it operated did not permit the use of risk assessment. But a Supreme Court decision regarding the agency's efforts to establish a permissible exposure limit for benzene caused OSHA to incorporate risk quantification (see below).

Those trends of the 1970s toward increasing the use of risk assessment in carcinogen regulation caused several regulatory agencies, working together as the Interagency Regulatory Liaison Group (IRLG), to develop and publicize a set of guidelines for the conduct of risk assessments (IRLG, 1979). The guidelines were said by the agencies to specify a common approach to risk assessment. No commitment was made by the agencies to use the methods for all possible carcinogens in all classes of regulated products, but, to the extent that an agency decided to use risk assessment, its approach would be that specified in the IRLG guidelines. The agencies also noted that the guidelines did not include an approach to what later came to be called risk management; such issues were said to remain the prerogative of the individual agencies.

The IRLG guidelines embodied several important scientific principles that originated in efforts of the WHO International Agency for Cancer Research (IARC) (IARC, 1972, 1982), NCI (Shubik, 1977), and the federal regulatory agencies (FDA, 1971; Albert et al., 1977; OSHA, 1982). Among them were principles concerning the appropriate uses of epidemiologic and animal data in identifying potential human carcinogens and the extrapolation of such data to humans. The IRLG guidelines did not explicitly incorporate the "default options" language described earlier (that came only after the 1983 NRC report), but it is clear that they do include science-policy choices (e.g., the generic adoption of a linearized, no-threshold model for carcinogen dose-response assessment).

By the early 1980s, risk assessment had begun to take on considerable importance within the regulatory agencies and to capture the attention of regulated industries. One important impetus to the development of risk-assessment techniques was the Supreme Court's decision in *Industrial Union Department, AFL-CIO v. American Petroleum Institute*, 448 U.S. 607 (1980), the "*Benzene*" decision. That decision struck down the OSHA standard for exposure to benzene in the workplace. The standard was based on OSHA's policy of trying to reduce concentrations of carcinogens in the workplace as far as technologically possible without consideration of whether existing concentrations posed a significant risk to health. There was no opinion for the majority of the Supreme Court in *Benzene,* but four justices concluded that, under the Occupational Safety and Health Act, OSHA could regulate only if it found that benzene posed a significant risk of harm. Although the plurality did not define *significant risk of harm* and stressed that the magnitude of the risk need not be determined precisely, the decision strongly signaled that some form of quantitative risk assessment was necessary as a prelude to deciding whether a risk was large enough to deserve regulation.

Under those circumstances, Congress instructed FDA to arrange for the National Research Council in 1981 to undertake a study of federal efforts to use risk assessment.

NRC STUDY OF RISK ASSESSMENT
IN THE FEDERAL GOVERNMENT

In 1983, NRC was asked to issue recommendations regarding the scientific basis of risk assessment and the institutional arrangements under which it was being conducted and used. In particular, NRC's charge involved a close examination of the possibility that risk assessment might be conducted by a separate, centralized scientific body that would serve all the relevant agencies. It was proposed that such an arrangement might reduce the influence of policy-makers on the conduct of risk assessment, so that there would be minimal opportunities for the results of risk assessments to be manipulated to meet predetermined policy objectives.

The NRC committee drew extensively on the earlier work of EPA, FDA, OSHA, IARC, and NCI, and much of its effort was directed at a synthesis of scientific principles and concepts first elucidated by these agencies. The NRC study did not, however, recommend specific methods for the conduct of risk assessment.

The risk assessment framework and specific definitions of risk assessment and its component steps from the 1983 NRC report have been widely adopted.

Many of the recommendations from the 1983 report have been implemented by EPA and other regulatory agencies. Two of the major recommendations of the committee, summarized below, are particularly relevant to this report:

• A clear conceptual distinction between risk assessment and risk management should be maintained. It is, however, not necessary—indeed, it is inadvisable—to provide for a physical separation of the two activities. (The committee rejected the proposal for the establishment of an independent scientific group that would perform risk assessments for the regulatory agencies.) Risk assessments should be undertaken with careful attention to the contexts in which those assessments will be used.

• Regulatory agencies should develop and use inference guidelines that detail the scientific basis for the conduct of risk assessment and that set forth the default options. The guidelines should be explicit about the steps of risk assessment that require such science-policy choices. The guidelines are necessary to avoid the appearance of case-by-case manipulation of assumptions to meet preset management goals. Guidelines should be flexible, however, and allow departures from defaults when data in specific cases show that a default option is not appropriate.

The NRC committee did not specify any particular methodologic approach to risk assessment, nor did it address the issue of which default options should be used by regulatory agencies. It did, however, note that provisions should be made for continuing review of the science underlying the guidelines and of the basis of the default options incorporated in them.

EVENTS AFTER RELEASE OF THE 1983 NRC REPORT

The Office of Science and Technology Policy (OSTP) brought together scientists from the regulatory agencies, the National Institutes of Health, and other federal agencies and, in 1985, issued a comprehensive review of the scientific basis of risk assessment of chemical carcinogens. The OSTP review adopted the framework for risk assessment proposed by the NRC committee and provided the individual regulatory agencies a basis for developing the type of guidelines recommended by that committee.

Alone among federal agencies, EPA adopted a set of guidelines for carcinogen risk assessment in 1986, as recommended by NRC. The EPA guidelines specify default options, note the distinction between risk assessment and risk management, and otherwise meet NRC's and OSTP's recommendations. EPA has issued guidelines for assessing risks associated with several other adverse health effects of toxic substances (without the benefit of OSTP review of the underlying science) and for the conduct of human exposure assessments. Beginning in 1984, it initiated work and published guidelines for evaluating mutage-

nicity, developmental toxicity, effects of chemical mixtures, and human exposure (EPA, 1986a, 1987a). It later published proposed guidelines on female
reproductive risk (EPA, 1988a), male reproductive risk (EPA, 1988b), and exposure-related measurements (EPA, 1988c). Final, revised guidelines on developmental toxicity were published in 1991 (EPA, 1991a). The agency is now in the
process of issuing revised guidelines on cancer risk assessment and has issued
revised guidelines for the assessment of human exposures (EPA, 1992a).

Increasing activity at the state level was first indicated by California's publication in 1985 of *Guidelines for Chemical Carcinogen Risk Assessments and
Their Scientific Rationale* (CDHS, 1985). The purpose of the guidelines was "to
clarify internal procedures which risk assessment staff of the California Department of Health Services will usually use to deal with certain decision points
which are characteristic of most risk assessments." The authors went on to state
why guidelines were thought necessary, in language consistent with earlier statements of IRLG (1979), NRC (1983a), OSTP (1985), and EPA (1987a):

> These California guidelines, while in harmony with recent federal statements
> on carcinogenic risk assessment, are more specific and practical. The Depart-
> ment of Health Services' staff believe that there are important advantages to the
> announcement of such flexible nonregulatory guidelines. First, the publishing
> of guidelines increases the likelihood of consistency in risk assessment among
> agencies and decreases the time spent repeatedly arguing risk assessment policy
> for each separate substance. Second, announcing guidelines in advance makes
> it clear that one is not tailoring risk assessment to justify some predetermined
> risk management decision. Third, specific guidelines allow the regulated com-
> munity to predict what emissions, food residues, or other exposures are apt to
> be of public health concern. Fourth, the publication and discussion of these
> guidelines should make the process more understandable to risk managers who
> have to make decisions that depend in part on risk assessment determinations.

The NRC, OSTP, EPA, and California documents were produced during a
time in which the uses of risk assessment to guide regulatory decision-making
were expanding rapidly. Particularly important was EPA's adoption of risk
assessment as a guide to decisions at Superfund and other hazardous-waste sites,
including those covered by the Resource Conservation and Recovery Act
(RCRA).

The agency also extended the uses of risk assessment to decisions regarding
pesticide residues in food, carcinogenic contaminants of drinking-water sup-
plies, industrial emissions of carcinogens to surface waters, and industrial chemicals subject to regulation under the Toxic Substances Control Act (TSCA).
Risk-management approaches varied according to the specific legal requirements
applicable to the sources of carcinogen exposure, but the EPA guidelines were
intended to ensure that the agency's approach to risk assessment was uniform
across the various programs.

USES OF RISK ASSESSMENT IN THE REGULATION
OF HAZARDOUS AIR POLLUTANTS

Section 112 of the Clean Air Act, as originally adopted in the Clean Air Act Amendments of 1970, required EPA to set emissions standards for hazardous air pollutants so as to protect public health with an "ample margin of safety." EPA was slow in carrying out that mandate, listing only eight chemicals as hazardous air pollutants in 20 years.[1] Standards were issued for only seven (there was no standard for coke ovens), and the standards that were issued covered only some of the sources that emit these pollutants. One major reason was the ambiguity of "ample margin of safety." Many commentators long thought that that term barred EPA from considering costs; EPA might well have to set a standard of zero for any pollutant for which no threshold could be defined (i.e., virtually all carcinogens).

That interpretation of the act (originally developed well before 1987), however, was unanimously rejected by the District of Columbia Circuit court in *Natural Resources Defense Council v. EPA* (824 F.2d 1146 [en banc] [D.C.Cir. 1987]). At the same time, the Court of Appeals also rejected EPA's position that it could use technologic or economic feasibility as the primary basis for standard-setting under Section 112. Instead, the court held that EPA had first to determine what concentration was "safe"—i.e., represented an acceptable degree of risk—and had then to select a margin of safety necessary to incorporate the uncertainties in scientific knowledge. In the latter step, but not the former, technological feasibility could be taken into account. In accordance with the plurality opinion in the Supreme Court's *Benzene* decision, the circuit court also held that EPA's standards did not have to eliminate all risk.

As in the *Benzene* case, the court did not define any particular method for EPA to use in determining what risks are acceptable. On remand, the agency, after taking comment on a number of possibilities, decided that it could not use any single metric as a measure of whether a risk is acceptable. Instead, it adopted a general presumption that a lifetime excess risk of cancer of approximately one in 10,000 (10^{-4}) for the most exposed person would constitute acceptable risk and that the margin of safety should reduce the risk for the greatest possible number of persons to an individual lifetime excess risk no higher than one in 1 million (10^{-6}). Such factors as incidence (e.g., the number of possible new cases of a disease in a population), the distribution of risks, and uncertainties would be taken into account in applying those benchmarks. The agency approach thus put primary emphasis on estimating individual lifetime risks through quantitative risk assessment.

Congress lessened the role of quantitative risk assessment for air-pollution regulation by rewriting Section 112 in Title III of the 1990 amendments. Congress defined 189 chemicals as hazardous (subject to possible deletion) and required technology-based controls on sources of those chemicals, as well as any others that might be added to the list by EPA. Sources that emit hazardous air pollutants will be regulated in two stages. In the first, technology-based emissions standards will be imposed. Each major source (defined, generally, as a stationary source having the potential to emit 10 tons per year of a single hazardous air pollutant or 25 tons per year of a combination of hazardous air pollutants) must meet an emission standard based on using the maximum available control technology (MACT) as defined by standards to be issued by EPA. Smaller sources, known as area sources, must meet emissions standards based on using generally available control technology.

Section 112 defines some contexts in which quantitative risk assessment will remain important. First, quantitative risk assessment will be relevant in determining which categories of sources will not be subject to technology-based regulation; EPA may delete a source category from regulation if no source in the category poses a risk of greater than 10^{-6} to the "individual most exposed to emissions." Even here, judging from the use of the word "may," EPA is not required to make the deletion; thus, the results of the quantitative risk assessment need not be decisive.

Quantitative risk assessment has a greater, but still limited, role in the second stage of standard-setting under Section 112(f), the "residual-risk" stage. That section requires EPA to set standards that protect public health with an ample margin of safety if it concludes that the first stage of technology-based standard-setting has not done so. Second-stage standards must be set for a category of "major sources" if the first stage allows a residual risk of greater than 10^{-6} to the individual most exposed to emissions. This requirement might seem a wholesale adoption of risk management based on the maximally exposed person, but two points must be noted. First, the 10^{-6} criterion for standard-setting need only be an upper-limit screening device. EPA is free, if it chooses, to set second-stage standards for source categories posing lesser risks. Second, the actual second-stage standard need not be expressed in terms of quantitative risk. Section 112(f)(2) authorizes EPA to continue the $10^{-4}/10^{-6}$ approach described earlier, but it does not require the agency to do so. Instead, any method is acceptable that comports with *NRDC v. EPA*'s requirement that the standards provide an "ample margin of safety" in addition to reducing risk to a level judged acceptable by EPA.

Such techniques as hazard assessment, hazard ranking, and risk ranking (discussed above), and in some cases quantitative risk assessment, can also play a role in the agency's decisions on questions such as these:

• *Should EPA modify the definition of "major source" to include sources emitting less than the statutory cutoffs?* Section 112(a) defines a major source as one with the potential to emit 10 tons per year of any single listed hazardous air pollutant or 25 tons of any combination of listed pollutants, but allows EPA to lower these thresholds for a pollutant on the basis of such factors as potency, persistence, and potential for bioaccumulation.

• *Should EPA list additional pollutants as hazardous or remove some pollutants from the list?* Section 112(b) establishes a list of 189 hazardous air pollutants and requires that EPA add a substance to the list on a determination, either on its own accord or in response to a petition, that the substance is "known to cause or may reasonably be anticipated to cause adverse effects to human health or adverse environmental effects." This standard represents a reaffirmation of the *Ethyl* decision (discussed later) that EPA may regulate in the face of scientific uncertainty about a substance's effects. EPA is required to delete a substance if it decides that data are adequate to show that the substance will not cause, or be reasonably anticipated to cause, an adverse effect. In deletions as well, the risks of uncertainty are put on the source.

• *Which sources of hazardous air pollutants ought EPA to regulate first?* Section 112 requires that EPA set technology-based standards for categories of major sources on a phased schedule beginning in 1992 and ending in 2000. In deciding the order in which standards will be set, EPA must consider known or expected adverse effects of the pollutants to be regulated, as well as the quantity and location of emissions, or reasonably anticipated emissions, of hazardous air pollutants in each category. EPA has completed this preliminary task (see EPA, 1992a).

• *What restrictions ought EPA to place on offsetting within plants?* Generally, a physical change at a plant that increases emissions of a hazardous air pollutant will subject the plant to special new-source requirements. Under Section 112(g), this will not be the case if the plant simultaneously decreases by an offsetting amount emissions of a more hazardous pollutant. Deciding which offsets, if any, qualify for Section 112(g) may require EPA to rank the relative potency of hazardous air pollutants.

• *What restrictions ought EPA to place on offsetting by sources seeking to qualify for the early-reduction program?* The "early-reduction" program will pose similar issues. Usually, a source will have up to 3 years to comply with an EPA standard for controlling hazardous air pollutants. A source can obtain a 6-year extension, however, if it shows that it has achieved by approximately the end of 1993 a reduction of at least 90% in emissions of hazardous air pollutants (95% for particulate hazardous air pollutants) from baseline emissions. EPA is required to disqualify reductions that were used to offset increases in emissions of pollutants for which high risks of adverse health effects might be associated with exposure even to small quantities. Here, too, EPA will have to grapple with the relative potency factors of hazardous air pollutants. These rules have already been issued (see EPA, 1992b).

• *Which substances should EPA attempt to control through its urban-area source program?* EPA is required to identify at least 30 hazardous air pollutants that, as the result of emissions from area sources (nonmajor sources other than vehicles or off-road engines), present the greatest threat to public health in the largest number of urban areas. The agency must also identify categories responsible for those emissions and develop a national strategy that accounts for over 90% of the emissions of the identified air pollutants and that reduces by at least 75% the incidence of cancer attributed to exposure to hazardous air pollutants emitted by major and area sources.

• *Which pollutants ought EPA control under its authority to protect against accidental releases?* EPA must promulgate a list of 100 substances that, in the event of accidental release, are known to cause or can reasonably be expected to cause death, injury, or serious adverse effects to human health or the environment. The agency must also establish a "threshold quantity" for each. Operators of sources at which a listed substance is present in more than a threshold quantity must prepare a risk-management plan to prevent accidental releases.

NONCANCER RISK ASSOCIATED WITH HAZARDOUS AIR POLLUTANTS

The current EPA approach to risk assessment for noncancer hazards posed by hazardous air pollutants, refined in several ways, is conceptually similar to the traditional approach to threshold agents described earlier. The agency identifies a so-called inhalation reference concentration (RfC). An RfC is defined by EPA as "an estimate (with uncertainty) of the concentration that is likely to be without appreciable risk of deleterious effects to the exposed population after continuous, lifetime exposure" (EPA, 1992b). RfCs are derived from chemical-specific toxicity data. The latter are used to identify the most sensitive indicator of a chemical's toxicity and the so-called no-observed-adverse-effect level (NOAEL) for that indicator effect. If the NOAEL is derived from an animal study, as is typically the case, it can be converted to a human equivalent concentration by taking into account species differences in respiratory physiology. Uncertainty factors, whose magnitudes depend on the nature of the toxic effect and the quantity and quality of the data on which the NOAEL is based, are applied to the human-equivalent NOAEL to estimate the RfC. That procedure is used for all forms of toxic hazard except carcinogenicity. The use of RfCs depends on the assumption that toxic effects will not occur until a threshold dose is exceeded (EPA, 1992b).

Another important provision of Title III of the 1990 Amendments was the requirement that environmental effects be included in the evaluation of a risk associated with a pollutant. An adverse environmental effect is defined in Section 112(a)(7) of the act as "any significant and widespread adverse effect, which may reasonably be anticipated, to wildlife, aquatic life, or other natural resources, including adverse impacts on populations of endangered species or significant degradation of environmental quality over broad areas." Appendix III of EPA's *Unfinished Business* report (EPA, 1987b) found that airborne toxic substances, toxic substances in surface waters, and pesticides and herbicides were in the second highest category of relative risk in the ecological and welfare categories. Of particular concern in this report was the transport by air and water of toxic substances (heavy metals and organics) that accumulate in ecological food chains. Such bioaccumulation has impacts on both ecological resources and the use by humans of specific ecological populations (e.g., fish consumption). Ecological risk assessment is not discussed in this report except to the extent that bioaccumulation affects the health of people who eat and drink contaminated ecological resources, but is discussed in another recent NRC report entitled *Issues in Risk Assessment* (NRC, 1993a).

PUBLIC CRITICISM OF CONDUCT AND
USES OF RISK ASSESSMENT

The development of risk-assessment methods and their expanding uses in the federal and state regulation of hazardous substances have been carefully scrutinized by interested parties in the regulated industries, environmental organizations, and academic institutions. That scrutiny has led to frequent and sharp criticisms of the methods used for assessing risk and of ways in which the results of risk assessment have been used to guide decision-making. The criticisms have not been directed solely at the use of risk assessment in regulation of hazardous air pollutants, but rather cover a range of uses.

We cite here some of the criticisms that have appeared in the literature or that have otherwise been presented to the committee, because they help to define the issues reviewed in this report. *We emphasize that our citation of these criticisms does not mean that we believe them to be valid. Nor is the order of their listing meant to suggest our opinion regarding their possible importance.*

Criticisms Pertaining to Conduct of Risk Assessment

(1) Some analysts have commented that the default options used by EPA (i.e., the science-policy components of risk assessment) are excessively "conservative" or are not consistent with current scientific knowledge. The cumulative and combined effect of the many conservative default options used by EPA might yield results that seriously overstate actual risks, and thus tend to overcontrol emissions.

(2) Some experts have noted that important aspects of risk are neglected by EPA. The agency does not appear to recognize the possibility of synergistic interactions when multiple chemical exposures occur, nor does it seem concerned that available data show extreme variability among individuals in their responses to toxic substances. The failure to deal with those issues can lead to serious underestimation of human risk, especially at very low exposures. A related issue is the overlooked problem of risk aggregation—how risks associated with multiple chemicals are to be combined.

(3) The default options used by EPA have, according to some, become ex-

cessively rigid. The barriers to using alternative assumptions by incorporating chemical-specific data are said to be in effect impassable, because the degree of scientific certainty has never been explicitly or implicitly defined by EPA. The too-rigid adherence to the preselected default options also impedes research, because there is little likelihood that novel data will be incorporated into EPA risk assessments.

(4) Many commentators have stated that insufficient attention has been paid to the issue of human exposure itself. In particular, EPA has not defined the terms of exposure assessment with sufficient clarity. How are populations and subpopulations of interest to be characterized? What is meant by such terms as "maximally exposed individual" and "reasonable maximal exposure"? How are multiple exposure pathways to be assessed in evaluating individual's total risk associated with a hazardous air pollutant?

(5) Some have noted that the uncertainties in the results of risk assessments are inadequately described. Risks are most often reported as "point estimates," single numbers that admit to no uncertainty. Large uncertainties are often overlooked, and descriptions of risk as "upper bounds" are misleading and simplistic.

(6) According to some, insufficient attention has been devoted to noncancer risks. The NOEL-safety factor approach, although useful, is not scientifically rigorous.

(7) Some believe that we do not have sufficient knowledge to make risk estimates. In addition, some believe that a risk assessor can make risk calculations come out high or low, depending on what answer is desired. Thus, some people believe that credible risk assessment might be impossible to obtain with the existing state of science and risk-assessment institutions.

Criticisms Pertaining to the Relationship Between Risk Assessment and Risk Management

(1) Several commentators have concluded that the conceptual separation of risk assessment and risk management called for in the 1983 NRC report has resulted in procedural separation to the detriment of the process. Some commentators have viewed the publication of toxicity values (cancer potency factors and reference doses) by one office of EPA for the use of other offices (those responsible for regulatory decision-making) as a prime example of undesirable separation.

(2) According to some analysts, upper-bound point estimates of risk, produced solely for screening or risk-ranking purposes, have too often been used inappropriately as a definitive basis for decision-making. Such use might be attractive to decision-makers, but it seriously distorts the intentions of risk assessors who produce the estimates. Managers need to consider scientific uncertainties more fully.

(3) Several commentators have expressed the view that risk assessment is

too resource-intensive and thus impedes action. Given the substantial uncertainties in the results of risk assessment, it seems inappropriate to devote so much effort to its conduct. Moreover, no good mechanisms exist to resolve controversies, so debates over the appropriateness of various risk-assessment outcomes can be endless.

(4) Some reviewers, particularly those with state governments, believe that more effort needs to be devoted to defining the uses to which a risk assessment is to be put before it is attempted. Such planning will help to deal with the problem of resource allocation, because the amount of effort needed for a risk assessment can be more appropriately matched to its ultimate uses.

(5) Some analysts have pointed out that the failure to pay sufficient attention to the results of risk assessment has resulted in misplaced priorities and regulatory actions that are driven by social forces, not by science. They note that the fact that risk assessment is imperfect does not justify the use of decision-making approaches that suffer from even greater imperfections.

(6) On the other hand, some commentators feel that risk assessment has been given too much weight, especially in light of its methodological limitations and inability to account for unquantifiable features of risk, such as voluntariness and fear.

(7) Some analysts also point out that far too little attention has been devoted to research to improve risk-assessment methods. It is unfair simply to criticize the methods without offering risk assessors the means to improve them.

Are any of those criticisms justified? If so, what responses can be made to them? Can improvements be made? If so, how will they affect the conduct of risk assessment and the use of risk-assessment results in regulatory decision-making? These and related issues are the primary focus of Chapters 6-12 of this report.

NOTE

1. The chemicals listed as hazardous air pollutants under the National Standards for Hazardous Air Pollutants (NESHAP) (with the date of public notice): asbestos (3/71); benzene (6/77); beryllium (3/71); coke-oven emissions (9/84); inorganic arsenic (6/80); mercury (3/71); radionuclides (12/79); and vinyl chloride (12/75).

3

Exposure Assessment

INTRODUCTION

Accurate information on human exposure to hazardous air pollutants emitted by various sources is crucial to assessing their potential health risks. This chapter describes methods used to assess exposure to hazardous air pollutants. Section 112 of the Clean Air Act Amendments of 1990 applies to major sources that either singly or in combination emit defined quantities of one or more of the 189 hazardous air pollutants. The sources to which the act applies emit pollutants both continuously and episodically, and the pollutants can move from air to water, soil, or food.

In the terminology of the Environmental Protection Agency (EPA) and Title III of the 1990 Amendments, a major source of pollution is considered to be

> any stationary source or group of stationary sources located within a contiguous area and under common control that emits or has the potential to emit considering controls, in the aggregate, 10 tons per year or more of any hazardous air pollutant or 25 tons per year of any combination of hazardous air pollutants. The [EPA] Administrator may establish a lesser quantity, or in the case of radionuclides different criteria, for a major source than that specified in the previous sentence, on the basis of the potency of the air pollutant, persistence, potential for bioaccumulation, other characteristics of the air pollutant, or other relevant factors.

A stationary source is "any building, structure, emission source, or installation which emits or may emit any air pollutant."

As part of determining the health threat of a pollution source to humans, EPA assesses how a pollutant moves from a source through the environment until it makes contact with humans in its original form or after conversion to other substances. For most airborne substances, inhalation is assumed to be the primary route of entry into the body. There has recently been an extensive review of advances in assessing human exposure to airborne constituents (NRC, 1991a). That review attempted to define exposure carefully as a part of the overall continuum that leads to illness brought about by environmental contaminants. The definition of exposure as a part of this continuum has been incorporated into the 1992 revised guidelines for exposure assessment developed by EPA (1992a).

Human exposure to a contaminant is an event consisting of contact with a specific contaminant concentration at a boundary between a human and the environment (e.g., skin or lung) for a specified interval; total exposure is determined by the integrated product of concentration and time. The amount of a substance that is absorbed or deposited in the body of an exposed person in a given period is the administered dose. Calculating the dose from the exposure depends on a number of factors, including the mode of entry into the body. For substances that move into the body through an opening—such as the mouth or nose via breathing, eating, or drinking—the dose depends on the amount of the carrier medium that enters the body. For airborne substances, the potential dose is the product of breathing rate (volume of air inhaled per unit of time), exposure concentration, and fractional deposition of the substance throughout the respiratory tract. However, an inhalation exposure will not lead to a dose if none of the substance is absorbed through the lung or deposited on the surface of the lung or other sections of the respiratory tract.

A pollutant can also enter the body through the skin or other exposed tissues, such as the eyes. The substance is then directly absorbed from the carrier medium into the tissue, often at a rate that is different from the rate of absorption of the carrier. The pollutant uptake rate is the amount of the pollutant absorbed per unit of time, and the dose is the product of exposure concentration and uptake rate at that concentration. The NRC report on exposure assessment (NRC, 1991a) provides a scientific framework to identify routes of entry and degree of contact and indicates how exposure assessment integrates data on emitted pollutants with biological effects.

Exposure assessment involves numerous techniques to identify a pollutant, pollutant sources, environmental media of exposure, transport through each medium, chemical and physical transformations, routes of entry to the body, intensity and frequency of contact, and spatial and temporal concentration patterns of the pollutant. Mathematical models that can be used to describe the relationships among emissions, exposures, and doses are shown in Appendix C.

Exposure to a contaminant can be estimated in three ways. It can be evaluated directly by having a person wear a device that measures the concentration of a pollutant when it comes into contact with the body. Environmental monitoring is an indirect method of determining exposure, in which a chemical's concentration is measured in an environmental medium at a particular site, and the extent to which a person is exposed to that medium is used to estimate exposure. Finally, exposure can be estimated from the chemical's actual dose to the body, if it manifests itself in some known way through a measurable internal indicator (biological marker), such as the concentration of the substance or its metabolite in a body tissue or excreted material (NRC, 1991a). This is a direct method of exposure estimation and, unlike the other two, accounts for the amount of contaminant absorbed by the body. Each of these methods provides an independent estimate of exposure; when it is possible to use more than one approach, comparison of results can be useful in validating exposure estimates.

EPA's air-pollution regulatory programs have relied primarily on mathematical models to predict the dispersion of emissions to air and the potential for human exposure under different emission-control scenarios (see Appendix C for a description of EPA's Human Exposure Model). Source-emission estimates and meteorologic data were used to calculate the expected long-term ambient concentrations at various distances and directions from the source. Census data were used to estimate the number and location of people living near the source. A high-exposure scenario was estimated for a person (e.g., maximally exposed individual, MEI) assumed to be living near the source and constantly exposed for 70 years to the highest estimated air-pollutant concentration. EPA does not modify exposure estimates by including mobility of the population, shielding due to indoor locations, or additional exposures from indoor or other community sources. EPA also used a modeling approach to estimate the exposure of the local population to an average concentration of pollutant emitted from a source (EPA, 1985a).

1992 Exposure-Assessment Guidelines

EPA has recently promulgated a new set of exposure-assessment guidelines to replace the previous (1986) version (EPA, 1992a). The approach in the new guidelines is very different from that in the previous version and generally follows many of the concepts of exposure assessment presented in the 1991 NRC report (NRC, 1991a). The guidelines explicitly consider the need to estimate the distribution of exposures of individuals and populations and discuss the need to incorporate uncertainty analysis into exposure assessment. This approach is consistent with the most recent NRC recommendations on exposure analysis (NRC, 1993e).

The guidelines discuss the roles of both analytic measurement and mathematical modeling in estimating concentrations and durations of exposure. They do not recommend specific models, but suggest that models match the objectives of the particular exposure assessment being conducted and that they have the accuracy needed to achieve those objectives. They also call for detailed explication of the choices and assumptions that often must be made in the face of incomplete data and insufficient resources.

Exposure Calculation and the Maximally Exposed Individual

EPA has traditionally characterized exposure according to two criteria: exposure of the total population and exposure of a specified, usually highly or maximally exposed individual. The MEI's exposure is estimated as the plausible upper bound of the distribution of individual exposures. The reason for finding the MEI, as well as population exposure, is to assess whether any individual exposure might occur above a particular threshold that, as a policy matter, is considered to be important. Because the MEI's exposure level is intended to represent a potential upper bound, its calculation has involved a variety of conservative assumptions. Among the more conservative, and more contentious, were that the MEI lived for 70 years at the location deemed by the dispersion model to receive the heaviest annual average concentration, that the person stayed there 24 hours/day, and that there is no difference between outdoor and indoor concentrations. In practice, it is straightforward to estimate the exposure of an immobile MEI with the air-quality models described below. However, estimating exposure for a more typical person requires much more information as to his or her activities during the assessment period. Usually, these activities include spending a majority of time inside (where pollutant concentrations can be attenuated) and time spent in travel away from the residence. The 70-year, 24-hour/day and no-indoor-attenuation assumptions are, in effect, bounding estimates. Some people do live in a small community for a whole lifetime. Some people do spend virtually their whole life at home. And for some pollutants, there is little attenuation of pollutant concentrations indoors. Nonetheless, the occurrence of these conditions is rare, and it is even rarer that all these are found together.

In the most recent exposure guidelines, EPA no longer uses the term MEI, noting the difficulty in estimating it and the variety of its uses. The MEI has been replaced with two other estimators of the upper end of the individual exposure distribution, a "high-end exposure estimate" (HEEE) and the theoretical upper-bounding estimate (TUBE). The HEEE is not specifically defined ("the Agency has not set policy on this matter" [EPA, 1992a]); rather, the new exposure guidelines discuss some of the issues and procedures that should be considered as part of the choice of the methods and criteria. The HEEE is "a plausible estimate of exposure of the individual exposure of those persons at the upper end of an exposure distribution." *High end* is stated conceptually as "above the 90th percentile of the population distribution, but not higher than the individual in the population who has the highest exposure." As is implied by those statements, the new guidelines have adopted the use of individual exposure distributions, and the HEEE is a value in the upper tail of that distribution. The exact percentile for the HEEE that should be picked from the exposure distribution is not specified, but, according to EPA, should be chosen to be consistent with the population size in the particular application. The TUBE is a "bounding calculation that can easily be calculated and is designed to estimate exposure, dose, and risk levels that are expected to exceed the levels experienced by all individuals in the actual distribution. The TUBE is calculated by assuming limits for all the variables used to calculate exposure and dose that, when combined, will result in mathematically highest exposure or dose. . . ." In addition, calculation of the TUBE includes using a limiting case for the exposure-dose and dose-response relationships in calculating risk.

To be responsive to the concerns raised in the NRC (1991a) report, EPA changed its approach to the MEI. The TUBE is to be used only for bounding purposes and is to be superseded by the HEEE in detailed risk characterizations. Although the exposure guidelines are ambiguous in details about the determination of the HEEE, the HEEE is based on the estimation of the distribution of exposures that people might actually encounter. From the individual exposures, it is possible to develop population exposure (and risk) distributions and include uncertainty estimation, and personal-activity patterns. The details of these approaches are discussed in the applicable sections of this report (Chapters 10, 11, and 12).

The calculation of the exposure distribution for an individual requires knowledge of both the distribution of hazardous-pollutant concentrations and the distribution of times that the individual spends in places for which the concentrations are measured or modeled (time-activity patterns). For estimates of population exposure, the individual time-activity patterns are estimated for the population of the individuals that might be exposed.

EMISSION CHARACTERIZATION

The first step in exposure assessment is estimation of the quantity of toxic materials emitted by a given source. Emission characterization involves identifying the chemical components of emissions and determining the rates at which they are emitted. Although emission characterization is a necessary part of the exposure-assessment process, it is often conducted separately from exposure assessment to determine whether a given operation falls into one or another regulatory category.

Sources of Emissions

The emission rate often is considered to be proportional to the type and magnitude of industrial activity at a source. Emissions from a source might occur from process vents, handling equipment such as valves, pumps, etc., storage tanks, transfer, and wastewater collection and treatment. Process-vent emissions are released to the atmosphere from the use, consumption, reaction, and production of chemicals. Fugitive emissions are produced when chemicals "escape" from handling equipment, such as pumps and valves. Storage-tank emissions are released from the locations where chemical feedstocks or products are stored. These emissions depend on the chemical properties of the product stored (e.g., the vapor pressure), the atmospheric conditions (e.g., temperature), the type of tank (e.g., fixed or floating roof), and the type of seal and venting used. Transfer emissions are produced as material is received from or loaded into storage tanks, tank trucks, rail cars, and marine vessels (e.g., barges and ships). When material is added to a storage tank, for example, it can displace contaminated air into the atmosphere. Wastewater collection and treatment emissions can be released into a plant's wastewater system when chemicals are processed and released from the wastewater treatment plant. In continuous processes, a malfunction (upset), startup, or shutdown of the process can result in a much greater emission than normal.

Emission Estimation Methods

EPA (1991c) has provided a detailed procedure for estimating the emissions from facilities that use hazardous chemicals. In estimating emissions, information is generally needed on the magnitude of use of given chemicals, the chemical characteristics of the chemicals, and the efficiency with which the emissions are controlled.

The EPA protocols (1991c) provide a tiered approach to emission estimations ranging from relatively simple emission factors to material balances and direct measurements. These approaches have varied accuracy in estimation and a wide range of costs.

An emission factor is a multiplication factor that allows determination of the average emissions likely to come from a facility on the basis of its level of activity (EPA, 1985b). Emission factors are calculated on the basis of average measured emissions at several facilities in a given industry (*Compilation of Air Pollutant Emission Factors,* commonly known as AP-42 [EPA, 1985b]).

A material balance is performed by assuming that the sum of the mass of chemical inputs minus the sum of the outputs, after all chemical changes and accumulation within the process or equipment have been accounted for, is the emission. In general, material balances produce information about emissions that depends on relatively small differences between the large numbers that characterize inputs (raw materials) and outputs (finished products, byproducts, and other wastes).

Emissions can be estimated with calculation methods presented in EPA (1988d) publications, such as *Protocols for Generating Unit-Specific Emission Estimates for Equipment Leaks of VOC and VHAP* (used for fugitive emissions). This emission-estimation method allows the development of site-specific emission factors based on testing a statistical number of sources at a facility. These site-specific emission factors can be used to develop emission estimates in the future.

Ideally, emissions from a source can be calculated on the basis of measured concentrations of the pollutant in the source and the emission rate of the source. This approach can be very expensive and is not often used. The emission rates, characteristics of the source facility (stack height, plume temperature, etc.), and local topography (flat or complex terrain) are used to estimate the ambient concentrations of the hazardous pollutants to which people can be exposed.

Measurement Methods

The concentration of a given pollutant can be measured in each microenvironment. A microenvironment is a three-dimensional space with defined boundaries of which contaminant concentration is approximately spatially uniform during some specific period (Sexton and Ryan, 1988). There have been substantial improvements in analytic methods to measure concentrations, as described in a 1991 NRC report (NRC, 1991a). Modern methods in computerization of instruments, data recording, and data processing also permit much greater capability to obtain detailed information on the temporal and spatial variability of contaminants over a range of microenvironments. Other substantial improvements have enhanced the utility of personal monitors, which are worn by subjects directly and record the concentration or collect time-integrated samples of specific pollutants with which the wearers come into contact for specific intervals. For example, assessment of exposure to radiation has long made use of inexpensive, accurate, integrating dosimeters that were first developed when research on radioactive materials and the use of radioactivity were expanding rapidly. There are often substantial variations in the spatial distribution of radiation within a microenvironment, so individual dosimeters have been thought to provide the best estimates of individual exposure. Individual monitoring and extensive microenvironmental measurements are not generally practical for assessing exposures of the general population, but because of cost and the unwillingness of individuals to participate in exposure assessments, new instruments, including passive dosimeters for airborne chemicals, are likely to permit a similar strategy. These methods have been used in the TEAM studies (Wallace, 1987) to examine the total exposure of individuals to a number of volatile organic compounds in several locations around the country. This approach to exposure assessment has been applied in other research studies. One important finding of the TEAM studies (and others) is that substantially greater exposures to many contaminants occur indoors, both because of the higher concentrations and because most people spend considerably more time inside.

Although field measurement studies are generally expensive and require careful planning, organization, and quality-assurance programs, measurement programs can provide the large amounts of high-quality data needed to characterize environmental systems, to estimate exposure, and to develop, test, and evaluate models for evaluating exposure. Documented reliable models can then be used in place of more expensive, direct measurements. Reliable measurements are generally needed to provide knowledge of emissions of chemicals that give rise to human exposures. However, measurements provide only information on the current status of the system. To allow for a broader range of meteorologic conditions, estimate the effects of changes in plant operating capabilities and procedures, or estimate the effects of an accident or upset condition, models are needed to estimate emissions and the transport of emitted materials in the atmosphere.

MODELING USED IN EXPOSURE ASSESSMENT

Mathematical models used in exposure assessment can be classified in two broad categories: models that predict exposure (in units of concentration multiplied by time) and models that predict concentration (in units of mass per volume). Exposure models can be used to estimate population exposures from small numbers of representative measurements. Although concentration (or air-quality) models are not truly exposure models, they can be combined with information on human time-activity patterns to estimate exposures.

Air-quality models are also used to predict the fate, such as deposition or chemical transformation, of atmospheric pollutants to which people can be exposed indirectly (e.g., through deposition of pollutants from air onto surface water followed by bioaccumulation in fish). Such models are central to risk assessment (see Figure 3-1). They constitute the only method of determining the total impact of diverse emissions on air quality and are key tools in assessing the impact of specific sources on future air-pollutant concentrations and deposition.

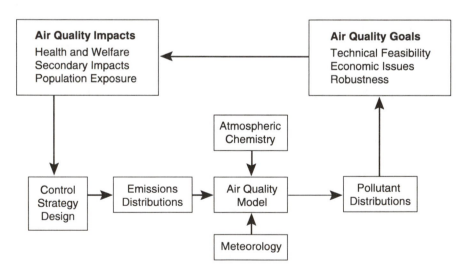

FIGURE 3-1 Air quality control strategy design process. SOURCE: Adapted from Russell et al., 1988.

Modeling Airborne Concentrations

Mathematical air-quality models used in air-pollution analysis are in two classes: empirical and analytic. The former type of model statistically relates observed air quality to the accompanying emission patterns, with chemistry and meteorology included only implicitly. Although they hold promise for use in some aspects of air-pollutant risk assessment, these models are not commonly used by EPA in its risk-assessment practice and will be discussed later. EPA and others more commonly use the form of analytical models, in which analytic or numerical expressions describe the complex transport processes and chemical reactions that affect air-pollutant concentrations. Pollutant concentrations are determined as explicit functions of meteorologic and topographic characteristics, chemical transformation, surface deposition, and source characteristics. In exposure assessments of air pollutants, the most widely used set of models has been the class called Gaussian-plume models. Gaussian-plume models are derived from atmospheric diffusion theory assuming stationary, homogeneous turbulence or, alternatively, by solution of the atmospheric-diffusion equation assuming simplified forms of the effective diffusivity (Seinfeld, 1986). Within the limits of the simplifications involved in their derivation, they can describe the individual processes that affect pollutant concentrations, such as diffusion, bulk transport by the wind, and deposition. These models are a type of a much broader family of models called dispersion or atmospheric-transport models. See Appendix C for more information.

Modeling Multimedia Exposure to Air Pollutants

In some cases, exposure to toxic pollutants emitted into the atmosphere occurs by pathways other than, or in addition to, inhalation. An example is deposition of metals like mercury in surface waters followed by the bioaccumulation of methyl mercury in fish and then ingestion of contaminated fish. Another is exposure of an infant ingesting the breast milk of a mother exposed to a toxic pollutant, such as polychlorinated biphenyls; this can be an important route for lipophilic compounds (NRC, 1993e), and EPA has investigated it in some exposure assessments. Recent studies (Travis and Hattemer-Frey, 1988; Bacci et al., 1990; Trapp et al., 1990) have also found significant bioaccumulation of chemicals from the atmosphere in plant tissues, particularly of nonionic organic compounds. These studies have found that the degree of bioaccumulation depends on solubility, and models for the uptake have been developed (Stevens, 1991). Such "indirect" pathways can concentrate pollutants and thus result in significant increases in exposure.

Multimedia exposure and indirect exposure have been considered more frequently in hazardous-waste site (e.g., Superfund) cleanup than in the management of exposure to industrial air pollutants. One example of multiple-path exposure to a source of primary air pollutants conducted by EPA is found in Cleverly et al. (1992). Multiple air pollutants, including heavy metals and organic chemicals, were followed after emission from a municipal-waste combustor. Atmospheric transport and deposition were modeled with a Gaussian-plume model modified to include wet and dry deposition. Other models were used to assess pollutant concentrations in nearby bodies of water; bioaccumulation; consumption of animal tissue, plants, and water; soil ingestion; and total potential dose.

Alternative Transport and Fate Models

The 1992 EPA guidelines for exposure assessment offer an approach to selection and use of models to estimate transport and fate, as well as exposure, so a variety of models can be used. For rapid screening analyses, Gaussian-plume models are adequate for limited distances around the source. However, for a more complete characterization of the distribution of concentrations downwind of a source, more refined modeling approaches may be needed.

In recent years, stochastic modeling of atmospheric dispersion has increased in popularity because of its relatively simple concept, its applicability to more complicated problems, and the improvements in computer capability and costs that make such models practical. Stochastic models can easily incorporate real physical phenomena, such as buoyancy, droplet evaporation, variations in the dispersity of released particles, and dry deposition. Stochastic modeling is typically implemented as a numerical Monte Carlo model in which the movement of a large number of air parcels is tracked in a Lagrangian reference frame. The concentration profile is then obtained from the air-parcel positions.

Boughton et al. (1987) described a Monte Carlo simulation of atmospheric dispersion based on treating either parcel displacement or parcel velocity as a continuous-time Markov process (a one-step-memory random process like Brownian movement). They simplified the problem by restricting themselves to crosswind-integrated point sources and assumed that dispersion in the mean wind direction is negligible. Thus, they reduced the problem to a one-dimensional model. Liljegren (1989) extended the model to incorporate both horizontal and vertical dispersion perpendicular to the mean wind direction. He found good agreement between the results of the three-dimensional stochastic model with concentration data found in the literature. Recent measurements of the dispersion of ground-released smokes and obscurants have shown excellent agreement of his stochastic model both with the average concentration values, including the profile across the plume, and with the time-varying concentrations observed (pers. comm., W. E. Dunn, U. of Illinois, 1988). It appears from those results that stochastic models offer considerable improvement over conventional Gaussian-plume models. Thus, there will soon be a substantially improved ability to predict average and time-varying ground-level concentrations.

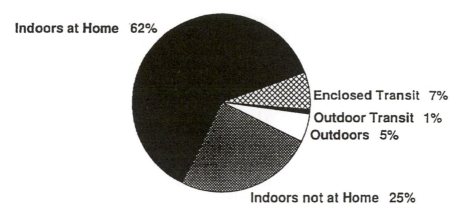

Indoors at Home 62%

Enclosed Transit 7%

Outdoor Transit 1%

Outdoors 5%

Indoors not at Home 25%

FIGURE 3-2 Percentage of day spent in different locations. Californians > 11 years of age. (Population Means). SOURCE: Jenkins et al., 1992. Reprinted with permission from *Atmospheric Environment*, copyright 1992 by Pergamon Press, Oxford, U.K.

Time-Activity Patterns

Exposure occurs when someone is in contact with a substance for some period. To estimate exposures, it is necessary to estimate the time spent in various activities that provide the opportunity for exposure. Figure 3-2 shows one such analysis. Various methods are available (NRC, 1991a), including recording of activities in a time-use diary (which might be automated to facilitate the recording of locations at specific times of the day and might use questionnaires to help reconstruct kinds and duration of activities). Some participants are careful in recording their activities; others might not provide accurate accounts, because of oversight or carelessness. The framing and wording of questionnaires can substantially affect the results of a survey and thus bias the resulting estimates of time spent in various activities and locations. Further work in the measurement and modeling of time and activity is needed; research recommendations were presented in an earlier report (NRC, 1991a).

Exposure-Assessment Models

The 1992 guidelines call for the development of distributions, instead of point estimates, for exposure parameters. It is the exposure-prediction models that combine microenvironmental concentration estimates with information on time-activity patterns of people to estimate individual exposures or the distribution of individual exposures in a typical population. Activity patterns and microenvironmental concentrations can both be measured or modeled. Microenvironmental concentrations and activity patterns can vary from person to person, and from period to period. Three types of models have been developed to estimate population exposures: simulation models, such as the simulation of human air pollution exposure (SHAPE) model (Ott, 1981, 1984; Ott et al., 1988) and National Ambient Air Quality Standards (NAAQS) Exposure Model (NEM) (Johnson and Paul, 1981, 1983, 1984), the convolution model of Duan (1981, 1987), and the variance-components model of Duan (1988) and Switzer (1988) (see Appendix C for additional information). The development of total-exposure models is one of the advances in modeling.

Several of the models for predicting exposures assume some correlation between measured contaminant concentrations in a microenvironment and the time spent by the exposed person in that space. Studies by Duan et al. (1985) suggested, on the basis of data from the Washington, D.C., carbon monoxide (CO) study (Akland et al., 1985), that there is no correlation between CO concentrations and time. However, there will be problems in existing models if occupancy times and concentrations of other contaminants correlate, as they might for irritating toxicants, such as formaldehyde.

Current exposure models use a variety of crude assumptions about the constancy of concentrations in microenvironments, the human activity patterns that determine the amount of time people spend in each microenvironment, and how representative the sampled population is to the total population that might be exposed to a contaminant.

Long-Term Exposure Modeling

Modeling very-long-term exposures, as is required for cancer risk assessment, presents several major difficulties. The current practice is to measure or model the concentration of a contaminant at one time and determine lifetime exposure by multiplying that concentration by a fixed number of years, e.g., the lifetime of an exposed person. However, the nature of exposure sources (e.g., changes in industrial processes) and activity patterns can change substantially over a lifetime. New sources or uses of sources can be introduced into the environment (e.g., the spreading use of wood-burning stoves), and old sources can be eliminated or modified (e.g., by the use of catalytic converters in motor vehicles). Typically, large facilities have a design life of 30 years, so considerable change can be anticipated in sources over the 70 years of a typical lifetime-exposure calculation.

Time-activity patterns of people can also vary substantially over very long periods. In the United States, people generally change their place of residence frequently, although some live in the same place over a lifetime. Population mobility can have a large impact on exposure assessments of agents, such as radon, that require reasonable estimates of long-term and highly variable exposure concentrations.

A person's activity pattern changes from childhood through young adulthood to middle and old age. Some efforts have addressed age-related differences in exposure that arise because of age. However, that aspect of variability in exposure over long periods has generally not received much attention in exposure modeling.

Short-Term Exposure Modeling

The typical steady-state airborne-concentration models are not able to provide estimates below 1-hour averages and have difficulty in modeling concentrations that vary widely over time and that can lead to short-term high exposures. If an exposure model is to estimate the effects of peak exposures on sensitive populations, the concentration model must provide reliable estimates for the time scales needed. There have been some important developments in stochastic models that could provide such estimates, but these developments have not yet been incorporated into the procedures for estimating exposure.

4

Assessment of Toxicity

INTRODUCTION

This chapter discusses the methods used to evaluate the toxicity of a substance for the purpose of health risk assessment. Evaluation of toxicity involves two steps: hazard identification and dose-response evaluation. Hazard identification includes a description of the specific forms of toxicity (neurotoxicity, carcinogenicity, etc.) that can be caused by a chemical and an evaluation of the conditions under which these forms of toxicity might appear in exposed humans. Data used in hazard identification typically are derived from animal studies and other types of experimental work, but can also come from epidemiologic studies. Dose-response evaluation is a more complex examination of the conditions under which the toxic properties of a chemical might be evidenced in exposed people, with particular emphasis on the quantitative relationship between dose and toxic response. This step also includes study of how response can vary from one population subgroup to another.

PRINCIPLES OF TOXICITY ASSESSMENT

The basic principles guiding the assessment of a substance's toxicity are outlined in the *Guidelines for Carcinogen Risk Assessment* (EPA, 1987a) (currently being updated), *Chemical Carcinogens: A Review of the Science and Its Associated Principles* (OSTP, 1985), *Guidelines for Developmental Toxicity Risk Assessment* (EPA, 1991a) and have recently been summarized by the NRC (1993a). In addition, guidelines for the assessment of acute toxicity have recently been developed by NRC (1993b). The developmental-toxicity guidelines are used in this chapter to illustrate EPA's approach to health effects that involve noncancer end points. They constitute the first completed noncancer risk-assessment guidelines in a series that EPA plans to issue.

Hazard Identification

The first of the two questions typically considered in the assessment of chemical toxicity concerns the types of toxic effects that the chemical can cause. Can it damage the liver, the kidney, the lung, or the reproductive system? Can it cause birth defects, neurotoxic effects, or cancer? This type of *hazard* information is obtained principally through studies in groups of people who happen to be exposed to the chemical (epidemiologic studies) and through controlled laboratory experiments involving various animal species. Several other types of experimental data can also be used to assist in identifying the toxic hazards of a chemical.

Epidemiologic Studies

Epidemiologic studies clearly provide the most relevant kind of information for hazard identification, simply because they involve observations of human beings, not laboratory animals. That obvious and substantial advantage is offset to various degrees by the difficulties associated with obtaining and interpreting epidemiologic information. It is often not possible to identify appropriate populations for study or to obtain the necessary medical information on the health status of individuals in them. Information on the magnitude and duration of chemical exposure, especially that experienced in the distant past, is often available in only qualitative or semiquantitative form (e.g., the number of years worked at low, medium, and high exposure). Identifying other factors that might influence the health status of a population is often not possible. Epidemiologic studies are not controlled experiments. The investigator identifies an exposure situation and attempts to identify appropriate "control" groups (i.e., unexposed parallel populations), but the ease with which this can be accomplished is largely beyond the investigator's control. For those and several other reasons, it is difficult or impossible to identify cause-effect relationships clearly with epidemiologic methods (OSTP, 1985).

It is rare that convincing causal relationships are identified with a single study. Epidemiologists usually weigh the results from several studies, ideally involving different populations and investigative methods, to determine whether there is a consistent pattern of responses among them. Some of the other factors that are often considered are the strength of the statistical association between a particular disease and exposure to the suspect chemical; whether the risk of the disease increases with increasing exposure to the suspect agent; and the degree to which other possible causative factors can be ruled out. Epidemiologists attempt to reach consensus regarding causality by weighing the evidence. Needless to say, different experts will weigh such data differently, and consensus typically is not easily achieved (IARC, 1987).

In the case of chemicals suspected of causing cancer in humans, expert groups ("working groups") are regularly convened by the International Agency for Research on Cancer (IARC) to consider and evaluate epidemiologic evidence. These groups have published their conclusions regarding the "degrees" of strength of the evidence on specific chemicals (sometimes chemical mixtures or even industrial processes when individual causative agents cannot be identified). The highest degree of evidence—sufficient evidence of carcinogenicity—is applied only when a working group agrees that the total body of evidence is convincing with respect to the issue of a cause-effect relationship.

No similar consensus-building procedure has been established regarding other forms of toxicity. Some epidemiologists disagree with IARC's cancer classification judgments in particular cases, and there seems to be even greater potential for scientific controversy regarding the strength of the epidemiologic evidence of non-cancer (e.g., reproductive, developmental, etc.) effects. There has been much less epidemiologic study of other toxic effects, in part because of lack of adequate medical documentation.

Animal Studies

When epidemiologic studies are not available or not suitable, risk assessment may be based on studies of laboratory animals. One advantage of animal studies is that they can be controlled, so establishing causation (assuming that the experiments are well conducted) is not in general difficult. Another advantage is that animals can be used to collect toxicity information on chemicals before their marketing, whereas epidemiologic data can be collected only after human exposure. Indeed, laws in many countries require that some classes of chemicals (e.g., pesticides, food additives, and drugs) be subjected to toxicity testing in animals before marketing. Other advantages of animal tests include the facts that

• The quantitative relationship between exposure (or dose) and extent of toxic response can be established.

• The animals and animal tissues can be thoroughly examined by toxicologists and pathologists, so the full range of toxic effects produced by a chemical can be identified.

• The exposure duration and routes can be designed to match those experienced by the human population of concern.

But laboratory animals are not human beings, and this obvious fact is one clear disadvantage of animal studies. Another is the relatively high cost of animal studies containing enough animals to detect an effect of interest. Thus, interpreting observations of toxicity in laboratory animals as generally applicable to humans usually requires two acts of extrapolation: interspecies extrapolation and extrapolation from high test doses to lower environmental doses. There are reasons based on both biologic principles and empirical observations to support the hypothesis that many forms of biologic responses, including toxic responses, can be extrapolated across mammalian species, including *Homo sapiens,* but the scientific basis of such extrapolation is not established with sufficient rigor to allow broad and definitive generalizations to be made (NRC, 1993b).

One of the most important reasons for species differences in response to chemical exposures is that toxicity is very often a function of chemical metabolism. Differences among animal species, or even among strains of the same species, in metabolic handling of a chemical, are not uncommon and can account for toxicity differences (NRC, 1986). Because in most cases information on a chemical's metabolic profile in humans is lacking (and often unobtainable), identifying the animal species and toxic response most likely to predict the human response accurately is generally not possible. It has become customary to assume, under these circumstances, that in the absence of clear evidence that a particular toxic response is not relevant to human beings, any observation of toxicity in an animal species is potentially predictive of response in at least some humans (EPA, 1987a). This is not unreasonable, given the great variation among humans in genetic composition, prior sensitizing events, and concurrent exposures to other agents.

As in the case of epidemiologic data, IARC expert panels rank evidence of carcinogenicity from animal studies. It is generally recognized by experts that evidence of carcinogenicity is most convincing when a chemical produces excess malignancies in several species and strains of laboratory animals and in both sexes. The observation that a much higher proportion of treated animals than untreated (control) animals develops malignancies adds weight to the evidence of carcinogenicity as a result of the exposure. At the other extreme, the observation that a chemical produces only a relatively small increase in incidence of mostly benign tumors, at a single site of the body, in a single species and sex of test animal does not make a very convincing case for carcinogenicity, although any excess of tumors raises some concern.

EPA combines human and animal evidence, as shown in Table 4-1, to categorize evidence of carcinogenicity; the agency's evaluations of data on individual carcinogens generally match those of IARC. For noncancer health effects, EPA uses categories like those outlined in Table 4-2. Animal data on other forms of toxicity are generally evaluated in the same way as carcinogenicity data, although this classification looks at hazard identification (qualitative) and dose-response relationships (quantitative) together. No risk or hazard ranking schemes similar to those used for carcinogens have been adopted.

The hazard-identification step of a risk assessment generally concludes with a qualitative narrative of the types of toxic responses, if any, that can be caused

TABLE 4-1   Categorization of Evidence of Carcinogenicity

Group	Criteria for Classification
A Human carcinogen	Sufficient evidence from epidemiologic studies
B Probable human carcinogen (two subgroups)	Limited evidence from epidemiologic studies and sufficient evidence from animal studies (B1); *or* inadequate evidence from epidemiologic studies (or no data) and sufficient evidence from animal studies (B2)
C Possible human carcinogen	Limited evidence from animal studies and no human data
D Not classifiable as to human carcinogenicity	Inadequate human and animal data or no data
E Evidence of noncarcinogenicity in humans	No evidence of carcinogenicity from adequate human and animal studies

SOURCE: Adapted from EPA, 1987a.

by the chemical under review, the strength of the supporting evidence, and the scientific merits of the data and their value for predicting human toxicity. In addition to the epidemiologic and animal data, information on metabolism and on the behavior of the chemical in tissues and cells (i.e., on its mechanism of toxic action) might be evaluated, because clues to the reliability of interspecies extrapolation can often be found here.

Identifying the potential of a chemical to cause particular forms of toxicity in humans does not reveal whether the substance poses a risk in specific exposed populations. The latter determination requires three further analytic steps: emission characterization and exposure assessment (discussed in Chapter 3), dose-response assessment (discussed next), and risk characterization (discussed in Chapter 5).

Dose-Response Assessment

In the United States and many other countries, two forms of dose-response assessment involving extrapolation to low doses are used, depending on the nature of the toxic effect under consideration. One form is used for cancer, the other for toxic effects other than cancer.

Toxic Effects Other Than Cancer

For all types of toxic effects other than cancer, the standard procedure used by regulatory agencies for evaluating the dose-response aspects of toxicity involves identifying the highest exposure among all the available experimental

TABLE 4-2   Weight-of-Evidence Classification Methods for Noncancer Health Effects

---

Sufficient Evidence

The sufficient-evidence category includes data that collectively provide enough information to judge whether a human developmental hazard could exist within the context of dose, duration, timing, and route of exposure. This category includes both human and experimental-animal evidence.

Sufficient Human Evidence: This category includes data from epidemiologic studies (e.g., case-control and cohort studies) that provide convincing evidence for the scientific community to judge that a causal relationship is or is not supported. A case series in conjunction with strong supporting evidence may also be used. Supporting animal data might or might not be available.

Sufficient Experimental Animal Evidence or Limited Human Data: This category includes data from experimental-animal studies or limited human data that provide convincing evidence for the scientific community to judge whether the potential for developmental toxicity exists. The minimal evidence necessary to judge that a potential hazard exists generally would be data demonstrating an adverse developmental effect in a single appropriate, well-conducted study in a single experimental-animal species. The minimal evidence needed to judge that a potential hazard does not exist would include data from appropriate, well-conducted laboratory-animal studies in several species (at least two) that evaluated a variety of the potential manifestations of developmental toxicity and showed no developmental effects at doses that were minimally toxic to adults.

Insufficient Evidence

This category includes situations for which there is less than the minimal sufficient evidence necessary for assessing the potential for developmental toxicity, such as when no data are available on developmental toxicity, when the available data are from studies in animals or humans that have a limited design (e.g., small numbers, inappropriate dose selection or exposure information, or other uncontrolled factors), when the data are from a single species reported to have no adverse developmental effects, or when the data are limited to information on structure/activity relationships, short-term tests, pharmacokinetics, or metabolic precursors.

SOURCE: EPA, 1987a.

studies at which no toxic effect was observed, the "no-observed-effect level" (NOEL) or "no-observed-adverse-effect level" (NOAEL). The difference between the two values is related to the definition of adverse effect. The NOAEL is the highest exposure at which there is no statistically or biologically significant increase in the frequency of an adverse effect when compared with a control group. A similar value used is the lowest-observed-adverse-effect level (LOAEL), which is the lowest exposure at which there is a significant increase in an observable effect. All are used in a similar fashion relative to the regulatory need. The NOAEL is more conservative than the LOAEL (NRC, 1986).

For example, if a chemical caused signs of liver damage in rats at a dosage of 5 mg/kg per day, but no observable effect at 1 mg/kg per day and no other study indicated adverse effects at 1 mg/kg per day or less, then 5 mg/kg per day would be the LOAEL and 1 mg/kg per day would be the NOAEL under the conditions tested in that study. For human risk assessment, the ratio of the NOAEL to the estimated human dose gives an indication of the margin of safety for the potential risk. In general, the smaller the ratio, the greater the likelihood that some people will be adversely affected by the exposure.

The uncertainty-factor approach is used to set exposure limits for a chemical when there is reason to believe that a safe exposure exists; that is, that its toxic effects are likely to be expressed in a person only if that person's exposure is above some minimum, or threshold. At exposures below the threshold, toxic effects are unlikely. The experimental NOAEL is assumed to approximate the threshold. To establish limits for human exposure, the experimental NOAEL is divided by one or more uncertainty factors, which are intended to account for the uncertainty associated with interspecies and intraspecies extrapolation and other factors. Depending on how close the experimental threshold is thought to be to the exposure of a human population, perhaps modified by the particular conditions of exposure, a larger or smaller uncertainty factor might be required to ensure adequate protection. For example, if the NOAEL is derived from high-quality data in (necessarily limited groups of) humans, even a small safety factor (10 or less) might ensure safety, provided that the NOAEL was derived under conditions of exposure similar to those in the exposed population of interest and the study is otherwise sound. If, however, the NOAEL was derived from a less similar or less reliable laboratory-animal study, a larger uncertainty factor would be required (NRC, 1986).

There is no strong scientific basis for using the same constant uncertainty factor for all situations, but there are strong precedents for the use of some values (NRC, 1986). The regulatory agencies usually require values of 10, 100, or 1,000 in different situations. For example, a factor of 100 is usually applied when the NOAEL is derived from chronic toxicity studies (typically 2-year studies) that are considered to be of high quality and when the purpose is to protect members of the general population who could be exposed daily for a full lifetime (10 to account for interspecies differences and 10 to account for intraspecies differences).

Using the NOAEL/LOAEL/uncertainty-factor procedure yields an estimate of an exposure that is thought to "have a reasonable certainty of no harm." Depending on the regulatory agency involved, the resulting estimate of "safe" exposure can be termed an acceptable daily intake, or ADI (Food and Drug Administration, FDA); a reference dose, or RfD (EPA); or a permissible exposure level, or PEL (Occupational Safety and Health Administration, OSHA). For risk assessments, the dose received by humans is compared with the ADI, RfD, or PEL to determine whether a health risk is likely.

The requirement for uncertainty factors stems in part from the belief that humans could be more sensitive to the toxic effects of a chemical than laboratory animals and the belief that variations in sensitivity are likely to exist within the human population (NRC, 1980a). Those beliefs are plausible, but the magnitudes of interspecies and intraspecies differences for every chemical and toxic end point are not often known. Uncertainty factors are intended to accommodate scientific uncertainty, as well as uncertainties about dose delivered, human variations in sensitivity, and other matters (Dourson and Stara, 1983).

EPA's approaches to risk assessment for chemically induced reproductive and developmental end points rely on the threshold assumption. The EPA (1987a) guidelines for health-risk assessment for suspected developmental toxicants states that, "owing primarily to a lack of understanding of the biological mechanisms underlying developmental toxicity, intra/interspecies differences in the types of developmental events, the influence of maternal effects on the dose-response curve, and whether or not a threshold exists below which no effect will be produced by an agent," many developmental toxicologists assume a threshold for most developmental effects, because "the embryo is known to have some capacity for repair of the damage or insult" and "most developmental deviations are probably multifactorial."

EPA (1988a,b) later proposed guidelines for assessing male and female reproductive risks that incorporate the threshold default assumption "usually assumed for noncarcinogenic/nonmutagenic health effects," as well as the agency's new RfD approach to deriving acceptable intakes. The RfD is obtained as described above. The total adjustment or uncertainty factor referred to in the proposed guidelines for use in obtaining an RfD from toxicity data "usually ranges" from 10 to 1,000. The adjustment incorporates (as needed) uncertainty factors ("often" 10) for "(1) situations in which the LOAEL must be used because a NOAEL was not established, (2) interspecies extrapolation, and (3) intraspecies adjustment for variable sensitivity among individuals." An additional modifying factor may be used to account for extrapolating between exposure durations (e.g., from acute to subchronic) or for NOAEL-LOAEL inadequacy due to scientific uncertainties in the available database.

EPA's 1992 revision of its guidelines for developmental-toxicity risk assessment state that "human data are preferred for risk assessment" and that the "most relevant information" is provided by good epidemiologic studies. When these data are not available, however, reproductive risk assessment and developmental-agent risk assessment, according to EPA, are based on four key assumptions:

- An agent that causes adverse developmental effects in animals will do so in humans, with sufficient exposure during development, although the types of effects might not be the same in humans as in animals.
- Any significant increase in any of the expressions of developmental tox-

icants (e.g., death, structural abnormalities, growth alterations, and functional deficits) indicates a likelihood that the agent is a developmental hazard.

• Although the types of effects in humans and animals might not be the same, the use of the most sensitive animal species to estimate human hazards is justified.

• A threshold is assumed in dose-response relationships on the basis of current knowledge, although some experts believe that current science does not fully support this position.

The new guidelines state that "the existence of a NOAEL in an animal study does not prove or disprove the existence or level of a biological threshold." The guidelines also address statistical deficiencies and improvements in the NOAEL-based uncertainty-factor approach (Crump, 1984; Kimmel and Gaylor, 1988; Brown and Erdreich, 1989; Chen and Kodell, 1989; Gaylor, 1989; Kodell et al., 1991a). The guidelines also discuss EPA's plans to move toward a more quantitative "benchmark dose" (BD) for risk assessment for developmental end points "when sufficient data are available"; the BD approach would be consistent with the uncertainty-factor approach now in use (EPA, 1991a). Like the NOAEL and LOAEL, the BD is based on the most sensitive developmental effect observed in the most appropriate or most sensitive mammalian species. It would be derived by modeling the data in the observed range, selecting an incidence rate at a preset low observed response (e.g., 1% or 10%), and determining the corresponding lower confidence limit on dose that would yield that level of excess response. A BD thus calculated would then be divided by uncertainty factors to derive corresponding acceptable intake (e.g., RfD) values (EPA, 1991a). Thus, the traditional uncertainty-factor approach is retained in the 1991 developmental-toxicity guidelines, as well as in the proposed BD approach. However, the new guidelines are unique, in that they emphasize both the possible effect of interindividual variability in the interpretation of acceptable exposures and the improvements that biologically based models could bring to developmental risk assessment (EPA, 1991a):

> It has generally been assumed that there is a biological threshold for developmental toxicity; however, a threshold for a population of individuals may or may not exist because of other endogenous or exogenous factors that may increase the sensitivity of some individuals in the population. Thus, the addition of a toxicant may result in an increased risk for the population, but not necessarily for all individuals in the population. . . . Models that are biologically based should provide a more accurate estimation of low-dose risk to humans. . . . The Agency is currently supporting several major efforts to develop biologically based dose-response models for developmental toxicity risk assessment that include the consideration of threshold.

Cancer

For some toxic effects, notably cancer, there are reasons to believe either that no threshold for dose-response relationships exists or that, if one does exist, it is very low and cannot be reliably identified (OSTP, 1985; NRC, 1986). This approach is taken on the basis not of human experience with chemical-induced cancer, but rather of radiation-induced cancer in humans and radiologic theory of tissue damage. Risk estimation for carcinogens therefore follows a different procedure from that for noncarcinogens: the relationship between cancer incidence and the dose of a chemical observed in an epidemiologic or experimental study is extrapolated to the lower doses at which humans (e.g., neighboring population) might be exposed (e.g., due to emissions from a plant) to predict an excess lifetime risk of cancer—that is, the added risk of cancer resulting from lifetime exposure to that chemical at a particular dose. In this procedure, there is no "safe" dose with a risk of zero (except at zero dose), although at sufficiently low doses the risk becomes very low and is generally regarded as without public-health significance.

The procedure used by EPA is typical of those used by the other regulatory agencies. The observed relationship between lifetime daily dose and observed tumor incidence is fitted to a mathematical model to predict the incidence at low doses. Several such models are in wide use. The so-called linearized multistage model (LMS) is favored by EPA for this purpose (EPA, 1987a). FDA uses a somewhat different procedure that nevertheless yields a similar result. An important feature of the LMS is that the dose-response curve is linear at low doses, even if it displays nonlinear behavior in the region of observation.

EPA applies a statistical confidence-limit procedure to the linear multistage no-threshold model to generate what is sometimes considered an upper bound on cancer risk. Although the actual risk cannot be known, it is thought that it will not exceed the upper bound, might be lower, and could be zero. The result of a dose-response assessment for a carcinogen is a potency factor. EPA also uses the term *unit risk factor* for cancer potency. This value is the plausible upper bound on excess lifetime risk of cancer per unit of dose. In the absence of strong evidence to the contrary, it is generally assumed that such a potency factor estimated from animal data can be applied to humans to estimate an upper bound on the human cancer risk associated with lifetime exposure to a specified dosage.

The dose-response step involves considerable uncertainty, because the shape of the dose-response curve at low doses is not derived from empirical observation, but must be inferred from theories that predict the shape of the curve at the low doses anticipated for human exposure. The adoption of linear models is based largely on the science-policy choice that calls for caution in the face of scientific uncertainty. Models that yield lower risks, indeed models incorporating a threshold dose, are plausible for many carcinogens, especially chemicals that do not directly interact with DNA and produce genetic alterations. For example, some chemicals, such as chloroform, are thought to produce cancers in laboratory animals as a result of their cell-killing effects and related stimulation of cell division. However, in the absence of compelling mechanistic data to support such models, regulators are reluctant to use them, because of a fear that risk will be understated. For other substances (e.g., vinyl chloride), evidence shows that the human cancer risk at low doses could be substantially higher than would be estimated by the usual procedures from animal data. Models that yield higher potency estimates at lower doses than the LMS model might also be plausible, but are rarely used (Bailar et al., 1988).

NEW TRENDS IN TOXICITY ASSESSMENT

With respect to carcinogenic agents, two types of information are beginning to influence the conduct of risk assessment.

For any given chemical, a multitude of steps can occur between intake and the occurrence of adverse effects. Those events can occur dynamically over an extended period, in some cases decades. One approach to understanding the complex interrelationships is to divide the overall scheme into two pieces, the linkages between exposure and dose and between dose and response. *Pharmacokinetics* has often been used to describe the linkage between exposure (or intake) and dose, and *pharmacodynamics* to describe the linkage between dose and response. Use of the root *pharmaco* (for drug) reflects the origin of those terms. When applied to the study and evaluation of toxic materials, the corresponding terms might more appropriately be *toxicokinetics* and *toxicodynamics.*

Exploration of the use of pharmacokinetic data is especially vigorous. Risk assessors are seeking to understand the quantitative relationships between chemical exposures and target-site doses over a wide range of doses. Because the target-site dose is the ultimate determinant of risk, any nonlinearity in the relationship between administered dose and target-site dose or any quantitative differences in the ratio of the two quantities between humans and test animals could greatly influence the outcome of a risk assessment (which now generally relies on an assumed proportional relationship between administered and target doses). The problem of obtaining adequate pharmacokinetic data in humans is being attacked by the construction of physiologically based pharmacokinetic (PBPK) models, whose forms depend on the physiology of humans and test animals, solubilities of chemicals in various tissues, and relative rates of metabolism (NRC, 1989). Several relatively successful attempts at predicting tissue dose in humans and other species have been made with PBPK modeling, and greater uses of this tool are being encouraged by the regulatory community (NRC, 1987).

A second major trend in risk assessment stems from investigations indicating that some chemicals that increase tumor incidence might do so only indirectly, either by causing first cell-killing and then compensatory cell proliferation or by increasing rates of cell proliferation through mitogenesis. In either case, increasing cell proliferation rates puts cells at increased risk of carcinogenesis from spontaneous mutation. Until a dose of such a carcinogen sufficient to cause the necessary toxicity or intracellular response is reached, no significant risk of cancer can exist. Such carcinogens, or their metabolites, show little or no propensity to damage genes (they are nongenotoxic).

5

Risk Characterization

INTRODUCTION

Characterization of risk is the final step in health risk assessment. This chapter discusses the methods used by the Environmental Protection Agency (EPA) to characterize the public-health risk associated with an emission source. In risk characterization, the assessor takes the exposure information from the exposure-assessment stage (discussed in Chapter 3) and combines it with information from the dose-response assessment stage (discussed in Chapter 4) to determine the likelihood that an emission could cause harm to nearby individuals and populations. The results of this risk characterization are then communicated to the risk manager with an overall assessment of the quality of the information in that analysis. The goal of risk characterization is to provide an understanding of the type and magnitude of an adverse effect that a particular chemical or emission could cause under particular circumstances. The risk manager then makes decisions on the basis of the public-health impact as determined by the risk characterization and other criteria outlined in the appropriate statute.

The elements of risk characterization are discussed here on the basis of several EPA documents, including EPA's *Risk Assessment Guidelines of 1986* (EPA, 1987a); *Guidelines for Exposure Assessment* (EPA, 1992a); a memorandum from Henry Habicht II, deputy administrator of EPA, dated February 26, 1992 (EPA, 1992c) (see Appendix B) (known hereafter as the "risk-characterization memorandum"); and *Risk Assessment Guidance for Superfund* (EPA, 1989a) (the "Superfund document").

ELEMENTS OF RISK CHARACTERIZATION

EPA's risk-characterization step has four elements: generation of a quantitative estimate of risk, qualitative description of uncertainty, presentation of the risk estimate, and communication of the results of risk analysis.

Quantitative Estimates of Risk

To determine the likelihood of an adverse effect in an exposed population, quantitative information on exposure—i.e., the dose (determined from the analysis in Chapter 3)—is combined with information on the dose-response relationship (determined from the analysis in Chapter 4). This process is different for carcinogens and for noncarcinogens. For noncarcinogens, the dose estimate is divided by the RfD to obtain a hazard index. If the hazard index is less than 1, the chemical exposure under consideration is regarded as unlikely to lead to adverse health effects. If the hazard index is greater than 1, adverse health effects are more likely and some remedial action is called for. The hazard index is thus not an actual measure of risk; it is a benchmark that can be used to estimate the likelihood of risk.

For carcinogens, excess lifetime risk is calculated by multiplying the dose estimate by a potency factor. The result is a value that represents an upper bound on the probability that lifetime exposure to an agent, under the specified conditions of exposure, will lead to excess cancer risk. This value is usually expressed as a population risk, such as 1×10^{-6}, which means that no more than one in 1 million exposed persons is expected to develop cancer. Risk estimates obtained in this way are *not* scientific estimates of actual cancer risk; they are upper bounds on actual cancer risk that are useful to regulators for setting priorities and for setting exposure limits.

When exposure to more than one agent occurs simultaneously, the cancer risk estimates obtained for each agent can be combined in an additive manner for each route of exposure. Hazard indexes for noncarcinogens may be combined when the agents of concern elicit similar end points of toxicity.

Sometimes, this risk-characterization technique is used to estimate an upper bound on excess lifetime cancer risk to exposed individuals, instead of populations. EPA's *Guidelines for Exposure Assessment* (EPA, 1992a) (not yet implemented) lists some of the questions that should be answered when considering individual versus population risk. These questions are stated by EPA as follows:

Individual Risk

• Are individuals at risk from exposure to the substances under study? Although for substances, such as carcinogens, that are assumed to have no threshold, only a zero dose would result in nonexcess risk for noncarcinogens, this question can often be addressed. In the case of the use of hazard indices, where exposure or doses are compared to a reference dose or some other acceptable level, the risk descriptor would be a statement based on the ratio between the dose incurred and the reference dose.

• To what risk levels are the persons at the highest risk subjected? Who are these people, what are they doing, where do they live, etc., and what might be putting them at this higher risk?

• Can people with a high degree of susceptibility be identified?

• What is the average individual risk?

Population Risk

• How many cases of a particular health effect might be probabilistically estimated for a population of interest during a specified time period?

• For noncarcinogens, what portion of the population exceed the reference dose (RfD), the reference concentration (RfC), or other health concern level? For carcinogens, how many persons are above a certain risk level such as 10^{-6} or a series of risk levels such as 10^{-5}, 10^{-4}, etc.

• How do various subgroups fall within the distributions of exposure, dose, and risk?

• What is the risk for a particular population segment?

• Do any particular subgroups experience a high exposure, dose, or risk?

Description of Uncertainty

Analysis of the uncertainty associated with a health risk estimate involves each step of the risk-assessment process: it brings together the uncertainty in emissions and exposure estimates with that of the toxicity dose-response assessment. Table 5-1 lists the uncertainty issues to be addressed at each step of a health risk assessment. Uncertainty analysis can take place at the time of each of those analyses, but because it affects the eventual risk estimate, it is considered part of the final step of risk assessment—risk characterization.

Several recent documents illustrate EPA's current approach to the analysis of uncertainty associated with health risk assessment, including the Superfund document (EPA, 1989a), the background information document for NESHAPS for radionuclides (EPA, 1989b), the *Guidelines for Exposure Assessment* (EPA, 1992a), and the risk-characterization memorandum (Appendix B).

Superfund Risk-Assessment Guidance

The Superfund document provides guidance to EPA and other government employees and contractors who are risk assessors, risk-assessment reviewers, remedial project managers, or risk managers involved in Superfund-site cleanup.

TABLE 5-1 Uncertainty Issues To Be Addressed in Each Risk Assessment Step

A. *Hazard Identification:* What do we know about the capacity of an environmental agent for causing cancer (or other adverse effects) in laboratory animals and in humans?

1. the nature, reliability, and consistency of the particular studies in humans and in laboratory animals;
2. the available information on the mechanistic basis for activity; and
3. experimental animal responses and their relevance to human outcomes.

B. *Dose-Response Assessment:* What do we know about the biological mechanisms and dose-response relationships underlying any effects observed in the laboratory or epidemiology studies providing data for the assessment?

1. relationship between extrapolation models selected and available information on biological mechanisms;
2. how appropriate data sets were selected from those that show the range of possible potencies both in laboratory animals and humans;
3. basis for selecting interspecies dose scaling factors to account for scaling dose from experimental animals to humans; and,
4. correspondence between the expected route(s) of exposure and the exposure route(s) utilized in the hazard studies, as well as the interrelationships of potential effects from different exposure routes.

C. *Exposure Assessment:* What do we know about the paths, patterns, and magnitudes of human exposure and number of persons likely to be exposed?

1. The basis for the values and input parameters used in each exposure scenario. If based on data, information on the quality, purpose, and representatives of the database is needed. If based on assumptions, the source and general logic used to develop the assumption (e.g., monitoring, modeling, analogy, professional judgment) should be described.
2. The major factor or factors (e.g., concentration, body uptake, duration/frequency of exposure) thought to account for the greatest uncertainty in the exposure estimate, due either to sensitivity or lack of data.
3. The link of the exposure information to the risk descriptors. These risk descriptors should include: (1) individual risk including the central tendency and high end portions of the risk distribution, (2) important subgroups of the population such as highly exposed or highly susceptible groups or individuals (if known), and (3) population risk. This issue includes the conservatism or non-conservatism of the scenarios, as indicated by the choice of descriptors. In addition, information that addresses the impact of possible low probability but possibly high consequence events should be addressed.

   For individual risk, information such as the people at highest risk, the risk levels these individuals are subject to, the activities putting them at higher risk, and the average risk for individuals in the population of interest should be addressed. For population risk, information as to the number of cases of a particular health effect that might be probabilistically estimated in this population for a specific time period, the portion of the population that are within a specified range of some benchmark level for non-carcinogens; and, for carcinogens, the number of persons above a certain risk level should be included. For subgroups, information as to how exposure and risk impact the various subgroups and the population risk of a particular subgroup should be provided.

continued on next page

TABLE 5-1   *Continued*

D. *Risk Characterization:* What do other assessors, decision-makers, and the public need to
know about the primary conclusions and assumptions, and about the balance between
confidence and uncertainty in the assessment? What are the strengths and limitations of
the assessment?

1. Numerical estimates should never be separated from the descriptive information that is
   integral to the risk assessment. For decisionmakers, a complete characterization (key
   descriptive elements along with numerical estimates) should be retained in all
   discussions and papers relating to an assessment used in decision-making. Differences
   in assumptions and uncertainties, coupled with non-scientific considerations called for
   in various environmental statutes, can clearly lead to different risk management
   decisions in cases with ostensibly identical quantitative risks; i.e., the "number" alone
   does not determine the decisions.
2. Consideration of alternative approaches involves examining selected plausible options
   for addressing a given uncertainty. The strengths and weaknesses of each alternative
   approach and as appropriate, estimates of central tendency and variability (e.g., mean,
   percentiles, range, variance). The description of the option chosen should include the
   rationale for the choice, the effect of option selected on the assessment, a comparison
   with other plausible options, and the potential impacts of new research.

SOURCE: Risk-characterization memorandum (Appendix B).

Section 8.4 of the document "discusses practical approaches to assessing uncertainty in Superfund site risk assessments and describes ways to present key information bearing on the level of confidence in quantitative risk estimates for a site." The document considers three categories of uncertainty associated with site risk assessments: selection of substances, toxicity values, and exposure assessments. Table 5-2 is EPA's uncertainty checklist for Superfund-site risk assessments. Risk assessors are to use the checklist to ensure that they describe adequately the uncertainty in a risk assessment. The document indicates that, although the uncertainty associated with each variable in a risk assessment would ideally be associated with the final risk estimate, a more practical approach is to describe qualitatively how the uncertainties might be magnified or the estimates of risk biased because of the risk models used. This document is being updated.

Uncertainty Analysis for Radionuclide Risk

EPA undertook a more comprehensive, integrated, quantitative approach to uncertainty characterization in the background document for its environmental impact statement on the National Emission Standards for Hazardous Air Pollutants (NESHAPS) for radionuclides (EPA, 1989b). This document includes an extensive presentation of estimates of fatal cancer risks associated with exposure to radionuclides. The estimates were "intended to be reasonable best estimates of risk; that is, to not significantly underestimate or overestimate risks and be of

TABLE 5-2 EPA Guidance for Uncertainty Analysis in Superfund Risk Assessments

---

LIST PHYSICAL SETTING DEFINITION UNCERTAINTIES

- For chemicals not included in the quantitative risk assessment, describe briefly:
  — reason for exclusion (e.g., quality control), and
  — possible consequences of exclusion on risk assessment (e.g., because of widespread contamination, underestimate of risk).
- For the *current land uses* describe:
  — sources and quality of information, and
  — qualitative confidence level.
- For the *future land uses* describe:
  — sources and quality of information, and
  — information related to the likelihood of occurrence.
- For *each exposure pathway,* describe why pathway was selected or not selected for evaluation.
- For *each combination of pathways,* describe any qualifications regarding the selection of exposure pathways considered to contribute to exposure of the same individual or group of individuals over the same period of time.

CHARACTERIZE MODEL UNCERTAINTIES

- List/summarize the key model assumptions.
- Indicate the potential impact of each on risk:
  — direction (i.e., may over- or underestimate risk); and
  — magnitude (e.g., order of magnitude).

CHARACTERIZE TOXICITY ASSESSMENT UNCERTAINTIES

For each substance carried through the quantitative risk assessment, list uncertainties related to:
- qualitative hazard findings (i.e., potential for human toxicity);
- derivation of toxicity values, e.g.,
  — human or animal data,
  — duration of study (e.g., chronic study used to set subchronic RfD), and
  — any special considerations;
- the potential for synergistic or antagonistic interactions with other substances affecting the same individuals; and
- calculation of lifetime cancer risks on the basis of less-than-lifetime exposures.

For each substance not included in the quantitative risk assessment because of inadequate toxicity information, list:
- possible health effects; and
- possible consequences of exclusion on final risk estimates.

RISK CHARACTERIZATION

- confidence that the key site-related contaminants were identified and discussion of contaminant concentrations relative to background concentration ranges;

continued on next page

TABLE 5-2   *Continued*

<hr>

RISK CHARACTERIZATION—*continued*

- a description of the various types of cancer and other health risks present at the site (e.g., liver toxicity, neurotoxicity), distinguishing between known effects in humans and those that are predicted to occur based on animal experiments;
- level of confidence in the quantitative toxicity information used to estimate risks and presentation of qualitative information on the toxicity of substances not included in the quantitative assessment;
- level of confidence in the exposure estimates for key exposure pathways and related exposure parameter assumptions;
- the magnitude of the cancer risks and noncancer hazard indices relative to the Superfund site remediation goals in the NCP (e.g., the cancer risk range of 10^{-4} to 10^{-7} and noncancer hazard index of 1.0);
- the major factors driving the site risks (e.g., substances, pathways, and pathway combinations);
- the major factors reducing the certainty in the results and the significance of these uncertainties (e.g., adding risks over several substances and pathways);
- exposed population characteristics; and
- comparison with site-specific health studies, when available.

<hr>

SOURCE: Adapted from EPA, 1989a.

sufficient accuracy to support decisionmaking" (EPA, 1989b). One chapter of the document, however, provides a detailed analysis of uncertainties in the calculated risks that was undertaken by EPA's Office of Radiation Programs for four selected exposure sites, such as a uranium-mill tailings pile in Washington and an elemental-phosphorus plant in Idaho. The stated reason for the uncertainty analysis was that "quantitative uncertainty analysis can provide results that indicate the likelihood of realizing different risk levels across the range of uncertainty. This type of information is very useful for incorporating acceptable and reasonable confidence levels into decisions" (EPA, 1989b).

The EPA uncertainty analysis for radionuclide risks focused on "parameter uncertainty," because it was felt that other sources of uncertainty involving alternative or additional exposure pathways and risk-model structures were "not readily amenable to explicit analysis" (EPA, 1989b). Parameter uncertainties were first modeled as particular probability distributions for each parameter involved in four key components of the radionuclide risk assessments: source terms, atmospheric-dispersion factors, environmental-transport and radionuclide-uptake factors, and risk-conversion (that is, radionuclide-potency) factors. All the distributions pertaining to exposure-related factors were intended to model uncertainty in factor values characteristic of a maximally exposed person. All the distributions pertaining to uptake-related factors were intended to model uncertainty in factor values characteristic of an average individual, except in a set of separate corresponding analyses in which census-based interindividual variability in home-residence time was incorporated into the analysis, where it was computationally treated as an uncertain parameter.

Monte Carlo methods were used to propagate uncertainty within contamination-uptake-risk models for calculating radionuclide-specific, increased lifetime risks of fatal cancer to an otherwise typical person who is maximally exposed over a lifetime (70 years) or over some shorter period sampled randomly from the distribution used to characterize home-residence time. The resulting characterization obtained for uncertainty in estimated total increased fatal-cancer risk associated with potential maximal exposure to all radionuclides for an exposure scenario involving a uranium-mill tailings pile is shown in Figure 5-1. The horizontal axis in that figure represents increased risk multiplied by 3.5×10^{-6}, which is the geometric mean of the distribution (shown as the solid curve) of risk to an individual maximally exposed for 70 years. (Normalization to the geometric mean value was done simply because all the risk distributions obtained were very close to lognormal.)

The vertical axis in Figure 5-1 represents cumulative probability expressed as a percentage, that is, the probability that the true (but certain) risk is less than or equal to a given, corresponding particular risk value shown on the horizontal axis. The solid horizontal line in the figure corresponds to cumulative probability equal to 50%. The dashed curve in the figure represents estimated risk accounting for less-than-lifetime home residence. In commenting on the substantial difference between the solid and dashed curves for the four types of exposure scenarios considered in this uncertainty analysis, EPA concluded that "it is clear . . . that many moves are to nearby locations," that "we do not believe that including a factor for exposure duration improves the assessment of maximum individual risk," and that "improper application of such a factor can easily lead to erroneous conclusions regarding uncertainties in the risk assessment" (EPA, 1989b).

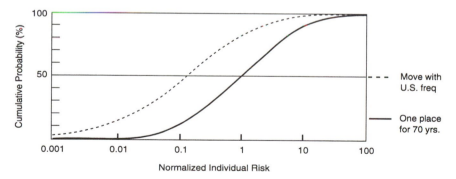

FIGURE 5-1 Uncertainty in estimated total increased fatal-cancer risk associated with potential maximal exposure to all radionuclides for an exposure scenario involving a uranium-mill tailings pile. SOURCE: Adapted from EPA, 1989b.

Presentation of Risk Estimates

Several methods can be used to display health risk estimates. Some of the terms used most often are listed in Table 5-2. The definitions are from the new 1992 exposure guidelines (EPA, 1992a). Any combination of them can be used to display the risk estimate to either the risk manager or the public. The choice of descriptors is often based on legal mandates. In general, the display includes a table indicating the risk estimated for the exposed population by route of exposure.

1992 Exposure-Assessment Guidelines

EPA's 1992 *Guidelines for Exposure Assessment* shows a clear presentation of hazard-identification, dose-response, and exposure-assessment information that might be useful in future risk assessments. Risk assessors are to examine the judgments made during the process, the constraints of available data, and the state of knowledge. According to EPA, the risk characterization should include (EPA, 1992a)

• the qualitative, weight-of-evidence conclusions about the likelihood that the chemical may pose a specific hazard (or hazards) to human health, the nature and severity of the observed effects, and by what route(s) these effects are seen to occur. These judgments affect both the dose-response and exposure assessments.

• for noncancer effects, a discussion of the dose-response behavior of the critical effect(s), data such as the shapes and slopes of the dose-response curves for the various other toxic end points, and how this information was used to determine the appropriate dose-response assessment techniques; and

• the estimates of the magnitude of the exposure, the route, duration and pattern of the exposure, relevant pharmacokinetics, and the number and characteristics of the population exposed. This information must be compatible with both the hazard identification and dose-response assessments.

The risk-characterization summary should highlight the key points of each step of the risk-assessment process.

Risk-Characterization Memorandum

EPA is in transition on risk characterization. Besides the exposure guidelines described above, the risk-characterization memorandum (Appendix B) provides guidance on risk characterization and uncertainty analysis for EPA risk managers and risk assessors. The memorandum

> addresses a problem that affects public perception regarding the reliability of EPA's scientific assessments and related regulatory decisions. . . Significant information is often omitted as the results of the assessment are passed along in the decision-making process. . . . Often, when risk information is presented to

the ultimate decision-maker and to the public, the results have been boiled down to a point estimate of risk. Such "short hand" approaches to risk assessment do not fully convey the range of information considered and used in developing the assessment. In short, informative risk characterizations clarified the scientific basis for EPA decisions, while numbers alone do not give a true picture of the assessment.

A statement attached to the memorandum from the Risk Assessment Council, made up of EPA senior managers, emphasized the following principles:

• *Full Characterization of Risk:* A full and open discussion of uncertainties in the body of each EPA risk assessment, including prominent display of critical uncertainties in the risk characterization. Numerical risk estimates should always be accompanied by descriptive information carefully selected to ensure an objective and balanced characterization of risk in risk assessment reports and regulatory documents.

• *Comparability and Consistency:* Confusion as to the comparability of similar looking (but quite different) risks, for example, the risk estimate for an average individual risk relative to the risk estimate for the most exposed individual, have led to misunderstandings about the relative significance of risks and the protectiveness of risk, reduction action. Therefore, several different descriptors of risk as outlined in the newly revised *Exposure Assessment Guidelines,* should be presented to provide a more complete picture of the risk than available from a single descriptor of risk.

• *Professional Judgment:* There are limits to the degree to which a full characterization of risk may be provided. The degree to which confidence and uncertainty are addressed depends largely on the scope of the assessment and available sources. So decision-makers and the public are not overwhelmed, only the most significant data and uncertainties need be presented. Further, when special circumstances (e. g., lack of data, extremely complex situations, resource limitations, statutory deadlines) preclude an assessment, such circumstances should be explained.

In implementing that guidance, EPA staff should:

1. Clearly present risk assessment information separate from any non-scientific risk management considerations.

2. Key scientific information on data and methods (e.g., use of animal or human data for extrapolating from high to low doses, use of pharmacokinetics data) must be highlighted, and a statement of confidence in the assessments that identifies all major uncertainties along with comment on their influence on the assessment must be provided.

3. The range of exposures derived from exposure scenarios and on the use of multiple risk descriptors (i.e., central tendency, high end of individual risk, population risk, important subgroups (if known) should be presented.

The risk-characterization memorandum goes through each step of risk assessment and outlines the questions to be answered. These are shown in Table 5-1, which suggests several issues that should be addressed to describe the information in each step fully.

Communication of Risk

Risk communication consists of two parts: communication between the risk assessor and the risk manager and communication between the risk-assessment management team and the public. The risk manager often receives the individual and population risk estimates (generally point estimates but occasionally ranges of these estimates) with only a qualitative description of the uncertainties in each. The general public often receives much less information—only the point estimate or range (without a description of the uncertainty) and the risk manager's decision—although far more is available from published sources or on request. In most regulatory situations, the manager's decision and supporting information are published in the *Federal Register*. In addition, extensive background documents that discuss the risk analysis in much more depth are often available to the public. The public is generally given an opportunity to comment within 30-60 days on the analysis and resulting decision. EPA may adjust a risk assessment on the basis of public comments.

PART
II

Strategies for Improving Risk Assessment

Previous chapters have examined the various steps of the health risk-assessment process in the sequence developed by the 1983 Red Book committee. In considering the various steps to risk assessment, the committee observed that several common themes cut across the various stages of risk assessment and arise in criticisms of each individual step. These themes are as follows:

- *Default options.* Is there a set of clear and consistent principles for choosing and departing from default options?
- *Validation.* Has the Environmental Protection Agency (EPA) made a sufficient case that its methods and models for carrying out risk assessments are consistent with current scientific information available?
- *Data needs.* Is enough information available to EPA to generate risk assessments that are protective of public health and are scientifically plausible? What types of information should EPA obtain and how should the information best be used?
- *Uncertainty.* Has EPA taken sufficient account of the need to consider, describe, and make decisions in light of the inevitable uncertainty in risk assessment?
- *Variability.* Has EPA sufficiently considered the extensive variation among individuals in their exposures to toxic substances and in their susceptibilities to cancer and other health effects?
- *Aggregation.* Is EPA appropriately addressing the possibility of interactions among pollutants in their effects on human health, and addressing the consideration of multiple exposure pathways and multiple adverse health effects?

The "Red Book" paradigm should be supplemented by applying a cross-cutting approach that uses those themes. Such an approach could ameliorate the following problems in risk assessment as it is currently practiced within the agency:

• The differing opinions in the scientific community on the merits of particular scientific evidence and the resulting lack of credibility caused by periodic revisions of particular "risk numbers" (e.g., those for dioxin).

• The reluctance to incorporate new scientific information into risk assessments when it might (erroneously) appear to increase uncertainty.

• The incompatibility of various inputs to risk characterization, e.g., dose estimates in units that cannot be combined with more sophisticated dose-response evaluations, or hazard-identification evidence that cannot readily be integrated into potency assessment.

• The emphasis on theoretical modeling over measurement.

• The production of risk assessments that are either insufficiently informative or too detailed for the needs of risk managers, and the related problem of lack of clear signals to guide risk-assessment research.

Considering the six cross-cutting themes in the planning and analysis of risk assessment will not solve the problems of risk assessment by itself. Indeed, too much emphasis on a cross-cutting vision of risk assessment might create unanticipated problems. On balance, however, the view of risk assessment proposed in Chapters 6-11 will serve two important purposes: it will give the individual cross-cutting themes a more prominent place in the risk-assessment process, and it will encourage the gradual evolution of attempts to improve risk assessment from its current, somewhat piecemeal orientation to a more holistic one, with the goal of improving the precision, comprehensibility, and usefulness for regulatory decision-making of the *entire* risk-assessment process. Whatever conceptual framework is used, the committee believes that EPA must develop principles for choosing default options and for judging when and how to depart from them. This controversial issue is described in the next section.

THE NEED FOR RISK-ASSESSMENT PRINCIPLES

Our scientific knowledge of hazardous air pollutants has numerous gaps. Hence, there are many uncertainties in the health risk assessments of those pollutants. Some of these can be referred to as model uncertainties—for example, uncertainties regarding dose-response model choices due to a lack of knowledge about the mechanisms by which hazardous air pollutants elicit toxicity. As discussed more fully in Chapter 6, EPA has developed "default options" to use when such uncertainties arise. These options are used in the absence of convincing scientific information on which of several competing models and theories is correct. The options are not rules that bind the agency; rather, they constitute guidelines from which the agency may depart when evaluating the risks posed by a specific substance. The agency may also change the guidelines as scientific knowledge accumulates.

The committee, as discussed in Chapter 6, believes that EPA has acted reasonably in electing to issue default options. Without uniform guidelines, there is a danger that the models used in risk assessment will be selected on an ad hoc basis, according to whether regulating a substance is thought to be politically feasible or according to other parochial concerns. In addition, guidelines can provide a predictable and consistent structure for risk assessment.

The committee believes that only the description of default options in a risk assessment is not adequate. We believe that EPA should have principles for choosing default options and for judging when and how to depart from them. Without such principles, departures from defaults could be ad hoc, thereby undercutting the purpose of the default options. Neither the agency nor interested parties would have any guidance about the quality or quantity of evidence necessary to persuade the agency to depart from the default options or the point(s) in the process at which to present that evidence.

Moreover, without an underlying set of principles, EPA and the public will have no way to judge the wisdom of the default options themselves. The individual default options inevitably vary in their scientific basis, foundation in empirical data, degree of conservatism, plausibility, simplicity, transparency, and other attributes. If defaults were chosen without conscious reference to these or other attributes, EPA would be unable to judge the extent to which they fulfill the desired attributes. Nor could the agency make intelligent and consistent judgment about when and how to add new default options when "missing defaults" are identified. In addition, the policies that underlie EPA's choice of risk-assessment methods would not be clear to the public and Congress—for example, it would be unclear whether EPA places the highest value on protecting public health, on generating scientifically accurate estimates, or on other concerns.

The committee has identified a number of objectives that should be taken into account when considering principles for choosing and departing from default options: protecting the public health, ensuring scientific validity, minimizing serious errors in estimating risks, maximizing incentives for research, creating an orderly and predictable process, and fostering openness and trustworthiness. There might be additional relevant criteria as well.

The choice of principles inevitably involves choosing how to balance such objectives. For instance, the most open process might not be the one that yields the result most likely to be scientifically valid. Similarly, the goal of minimizing errors in estimation might conflict with that of protecting the public health, inasmuch as (given the pervasiveness of uncertainty) achievement of the latter objective might involve accepting the possibility that a given risk assessment will overestimate the risk.

The committee therefore found it difficult to agree on what principles EPA should adopt. For example, the committee debated whether EPA should base its practices on "plausible conservatism"—that is, on attempting to use models that have support in the scientific community and that tend to minimize the possibility that risk estimates generated by these models will significantly underestimate true risks. The committee also discussed whether EPA instead should attempt as much as possible to base its practices on calculating the risk estimate most likely to be true in the light of current scientific knowledge. After extensive discussion, no consensus was reached on this issue.

The committee also concluded that the choice of principles to guide risk assessment, although it requires a knowledge of science and scientific judgment, ultimately depends on policy judgments, and thus is not an issue for specific consideration by the committee, even if it could agree on the substance of specific recommendations. The choice reflects decisions about how scientific data and inferences should be used in the risk-assessment process, not about which data are correct or about what inferences should be drawn from those data. Thus, the selection of principles inevitably involves choices among competing values and among competing judgments about how best to respond to uncertainty.

Many members contended that the committee ought not attempt to recommend principles, but should leave their formulation to the policy process. They concluded that weighing societal values is properly left to those who have been chosen, directly or indirectly, to represent the public. Indeed, in the view of these members, any recommendation by the committee would give the false impression that the choice of principles is ultimately an issue of science; noting the sharp differentiation that Congress made between the tasks of this committee and those of the Risk Assessment and Management Commission established by Section 303 of the Clean Air Act Amendments of 1990. That commission, rather than this committee, appears to have been intended to address issues of policy.

Other members contended that the committee should attempt to recommend principles. They urged that the choice of risk-assessment principles is one of the most important decisions to be made in risk assessment and one on which risk assessment experts, because of their expertise on the scientific issues related to the choice, ought to make themselves heard. They believe that the choice of principles is no more policy-laden than many other issues addressed by the committee, and that the decision not to recommend principles is itself a policy choice. They also note that the scientific elements involved in making the choice distinguish the selection of principles from other pure "policy" issues that the committee agreed not to address such as the use of cost-benefit methods or the implications of the psychosocial dimensions of risk perception.

The committee has decided not to recommend principles in its report. Instead, it has included in Appendix N papers by three of its members that offer various perspectives on the issue. One paper, by Adam Finkel, urges that EPA should strive to advance scientific consensus while minimizing serious errors of risk underestimation, by adopting an approach of "plausible conservatism." The other, by Roger McClellan and Warner North, argues that EPA should promote risk assessments that reflect current scientific understanding. Those perspectives are not intended to reflect the total range of opinion among committee members on the subject, but are presented to illustrate the issues involved.

REPORTING RISK ASSESSMENTS

As already mentioned, uncertainties are pervasive in risk assessment. When uncertainty concerns the magnitude of a physical quantity that can be measured or inferred from assumptions (e.g., ambient concentration), it can often be quantified, as Chapter 9 suggests.

Model uncertainties result from an inability to determine which scientific theory is correct or what assumptions should be used to derive risk estimates. Such uncertainties cannot be quantified on the basis of data. Any expression of probability, whether qualitative (e.g., a scientist's statement that a threshold is likely) or quantitative (e.g., a scientist's statement that there is a 90% probability of a threshold), is likely to be subjective. Subjective quantitative probabilities could be useful in conveying the judgments of individual scientists to risk managers and to the public, but the process of assessing subjective probabilities is difficult and essentially untried in a regulatory context. Substantial disagreement and misunderstanding about the reliability of quantitative probabilities could occur, especially if their basis is not set forth clearly and in detail.

In the face of important model uncertainties, it may still be undesirable to reduce a risk characterization to a single number, or even to a range of numbers intended to portray uncertainty. Instead, EPA should consider giving risk managers risk characterizations that are both qualitative and quantitative and both verbal and mathematical.

If EPA takes this route, quantitative assessments provided to risk managers should be based on the principles selected by EPA. EPA might choose to require that a risk assessment be accompanied by a statement describing alternative assumptions presented to the agency that, although they do not meet the principles selected by EPA for use in the risk characterization, satisfy some lesser test (e.g., plausibility). For example, EPA generally assumes that no threshold exists for carcinogenicity and calculates cancer potency using the linearized multistage model as the default. Commenters to the agency on a specific substance might attempt to show that there is a threshold for that substance on the basis of what is known about its mechanism of action. If the threshold can be demonstrated in a manner that is satisfactory under the agency's risk-assessment principles, the risk characterization would be based on the threshold assumption. If such a demonstration cannot be made, then the risk characterization would be based on the no-threshold assumption; but if the threshold assumption were found to be plausible, the risk manager might be informed of its existence as a plausible assumption, its rationale, and its effect on the risk estimate. In this way, risk assessors would receive both qualitative and quantitative information relevant to characterizing the uncertainty associated with the risk estimate.

THE ITERATIVE APPROACH

One strategy component that deserves emphasis is the need for iteration. Neither the resources nor the necessary scientific data exist to perform a full-scale risk assessment on each of the 189 chemicals listed as hazardous air pollutants by Section 112 of the Clean Air Act. Nor, in many cases, is such an assessment needed. Some of the chemicals are unlikely to pose more than a de minimis (trivial) risk once the maximum available control technology is applied to their sources as required by Section 112. Moreover, most sources of Section 112 pollutants emit more than one such pollutant, and control technology for Section 112 pollutants is rarely pollutant-specific. Therefore, there might not be much incentive for industry to petition EPA to remove substances from Section 112's list (or much need for EPA to devote its resources to carrying out risk assessments in response to such petitions).

An iterative approach to risk assessment would start with relatively inexpensive screening techniques and move to more resource-intensive levels of data-gathering, model construction, and model application as the particular situation warranted. To guard against the possibility of underestimating risk, screening techniques must be constructed that err on the side of caution when there is uncertainty. (As discussed in Chapter 12, the committee has some doubts about whether EPA's current screening techniques are so constructed.) The results of such screening should be used to set priorities for gathering further data and applying successively more complex techniques. These techniques should then be used to the extent necessary to make a judgment. In Chapter 7, the kinds of data that should be obtained at each stage of such an iterative process are described. The result would be a process that yields the risk-management decisions required by the Clean Air Act and that provides incentives for further research without the need for costly case-by-case evaluations of individual chemicals. Use of an iterative approach can improve the scientific basis of risk-assessment decisions and account for risk-management concerns, such as the level of protection and resource constraints.

6

Default Options

EPA's risk-assessment practices rest heavily on "inference guidelines" or, as they are often called, "default options." These options are generic approaches, based on general scientific knowledge and policy judgment, that are applied to various elements of the risk-assessment process when the correct scientific model is unknown or uncertain. The 1983 NRC report *Risk Assessment in the Federal Government: Managing the Process* defined *default option* as "the option chosen on the basis of risk assessment policy that appears to be the best choice in the absence of data to the contrary" (NRC, 1983a, p. 63). Default options are not rules that bind the agency; rather, as the alternative term *inference guidelines* implies, the agency may depart from them in evaluating the risks posed by a specific substance when it believes this to be appropriate. In this chapter, we discuss EPA's practice of adopting guidelines containing default options and departing from them in specific cases.

ADOPTION OF GUIDELINES

As our discussion of risk assessment has made clear, current knowledge of carcinogenesis, although rapidly advancing, still contains many important gaps. For instance, for most carcinogens, we do not know the complete relationship between the dose of a carcinogen and the risk it poses. Thus, when there is evidence of a carcinogenic effect at a high concentration (for instance, in the workplace or in animal testing), we do not know for certain how strong the effect (if any) would be at the lower concentrations typically found in the environment. Similarly, we do not know how much importance to attach to experiments that show that exposure to a substance causes only benign tumors in animals or how to adjust for metabolic differences between animals and humans in calculating the carcinogenic potency of a chemical.

Other uncertainties are not peculiar to carcinogenesis, but are characteristic of many aspects of risk assessment. For example, calculating the doses received by individuals might require knowledge of the relationship between emission of a substance by a source and the ambient concentration of that substance at a particular place and time. It is impossible to install a monitor at every place where people might be exposed; moreover, monitoring results are subject to error. Thus, regulators attempt to use air-quality models to predict ambient concentrations. But because our knowledge of atmospheric processes is imperfect and the data needed to use the models cannot always be obtained, the predictions from atmospheric-transport models can differ substantially from measured ambient concentrations (NRC, 1991a).

In time, we hope, our knowledge and data will improve. Indeed, we believe that EPA and other government agencies must engage in scientific research and be receptive to the results of sound scientific research conducted by others. In the meantime, decisions about regulating hazardous air pollutants must be made under conditions of uncertainty. It is vital that the risk-assessment process handle uncertainties in a predictable way that is scientifically defensible, consistent with the agency's statutory mission, and responsive to the needs of decision-makers.

These uncertainties, as we explain further in Chapter 9, are of two major types. One type, which we call *parameter uncertainty*, is caused by our inability to determine accurately the values of key inputs to scientific models, such as emissions, ambient concentrations, and rates of metabolic action. The second type, *model uncertainty*, is caused by gaps in our knowledge of mechanisms of exposure and toxicity—gaps that make it impossible to know for certain which of several competing models is correct. For instance, as mentioned above, we often do not know whether a threshold may exist below which a dose of a carcinogen will not result in an adverse effect. As we discuss in Chapter 9, model uncertainties, unlike parameter uncertainties, are often difficult to quantify.

The Red Book recommended that model uncertainties be handled through the development of uniform inference guidelines for the use of federal regulatory agencies in the risk-assessment process. Such guidelines would structure the interpretation of scientific and technical information relevant to the assessment of health risks. The guidelines, the report urged, should not be rigid, but instead should allow flexibility to consider unique scientific evidence in particular instances.

The Red Book described the advantages of such guidelines as follows (pp. 7-8):

The use of uniform guidelines would promote clarity, completeness, and consistency in risk assessment; would clarify the relative roles of scientific and other factors in risk assessment policy; would help to ensure that assessments reflect the latest scientific understanding; and would enable regulated parties to anticipate government decisions. In addition, adherence to inference guidelines will aid in maintaining the distinction between risk assessment and risk management.

This committee believes that those considerations continue to be valid. In particular, we stress the importance of inference guidelines as a way of keeping risk assessment and risk management from unduly influencing each other. Without uniform guidelines, risk assessments might be manipulated on an ad hoc basis according to whether regulating a substance is thought to be politically feasible. In addition, we believe that inference guidelines can provide a predictable and consistent structure for risk assessment and that a statement of guidelines forces an agency to articulate publicly its approach to model uncertainty.

Like the committee that produced the 1983 NRC report, we recognize that there is an inevitable interplay between risk assessment and risk management. As the 1983 report states (pp. 76, 81), "risk assessment must always include policy, as well as science," and "guidelines must include both scientific knowledge and policy judgments." Any choice of defaults, or the decision not to have defaults at all, therefore amounts to a policy decision. Indeed, without a policy decision, the report stated, risk-assessment guidelines could do no more than "state the scientifically plausible inference options for each risk assessment component without attempting to select or even suggest a preferred inference option" (NRC, 1983a, p. 77). Such guidelines would be virtually useless. The report urged that risk-assessment guidelines include risk-assessment policy and explicitly distinguish between scientific knowledge and risk-assessment policy to keep policy decisions from being disguised as scientific conclusions (NRC, 1983a, p. 7). That report urged that for consistency, policy judgments related to risk assessment ought to be based on a common principle or principles.

We believe that EPA acted reasonably in electing to issue *Guidelines for Carcinogen Risk Assessment* (EPA, 1986a). Those guidelines set out policy judgments about the accommodation of model uncertainties that are used to assess risk in the absence of a clear demonstration that a particular theory or model should be used.

For instance, the default options indicate that, in assessing the magnitude of risk to humans associated with low doses of a substance, "in the absence of adequate information to the contrary, the linearized multistage procedure will be employed" (EPA, 1986a, p. 33997). The linearized multistage procedure implies low-dose linearity. At low doses, if the dose is reduced by, say, a factor of 1,000, the risk is also reduced by a factor of 1,000; dose is linearly related to risk. Departure from this default option is allowed, under EPA's current guidelines, if there is "adequate evidence" that the mechanism through which the substance is carcinogenic is more consistent with a different model—for instance, that there is a threshold below which exposure is not associated with a risk. Thus, the default option in guiding a decision-maker, in the absence of evidence to the contrary, assigns the burden of persuasion to those who wish to show that the linearized multistage procedure should not be used. Similar default options cover such important issues as the calculation of effective dose, the treatment of benign tumors, and the procedure for scaling animal-test results to estimates of potency in humans.

Some default options are concerned with issues of extrapolation—from laboratory animals to humans, from large to small exposures (or doses), from intermittent to chronic lifetime exposures, and from route to route (as from ingestion to inhalation). That is because few chemicals have been shown in epidemiologic studies to cause measurable numbers of human cancers directly, and epidemiologic data on only a few of these are sufficient to support quantitative estimates of human epidemiologic cancer risk. In the absence of adequate human data, it is necessary to use laboratory animals as surrogates for humans.

One advantage of guidelines, as already noted, is that they can articulate both the agency's choice of individual default options and its rationale for choosing all of the options. EPA's guidelines set out individual options but do not do so with ideal clarity. Nor has the agency explicitly articulated the scientific and policy bases for its options. Hence, there might be disagreement about precisely what the agency's default options are and the rationales for these options. We attempt here to identify the most important of the options (numbered points in the 1986 guidelines are cited):

• Laboratory animals are a surrogate for humans in assessing cancer risks; positive cancer-bioassay results in laboratory animals are taken as evidence of a chemical's cancer-causing potential in humans (IV).

• Humans are as sensitive as the most sensitive animal species, strain, or sex evaluated in a bioassay with appropriate study-design characteristics (III.A.1).

• Agents that are positive in long-term animal experiments and also show evidence of promoting or cocarcinogenic activity should be considered as complete carcinogens (II.B.6).

• Benign tumors are surrogates for malignant tumors, so benign and malignant tumors are added in evaluating whether a chemical is carcinogenic and in assessing its potency (III.A.1 and IV.B.1).

• Chemicals act like radiation at low exposures (doses) in inducing cancer; i.e., intake of even one molecule of a chemical has an associated probability for cancer induction that can be calculated, so the appropriate model for relating exposure-response relationships is the linearized multistage model (III.A.2).

• Important biological parameters, including the rate of metabolism of chemicals, in humans and laboratory animals are related to body surface area. When extrapolating metabolic data from laboratory animals to humans, one may use the relationship of surface area in the test species to that in humans in modifying the laboratory animal data (III.A.3).

• A given unit of intake of a chemical has the same effect, regardless of the time of its intake; chemical intake is integrated over time, irrespective of intake rate and duration (III.B).

• Individual chemicals act independently of other chemicals in inducing cancer when multiple chemicals are taken into the body; when assessing the risks associated with exposures to mixtures of chemicals, one treats the risks additively (III.C.2).

EPA has never articulated the policy basis for those options. As we discuss in the previous introductory section (Part II), the agency should choose and explain the principles underlying its choices to avoid the dangers of ad hoc decision-making. The agency's choices are for the most part intended to be conservative—that is, they represent an implicit choice by the agency, in dealing with competing plausible assumptions, to use (as default options) the assumptions that lead to risk estimates that, although plausible, are believed to be more likely to overestimate than to underestimate the risk to human health and the environment. EPA's risk estimates thus are intended to reflect the upper region of the range of risks suggested by current scientific knowledge.

EPA appears to use conservative assumptions to implement Congress's authorization in several statutes, including the Clean Air Act, for the agency to undertake preventive action in the face of scientific uncertainty (see, e.g., *Ethyl v. EPA*, 541 F.2d 1 (D.C. Cir.) (en banc), *certiorari denied* 426 U.S. 941 (1976), ratified by Section 401 of the Clean Air Act Amendments of 1977) and to set standards that include a precautionary margin of safety against unknown effects and errors in calculating risks (see *Environmental Defense Fund v. EPA*, 598 F.2d 62, 70 (D.C. Cir. 1978) and *Natural Resources Defense Council v. EPA*, 824 F.2d 1146, 1165 (en banc) (D.C. Cir. 1987)).

EPA's choice of defaults has been controversial. We note, though, that some of the arguments about EPA's practices are directed less at conservatism than at the means of implementation that the agency has adopted. We believe that the iterative approach recommended in the previous chapter combined with quantitative uncertainty analysis will improve the agency's practices regardless of the degree of conservatism chosen by the agency. We also note that with an iterative approach, the agency must use relatively conservative models in performing screening estimates designed to indicate whether a pollutant is worthy of further analysis and comprehensive risk assessment. Such estimates are intended to obviate the detailed assessment of risks that can with a high degree of confidence be deemed acceptable or de minimis (trivial). By definition, therefore, screening analyses must be sufficiently conservative to make sure that a pollutant that could pose dangers to health or welfare will receive full scrutiny.

Over time, the choice of defaults should have decreasing impact on regulatory decision-making. As scientific knowledge increases, uncertainty diminishes. Better data and increased understanding of biological mechanisms should enable risk assessments that are less dependent on default assumptions and more accurate as predictions of human risk.

In evaluating EPA's risk-assessment methods, we are aware that the agency's guidelines, to use the terminology of the earlier NRC report, are in part statements of science policy, rather than purely statements of scientific fact. The guideline cited above dealing with extrapolation of high doses to low doses is illustrative. The guideline is not a claim that it is known that the relationship between dose and response is linear; that the true relationship between dose and response is uncertain and could be nonlinear is readily acknowledged. Rather, the guideline is based (1) on the scientific conclusion that the linear model has substantial support in current data and biologic theory and that no alternative model has sufficient support to warrant departure from the linear model for most chemicals identified as carcinogens; (2) on the further scientific conclusion that the linear model is more conservative than most alternative plausible models; and (3) on the policy judgment that a conservative model should be chosen when there is model uncertainty.

DEPARTURES FROM DEFAULT OPTIONS

Agency policies should encourage further scientific research. Risk assessors and managers must be receptive to new scientific information about the character and magnitude of the toxic effects of a chemical substance. Putting this receptivity into practice, though, has proved difficult. The 1983 NRC report criticized how agencies had implemented their guidelines. The report noted that "the application of inference options to specific risk assessments has been marked by a general lack of explicitness" and that that made it "difficult to know whether assessors adhere to guidelines" (NRC, 1983a, p. 79). The NRC report recognized the need to prevent ad hoc and undocumented departures from guidelines in specific risk assessments. But the NRC report made it clear that well-designed guidelines "should permit acceptance of new evidence that differs from what was previously perceived as the general case, when scientifically justifiable." NRC urged a recognition of the need for a tradeoff between flexibility on the one hand and predictability and consistency on the other (NRC, 1983a, p. 81).

The NRC advocated that agencies seek a middle path between inflexibility and ad hoc judgments, but steering this course is difficult. Consistency and predictability are served if an agency sets out criteria for departing from its guidelines. If such criteria are themselves too rigidly applied, the guidelines could ossify into inflexible rules; but without such criteria, the guidelines could be subverted at will with the potential for political manipulation of risk assessment.

NRC's approach requires that agencies regard their inference options not as binding rules, but rather as guidelines that are to be followed unless a sufficient showing is made. In the decade since the NRC report, EPA has never articulated clearly its criteria for a departure. We believe that a structured approach would give better guidance to the scientific community and to the public and would ensure both that the default options are set aside only when there is a valid scientific reason for doing so and that decisions to set aside defaults are scientifically credible and receive public acceptance.

EPA's practice appears to be to allow departure in a specific case when it ascertains that there is a consensus among knowledgeable scientists that the available scientific evidence justifies departure from the default option. The agency apparently considers both the quality of the data submitted and the robustness of the theory that is used to justify the departure.

EPA needs to be more precise in describing the kind and strength of evidence that it will require to depart from a default option. Because the decision as to the evidentiary burden to be required is ultimately one of policy, and because we could not reach agreement on proposed language to implement such a standard (see Appendixes N-1 and N-2), we do not urge any particular standard; moreover, we are conscious of the difficulties of capturing the nuances of judgment in any verbal formula that will not be open to misinterpretation.

We believe that the agency must continue to rely on its Science Advisory Board (SAB) and other expert bodies to determine when departing from a default option is warranted according to default options EPA will develop. EPA has increasingly used peer review and workshops as a way to ensure that it carefully considers the propriety of departing from a default. These and other devices should continue to ensure broad peer and scientific participation to guarantee, as much as possible, that the agency's risk-assessment decisions are made with access to the best science available.

We note that here, too, EPA has a difficult path to tread. EPA has been criticized for delay in deciding whether to depart from default options. Increased procedural formality raises the possibility of further delays, especially in a period of budgetary stringency such as EPA can expect to face for some time. It is likely that EPA will be cutting back on hiring personnel at the salary ranks necessary to attract scientists with the needed experience and training to judge whether departure from a default option is justifiable. Congress ought to be aware of the need for greater agency resources to carry out the mandates of the Clean Air Act and similar legislation.

Even if a default option is not set aside, we believe that decision-makers ought to be informed in a narrative way of any specific information suggesting that, in specific cases, alternatives to the default options might have equal or greater scientific support, and believe that the characterization of risk should include a discussion of the effect of the alternative options on risk estimates.

CURRENT EPA PRACTICE IN DEPARTING
FROM DEFAULT OPTIONS

As discussed above, EPA needs simultaneously to be receptive to evidence indicating the need to depart from a default option and to be careful that it departs from a default in a specific case only when a departure is justifiable. In addition, the agency needs to follow a process that allows peer participation and review.

We discuss below some of the cases in which EPA has addressed the issue of whether to depart from default options. In each of these cases, EPA decisions to depart from default options lessened its estimate of the risk; however, it is important to note that new scientific data could also increase the estimate of risk above that reached by using the default options.

Example 1: Use of Animal-Cancer Bioassay Data

The example that follows illustrates a departure from the two default options that: (1) positive animal-bioassay results for cancer induction are sufficient proof of cancer hazard in humans; and (2) humans are at least as sensitive as the most sensitive responding animal species. It involves induction of kidney cancer in male laboratory rats by a number of chemicals—most important, 1,4-dichlorobenzene, hexachloroethane, isophorone, tetrachloroethylene, dimethyl methyl phosphorate, d-limonene, pentachloroethane, and unleaded gasoline (EPA, 1991d). The first four have been classified as hazardous air pollutants by the Clean Air Act Amendments of 1990.

Male rats exposed to those chemicals develop dose-related kidney cancer; the highest incidence is usually 25% or less. The tumors do not occur in other organs or other species or in female rats. Because of the economic importance of several of the compounds and unleaded gasoline, extensive studies were conducted to understand the mechanisms involved in the development of the tumors. The studies suggested that a special mechanism was responsible for the tumors in male rats. When the chemicals in question are inhaled by male rats, the chemicals, or products of their metabolism, reach the bloodstream and form complexes with a specific protein, alpha-2μ-globulin, that is produced in the male rat liver and removed from the blood by the kidneys. As the complex is cleared from the blood by the kidneys, it accumulates there in the form of hyaline droplets, which lead to the development of kidney disease characterized by cell death, cast formation, mineralization, and hyperplasia. This accumulation, as well as statistically significant increases in tumors that result from exposure to the chemicals, occurs only in male rats.

In contrast, female rats, which do not have the same concentrations of alpha-2μ-globulin protein, do not develop statistically significant increases tumors as a result of exposure. Similarly, the protein is not present in detectable quantities in humans, so no risk of kidney-cancer development by this mechanism would be expected in humans exposed to the chemicals in question. It was therefore suggested that, inasmuch as a special mechanism not found in humans seemed to be responsible for the tumors, EPA ought to depart in this case from its default option that a substance that is carcinogenic in animals is also a human carcinogen. In response, EPA (1991d) evaluated the evidence of production of kidney tumors in male rats by chemicals inducing alpha-2μ-globulin accumulation (CIGAs), such as those in question. EPA's review suggested that kidney cancer in male rats from exposure to CIGAs is due only to the kidney disease that CIGAs cause through accumulation of alpha 2μ-globulin. For instance, EPA noted, the CIGAs are not known to react with DNA and are generally negative in short-term tests for genotoxicity. In contrast, classical kidney carcinogens (or their active metabolites) are usually electrophilic species that bind covalently to macromolecules and form DNA adducts. With the classical kidney carcinogens, which presumably are carcinogenic in both laboratory animals and humans, the kidney carcinogenesis is presumed to result from the interaction of the compounds or their metabolites with DNA. Classical kidney carcinogens, such as dimethylnitrosamine, induce renal tubule cancer in laboratory animals at a high incidence in both sexes after short periods of exposure, with a clear increase in kidney tumor incidence with increased dose. Thus, the classical kidney carcinogens and CIGAs appear to act via different mechanisms.

After reviewing the data, EPA (1991d) provided specific decision criteria for categorizing a chemical as a CIGA. A substance may be so classified only if it meets all the decision criteria, and classification of a chemical as a CIGA does not keep it from being considered as a carcinogen because of other modes of action. In that way, the agency precisely tailored its proposed departure from default options. EPA concluded that renal tubule tumors in male rats attributable solely to chemically induced alpha-2μ-globulin accumulation should not be used for human-cancer hazard identification or for dose-response extrapolations. Furthermore, EPA noted that even in the absence of renal tubule tumors in the male rat, if the lesions of alpha-2μ-globulin syndrome are present, the associated nephropathy in male rats should not contribute to determinations of noncarcinogenic hazard or risk.

EPA's documents reviewed and synthesized the available scientific information in a document that was then presented to peers in a public meeting, reviewed by the SAB's Environmental Health Committee and later endorsed by the SAB Executive Committee, and transmitted to the administrator (EPA, 1991d). Transmission to the administrator was accompanied by endorsement by the SAB that the document outlined a scientifically sound policy for departing from the default option for this specific class of compounds. This policy has been generally supported by the scientific community. However, it is noteworthy that some researchers (see, e.g., Melnick, 1992) believe that another mechanism to explain all of the observed data is equally or more plausible than the one

EPA endorsed. Alpha-2µ-globulin may be a carrier protein that transports certain chemicals to the kidney, where toxic metabolites can be released; this mechanism defines alpha-2µ-globulin accumulation as an *indicator*, rather than the *cause* of renal toxicity. If so, humans may have other carrier proteins that could transport toxins to the kidney and cause toxicity or carcinogenicity in the absence of protein droplet information, and the assumption that the rat studies are irrevelant to humans might therefore be erroneous.

Example 2: Linkages Between Exposure, Dose, and Response

In the previous example, a departure from default options occurred at the hazard-identification stage. As discussed in examples 2 and 3, such departures can also be used to refine the unit risk estimate of a carcinogen.

Calculating the unit risk through quantitative risk assessment requires an understanding of the relationship between exposure to a substance and response. One part of this relationship involves the link between exposure (that is, intake of a substance) and dose (that is, the amount of the substance, or harmful metabolites, that is taken up by bodily organs). However, that understanding is incomplete. EPA's default options assume that all species are equally sensitive to a given target-tissue dose of the toxicant or its metabolites. The surface-to-area ratios in the test species and humans are used as the key to relating the dose received by the test species to the dose that would cause similar effects in humans (see pp. 6-7, III.A.3). As the following examples show, however, evidence can sometimes support departing from this default option.

Methylene Chloride

Epidemiological studies on whether exposure to methylene chloride causes cancer in humans have produced equivocal results. Thus, assessment of methylene chloride's carcinogenic risk depends on use of laboratory animal data and especially on several long-term bioassays. Syrian hamsters did not show a tumor response at any site at exposures up to 3,500 ppm for 6 hr/day 5 days/week, but mice and rats exposed at up to 4,000 ppm for 6 hr/day 5 days/week had treatment-related tumorigenic effects. EPA, after evaluating the data, classified methylene chloride as a probable human carcinogen (B2).

In accord with the default options of EPA's guidelines, the carcinogenic potency of methylene chloride was estimated by scaling the laboratory animal data to humans with a body surface-area conversion factor. The resulting cancer risk estimate was 4.1×10^{-6} for exposure at 1 µg/m^3 (Table 6-1). After further consideration, EPA has decreased this estimate by an order of magnitude (EPA, 1991d). The reduction is based on research on the pathways through which methylene chloride is metabolized. As with some other carcinogens, the risk of cancer arises not from methylene chloride itself, but rather from its metabolites.

TABLE 6-1 Cancer Incidence in B6C3F1 Female Mice Exposed to Methylene Chloride and Human Cancer Risk Estimates Derived from Animal Data

Animal Data				
Concentration, Administered mg/kg, day	Transformed Animal mg/kg, day	Human Equivalent mg/kg, day	Incidence of Liver Tumors	Incidence of Lung Tumors
4000	3162	712	40/46	41/46
2000	1582	356	16/46	16/46
0	0	0	3/45	3/45

Human Risk Estimates	
Extrapolation Model	Cancer Risk[b] for 1 $\mu g/m^3$
LMS[a], surface area	4.1×10^{-6}
LMS, PB-PK[c]	3.7×10^{-8}
Logit	2.1×10^{-13}
Weibull	9.8×10^{-8}
Probit	$<10^{-15}$
LMS-PB-PK with scaling for sensitivity	4.7×10^{-7}

[a]LMS = linearized multistage model.
[b]Upper 95% confidence limit.
[c]PB-PK = physiologically based pharmacokinetic.

SOURCE: Modified from Reitz et al., 1989.

A correct calculation of the risk posed by methylene chloride therefore rests on understanding the human body's processes for metabolizing this chemical.

Research with animal species used in the bioassays and human tissue has shed light on the metabolism of methylene chloride. Much of the research was conducted with the goal of providing input for physiologically based pharmaco-kinetic (PBPK) models (Andersen et al., 1987, 1991; Reitz et al., 1989). The data were modeled in various ways, including consideration of two metabolic pathways. One involves oxidation by mixed-function oxidase (MFO) enzymes, and the other involves a glutathione-S-transferase (GST). Both pathways in-volve the formation of potentially reactive intermediates: formyl chloride in the MFO pathway and chloromethyl glutathione in the GST-mediated pathway. The MFO pathway was modeled as having saturable, or Michaelis-Menten, kinetics, and the GST pathway as a first-order reaction, i.e., proportional to concentration. The analyses suggested that a reactive metabolite formed in the GST pathway was responsible for tumor formation. This pathway, according to the analyses, contributes importantly to the disposition of methylene chloride only at exposures that saturate the primary MFO pathway. The analyses further indicated that the GST pathway is less active in human tissues than in mice. This suggests that the default option of scaling for surface area yields a human risk estimate that is too high to be plausible. EPA incorporated the data on pharmacokinetics and metabolism into its most recent risk assessment for methylene chloride, although it retained a surface-area correction factor—now identifying it as a correction for interspecies differences in sensitivity. The new risk estimate is 4.7 $\times 10^{-7}$ for continuous exposure at 1 $\mu g/m^3$ (Table 6-1).

The process by which EPA arrived at the current risk estimate for methylene chloride with PBPK modeling involved use of peer-review groups and SAB review to achieve a scientifically acceptable consensus position on the validity of the alternative model. After EPA's re-evaluation, however, articles in the peer-reviewed literature began to focus attention on parameter uncertainties in PBPK modeling, which neither EPA nor the original researchers in the methylene chloride case had considered. In the specific case of methylene chloride, at least one of the analyses (Portier and Kaplan, 1989) suggested that according to the new PBPK information EPA should have raised, rather than lowered, its original unit risk estimate if it wanted to continue to take a conservative stance. The more general point, which we discuss in Chapter 9, is that EPA must simultaneously consider both the evidence for departing from default models and the need to generate or modify the parameters that drive both the alternative and default models.

Formaldehyde

The toxicity and carcinogenicity of formaldehyde, a widely used commodity chemical, have been intensely studied and recently reviewed (Heck et al., 1990; EPA, 1991e). Concern for the potential human carcinogenicity of formaldehyde was heightened by the observation that exposure of rats at high concentrations (14.3 ppm) resulted in a very large increase in the incidence of nasal cancer. That observation gave impetus to the conduct and interpretation of epidemiologic studies of formaldehyde-exposed human populations. In the aggregate, the 28 studies that have been reported provide limited evidence of human carcinogenicity (EPA, 1991e). The "limited" classification is used primarily because the incidence of cancers of the upper respiratory tract has been confounded by exposure to other agents known to increase the rate of cancer, such as cigarette smoke and wood dusts.

The effects of chronic inhalation of formaldehyde have been investigated in rats, mice, hamsters, and monkeys. The principal evidence of carcinogenicity comes from studies in both sexes and two strains of rats and the males of one strain of mice, all showing squamous cell carcinomas of the nasal cavity.

The results of the rat bioassay have been used to derive quantitative risk estimates for cancer induction in humans (Kerns et al., 1983). Table 6-2 shows these animal data and the estimates of human cancer risk based on different exposure-dose models. (The table uses the inhalation cancer unit risk—the lifetime risk of developing cancer from continuous exposure at 1 ppm.) The 1987 EPA risk estimate (EPA, 1987c) measured exposure as the airborne concentration of formaldehyde. The rat bioassay shows a steep nonlinear exposure-response relationship for nasal-tumor induction. For example, two tumors were observed at 5.6 ppm, whereas 37 would have been expected from linear extrapolation from 14.3 ppm. Similarly, no tumors were observed at 2 ppm, whereas linear extrapolation from 14.3 ppm would have predicted 15.

The key issue became whether the same exposure-response relationship exists in people as in rats. To determine the answer, researchers directed substantial effort toward investigating the mechanisms by which formaldehyde exerted a carcinogenic effect. One avenue of investigation was directed toward character-

TABLE 6-2 Incidence of Nasal Tumors in F344 Rats Exposed to Formaldehyde and Comparison of EPA Estimates of Human Cancer Risk Associated with Continuous Exposure to Formaldehyde

	Exposure rate, ppm[a]	Incidence of Rat Nasal Tumors	
	14.3	94/140	
	5.6	2/153	
	2.0	0/159	
	0	0/156	
	Upper 95% Confidence Limit Estimates		
Exposure Concentration, ppm	1987 Risk Estimates[b]	1991 Risk Estimates[c]	
		Monkey-Based	Rat-Based
1.0	2×10^{-2}	7×10^{-4}	1×10^{-2}
0.5	8×10^{-3}	2×10^{-4}	3×10^{-3}
0.1	2×10^{-3}	3×10^{-5}	3×10^{-4}
	Maximum Likelihood Estimates		
1.0	1×10^{-2}	1×10^{-4}	1×10^{-2}
0.5	5×10^{-4}	1×10^{-5}	1×10^{-3}
0.1	5×10^{-7}	4×10^{-7}	3×10^{-5}

[a]Exposed 6 hr/day, 5 days/week for 2 years.

[b]Estimated with 1987 inhalation cancer unit risk of 1.6×10^{-2} per ppm, which used airborne concentration as measure of exposure.

[c]Estimated with 1991 inhalation cancer unit risks of 2.8×10^{-3} per ppm (rat) and 3.3×10^{-4} per ppm (monkey), which used DNA-protein cross-links as measure of exposure.

SOURCE: Adapted from EPA, 1991b.

izing DNA-protein cross-links as a measure of internal dose of formaldehyde (Heck et al., 1990). That work, initially conducted in rats, demonstrated a steep nonlinear relationship between formaldehyde concentration and formation of DNA-protein cross-links in nasal tissue, where most inhaled formaldehyde is deposited in rats. This suggested a correlation between such cross-links and tumors.

When the studies were extended to monkeys, a similar nonlinear relationship was observed between exposure concentration and DNA-protein cross-links in nasal tissue, but the concentration of DNA-protein cross-links per unit of exposure concentration was substantially lower than in the rat. Because the breathing patterns of humans more closely resemble those of monkeys than those of rats, the results of these studies suggested that using rats as a surrogate for humans might overestimate doses to humans, and hence the risk presented to humans by formaldehyde. EPA's most recent risk assessment (EPA, 1991e) used DNA-protein cross-links as the exposure indicator and estimated the human cancer risk (Table 6-2). EPA noted that the cross-links were being used only as a measure of delivered dose and that present knowledge was insufficient to ascribe a mechanistic role to the DNA-protein cross-links in the carcinogenic process.

The EPA risk estimates for formaldehyde have been the subject of extensive peer review and review by the SAB. The 1992 update was reviewed by the SAB Environmental Health Committee and Executive Committee. The SAB recommended that the agency attempt to develop an additional risk estimate using the epidemiological data and prepare a revised document reporting all the risk estimates developed by the alternative approaches with their associated uncertainties. The two examples just discussed used mechanistic data and modeling to improve the characterization of the exposure-dose link. It is possible that as knowledge increases, models can be developed that link dose to response; the possibility is further discussed in Chapter 7.

The same is true of the linearized multistage model. As noted earlier, this model assumes that risk is linear in dose. As noted earlier, however, rats exposed to formaldehyde show a steep nonlinear exposure-response relationship. This raises the possibility that the linearized multistage model might be inappropriate for at least some chemicals. It is possible that advances in knowledge of the molecular and cellular mechanisms of carcinogenesis will show a need to use other models either case by case or generically. More discussion of this matter can be found in Chapter 7.

The strategy advocated for formaldehyde would build on multistage models of the carcinogenic process that describe the accumulation of procarcinogenic mutations in target cells and the consequent malignant conversion of these cells (Figure 6-1). The Moolgavkar-Venzon-Knudson model substantially oversimplifies the carcinogenic process but provides structural framework for integrating and examining data on the role of DNA-protein cross-links, cell replication, and other biologic phenomena in formaldehyde-induced carcinogenesis (Mool-

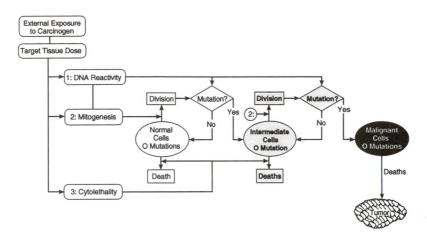

FIGURE 6-1 Model of chemical carcinogenesis built on multi-stage carcinogenesis model of Moolgavkar-Venzon-Knudson. SOURCE: Conolly et al., 1992. Reprinted with permission, copyright 1992 by Gordon & Breach, London.

gavkar and Venzon, 1979; Moolgavkar and Knudson, 1981; Moolgavkar et al., 1988; NRC, 1993b). Key features of this model are definition of the relationship of target-tissue dose to exposure and the use of that dose as a determinant of three outcomes: reactivity with DNA, mitogenic alterations, and cytolethality. These, in turn, cause further biologic effects: DNA reactivity leads to mutations, the mitogenic stimuli increase the rate of cell division, and cells die (cell death stimulates compensatory cell proliferation). Models like that shown provide a structured approach for integrating data on a toxicant, such as formaldehyde. It is anticipated that modeling will provide insight into the relative importance, at various exposure concentrations, of the two mechanisms that appear to have a dominant role in formaldehyde carcinogenesis: mutation and cell proliferation. Improved insight into their role could provide a mechanistic basis for selecting between the linearized multistage mathematical model now used for extrapolation from high to low doses and alternative models that might have more biologic plausibility.

Trichloroethylene

Trichloroethylene (TCE) is a chlorinated solvent that has been widely used in the industrial degreasing of metals. TCE is a concern to EPA as an air pollutant, a water pollutant, and a substance frequently present in ground water at Superfund sites. EPA carried out a risk assessment for TCE documented in a health assessment document (HAD) (EPA, 1985d) and a draft addendum incorporating additional inhalation-bioassay data (EPA, 1987e). Both documents were reviewed by the SAB (EPA, 1984a; EPA, 1988j,k). The second document has not been issued in final form, and no further revision of EPA's risk assessment on TCE has been made since 1987.

The carcinogenic potency of TCE is based on the liver-tumor response in B6C3F1 mice, a strain particularly prone to liver tumors. The carcinogenicity of TCE might result from trichloroacetic acid (TCA), a metabolite of TCE that is itself known to cause liver tumors in mice. TCA is one of a number of chemicals that cause proliferation of peroxisomes, an intracellular organelle, in liver cells. Peroxisome proliferation has been proposed as a causal mechanism for the liver tumors, and proponents have asserted that such tumors should receive treatment in risk assessments different from evaluation under EPA's default assumptions. In particular, human liver cells might be much less sensitive than mouse liver cells to tumor formation from this mechanism, and the dose-response relationship might be nonlinear at low doses.

The SAB held a workshop in 1987 on peroxisome proliferation as part of its reviews on risk assessments for TCE and other chlorinated solvents. While endorsing a departure from the default on the alpha-2μ-globulin mechanism described in example 1 above, the SAB declined to endorse such a departure for peroxisome proliferation, noting that a causal relationship for this mechanism was "plausible but unproven." The SAB strongly encouraged further research, describing this mechanism for mouse liver tumors as "most promising for immediate application to risk assessment" (EPA, 1988k). The SAB criticized EPA on the draft Addendum on TCE (EPA, 1987e) for not adequately presenting uncertainties and for not seriously evaluating recent studies on the role of peroxisome proliferation (EPA, 1988l).

In the TCE case, departure from the defaults was rejected after an SAB review that recognized the peroxisome proliferation mechanism as plausible. Controversy over the interpretation of liver tumors in B6C3F1 mice continues. Some scientists assert that EPA's use of the tumor-response data from this particularly sensitive strain has been inappropriate (Abelson, 1993; ILSI, 1992). In the TCE example, departure from the defaults might become appropriate, on the basis of improved understanding of mouse liver tumors and their implications for human cancer. Although the SAB declined to endorse such a departure in 1987, it strongly encouraged further research as appropriate for supporting improved risk assessment.

Cadmium

Cadmium compounds are naturally present at trace levels in most environmental media, including air, water, soil, and food. Substantial additional amounts might result from human activities, including mining, electroplating, and disposal of municipal wastes. EPA produced an HAD on cadmium (EPA, 1981b) and later an updated mutagenicity and carcinogenicity assessment (EPA, 1985e). The latter went through SAB review (EPA, 1984b), which pointed out many weaknesses and research needs for improving the risk assessment. No revision of the risk assessment on cadmium has occurred since 1985.

EPA used epidemiological data for developing a single unit risk estimate for all cadmium compounds. Use of the estimate from the best available bioassay would have given a unit risk for cadmium compounds higher by a factor of 50. The SAB and EPA in its response to SAB comments (EPA, 1985f) agreed that the solubility and bioavailability of different cadmium compounds were important in determining the risk associated with different cadmium compounds and that such differences might explain the discrepancy between the epidemiological data and the bioassay data. No implementation of the principle that cadmium compounds should be evaluated on the basis of bioavailability has yet been devised, although its importance to risk assessment for some air pollutants that contain cadmium is clearly set forth in EPA's response to the SAB (EPA, 1985f).

EPA's existing risk assessment for cadmium might be judged adequate for screening purposes. But the SAB review and the EPA response to it suggest that the carcinogenic risk associated with a specific cadmium compound could be overestimated or underestimated, because bioavailability has not been included in the risk assessment. A refined version of the risk assessment that includes bioavailability might be appropriate, especially if residual risks for cadmium compounds appear to be important under the Clean Air Act Amendments of 1990.

Nickel

Nickel compounds are found at detectable levels in air, water, food, and soil. Increased concentrations of airborne nickel result from mining and smelting and from combustion of fuel that contains nickel as a trace element. Nickel compounds present in smelters that use the pyrometallurgical refining process are clearly implicated as human carcinogens. EPA's HAD on nickel (EPA, 1986b) lists dust from such refineries and nickel subsulfide as category A (known human) carcinogens. A rare nickel compound, nickel carbonyl, is listed, on the basis of sufficient evidence in animals, as category B2. Other nickel compounds are not listed as carcinogens, although EPA states (EPA, 1986b, p. 2-11) :

> The carcinogenic potential of other nickel compounds remains an important area for further investigation. Some biochemical and *in vitro* toxicological studies seem to indicate the nickel ion as a potentially carcinogenic form of nickel and nickel compounds. If this is true, all nickel compounds might be potentially carcinogenic with potency differences related to their ability to enter and to make the carcinogenic form of nickel available to a susceptible cell. However, at the present time, neither the bioavailability nor the carcinogenesis mechanism of nickel compounds is well understood.

The SAB reviewed the nickel HAD and concurred with EPA's listing of only the three rare nickel species as category A and B2 carcinogens (EPA, 1986c).

The results of bioassays on three nickel species by the National Toxicology Program are due to be released soon, and these results should provide a basis for revision of risk assessments for nickel compounds.

The cadmium and nickel examples point out an important additional default option: Which compounds should be listed as carcinogens when it is suspected that a class of chemical compounds is carcinogenic? Neither the cadmium risk assessment, the nickel risk assessment, or EPA's *Guidelines for Carcinogen Risk Assessment* (EPA, 1986a) provide specific guidance on this issue.

Dioxins

Dioxins is a commonly used name for a class of organochlorine compounds that can form as the result of the combustion or synthesis of hydrocarbons and chlorine-containing substances. One isomer, 2,3,7,8-tetrachlorodibenzo-*p*-dioxin (TCDD), is one of the most potent carcinogens ever tested in bioassays. EPA issued an HAD for dioxins (EPA, 1985g), which the SAB criticized for its treatment of the non-TCDD isomers that may contribute substantially to the overall toxicity of a mixture of dioxins (EPA, 1985h).

The potency calculation for TCDD has continued to be a subject of controversy. Research indicates that the toxic effects of TCDD may result from the binding of TCDD to the Ah (aromatic hydrocarbon) receptor. In 1988, EPA asked the SAB to review a proposal to revise its risk estimate for TCDD. SAB agreed with EPA's criticism of the linearized multistage model and its assessment of the promise of alternative models based on the receptor mechanism. But SAB did not agree that there was adequate scientific support for a change in the risk estimate. SAB carefully distinguished its recommendation from a change that EPA might wish to make as part of risk management (EPA, 1989f)

> The Panel thus concluded that at the present time the important new scientific information about 2,3,7,8-TCDD does not compel a change in the current assessment of the carcinogenic risk of 2,3,7,8-TCDD to humans. EPA may for policy reasons set a different risk-specific dose number for the cancer risk of 2,3,7,8-TCDD, but the Panel finds no scientific basis for such a change at this time. The Panel does not exclude the possibility that the actual risks of dioxin-induced cancer may be less than or greater than those currently estimated using a linear extrapolation approach.

A recent conference affirmed the scientific consensus on the receptor mechanism for TCDD, but there was not a consensus that this mechanism implied a basis for departure from low-dose linearity (Roberts, 1991). After the conference, and after the recommendations of the SAB (EPA, 1989f), EPA initiated a new study to reassess the risk for TCDD. That study is now in draft form and scheduled for SAB review in 1994.

The potencies of other dioxin isomers and isomers of a closely related chemical class, dibenzofurans, have been estimated by EPA with a toxic-equivalency-factor (TEF) method (EPA, 1986d). The TEF method was endorsed by the SAB as a reasonable *interim* approach in the absence of data on these other isomers (EPA, 1986e). The SAB urged additional research to collect such data. Municipal incinerator fly ash was used as an example of a mixture of isomers of regulatory importance that might be appropriate for long-term animal testing.

The EPA initiative for a review of TCDD is one of the few instances in which the agency has initiated revision of a carcinogen risk assessment on the basis of new scientific information. Dioxins and dibenzofurans are unique in that potency differences within this class of closely related chemical isomers are dealt with through a formal method that has undergone peer review by the SAB.

Example 3: Modeling Exposure-Response Relationship

If chemicals act like radiation at low exposures (doses) inducing cancer—i.e., if intake of even one molecule of a chemical has an associated probability for cancer induction that can be calculated—the appropriate model for relating exposure-response relationships is a linearized multistage model.

Of the 189 hazardous air pollutants, unit risk estimates are available for only 51: 38 with inhalation unit risks, which are applicable to airborne materials, and 13 with oral unit risks. The latter probably have less applicability to estimating the health risks associated with airborne materials. All 38 inhalation unit risk values have been derived with a linearized multistage model; i.e., it is assumed that the chemicals act like radiation. That might be an appropriate assumption for chemicals known to affect DNA directly in a manner analogous to that of radiation. For other chemicals—e.g., such nongenotoxic chemicals as chloroform—the assumption of a mode of action similar to that of radiation might be erroneous, and it would be appropriate to consider the use of biologically-based exposure-response models other than the linearized multistage model.

The process of choosing between alternative exposure-response models is difficult because the models cannot be validated directly for their applicability for estimating lifetime cancer risks at exposures of regulatory concern. Indeed, it is possible to obtain cancer incidence data on exposed laboratory animals and distinguish them from the control incidence only over a narrow range, from some value over 1% (10^{-2}) to about 50% (5×10^{-1}) cancer incidence. In regulation of chemicals, the extrapolation may be over a range of up to 4 orders of magnitude (from 10^{-2} to 10^{-6}), going from experimental observations to estimated risks of cancer incidence at exposures of regulatory concern. One approach to increasing the accuracy with which comparisons between measured outcome and model projections can be made involves increasing the size of the experimental populations. However, statistical considerations, the cost of studying large numbers of animals, and the greater difficulty of experimental control in larger studies put narrow limitations on the use of this approach. Similar problems exist in conducting epidemiological studies.

An attractive alternative is to use advances in knowledge of the molecular and cellular mechanisms of carcinogenesis. Identification of events (e.g., cell proliferation) and markers (e.g., DNA adducts, suppressor genes, oncogenes, and gene products) associated with various steps in the multistep process of carcinogenesis creates a potential for modeling these events and products at low exposure. Direct tests of the validity of exposure-response models at risks of around 10^{-6} are not likely in the near future. However, with an order-of-magnitude improvement in sensitivity of detection of precancerous events with a probability of occurrence down to around 10^{-3}-10^{-2}, the opportunity will be available to evaluate alternative modes of action and related exposure-response models at substantially lower exposure concentrations than has been possible in the past. For example, it should soon be possible to evaluate compounds that are presumed to have different modes of action (direct interaction with DNA and genotoxicity versus cytotoxicity) and alternative models (linearized multistage versus nonthreshold) that might yield markedly different risks when extrapolated to realistic exposures and low risks.

FINDINGS AND RECOMMENDATIONS

Use of Default Options

FINDING: EPA's practice of using default options when there is doubt about the choice of appropriate models or theory is reasonable. EPA should have a means of filling the gap when scientific theory is not sufficiently advanced to ascertain the correct answer, e.g., in extrapolating from animal data to responses in humans.

RECOMMENDATION: EPA should continue to regard the use of default options as a reasonable way to cope with uncertainty about the choice of appropriate models or theory.

Articulation of Defaults

FINDING: EPA does not clearly articulate in its risk-assessment guidelines that a specific assumption is a default option.

RECOMMENDATION: EPA should clearly identify each use of a default option in future guidelines.

Justification for Defaults

FINDING: EPA does not fully explain in its guidelines the basis for each default option.

RECOMMENDATION: EPA should clearly state the scientific and policy basis for each default option.

Alternatives to Default Options

FINDING: EPA's practice appears to be to allow departure from a default option in a specific case when it ascertains that there is a consensus among knowledgeable scientists that the available scientific evidence justifies departure from the default option. EPA, though, has not articulated criteria for allowing departures.

RECOMMENDATION: The agency should consider attempting to give greater formality to its criteria for a departure, to give greater guidance to the public and to lessen the possibility of ad hoc, undocumented departures from default options that would undercut the scientific credibility of the agency's risk assessments. At the same time, the agency should be aware of the undesirability of having its guidelines evolve into inflexible rules.

PROCESS FOR DEPARTURES

FINDING: EPA has relied on its Science Advisory Board and other expert bodies to determine when a consensus among knowledgeable scientists exists.

RECOMMENDATION: EPA should continue to use the Science Advisory Board and other expert bodies. In particular, the agency should continue to make the greatest possible use of peer review, workshops, and other devices to ensure broad peer and scientific participation to guarantee that its risk-assessment decisions will have access to the best science available through a process that allows full public discussion and peer participation by the scientific community,

Missing Defaults

FINDING: EPA has not stated all the default options in each step in the risk-assessment process, nor the steps used when there is no default. Chapters 7 and 10 elaborate on this matter and identify several possible "missing defaults."

RECOMMENDATION: EPA should explicitly identify each generic default option in the risk-assessment process.

7

Models, Methods, and Data

INTRODUCTION

Health risk assessment is a multifaceted process that relies on an assortment of methods, data, and models. The overall accuracy of a risk assessment hinges on the validity of the various methods and models chosen, which in turn are governed by the scope and quality of data. The degree of confidence that one can place in a risk assessment depends on the reliability of the models chosen and their input parameters (i.e., variables) and on how well the boundaries of uncertainty have been quantified for the input parameters, for the models as a whole, and for the entire risk-assessment process.

Quantitative assessment of data quality, verification of method, and validation of model performance are paramount for securing confidence in their use in risk assessment. Before a data base is used, the validity of its use must be established for its intended application. Such validation generally encompasses both the characterization and documentation of data quality and the procedures used to develop the data. Some characteristics of data quality are overall robustness, the scope of coverage, spatial and temporal representativeness, and the quality-control and quality-assurance protocols implemented during data collection. More specific considerations include the definition and display of the accuracy and precision of measurements, the treatment of missing information, and the identification and analysis of outliers. Those and similar issues are critical in delineating the scope and limitations of a data set for an intended application.

The performance of methods and models, like that of data bases, must be characterized and verified to establish their credibility. Evaluation and validation procedures for a model might include sensitivity testing to identify the parameters having the greatest influence on the output values and assessment of its accuracy, precision, and predictive power. Validation of a model also requires an appropriate data base.

This chapter discusses the evaluation and validation of data and models used in risk assessment. In cases where there has been an insufficient assessment of performance or quality, research recommendations are made. Although in this chapter we consider validation issues sequentially, according to each of the stages in the (modified) Red Book paradigm, our goal here is to make the assessment of data and model quality an iterative, interactive component of the entire risk-assessment and risk-characterization process.

EMISSION CHARACTERIZATION

As described in Chapter 3, emissions are characterized on the basis of emission factors, material balance, engineering calculations, established Environmental Protection Agency (EPA) protocols, and measurement. In each case, this characterization takes the structural forms of a linearly additive process (i.e., emissions equals product – [feedstock + accumulations]), a multiplicative model (i.e., emissions equals [emission factor][process rate]), or an exponential relationship (e.g., emission equals intercept + [(emission factor) (measurement)^{exp}]).

The additive form is based on the mass-balance concept. An estimate is made by measuring the feedstock and product to determine an equipment-specific or process-specific transfer coefficient. This coefficient is used to estimate emissions to the atmosphere. The measurements available for the additive form are often not sufficiently precise and accurate to yield complete information on inputs and outputs (NRC, 1990a). For example, an NRC committee (NRC, 1990a) considered a plant that produced 5 million pounds of ethylene per day and used more than 200 monitoring points to report production with a measurement accuracy of 1%, equivalent to 50,000 lb of ethylene per day. The uncertainty in this estimate (50,000 lb) greatly exceeded a separate estimate of emissions, 191 lb, which was calculated by the plant and was confirmed by monitoring of the emission points. Thus, despite the apparently good precision of estimates within 1%, the additive method was not reliable. This seems to be generally true for complicated processes or multiple processing steps.

The other forms are based on exponential and multiplicative models. Each may be deterministic or stochastic. For example, emissions from a well-defined sample of similar sources may be tested to develop an emission factor that is meant to be representative of the whole population of sources. A general difficulty with such fits that use these functional (linear or one of several nonlinear forms) forms is that the choice of form may be critical but hard to validate. In addition, it must be assumed that data from the sources used in the calculations are directly applicable to the sources tested in process design and in the management and maintenance approaches of the organizations that run them are the same in all cases.

An example of an exponential form of an emission calculation is shown in Figure 7-1. This figure shows the correlation between screening value (the measurement) and leak rate (the emission rate) for fugitive emissions from a valve. The screening value is determined by measuring the hydrocarbons emitted by a piece of equipment (in this case, a valve in gas service) with an instrument like an OVA (organic-vapor analyzer). The leak rate (i.e., emission) is then determined by reading the value on the y axis corresponding to that screening value. Note that the plot is on a log-log scale, so that a "3" on the x axis indicates that a 1,000-ppm screening value corresponds to a "–3.4" on the y axis, or 0.001 lb/hr for each valve in gas service at that screening value. The observations here are based on an analysis conducted for 24 synthetic organic chemical manufacturing industry (SOCMI) units representing a cross-section of this industry (EPA, 1981a).

As part of this analysis, a six-unit maintenance study (EPA, 1981a) was used to determine the impact of equipment monitoring and maintenance using an OVA instrument on emission reduction. The equation derived for the valve

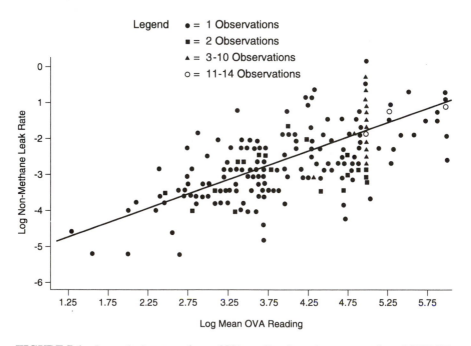

FIGURE 7-1 Log_{10} leak rate vs. log_{10} OVA reading for values-gas service. SOURCE: EPA, 1981a.

emissions in gas service explains only 44% (square of the correlation coefficient) of the variance in the points shown in Figure 7-1. Similar results were obtained from other possible emission points.

The facilities in this SOCMI study could reduce the estimate of their emissions by 29-99% by determining plant-specific emission factors, indicating the difficulties in using industry-wide average to represent specific plant behavior.

The multiplicative form improves on the emission-factor approach, in that it incorporates more features of the process, attempting to accommodate the types of equipment being used, the physical properties of the chemical, and the activity of the equipment as a whole. The deterministic form of the multiplicative model is based on the chemical and physical laws that determine the emission rate. The variables measured—vapor pressure, molecular weight, temperature, etc.—are chemical physical properties that are related to the emission rate. The multiplicative form provides some scientific basis for the estimate beyond the simple curve-fitting. However, it has difficulties, because some of the properties are not constant. For example, the ambient air temperature, one factor in determining the emission rate, can vary quite widely within a day. The average temperature for a given period, such as a month, is used for ease in calculation, but this practice introduces some error. EPA might want to consider a more detailed analysis in which the emissions that occur during the period are stratified into groups with smaller variations in variables such as ambient temperature. The emissions in the strata could be estimated and weighted sums calculated to provide a better estimate.

Probably the most accurate procedure is to use none of those "forms" to determine emissions, but rather to sample stack and vent emissions at each source. However, such sampling can be quite expensive, and the costs could overburden owners of small sources. Apart from costs, the primary difficulty with this procedure is that it yields an estimate for one site on one occasion. Emissions could change because of a variety of factors. An alternative to testing is to estimate emissions from monitoring data. Continuous emission monitors (CEMs), which are available for a small number of chemicals, are placed in stacks or near fugitive-emission points to measure the concentration of a chemical being released; concentrations can then be converted to amounts. However, CEMs can be expensive and difficult to maintain, and they may produce incomplete or inaccurate measurements. When such testing is conducted, however, they may show that other kinds of estimates are seriously in error. For example, a study (Amoco/EPA, 1992) compared emissions estimated primarily from emission factors with those determined during testing. The measured overall actual estimate of emissions was more than twice as high as the TRI estimate for a variety of reasons, including identification of new sources, overestimation or underestimation of the importance of some sources, and the lack of a requirement to report source emissions under a particular regulation.

Evaluation of EPA Practice   EPA has worked diligently to help members of the public who are required to provide emission estimates for regulatory purposes. This 20-year effort has provided documents that are used to estimate air-pollutant emissions throughout the world. However, in some cases, EPA has had to provide emission estimation factors based on very little information about the process involved; it was difficult to check the assumption that the process for which the calculation is being used is similar to the process that was tested in the development of the emission factor.

There are two basic difficulties with the way EPA applies its emission estimation techniques. First, most estimates are made by using the emission factors or by fitting the linear or exponential forms. As discussed previously, the accuracy of emission estimates using these techniques might not be high.

Second, the information is generated in such a way that only point estimates are presented. Although it is clear from the earlier discussion that there can be uncertainty in the estimates, EPA has extensive files on how the emission factors were determined, and this information presumably contains enough points to generate distribution of emissions rather than just a point estimate. EPA provides only qualitative ratings of the accuracy of the emission method. The ratings are not based on the variance in the estimate, but just on the number of emission points used to generate the data. If there are enough points to generate an emission factor, it is possible to estimate the distribution of emission factors from which an estimate can be chosen to solve a particular exposure-risk estimation problem.

However, the emission factors are given only a "grade" from A (best) to E relative to the quality and amount of data on which estimates are based. An emission factor based on 10 or more plants would likely get an "A" grade, whereas a factor based on a single observation of questionable quality or one extrapolated from another factor for a similar process would probably get a D or E. The grades are subjective and do not consider the variance in the data used to calculate factors. According to EPA (1988e), the grades should "be used only as approximations, to infer error bounds or confidence intervals about each emission factor. At most, a [grade] should be considered an indicator of the accuracy and precision of a given factor used to estimate emissions from a large number of sources." The uncertainty in the estimates is such that EPA is not comfortable with the A-E system and is developing a new qualitative system to indicate uncertainty. EPA is attempting to generate estimation factors for hazardous air pollutants industry by industry, but it is still hesitant to ascribe any sort of uncertainty to emission factors.

A single disruption in operation of a plant can increase the release rate for some interval (hour or day). An extreme example is the dioxin release from a manufacturing plant in Seveso, Italy. Such disruptions are not incorporated into any of the emission characterizations, except for the few cases where emission monitoring is available. However, in those cases, emissions might be so high that they exceed the maximum reading of a monitor and thereby lead to just a lower bound (if this problem is recognized) or even to a serious underestimate of the actual emission. Furthermore, the frequency and duration of such episodes are unpredictable.

Therefore, EPA should also attempt to make some sort of quantitative estimates of the variability of measured emissions among sources within a category and of the uncertainty in its overall emission estimates for individual sources and the source category as a whole. This issue is discussed in more depth in Chapter 10, but could involve analyzing the four kinds of circumstances as appropriate for a particular source type—routine, regular maintenance, upsets and breakdowns, and rare catastrophic failures. EPA could also note the implications of the dynamics of causation of different effects for emission estimation, and the resulting need for estimates of exposure and exposure variability over different averaging times.

The itemization of emissions by chemical constituent also raises problems. Emission characterization methods often provide only the amount of VOCs (volatile organic compounds) that is emitted. The amounts of particular compounds (benzene, toluene, xylene, etc.) within these VOC emissions are often not individually reported. Without the emission data on particular compounds, it is impossible to provide the information needed for exposure modeling in the risk-assessment process.

EPA does not appear to be making major strides toward improving the methods used to evaluate emissions. Although EPA is making extensive efforts to distribute the emission factors it has generated, the committee has found insufficient effort either to evaluate the accuracy of the underlying method used to derive the emission estimates or to portray the uncertainty in the emission factors. The primary exception is a joint effort of the Chemical Manufacturers Association (CMA) and EPA on fugitive emissions called Plant Organization Software System for Emission Estimation or POSSEE (CMA, 1989). In this case, companies are testing fugitive emissions within plants and collecting data on chemical and physical variables to derive emission estimates based on deterministic models (which use physical and chemical properties), rather than stochastic models. There have been efforts to increase the scientific justification of estimates of emissions from storage tanks: the American Petroleum Institute has developed data that have been used for developing the estimation method shown in the multiplicative form described above. The question then arises as to how to approach emission estimates in exposure assessments and risk assessments. The uncertainty in the mass-balance approach (additive form) can be so large that its use should be discouraged for any purposes other than for a very general screening. It is unlikely that an emission estimate derived with this method would be appropriate for risk assessment.

The linear emission-factor approach could be used as a general screening tool in an exposure assessment. As indicated by EPA in response to a question from this committee:

While emission factor-based estimates can be useful in providing a general picture of emissions across an entire industrial category, use of such factors to provide inputs to a site-specific risk assessment may introduce a great deal of uncertainty into that assessment.

If such an approach is used for an entire industrial category, then at least the uncertainty of each emission factor should be determined. If there is enough information to derive an emission factor, then a probability distribution could be calculated. There may then be disagreement about where on the probability distribution the emission estimate should be chosen. However, it is better to make the choice explicitly, as discussed in Chapter 9. The same situation is true for emissions estimated with the exponential and multiplicative approaches. EPA should include a probability distribution in all its emission estimates.

One method to determine the uncertainty in an emission estimate more easily would be to require each person submitting an emission estimate (for SARA 313 requirements, permitting, etc.) to include an evaluation of the uncertainty in the estimate. EPA could then evaluate the uncertainty in the estimation methods to determine whether the estimation was done properly. Although that might increase the costs of developing submissions slightly, the organization submitting the estimate might benefit from the results. Small sources unable to afford such analysis could instead define a range that is consistent with known or readily determined factors in their operation (e.g., for a dry cleaner, the pounds of clothes per week and gallons of solvent purchased each month).

EPA is reviewing, revising, and developing emission estimation methods for sources of the 189 chemicals. It is focusing on adding data, rather than evaluating its basic approach—the use of a descriptive model, instead of a model based on processes, for emission estimation. It appears from the examples given above that the uncertainties in emissions can dominate an exposure assessment and that a concerted effort to improve emission estimation could serve to substantially reduce the uncertainty in many risk estimates. Combined industry efforts to improve the techniques used to estimate fugitive emissions on the basis of physical and chemical properties (not just curve-fitting) should be encouraged.

EXPOSURE ASSESSMENT

Once an emission characterization is developed, it becomes one of the inputs into an air-quality model to determine the amount of a pollutant in ambient air at a given location. A population-exposure model is then used to determine how much of a pollutant reaches people at that location.

Population

The size of the population that might be exposed to an emission must be determined. Population data have been collected, published, and scrutinized for centuries. Many such data refer to entire populations or subpopulations, so questions of representation and statistical aspects of sampling do not arise in their usual form. Even where sampling is used, a large background of technique and experience allows complex estimation and other kinds of modeling to proceed without the large uncertainties inherent in, for example, extrapolation from high to low doses of toxic agents or from rodents to humans.

Population data are almost always affected to some degree by nonsampling error (bias), but this is well categorized, understood, and not a serious problem in the context of risk assessment. For example, terminal-digit preference (e.g., a tendency to report ages that end in zero or five) has been minimal since the attainment of nearly universal literacy and especially since the adoption of birth certification. Attainment of advanced ages (i.e., over 80 years) is still overstated, but this is not quantitatively serious in age estimation for purposes of risk assessment (because EPA still assumes that 70 years is the upper-bound value of the length of a lifetime). Population undercounts in the U.S. census of 1990 averaged about 2.1% and were substantially higher for some subgroups, perhaps up to 30%; however, even 30% uncertainty is smaller than many other sources of error that are encountered in risk assessment. The largest proportionate claim of uncertainty seems to be in the number of homeless persons in the United States; estimated uncertainty is less than a factor of 10.

Estimation of characteristics in groups or subgroups not examined directly is subject to additional uncertainty. For example, the 1992 population is not directly counted, but standard techniques are used to extrapolate from the census of 1990, which was a nearly complete counting of the population. Investigators have found earlier years estimates to be generally quite accurate, whether the extrapolations were strictly mathematical (e.g., based on linear extrapolation) or demographic (based on accounting for the addition of 3 years between 1990 and 1993, with adjustments for deaths, for births of the population under age 3, deaths, and net migration). The problems are greater for states and smaller areas, because data on migration (including internal migration) are not generally available.

Error tends to increase as subgroups get smaller, partly because statistical variability increases (i.e., small sample size leads to less precision in the estimate of the central tendency with any distributed measurement), but also because individual small segments are not as well characterized and as well understood as larger aggregates and because population data are generally collected according to a single nationwide protocol that allows for little deviation to accommodate special problems.

The committee is comfortable about using published population data for nearly all population characteristics and subgroups. Where adjustment to reduce errors is feasible, it should be used; but in the overall context of risk assessment, error in population assessment contributes little to uncertainty.

In some cases, a research study must define and identify its own population without help from official census and surveys. An example is a long-term followup study of workers employed in a specific manufacturing plant. When such studies are done by skilled epidemiologists, total counts, ages, and other demographic items tend to be accurate to within a factor of 2 or 3. The largest uncertainties are likely to be in the estimation of exposure to some toxic agent; these are often dealt with by the use of rough categories (high, medium, and low exposure) or surrogate measures (e.g., years employed in a plant, rather than magnitude of exposure). Errors in such work are of great concern, but they tend to be peculiar to each study and hence lead to study-specific remedies in design, performance, or analysis. They tend to be smaller than other kinds of uncertainties, but can still be of concern if a putative effect is also small.

As indicated, population data derived from a census and fortified with estimation methods are regarded as accurate and valid, and uncertainties introduced into risk assessment are relatively small. There is a need, however, for information on additional population characteristics that are not included in the census. There is a paucity of activity-pattern information, and population-exposure models or individual-exposure-personal-exposure models have not been adequately tested or validated, because they use people's activity to estimate exposure to chemicals in air. Only a few small efforts have been undertaken to develop such a data base, namely, EPA's Total Exposure and Assessment Methodology (TEAM) program and the California EPA's State Activity Pattern Study. Those programs have acquired information about people's activities that cause the emission of air pollutants or place people in microenvironments containing air pollutants that potentially lead to exposure. There is a need to develop a national data base on activity patterns that can be used to validate models that estimate personal exposure to airborne toxic chemicals. Accurately described activity patterns coupled with demographic characteristics (e.g., socioeconomic) can be used for making a risk assessment and assessing the environmental equity of risk across socioeconomic groups and races.

When exposure-characterization models are developed for use in risk assessment, the bias and uncertainty that they yield in the calculation of exposure estimates should be clearly defined and stated, regardless of whether activity patterns are included. Later, the choice of an appropriate model from an array of possibilities should be based on, but not necessarily limited to, its quantitative measure of performance and its rationale should be included with a statement of the criteria for its selection.

Air-Quality Model Evaluation

Air-quality models are powerful tools for relating pollutant emissions to ambient air quality. Most air-quality models used in assessing exposure to toxic air pollutants have been extensively evaluated with specific data sets, and their underlying mathematical formulations have been critically reviewed. Relative to some of the other models for risk assessment of air pollutants, air-quality models probably enjoy the longest history of model evaluation, refinement, and re-evaluation. For example, the original Gaussian-plume models were formulated and tested in the 1950s. That does not mean, however, that model evaluation does not still continue or that the model evaluation should be dismissed in assessing air-pollutant exposure; in fact, previous studies have shown the benefits of model evaluation in every application.

Evaluation of the air-quality models and other components of air-pollutant risk assessment is intended to determine accuracy for providing the details required in a given application and to provide confidence in the results. In air-quality modeling, that is particularly important. A Gaussian-plume model, when used with the input data generally available, might not correctly predict where maximal concentrations will be realized (e.g., because winds at the nearest station, such as an airport, might differ in direction from winds near the source of interest), but should provide a reasonable estimate of the distribution of pollutant concentrations around the site. That might be sufficient for some applications, but not others. Model evaluation can also add insight as to whether a tool is "conservative" or the opposite, and it can provide a quantitative estimate of uncertainty.

Of particular concern are the more demanding applications of models, such as in areas of complex terrain (e.g., hills, valleys, mountains, and over water), when deposition is important, and when atmospheric transformation occurs. As discussed below, it is difficult enough to use models in the simple situations for which they were specifically designed. One should always try to ascertain the level of accuracy that can be expected from a given model in a given application. Sufficient studies have been performed on most air-quality models to address that question.

Zannetti (1990) reviews evaluations of many air-quality models, including Gaussian-plume models. Evaluation procedures have recently been reviewed for photochemical air-quality models (NRC, 1991a). Similar procedures are applicable to other models. In essence, the models should be pushed to their limits, to define the range in which potential errors in either the models themselves or their inputs still lead to acceptable model performances and so that compensatory errors in the models and their inputs (e.g., meteorology, emissions, population distributions, routes of exposure, etc.) will be identified. That should lead to a quantitative assessment of model uncertainties and key weaknesses. As pointed out in the NRC (1991a) report, model evaluation includes evaluation of input data. The greatest limitation in many cases is in the availability and integrity of the input data; for the most part, many models can give acceptable results when good-quality input data are available.

A key motivation in model evaluation is to achieve a high degree of confidence in the eventual risk assessment. Pollutant-transport model evaluation, as it pertains to estimating air-pollutant emissions, has been somewhat neglected and is used without adequate discussion and analysis. For example, the modeling of emissions from the ASARCO smelter (EPA, 1985b) showed significant bias. However, the reasons for both the bias and errors were not fully identified. A major plume-model validation study was mounted in the early 1980s with support of the Electric Power Research Institute (EPRI); it was the first study of a large coal-fired power plant situated in relatively simple terrain. The study compared three Gaussian-plume models and three first-order closure numerical (stochastic) models, and an experimental, second-order closure model, for which ground-level concentrations were obtained with both routine and intensive measurement programs (Bowne and Londergan, 1983). (*First-order closure* and *second-order closure* refer to how the effects of turbulence are treated.) The authors conclude that

• The models were poor in predicting the magnitude or location of concentration patterns for a given event.
• The models performed unevenly in estimating peak concentrations as a function of averaging time; none provided good agreement for 1-, 3-, and 24-hour averaging periods.
• The cumulative distribution of hourly concentrations predicted by the models did not match the observed distribution over the full range of concentration values.
• The variation of peak concentration values with atmospheric stability and distance predicted by the Gaussian models did not match the pattern of observed peak values.
• One of the first-order closure models performed better than the Gaussian models in estimating peak concentration as a function of meteorological characteristics, but its predictive capacity was poorer than desirable for detailed risk assessments, and it systematically overpredicted the distance to the maximal concentrations.
• One of the other first-order closure models systematically underpredicted plume impacts, but its predictive capacity was otherwise superior to that of the Gaussian models.
• An experimental second-order closure model did not provide better estimates of ground-level concentrations than the operational models.

Predictions and observed pollutant concentrations often differed by factors of 2-10. It is clear from the study—in which there was no effect of complex terrain, heat islands, or other complicating effects—that the dispersion models had serious deficiencies. Dispersion models have been developed since then, but they require further development and improvement and they warrant evaluation when applied to new locations or periods.

Larger-scale urban air-quality models perform better in predicting concentrations of secondary species—such as ozone, nitrogen dioxide, and formaldehyde—even though the complex chemical reactions might seem to make the task harder. Prediction accuracy, on the average, is usually within about 10% (NRC, 1990a). This performance is due in part to the coarser spatial resolution used by the model, the chemical transformation times allowing the dispersion from the original sources, and better spatial separation of the sources. The lower spatial resolution, with increased chemical detail and performance, leads back to a consideration of model choice and evaluation: What type of detail is required from a particular model application and what level of performance can be expected?

In summary, model evaluation is an integral part of any risk assessment and is crucial for providing confidence in models. Evaluation procedures have been developed for various classes of air-quality models. Studies have shown that air-quality models can give reasonable predictions, but do not always (or often) do so. Results of a model evaluation can be used in an uncertainty analysis of predicted risk.

Evaluation of EPA Practice   The validity of the population-exposure models used by EPA remains largely untested. Ott et al. (1988) used data from EPA's TEAM studies of carbon monoxide (CO) of Denver and Washington, D.C., to examine the validity of the SHAPE model and compared the estimated co-exposure distribution based on the SHAPE model with the distribution based on direct measurement (personal monitoring). They found the estimated average exposure to be similar with the two approaches, but the ranges in estimated exposure distributions were quite different. The SHAPE exposure model predicted median values well, but there were substantial discrepancies in the tails of the distribution.

Duan (1991) also using data from EPA's TEAM study of carbon monoxide in Washington, D.C., found that the concentrations and time intervals were independent and tested the effectiveness of a "variance-components exposure" model in comparison with SHAPE. Both the long-term average concentrations and short-term fluctuations in concentration were important in predicting exposure. Duan (1988) and Thomas (1988) examined several statistical parameters for several microenvironments and found the time-invariant component (i.e., a component that does not vary with time, often taken as a background level) to be dominant. Thus, there has been some effort to validate the exposure models developed for research purposes.

There have been no systematic attempts, however, to validate either of the exposure models used for regulatory purposes, the Human Exposure Model (HEM) and the National Ambient Air Quality Standard Exposure Model (NEM). The dispersion-model portion of HEM was compared with other simple Gaussian-plume models, and the results were similar. However, neither actual airborne concentrations nor measured integrated exposures to any airborne constituent were compared with the model results to test its utility in estimating individual or population exposures. Comparison of the site-specific model used to evaluate the health impact of arsenic from the ASARCO smelter in Tacoma,

Washington, from the few available data proved to have low marginal accuracy, and arsenic in the exposed human urine samples did not correlate well with estimated exposures, as discussed in Chapter 3. Thus, the effectiveness of these models is essentially unknown, although it will be important to understand their strengths and limitations, including prediction accuracy and the associated uncertainty, when residual risk must be estimated after installation of Maximum Achievable Control Technology (MACT).

When EPA conducts a risk assessment of a hazardous air pollutant, it generally relies on Gaussian-plume models. Gaussian-plume models are inadequately formulated, so inaccuracies appear in predicted pollutant concentrations (e.g., Gaussian-plume models generally are not applicable for nonlinear chemistry or particle dynamics). Furthermore, the inputs to these models are often inaccurate and not directly appropriate for a given application. In practice, application of Gaussian-plume models has not been adequately evaluated, and some evaluations have shown substantial discrepancies. More comprehensive and robust pollutant-transport models (i.e., those more directly applicable to a wider variety of situations) are available, including stochastic Lagrangian and photochemical models, and evaluations have shown good agreement with direct observations. In specific applications, model evaluation (via pollutant monitoring and assessment of model inputs and theory) should be undertaken and ranges of applicability determined. Demonstrations should include, but not be restricted to, showing that the model assumptions reasonably represent physical-chemical behavior of the contaminant, source configuration, and atmospheric dispersion. For environmental conditions for which the performance of Gaussian-plume models are demonstrated to be unsatisfactory, more comprehensive models should be considered; however, their superior performance should be documented and clearly evident when they are considered as an alternative in a risk assessment.

EPA has generally not included population activity, mobility, and demographics and has not adequately evaluated the use of population averages (as used by default in HEM) in its exposure assessments. Exposure models, such as NEM and SHAPE, have been developed to account for personal activity. Population-activity models should be used in exposure assessments; however, their accuracy should be clearly demonstrated before considering them as alternatives to the default approach. Demographics might also play a role in determining risk. Further evaluation of some simple methods (e.g., use of population centroids), compared with more comprehensive tools (e.g., NEM and SHAPE), is warranted, before they are considered in lieu of the default option.

EPA currently uses HEM to screen exposure associated with HAP releases from stationary sources. The HEM-II model uses a standardized EPA Gaussian-plume dispersion model and assumes nonmobile populations residing outdoors at specific locations. The HEM construct is not designed to provide accurate estimates of exposure in specific locations and for specific sources and contaminants when conditions are not represented by the simplified exposure- and dis-person-model assumptions inherent in the standard HEM components. Alternative models for transport and for personal activity and mobility can be adopted in an exposure-modeling system to provide more accurate, scientifically founded, and robust estimates of pollutant-exposure distributions (including variability, uncertainty, and demographic information). Those models can be linked to geographic data bases to provide both geographic and demographic information for exposure-modeling systems.

Application of HEM generally does not include noninhalation exposures to hazardous air pollutants (HAPs) (e.g., dermal exposure), but these routes can be important. Modeling systems similar to extensions of HEM have been developed to account for the other pathways. Unless there is good evidence to the contrary, the contribution of alternative pathways of exposure to HAPs should be considered explicitly and quantified in a risk assessment.

Relatively simple models for exposure assessments, such as HEM, can provide valuable information for setting priorities and determining what additional data should be developed. However, exposure estimates that use this model can have large uncertainties (e.g., a factor of 2-10 due to the Gaussian-plume dispersion model used in HEM alone). Furthermore, Gaussian-plume models, in general, have not been validated for pollutants that are reactive and easily transformed to other chemicals such as organic gases (e.g., formaldehyde), particles, and acids (e.g., nitric and sulfuric acids). Multiple exposure routes can add still more uncertainty as to actual exposure. Uncertainty can be used as a tool for assessing the performance of a model like HEM. This is because HEM is based on very simplified descriptions of pollutant dynamics and was designed for use as a screening tool for estimating human exposure via inhalation.

The predictive accuracy and uncertainty associated with the use of the HEM should be clearly stated with each exposure assessment. The underlying assumption that the calculated exposure estimate is a conservative one should be reaffirmed; if not, alternative models whose performance has been demonstrated to be superior should be used in exposure assessment.

ASSESSMENT OF TOXICITY

The first step in assessing human toxicity based on animal experiments is the extrapolation of observations from studies in rats, mice, monkeys, and other laboratory animals to humans. The extrapolation procedure used in risk assessment to assess the toxicity of a substance is both an intellectual exercise and a tool for making practical decisions. It is based on two assumptions: that the biological response to an external stimulus in one species will occur in a different species that is subjected to the same stimulus and that the biological response is proportional to the size of the stimulus (except that a very small stimulus will often result in only a transient response or no immediate response at all). Those two assumptions are invoked whenever extrapolation from animals to humans and from high doses to low doses is performed. Cancer and other end points are discussed separately here because considerations related to extrapolation can differ.

Cancer

Qualitative Considerations

Cancer, defined as abnormal and uncontrolled growth, is ubiquitous among higher organisms; it occurs in plants, animals, and humans. In some cases, carcinogens can be identified as physical or chemical agents or self-replicating infectious agents. Many epidemiological studies have documented an association between exposure to particular chemicals and an increased incidence of particular malignancies in humans (Doll and Peto, 1981). Examples are cancers related to exposure to industrial agents—such as aniline dyes, mustard gas, some metal compounds, and vinyl chloride—and, in the general population, tobacco and tobacco smoke. Perhaps most convincing in this context is the repeated observation that cessation of exposure to a given chemical (e.g., cessation of smoking or introduction of appropriate mitigation or hygienic measures) results in a decrease in cancer incidence. When tested in animal studies, almost all known human carcinogens have been found to produce cancer in other mammals. There are a few exceptions to that rule, e.g. tobacco smoke in laboratory animals. Recent advances in the understanding of basic mechanisms of carcinogenesis, often very similar in laboratory animals and humans, lend credibility to a relationship between animal carcinogenesis and human carcinogenesis, particularly when mutagenicity is involved (OSTP, 1985; Barbacid, 1986; Bishop, 1987); in other cases, advances in the understanding of species-specific mechanisms of carcinogenesis do not support a relationship between humans and specific laboratory animals studied to date (Ellwein and Cohen, 1992). Current long-term carcinogenicity bioassays are conducted with rodents using, among other doses, the highest dose that does not reduce survival as a result of causes other than cancer, known as the maximum tolerated dose (MTD). Information acquired from rodent bioassays conducted at the MTD might yield information on whether a chemical can produce tumors in humans, but it generally cannot provide information on whether it produces tumors through generalized, indirect mechanisms or directly as a result of its specific properties. Mechanistic data could resolve the question of whether it is valid to extrapolate the results of a bioassay to humans (see NRC, 1993b). Current regulatory practice takes the view that in the absence of information to the contrary, animal carcinogens are human carcinogens; however, the data base supporting this assumption is not complete.

Obtaining more information on the biological mechanisms of carcinogenesis, their dose dependence, and their interspecies relevance will permit better and more valid qualitative and quantitative extrapolations. For example, there is a tendency to give more weight to an observation when it relates chemical exposure to development of malignant tumors and to place less emphasis on an observation that suggests that a given chemical induces benign tumors. It might be an oversimplification to consider one category of abnormal growth as invariably detrimental and another as comparatively harmless. Tumor biology is much more complicated. Most, if not all, bronchial adenocarcinomas will kill when they run their course, whereas subcutaneous lipomas will not; however, excision of a malignant basal cell skin tumor is considered a cure, whereas a benign tumor of the VIIIth cranial nerve or of the pituitary gland can be lethal. Available knowledge on causes of cancer and on the biological behavior of tumors does not permit us to ascertain whether a compound that produces a benign tumor in laboratory animals would be either capable or incapable of producing a malignant tumor in humans. In the absence of information to the contrary, the conservative view equates abnormal growth with carcinogenicity. Circumstances that produce benign tumors in animal systems might have the potential for producing abnormal growth in humans, depending on the mechanism involved. Many benign tumors are most easily produced in animal strains that already have an inherently high spontaneous incidence of such tumors (e.g., liver and lung adenomas in mice and mammary tumors in rats). Studies of the genetic, biochemical, hormonal, and other factors that determine development of such tumors might improve the validity of human risk assessments based on animal studies, and should be pursued more vigorously.

The assumption that the organ or tissue affected by a chemical in animals is also the site of greatest risk in humans should also be made cautiously. It is likely that the site of tumor formation is related to the route of exposure and to numerous pharmacokinetic and pharmacodynamic factors. Each route of exposure might result in carcinogenicity and should be considered separately. It probably is reasonable to assume that in some cases, animal models of carcinogenesis can be used to predict the development of human tumors at specific sites, provided that conditions of exposure are comparable. However, if exposure conditions are not similar, that might not be true. For example, it might well be incorrect to assume that agents that produce sarcomas in laboratory animals after subcutaneous injection will induce sarcomas in humans after inhalation. Animal models can be used to detect potential carcinogenicity; however, extrapolating from animal models to particular human organs is not valid without a great deal of additional mechanistic information, such as information on the effects of exposure route, dose, and many other factors, including the metabolism of the agent in question.

Evaluation of EPA Practice   Experience has shown that, in a broad sense, extrapolation from species to species is justifiable (Allen et al., 1988; Crump, 1989; Dedrick and Morrison, 1992). It is prudent to assume that agents that cause abnormal growth of tissue components in laboratory animals will do so in humans. The animal species (mice and rats) most commonly used in the National Toxicology Program (NTP) to make predictions about human carcinogenesis were selected for convenience, not because they have been demonstrated to predict human risks accurately. For example, the risk of inhaled particles for humans might be underestimated in animal assays that use rats and mice, which are obligatory nose breathers and thus might filter out much of the coarser dust. Conversely, some believe that rodents might overpredict human risk when mechanisms of carcinogenesis that are operative in rodents do not occur in humans (Cohen et al., 1992). It appears that NTP has not seriously explored alternatives to rats and mice in carcinogenesis testing, except perhaps for the use of hamsters in inhalation studies.

In principle, selection of data for estimation of carcinogenic potential from the most sensitive strain or species of animals tested is designed to be conservative; whether it is actually conservative and accurate is unknown. This default assumption contributes to the uncertainty in risk assessment, and research designed to investigate the biological mechanisms of carcinogenesis in both rodents and humans should be vigorously pursued so that more accurate risk assessments can be conducted.

Quantitative Considerations

Key terms in quantitative cancer risk characterization are *unit cancer risk* and *potency*. As currently estimated by EPA, potency is a statistical upper bound on the slope of the linear portion of a dose-response curve at low doses as calculated with a mathematical dose-response model. The unit cancer risk is based on potency and is an upper-bound estimate of the probability of cancer development due to continuous lifetime exposure to one unit of carcinogen. For airborne agents, that unit is commonly defined as exposure to 1 μg of agent per cubic meter of air over a 70-year lifetime.

Cancer potencies are generally based on dose-response relationships generated from cancer bioassays performed with rodents exposed to doses that are several orders of magnitude greater than those for which risk must be estimated. Bioassays typically include two, and to a lesser extent three or more, doses in addition to controls, and are rarely repeated. Often, positive results are obtained at only one dose. Therefore, for most carcinogens, few unequivocal data points are available for potency calculation. In addition, several assumptions often enter into calculations of potency, such as considerations related to tissue dosimetry, in which metabolism data obtained from different experimental systems and used in PBPK modeling might be used in place of bioassay exposure levels. It is not unusual for potency estimates based on the same bioassay data to vary substantially from one risk assessment to another, depending on these additional assumptions and the dose-response model used. Accordingly, potency values are often fraught with as much uncertainty as other aspects of quantitative risk assessment.

To estimate cancer potencies, EPA currently uses the linearized multistage model (EPA, 1987a). This model uses what is essentially an empirical curve-fitting procedure to describe the relationship between bioassay dose and response and to extrapolate the relationship to exposures below the experimental range. A statistical upper bound on the slope of the low-dose linear portion of the curve is considered to represent an upper bound on a chemical's carcinogenic potency. The multistage model is based on a theory of carcinogenic mechanism proposed in the early 1950s by Armitage and Doll. In essence, normal cells in a target organ are envisioned as undergoing a sequence of irreversible genetic transformations culminating in malignancy. Each transformation to a new stage is assumed to occur at some nonzero background rate. Exposure to a carcinogen is presumed simply to increase one or more of the transformation rates in proportion to the magnitude of the exposure (technically, dose at the target site). However, actual exposure circumstances are more complicated than can be briefly described here. No other potential effects of exposure or alternative mechanisms of carcinogenesis, such as induced cell proliferation or receptor-mediated alterations in gene expression, are included in the Armitage-Doll model. One important consequence of this assumption about how exposure influences transformations is the linearity of risk at low doses, i.e., risk increases and decreases in direct proportion to the delivered dose. That result arises in part because the model assumes that the number of cells at risk of undergoing the first transformation (the susceptible target-cell population) is constant and independent of age, magnitude of exposure, and exposure duration. Thus, the normal processes of cell division, differentiation, and death are not taken into account by the model.

Another cancer dose-response model that has been developed to estimate cancer potencies for risk assessment, but that is not used routinely for regulatory purposes, is the two-stage model. The two-stage model was developed by Moolgavkar, Venzon, and Knudson (Moolgavkar and Venzon, 1979; Moolgavkar and Knudson, 1981; Moolgavkar, 1988; Moolgavkar et al., 1988; Moolgavkar and Luebeck, 1990) and postulates that two critical mutations are required to produce a cancer cell. The model presupposes three cell compartments: normal stem cells, intermediate cells that have been altered by one genetic event, and malignant cells that have been altered by two genetic events. The size of each compartment is affected by cell birth, death, and differentiation processes and by the rates of transition between cell compartments. The model can accommodate some current concepts regarding the roles of inactivated tumor-suppressor genes and activated oncogenes in carcinogenesis. Unlike the Armitage-Doll model, it can explicitly account for many processes considered important in carcinogenesis, including cell division, mutation, differentiation, and death and the clonal expansion of populations of cells. Some knowledge of a chemical's mechanism of action and dose-response data for that mechanism are required to apply the two-stage model, however, and such data on most chemicals are scanty.

Potency estimates are generally based on the assumption that exposure to a particular agent occurs over a 70-year lifetime under constant conditions. That assumption is not likely to apply to the entire exposed population, however, and might produce a conservative estimate of risk. Use of a single potency number implies that the biological response of concern, such as carcinogenesis, depends only on total dose and therefore is independent of dose rate (the quantity of the agent received per unit time). This assumption might be invalid in some cases; for example, studies of low-energy-transfer radiation carcinogenesis show that low-dose-rate exposures are less effective than high-dose-rate exposures (NRC, 1990b). Other studies of radiation have differing results.

Potency estimates provide a means for comparing animal data with human data and for ranking potential carcinogens. Analysis of data available for some 20 known human carcinogens has shown that, in general, potency values derived from carcinogenicity bioassays in animals agree reasonably well with values calculated for humans from epidemiological studies (Allen et al., 1988). However, ranking of chemicals according to potency should not necessarily be used to make conclusions on the ranking of the corresponding hazards or risk. It is only multiplication of potency (unit risk) with exposure (dose) that yields an estimate of risk. Where there is no exposure, there might be little practical need for information on potency.

Evaluation of EPA Practice   The selection by EPA of a mathematical model to estimate potency is a critical step in quantitative risk assessment, in which alternative assumptions can lead to large differences in estimated risks. Such a model provides explicit, objective rules for extrapolating from the risks observed in controlled, high-dose laboratory experiments to those associated with the far lower doses that people might receive through inhalation. However, all dose-response models are simplified characterizations of the underlying biological reality. That is due, in part, to the incomplete scientific understanding of toxic mechanisms and to the requirement that the models be usable in a broad array of cases.

The challenge for EPA is to incorporate the expanding knowledge of mechanisms into the design of extrapolation models. The models would then depict more accurately the dose-response relationship at the low doses that are of concern to regulators, but are too low for toxic effects to be directly observed in whole animal studies or, often, any feasible human studies. The challenge can be illustrated by examining the simplified mechanistic assumptions that are included in the multistage model used by EPA in light of new understanding of mechanisms, which is not included in that model.

As long as exposure to a chemical has no substantial effect on cell processes other than genetic change, one would not expect the exclusion of these processes from the multistage model to compromise the resulting cancer risk estimates. The model would likely be appropriate for "direct-acting" carcinogens—ones, such as radiation, that act by directly attacking cellular DNA and thereby causing genetic transformation. In recent years, however, it has become apparent that many substances alter the pharmacodynamics of cells and can be carcinogenic by mechanisms that do not involve direct covalent interaction with DNA at all, but involve indirectly caused alterations in gene expression. One consequence of such a change could be altered cellular dynamics in the target organ. Because genetic transformations can occur spontaneously, many target organs contain a background of continuing steps in the multistep carcinogenic process. Exposure to a chemical could augment those background carcinogenic processes by simply increasing the pool of cells that are susceptible to further transformation. Such augmentation might occur as a regenerative response to cellular injury among surviving cells or to the cell-killing that occurs after exposure to highly toxic substances. The augmentation of background carcinogenic processes could also occur as an indirect response to alterations in hormonal balances induced by exposure or as a response to a directly mitogenic substance, i.e., one that stimulates normal cell division. By increasing the rate of cell division, such substances can increase the overall probability of generating a mutation, even though they have no direct effect on the transformation probability per cell division.

Similarly, exposure to substances classified as nongenotoxic carcinogens or "promoters" can create physiologic conditions within a target organ that favor the growth of "initiated" cells, i.e., cells that have already sustained at least one irreversible change from normal cells. Clonal expansions of initiated cell populations can be induced by exposure to promoters, thus increasing the probability of cell transformation and malignancy without directly affecting DNA.

Critical to effective regulatory use of biologically based models such as the two-stage model is accurate determination of the dose-response and time-response relationships for agent-induced cell death, differentiation, transformation and division, if any, in target tissues. Those processes might exhibit threshold-like dose-response relationships, in contrast to the presumed low-dose linear response of conventional multistage model transformation rates. Conversely, better understanding might show supralinear relations. Thus, use of a two-stage pharmacodynamic model might predict low-dose risks that are lower or higher than those predicted by the linearized multistage model.

Successful use of biologically based models in the risk assessment process will require a greater variety and amount of information on and understanding of carcinogenic mechanisms than is typically available for most chemicals. In the near term, such a data-intensive approach might be applied only to substances that have great economic value. In the long run, as knowledge and experience accrue, the use of models that incorporate relevant pharmacodynamic data should become more routine. Those models, used in conjunction with pharmacokinetic models for determining delivered doses, will increase the accuracy of quantitative risk assessment. For that reason, EPA should intensify their incorporation into the cancer-risk assessment process. For more information on two-stage models, see the NRC (1993c) report on this topic.

Carcinogen Classification

As noted in Chapter 4 (Table 4-1), EPA, following the lead of the International Agency for Research on Cancer (IARC), provides an evaluation of the available evidence of carcinogenicity of individual substances. The direction and strength of evidence are summarized by a letter: A, B_1, B_2, C, D, or E (see Table 4-1). The assignment of a substance to a class (actually, the assignment of available evidence to a class) depends almost entirely on epidemiological evidence and evidence derived from animal studies. The evidence for each of these is classified by EPA as "sufficient," "inadequate," or "limited." Some other types of experimental evidence (e.g., on genotoxicity) might sometimes play a role in the classification, but the epidemiological and bioassay data are generally of overriding importance.

The EPA classification scheme is intended to provide information on hazard—not to provide information about potential human risk; the latter cannot be assessed without the additional evaluation of dose-response and exposure information. The assignment of evidence to a class is intended by EPA only to suggest how convinced we should be that a substance poses a carcinogenic hazard to people. The classification is thus meant to depict the state of our knowledge regarding human carcinogenic hazard.

The difference between hazard and risk needs to be further emphasized here. As conceived in EPA's current four-step approach, identifying a substance as a possible, probable, or known carcinogenic hazard to humans means only that, under some unspecified conditions, the substance could cause excess cancers to occur in people. Evaluation of potency and of the exposures incurred by specific populations provides the information needed to assess the probability (risk) that the substance will cause cancer in the specified population. EPA developed the categorization scheme because it believes that, in addition to the risk estimate, decision-makers should have some sense of the strength of the evidence supporting identification of a substance as a carcinogen. There has been some confusion regarding the terms *strength of evidence*, as used by EPA, and *weight of evidence*. Some interpret *strength* to only describe the degree of positive evidence and *weight* to apply when *all* evidence—positive, negative, and evidence on relevance to humans—is considered. The committee adopts those uses of the terms. In many cases, substances for which the evidence of human carcinogenicity is strong (classification A) will, in specific circumstances, pose relatively small risks (because of low potency or low exposure), whereas substances for which the evidence of human carcinogenicity is much less convincing (classification B_2, for example) are likely to pose large risks (because of high potency or exposure). The typical question faced by a decision-maker is whether, for example, more restrictive controls should be placed on substances in class A that pose relatively small risks or on substances in lower classes that pose equal or greater risks. Stated in other terms, the issue concerns the justification for placing different degrees of regulatory restriction on substances that pose equal risks but which are differently classified. Should we control more carefully substances for which the state of our knowledge regarding human carcinogenicity is highly certain than we do substances for which the state of our knowledge is relatively weak? Although EPA includes a strength-of-evidence classification with each risk characterization, there is no clear indication of whether and how the classification influences ultimate agency decision-making.

Evaluation of EPA Practice   Does EPA's approach accurately portray the state of knowledge regarding human carcinogenic hazard? It is certainly the case that the state of scientific knowledge regarding the potential for various substances to contribute to the development of human cancers is highly variable among them. It also seems reasonable that risk assessors should have available a means to express that knowledge in a relatively simple way. It is for this reason that any such scheme should be examined carefully to ensure that it expresses as closely as possible what it is intended to express and that it summarizes all the relevant and appropriate findings derived from data, with no extraneous data.

Because two conclusions (that the substance might pose a carcinogenic hazard to humans under some conditions of exposure and that animal data can be unconditionally extrapolated to humans) are implicitly contained in the current EPA classification system, it could be conceived as misleading in some cases in which the scientific evidence does not support one or more of the typical default assumptions (for example, on route-to-route, high to low dose, or animal-to-human extrapolation). Such a situation could arise when, for example, data are available to show clearly and convincingly that some types of animal tumors would not likely be produced in humans or when mechanistic data show that results obtained at high doses are not relevant to low doses. Although different in kind, classification of substances at EPA's D or E level could also be misleading. If, for example, a substance were classified at level E on the basis of negative chemical bioassays in two species, but additional data suggested that neither animal species metabolized the substance in the way humans did, then the absence of potential human hazard would be improperly inferred.

The present EPA system might also be misleading because it is too susceptible to "accidents of fate." The carcinogenicity of a substance that happens to cause very rare tumors in humans (e.g., vinyl chloride, which causes angiosarcoma of the liver) is much easier to detect in epidemiological studies than is the carcinogenicity of a substance that causes very common human cancers, such as colorectal carcinoma. Although the available animal data on the latter substance might be very convincing with respect to carcinogenicity and there might be every reason to believe that it will be as hazardous to humans as the former (i.e., the "known" human, category A carcinogen), it will usually end up in category B, which may be interpreted as suggesting a lesser likelihood of hazard. Such a distinction might be due only to differences in our ability to detect the carcinogenic properties of substances that produce different types of cancers, and not to any true differences in human hazard.

Possible Improvements in EPA Practice   Before turning to the issue of improvements in EPA's carcinogen classification scheme, the committee first considered whether any such scheme should be used at all. As noted above, the current scheme can easily be misinterpreted—unfamiliar users might be led to believe that all substances in a specific category are equally hazardous or nonhazardous. Moreover, it is impossible to capture in any simple categorization scheme the completeness and complexity of the information that supports scientific judgments about the nature of a human carcinogenic hazard and the conditions under which it can exist. The quality, nature, and extent of such information vary greatly among carcinogens, and it is not an exaggeration to state that every substance is unique with respect to the scientific evidence bearing on its hazards.

It is for these reasons that the committee strongly recommends that EPA include in each hazard-identification portion of a risk assessment a *narrative* evaluation of the evidence of carcinogenicity. Such a narrative should contain at least the following:

• An evaluation of the strength of the available human and animal evidence.

• A weight-of-evidence evaluation of any available information on the relevance to humans of the animal models used and the results obtained from them and on the conditions of exposure (route, dose, duration, and timing) under which carcinogenic responses to other conditions of exposure (usually conditions that could exist in human populations exposed environmentally) have been measured (either in human populations or in laboratory animals).

Such a narrative seems to be the best way to describe the type of information typically available to evaluate carcinogenic hazards and should be used by EPA when it undertakes full-scale risk assessments.

Although the committee agreed that such narrative descriptions are the preferred way to express scientific evidence, it also recognized that there are important practical needs for some type of simple categorization of evidence. The committee recognized, for example, that many regulatory actions or plans for action require, for practical reasons, the creation of lists of carcinogens and that narrative statements are not likely to be included in such lists. Without some simple categorization scheme, such lists are likely to be completely undiscriminating with respect to the potential human hazards of the substances on them. When any such lists are used, for example, to create priorities for full risk assessment or for some type of regulation, the results could be seriously misleading to decision-makers and the public.

As already noted, however, the committee believes that the current EPA categorization scheme is inadequate. Substantial improvements could be made if the scheme incorporated not only "strength-of-evidence" information, but also some of the information we have called for in the narrative description.

It will not be easy to create a categorization scheme for carcinogens that incorporates both strength of evidence and the two "relevance" considerations. Moreover, EPA is not the only agency for which such a categorization scheme is useful. Indeed, there is a strong need for international agreement on a single classification. It would be highly desirable for EPA to convene a workshop on the matter and involve other agencies of federal and state governments, IARC, and other national and international bodies to develop a scheme that would have worldwide acceptance. IARC has recently moved to include information on mechanisms of carcinogenic action in its evaluation of carcinogens. Such an effort seems essential to eliminating the deficiencies of current schemes and the confusion that exists because of differences in approaches to categorization around the globe.

The committee suggests the scheme in Table 7-1 as a draft or prototype to avoid the difficulties of the current EPA scheme. The proposal in this table incorporates both strength-of-evidence considerations (as in the current EPA and IARC schemes) and "relevance" information, as specified in the two points mentioned above. The example also reduces the susceptibility of current classification schemes to the "accidents of fate" that can artificially influence the availability of evidence for different substances.

The classification in Table 7-1 takes place in two steps. In Step 1, a classification is made (into Categories I-IV) according to the two relevance criteria mentioned above. Note also that Category I is used for all substances on which positive carcinogenicity data are available and on which there are no substantive data to support conclusions that would place them in Category II or III—i.e., *Category I is the default option that applies when data related to relevance are weak or absent.* Step 2 of the classification involves evaluation of the strength of the available evidence.

Such a categorization scheme can provide guidance on priorities for both risk assessment and a variety of regulatory efforts. Substances placed in Category I, for example, would generally receive greater attention with respect to their carcinogenic properties than those in Category II; and within Category I, the nature of the attention received might be further influenced by the strength of available evidence (i.e., Ia > b > c > d). A Ia substance, for example, might be a prime candidate for immediate and stringent regulation, whereas a Id substance might be a prime candidate for high-priority information-gathering.

TABLE 7-1   Possible Scheme for Categorizing Carcinogens

Step 1: Categorization according to relevance of findings to humans

Category	Nature of Evidence
Category I Might pose a carcinogenic hazard to humans under any conditions of exposure. Magnitude of risk depends on dose-response relationship and extent of human exposure.	• Evidence of carcinogenicity in either human or animal studies (strength of evidence varies; see Step 2) • No information available to raise doubts about the relevance to humans of animal model or results • No information available to raise doubts about relevance of conditions of exposure (route, dose, timing, duration, etc.) under which carcinogenic effects were observed to conditions of exposure likely to be experienced by human populations exposed environmentally.
Category II Might pose a carcinogenic hazard to humans, but only under limited conditions. Whether a risk exists in specific circumstances depends on whether those conditions exist. Dose-response and exposure assessments must be completed to identify conditions under which risk exists.	• Evidence of carcinogenicity in either human or animal studies (strength of evidence varies; see Step 2) • Scientific information available to show that there are *limitations* in the conditions under which carcinogenicity might be expressed, owing to questions about the relevance to humans of the animal models or results or relevance of the conditions of exposure (route, dose, timing, duration, etc.) under which carcinogenic effects were observed to conditions of exposure likely to be experienced by human populations exposed environmentally.
Category III Notwithstanding the evidence of carcinogenicity in animals, not likely to pose a carcinogenic hazard to humans under any conditions.	• Evidence of carcinogenicity in animal studies • Scientific information available to show that the animal models or results are not relevant to humans under any conditions.
Category IV Evidence available to demonstrate lack of carcinogenicity or no evidence available.	• No evidence of carcinogenicity or evidence of non-carcinogenicity (weight of negative evidence varies; see Step 2)

TABLE 7-1 *Continued*

Step 2: Categorization according to strength of evidence (a through d, in decreasing order of strength)

Category	Data Source	Subcategory			
		a	b	c	d
I	Epidemiology	S	L	NI	NI
	Animal Studies	S/L/NI	S	S	L
II	Epidemiology	s	l	NI	NI
	Animal Studies	s/l/NI	s	s	l
III	Epidemiology	L/NI	NI		
	Animal Studies	s_i	l_i		
IV	Epidemiology	NI	NI/NA		
	Animal Studies	NI	NA		

S = sufficient evidence, high relevance. s = sufficient evidence, limited relevance.
L = limited evidence, high relevance. l = limited evidence, limited relevance.
NI = no or inadequate evidence. s_i = sufficient evidence, low relevance.
NA = no evidence in adequate studies. l_i = limited evidence, no relevance.

Placement of a substance in Category II does not mean that regulatory efforts should not be undertaken. For example, there might be reason to determine whether potentially risky conditions of exposure exist in any situations. The categories do not influence ultimate actions, but only priorities and the relative, inherent degrees of concern associated with different substances.

Although the committee recommends that any categorization scheme adopted by EPA include the elements associated with the above example, it also recognizes that there might be other ways to capture and express the same information. Some members suggested, for example, that substances listed as carcinogens simply be accompanied by a set of codes that specify both the strength of supporting evidence and the conditions and limitations, if any, that might pertain to the interpretation of that evidence (e.g., an asterisk next to a chemical might mean "assumed to be carcinogenic in humans only when inhaled").

Other End Points of Toxicity

The standard approach to regulating chemicals that are associated with non-cancer end points of toxicity has been based on the theory of homeostasis. According to that theory, biological processes that maintain homeostasis exist in an interdependent web of adaptive responses that automatically react to and compensate for stimuli that alter optimal conditions. An optimal condition is maintained as long as none of the stimuli that regulate it is pushed beyond some limit or "threshold." For the purposes of regulation, end points of toxicity other than cancer are lumped together under a toxicological paradigm that presumes a dose threshold for any chemical capable of inducing an adverse effect: there is an exposure below which the adverse effect would not be expected to occur. The current approach—no-observed-adverse-effect level (NOAEL) and uncertainty factor—is only a semiquantitative method designed to prevent exposures that are likely to result in an adverse effect, not a mechanistically based quantitative method for assessing the likely incidence and severity of effects in an exposed population. Moving beyond the current simplistic regulatory method will require, as is the case for carcinogenesis, a greater understanding of the mechanisms of disease causation, of pharmacokinetics, and of interindividual variation in each. Such improved understanding will permit final abandonment of the obsolete "threshold versus nonthreshold" paradigm for regulating carcinogens and noncarcinogens.

Evaluation of EPA Practice   The methodology now used by EPA to regulate human exposure to noncarcinogens is in a state of flux. That used by EPA in the past was not sufficiently rigorous. It was not based on evaluations of biological mechanisms of action or on differences in susceptibility between and within exposed populations. In addition, it incorporated risk management, not scientifically based risk-assessment techniques; and it did not permit incorporation of newer and better scientific information as it was obtained. The NOAEL-uncertainty factor approach might be adequate for the immediate future as a screening technique and for setting priorities, but its empirical and scientific basis is meager. EPA appears to be continuing to pursue simplistic, empirical techniques by adding to the list of uncertainty factors in use.

Impact of Pharmacokinetic Information in Risk Assessment

One of the critical steps in risk assessment is the selection of the measure of exposure to be used in defining the dose-response relationship. It is common today to calculate exposure on the basis of the "administered dose" of a chemical—the dose or amount fed to animals in toxicity studies or ingested by humans in food or water or inhaled in air. That dose can usually be accurately measured.

The dose that is of interest for risk assessment, however, is the amount of the biologically active form of a substance that reaches specific target tissues. This target-tissue dose is the "delivered dose," and its biologically active derivative, if any, is the "biologically active dose." The biologically active dose causes the events that culminate in toxicity to target cells and organs, and ideally it is used as the basis for defining the dose-response relationship and for assessing risk. The science of pharmacokinetics seeks to replace the current operating assumption—that administered dose and delivered dose are always directly proportional and that the administered dose is therefore an appropriate basis for risk assessment—with direct, accurate information about the delivered or biologically active dose.

Pharmacokinetic models are used to study the quantitative relationship between administered and delivered or biologically active doses. The relationship reflects the spectrum of biological responses to exposure, from physiological responses of a whole organism to biochemical responses within specific cells of a target organ. Pharmacokinetic models explicitly characterize biologic processes and permit accurate predictions of the doses of an agent's active metabolites that reach target tissues in exposed humans. As a consequence, the use of pharmacokinetic models to provide inputs to dose-response models reduces the uncertainty associated with the dose parameter and can result in more accurate estimates of potential cancer risks in humans.

The relationship between administered and delivered doses often differs among individuals: because of such differences, some people might be acutely sensitive and others insensitive to the same administered dose. The relationship between administered and delivered doses can also differ between large and small exposures and between continuous and intermittent exposures, and it can differ among species, some species being more or less efficient than humans in the transport of an administered dose to tissues or in its metabolism to a biologically active or inactive derivative. Those differences in the relationship between administered and delivered or biologically active doses can dramatically affect the validity of the predictions of dose-response models; failure to incorporate the difference into the models contributes to the uncertainty in risk assessment.

Differences between administered and biologically active doses occur because specialized organ systems intervene to modulate the body's responses to inhaled, ingested, or otherwise absorbed toxic materials. For example, the liver can detoxify materials circulating in the blood by producing enzymes to accelerate chemical reactions that break the materials down into harmless components (metabolic deactivation, or "detoxification"). Conversely, some substances can be activated by metabolism into more toxic reaction products. Activation and detoxification might occur at the same time and can occur in the same or different organ systems.

Furthermore, the rates at which activation and detoxification take place might have natural limits. Metabolic deactivation might thus be overwhelmed by high exposure concentrations, as seems to be the case with formaldehyde: the biologically active dose and the risk of nasal-tumor development rise rapidly in exposed rats only at high airborne concentrations. The assumption of a simple linear relationship between administered and biologically active doses of formaldehyde is believed by many to result in exaggerated estimates of cancer risk at low exposure concentrations. In contrast, metabolic activation of vinyl chloride occurs more and more slowly with increasing administered dose, because a critical enzyme system becomes overloaded; the biologically active dose and the resulting liver-tumor response increase more and more slowly as the administered dose increases. The assumption of a linear relationship between administered and delivered doses in the case of vinyl chloride could result in underestimation of the cancer risk associated with low doses. These examples illustrate how using pharmacokinetic models can reduce the uncertainty in risk estimation by modifying the dose values used in dose-response modeling to reflect the nonlinearity of metabolism.

Although most pharmacokinetic models are derived from laboratory-animal data, they provide a biological framework that is useful for extrapolating to human pharmacokinetic behavior. Anatomical and physiological differences among species are well documented and easily scaled by altering model parameters for the species in question. This aspect of pharmacokinetic modeling reduces the uncertainty associated with extrapolating from animal experiments to human cancer risk. For example, considerable effort has been devoted to the development of pharmacokinetic models for methylene chloride, which is considered a rodent carcinogen. The model was initially developed on the basis of rat data, then scaled to predict human behavior. Predictions in humans were compared with published data and with the results of experiments in human volunteers. The model was shown to predict accurately the pharmacokinetic behavior of inhaled methylene chloride and its metabolite carbon monoxide in both species (Andersen et al., 1991). Use of a particular pharmacokinetic model for methylene chloride in cancer risk assessment reduces human risk estimates for exposure to methylene chloride in drinking water by a factor of 50-210, compared with estimates derived by conventional linear extrapolation and body surface-area conversions (Andersen et al., 1987). Other analyses show different results (Portier and Kaplan, 1989). What pharmacokinetic models for methylene chloride do not predict, however, is whether methylene chloride is a human carcinogen. Thus, although use of the model might improve confidence in dose estimation by replacing the conventional scaling-factor approach, it cannot predict the outcome of exposure in humans.

Another way to reduce uncertainty would be to use pharmacokinetic models to extrapolate between exposure routes. If information on the disposition of an agent were available only as a result of its inhalation in the workplace, for example, and a risk assessment were required for its consumption in drinking water, appropriate models could be constructed to relate the delivered dose after inhalation to that expected after ingestion. To the committee's knowledge, pharmacokinetic models have not yet been used in a risk assessment for such regulatory purposes.

Failure to include pharmacokinetic considerations in dose-response modeling contributes to the overall uncertainty in a risk assessment, but uncertainty is associated with their use as well. This uncertainty comes from several sources. First, uncertainty is associated with the pharmacokinetic model parameters themselves. Parameter values are usually estimated from animal data and can come from a variety of experimental sources and conditions. Quantities can be measured indirectly, they can be measured *in vitro*, and they can vary among individuals. Different data sets might be available to estimate values of the same parameters. Hattis et al. (1990) evaluated seven pharmacokinetic models for tetrachloroethylene (perchloroethylene) metabolism and found that their predictions varied considerably, primarily because of the differences in choice of data sets used to estimate values of model parameters. Moreover, analogous parameter values are also needed for humans—although some values, such as organ weights, are amenable to direct measurement and do not vary widely among humans, others, such as rate constants for enzymatic detoxification and activation, are both difficult to measure and highly variable.

Second, there is uncertainty in the selection of the appropriate tissue dose available to model. For example, information might be available on the blood concentration of an agent, on its concentration in a tissue, or on the concentrations of its metabolites in the tissue. Tissue concentrations of one metabolite might be inappropriate if another metabolite is responsible for the biologic effects. Total tissue concentrations might not accurately reflect the biologically active dose if only one type of cell within the tissue is affected.

Choice of an appropriate measure of tissue dose can have an effect on cancer risk estimates. Farrar et al. (1989) considered three measures of tissue dose for tetrachloroethylene: tetrachloroethylene in liver, tetrachloroethylene metabolites in liver, and tetrachloroethylene in arterial blood. Using EPA's pharmacokinetic model for tetrachloroethylene and cancer bioassay data in mice, they found that human cancer risk estimates varied by a factor of about 10,000, depending on the dose surrogate used. Interestingly, the estimates bracketed that obtained in the absence of any pharmacokinetic transformation of dose as shown in Table 7-2.

This example illustrates the variation in dose and risk estimates that can be obtained under different assumptions, but it does not help to evaluate of the

TABLE 7-2   Risk Estimates Based on EPA's Pharmocokinetic Model for Tetrachloroethylene and Cancer Bioassay Data in Mice

Dose Surrogate	Risk Estimate[a]
Administered dose	5.57×10^{-3}
Dose to liver	425×10^{-3}
Dose of metabolites to liver	0.0195×10^{-3}
Dose in blood	126×10^{-3}

[a]Maximum-likelihood estimate.

SOURCE: Adapted from Farrar et al., 1989.

validity of any of the estimates in the absence of knowledge of the biologic mechanism of action of tetrachloroethylene as a rodent carcinogen and in the absence of knowledge of whether it is a human carcinogen. Although the dose of metabolites to the liver appears to be the most appropriate choice of dose surrogate, there is a high degree of nonlinearity between this dose and the tumor incidence in mice. The nonlinearity indicates either that this dose surrogate does not represent the actual biologically active dose for the particular sex-species combination analyzed by these authors or that the model does not adequately describe tetrachloroethylene pharmacokinetics.

The science of pharmacokinetics seeks to gain a clear understanding of all the biological processes that affect the disposition of a substance once it enters the body. It includes the study of many active biological processes, such as absorption, distribution, metabolism (whether activation or deactivation), and excretion. Accurate prediction of delivered and biologically active doses requires comprehensive, physiologically based computer models of those linked processes. Because the science of pharmacokinetics aims to replace general assumptions with a more refined model based on the specific relationship between administered and delivered or biologically active doses, its use in risk assessment will help to reduce the uncertainties in the process and the related bias in risk estimation. Advances will come slowly and at considerable cost, because detailed knowledge of the biologically active dose of many materials must be acquired before generalizations can be confidently exploited. Nevertheless, EPA increasingly incorporates pharmacokinetic data into the risk-assessment process, and its use represents one of the clearest opportunities for improving the accuracy of risk assessments.

Conclusions

Developing improved methods for assessing the long-term health impacts of chemicals will depend on improved understanding of the underlying science and on more effective coordination, validation, and integration of the relevant environmental, clinical, epidemiological, and laboratory data, each of which is limited by various kinds of error and uncertainty. Goodman and Wilson (1991) have demonstrated that, for 18 of 22 chemicals studied, there is good agreement between risk estimates based on rodent data and on epidemiologic studies. Their quantitative assessment, which can be compared to the Ennever et al. (1987) qualitative evaluation of the same issue, provides stronger evidence that current risk-assessment strategies produce reasonable estimates of human experience for known human carcinogens (Allen et al., 1988).

The reliability of a given health-risk assessment can be determined only by evaluating both the validity of the overall assessment and the validity of its components. Because the validity of a risk assessment depends on how well it predicts health effects in the human population, epidemiologic data are required for testing the predictions. To the extent that the requisite data are not already available, epidemiologic research will be necessary. An example is the study in which the New York Department of Health conducted biological monitoring for arsenic in schoolchildren (New York Department of Health, 1987). The researchers compared their findings with the arsenic concentrations predicted by the risk assessment conducted by EPA. The good agreement between the estimates and actual urinary arsenic concentrations in the children provided support for the EPA risk model.

The committee believes that substantial research is warranted to validate methods, models, and data that are used in risk assessment. In some instances the magnitude of uncertainty is not well understood, because information on the accuracy of the prediction process for each model used in risk assessment is insufficient. We also note that the uncertainties tend to vary considerably; for example, uncertainties are relatively low for estimation of population characteristics, compared with those associated with extrapolation from rodents to human beings.

The quality of risk analysis will improve as the quality of input improves. As we learn more about biology, chemistry, physics, and demography, we can make progressively better assessments of the risks involved. Risk assessment evolves continually, with re-evaluation as new models and data become available. In many cases, new information confirms previous assessments; in others, it necessitates changes, sometimes large. In either case, public confidence in the process demands that EPA make the best judgments possible. That an estimate of risk is subject to change is not a criticism of the process or of the assessors. Rather, it is a natural consequence of increasing knowledge and understanding. Re-evaluating risk assessments and making changes should be expected, embraced, and applauded, rather than criticized.

FINDINGS AND RECOMMENDATIONS

The following is a compilation of findings and recommendations related to evaluation of methods, data, and models for risk assessment.

Predictive Accuracy and Uncertainty of Models

Various methods and models are available to EPA and other organizations for conducting emission characterization, exposure assessment, and toxicity assessments. They include those used as default options and their corresponding alternatives, which represent deviations from the defaults. The predictive accuracy and uncertainty of the methods and models used for risk assessment are not clearly understood or fully disclosed in all cases.

- EPA should establish the predictive accuracy and uncertainty of the methods and models and the quality of data used in risk assessment with the high priority given to those which support the default options. EPA and other organizations should also conduct research on alternative methods and models that might represent deviations from the default options to the extent that they can provide superior performance and thus more accurate risk assessments in a clear and convincing manner.

Emission Characterization

Guidelines

EPA does not have a set of guidelines for emission characterization to be used in risk assessment.

• EPA should develop guidelines that require a given quality and amount of emission information relative to a given risk-assessment need.

Uncertainty

EPA does not adequately evaluate the uncertainty in the emission estimates used in risk assessments.

• Because of the wide variety of processes and differing maintenance of those sources, EPA should develop guidelines for the estimation and reporting of uncertainty in emission estimates; these guidelines may depend on the level of risk assessment.

External Collaboration

EPA has worked with outside parties to design emission characterization studies that have moved the agency from crude to more refined emission characterization.

• EPA should conduct more collaborative efforts with outside parties to improve the overall risk-assessment process, and each step within that process.

Exposure Assessment

Gaussian-Plume Models

In its regulatory practice, EPA has relied on Gaussian-plume models to estimate the concentrations of hazardous pollutants to which people are exposed. However, Gaussian-plume models are crude representations of airborne transport processes; because they are not always accurate, they lead to either underestimation or overestimation of concentrations. Stochastic Lagrangian and photochemical models exist, and evaluations have shown good agreement with observations. Also, EPA has typically evaluated its Gaussian-plume models for release and dispersion of criteria pollutants from plants with good dispersion characteristics (i.e., high thermal buoyancy, high exit velocity, and tall stacks). EPA has not fully evaluated the Gaussian-plume models for hazardous air pollutants with realistic plant parameters and locations; thus, their potential for underestimation or overestimation has not been fully disclosed.

• EPA should evaluate the existing Gaussian-plume models under more realistic conditions of small distances to the site boundaries, complex terrain, poor plant dispersion characteristics (i.e., low plume buoyancy, low stack exit momentum, and short stacks), and presence of other structures in the plant vicinity. When there is clear and convincing evidence that the use of Gaussian-plume models leads to underestimation or overestimation of concentrations (e.g., according to monitoring data), EPA should consider incorporating state-of-the-art models, such as stochastic-dispersion models, into its set of concentration-estimation models and include a statement of criteria for their selection and for departure from the default option.

Exposure Models

EPA has not adequately evaluated HEM-II for estimation of exposures, and prior evaluations of exposure models have shown substantial discrepancies between measured and predicted exposures, i.e., yielding under prediction of exposures.

• EPA should undertake a careful evaluation of all its exposure models to demonstrate their predictive accuracy (via pollutant monitoring and assessment of model input and theory) for estimating the distribution of exposures around plants that limit hazardous air pollutants. EPA should particularly ensure that, although exposure estimates are as accurate as possible, the exposure to the surrounding population is not underestimated.

Population Data

EPA has not previously used population activity, population mobility, and demographics in modeling exposure to hazardous air pollutants and has not adequately evaluated the effects of assuming that the population of a census enumeration district is all at the location of the district's population center.

• EPA should use population-activity models in exposure assessments when there is reason to believe that the exposure estimate might be inaccurate (e.g., as indicated by monitoring data) if the default option is applied. This is particularly important in the case of potential underestimation of risk. Population mobility and demographics will also play a role in determining risk and lifetime exposures. EPA should conduct further evaluation of the use of both simple methods (e.g., use of center of the population examined) and more comprehensive tools (e.g., NEM and SHAPE exposure models).

Human-Exposure Model

EPA uses the Human-Exposure Model (HEM) to evaluate exposure associated with hazardous air-pollutant releases from stationary sources. This model generally uses a standardized EPA Gaussian-plume dispersion model and assumes nonmobile populations residing outdoors at specific locations. The HEM construct will not provide accurate estimates of exposure in specific locations and for specific sources and contaminants where conditions do not match the simplified exposure and dispersion-model assumptions inherent in the standard HEM components.

• EPA should provide a statement on the predictive accuracy and uncertainty associated with the use of the HEM in each exposure assessment. The underlying assumption that the calculated exposure estimate based on the HEM is a conservative one should be reaffirmed; if not, alternative models whose performance has been clearly demonstrated to be superior should be used in exposure assessment. These alternative models should be adapted to include both transport and personal activity and mobility into an exposure-modeling system to provide more accurate, scientifically founded, and robust estimates of pollutant exposure distributions (including variability, uncertainty, and demographic information). Consideration may be given to linking these models to geographic information systems to provide both geographic and demographic information for exposure modeling.

EPA generally does not include non-inhalation exposures to hazardous air pollutants (e.g., dermal exposure and bioaccumulation); its procedure can lead to underestimation of exposure. Alternative routes can be an important source of exposure. Modeling systems similar to extensions of HEM have been developed to account for the other pathways.

• EPA should explicitly consider the inclusion of noninhalation pathways, except where there is prevailing evidence that noninhalation routes—such as deposition, bioaccumulation, and soil and water uptake—are negligible.

Assessment of Toxicity

Extrapolation from Animal Data for Carcinogens

EPA uses laboratory-animal tumor induction data, as well as human data, for predicting the carcinogenicity of chemicals in humans. It is prudent and reasonable to use animal models to predict potential carcinogenicity; however, additional information would enhance the quantitative extrapolation from animal models to human risks.

• In the absence of human evidence for or against carcinogenicity, EPA should continue to depend on laboratory-animal data for estimating the carcinogenicity of chemicals. However, laboratory-animal tumor data should not be used as the exclusive evidence to classify chemicals as to their human carcinogenicity if the mechanisms operative in laboratory animals are unlikely to be operative in humans; EPA should develop criteria for determining when this is the case for validating this assumption and for gathering additional data when the finding is made that the species tested are irrelevant to humans.

EPA uses data that generally assume that exposure of rats and mice after weaning and until the age of 24 months is the most sensitive and appropriate test system for conservatively predicting carcinogenicity in humans. These doses miss exposure of animals before they are weaned including newborns. Furthermore, the sacrifice of animals at the age of 2 years makes it difficult to estimate accurately the health affects of a disease whose incidence increases with age (as does that of cancer).

• EPA should continue to use the results of studies in mice and rats to evaluate the possibility of chemical carcinogenicity in humans. EPA and NTP are encouraged to explore the use of alternative species to test the hypothesis that results obtained in mice and rats are relevant to human carcinogenesis, the use of younger animals when unique sensitivity might exist for specific chemicals, and the age-dependent effects of exposure.

EPA typically extrapolates data from laboratory animals to humans by assuming that the delivered dose is proportional to the administered dose, as a default option. Alternative pharmacokinetic models are used less often to link exposure (applied dose) to effective dose.

• EPA should be encouraged to continue to explore and, when it is scientifically appropriate, incorporate mechanism-based pharmacokinetic models that link exposure and biologically effective dose.

The location of tumor formation in humans is related to route of exposure, chemical properties, and pharmacokinetic and pharmacodynamic factors, including systemic distribution of chemicals throughout the body. Thus, tumors might be found at different sites in humans and laboratory animals exposed to the same chemical. EPA has accepted evidence of carcinogenicity in tissues of laboratory animals as evidence of human carcinogenicity without necessarily assuming correspondence on a tumor-type or tissue-of-origin basis. EPA has extrapolated evidence of tumorigenicity by one route to another route where route-specific characteristics of disposition of the chemical are taken into account. EPA has traditionally treated almost all chemicals that induce cancer in a similar manner, using a linearized multistage nonthreshold model to extrapolate from large exposures and associated measured responses in laboratory animals to small exposures and low estimated rates of cancer in humans.

• Pharmacokinetic and pharmacodynamic data and models should be validated, and quantitative extrapolation from animal bioassays to humans should continue to be evaluated and used in risk assessments. EPA should continue to use the linearized multistage model as the default for extrapolating from high to low doses. If information on the mechanism of cancer induction suggests that the slope of the linearized multistage model is not appropriate for extrapolation, this information should be made an explicit part of the risk assessment. If sufficient information is available for an alternative extrapolation, a quantitative estimate should be made. EPA should develop criteria for determining what constitutes sufficient information to support an alternative extrapolation. The evidence for both estimates should be made available to the risk manager.

Extrapolation of Animal Data on Noncarcinogens

EPA uses a semiquantitative NOAEL-uncertainty factor approach to regulating human exposure to noncarcinogens.

• EPA should develop biologically based quantitative methods for assessing the incidence and likelihood of noncancer effects in an exposed population. These methods should permit the incorporation of information on mechanisms of action, as well as on differences in population and individual characteristics that affect susceptibility. The most sensitive end point of toxicity should continue to be used for establishing the reference dose.

Classification of Evidence of Carcinogenicity

EPA's narrative descriptions of the evidence of carcinogenic hazards are appropriate, but a simple classification scheme is also needed for decision-making purposes. The current EPA classification scheme does not capture information regarding the relevance to humans of animal data, any limitations regarding the applicability of observations, or any limitations regarding the range of carcinogenicity outside the range of observation. The current system might thus understate or overstate the degree of hazard for some substances.

• EPA should provide comprehensive narrative statements regarding the hazards posed by carcinogens, to include qualitative descriptions of both: 1) the strength of evidence about the risks of a substance; and 2) the relevance to humans of the animal models and results and of the conditions of exposure (route, dose, timing, duration, etc.) under which carcinogenicity was observed to the conditions under which people are likely to be exposed environmentally. EPA should develop a simple classification scheme that incorporates both these elements. A similar scheme to that set forth in Table 7-1 is recommended. The agency should seek international agreement on a classification system.

Potency Estimates

EPA uses estimates of a chemical's potency, derived from the slope of the dose-response curve, as a single value in the risk-assessment process.

• EPA should continue to use potency estimates—i.e., unit cancer risk—to estimate an upper bound on the probability of developing cancer due to lifetime exposure to one unit of a carcinogen. However, uncertainty about the potency estimate should be described as recommended in Chapter 9.

Although EPA routinely cites available human evidence, it does not always rigorously compare the quantitative risk-assessment model based on rodent data with available information on molecular mechanisms of carcinogenesis or with available human evidence from epidemiological studies.

• Because the validity of the overall risk-assessment model depends on how well it predicts health effects in the human population, EPA should acquire additional expertise in areas germane to molecular and mechanistic toxicology. In addition, EPA should also acquire additional epidemiological data to assess the validity of its estimates of risk. These data might be acquired in part by formalizing a relationship with the National Institute for Occupational Safety and Health to facilitate access to data from occupational exposures.

8

Data Needs

This chapter discusses the quantity, quality, and availability of data needed for conducting an adequate risk assessment in the context of the Clean Air Act Amendments of 1990 (CAAA-90). It begins by discussing the need for a priority-setting process, and the need for an iterative data-collection process. It then indicates the proper prioritization for data collection and the availability of data in each of the key risk-assessment steps. It concludes with a discussion of how data should be managed.

CONTEXT OF DATA NEEDS

Most would agree that, given the best available model, additional relevant data will lead to a more accurate and precise risk assessment. The quality of the data is critical, no matter how excellent the model chosen, to avoid the classic "garbage in, garbage out" problem. In the gathering of data, tradeoffs must often be made among data that are necessary, data that are desirable, and data that are affordable. Desirability must be defined in the context of the risk-management goals to be achieved, which might be the development of regulations, the setting of standards, or the screening of chemicals to set priorities.

The more precisely the risk manager frames the questions to be addressed by the risk assessment at the outset, the less ambiguity there will be as to what data are required to answer the questions, the less need for judgment in data-gathering, and the lower the likelihood that inappropriate or insufficient data will be gathered. As a corollary, public input into the framing of goals and questions can help to avoid public criticism and distrust of the process of risk assessment, including the gathering of exposure and toxicity data. Public confidence that risk managers are addressing real concerns, as opposed to going through a process perfunctorily, is critical to the future of risk assessment as an activity capable of improving the quality of life. Risk managers need to articulate clearly from the beginning who is to be protected from what, when and where, and at what cost (including how much effort and funds are to be expended to collect appropriate data), so that risk assessors can provide relevant information.

IMPLICATIONS FOR PRIORITY-SETTING

It is not necessary, nor would it be cost-effective, to collect all the data needed for a complete health-hazard assessment on all the 189 chemicals (or mixtures) listed in CAAA-90. It is important, however, that the entire list be examined to identify chemicals that are potentially hazardous and that the later full-scale evaluation of each chemical selected for further scrutiny proceed as effectively as possible. An overall strategy is essential for setting priorities among the steps in the information-gathering process and for determining the extent of assessment needed.

Because risk is a function of exposure, as well as toxicity, determining both that a chemical is of low toxicity to all humans and that all humans have only small exposures to it would lead to an overall low priority for a full-scale risk assessment. Obviously, assigning a high priority to both would lead to an overall high priority for such assessment and argue for collection of a complete data set in all categories of exposure and toxicity. There will be various intermediate levels between low and high overall priority.

In the absence of pertinent human data, toxicological evaluation should begin with the simplest, most rapid, and most economical tests and proceed to more complex, time-consuming, and more expensive tests only as warranted by the initial steps. Similarly, emission, transport, and exposure data might be used to rank chemicals for testing, from those with relatively large exposure potential down to those with a very low likelihood of significant exposure, either for the population at large or for any substantial subset of the population. What is "substantial" in this context will of course depend on concurrent assessments of toxicity. Ordering can then be based on an evaluation of a relatively modest or limited data set.

To assess whether there is a potential for exposure, and to gauge the magnitude and duration of exposure, one needs to know:

1. Is the chemical emitted into the air?
2. Is the chemical stable enough to be transported from its source to a population?

If the chemical is not emitted or is so unstable that it breaks down into innocuous products before reaching a population, no further data need be collected and further risk assessment is not warranted. But if it is emitted and can be transported to a population, one needs to ask:

3. Who is exposed, to how much, and for how long?

4. What is the relationship between exposure (dose) and response (effect) for humans and for animals?

In an iterative data-collection process, one works through data related to questions 1-4, first collecting the most critical data within each category, then judging needs for more data within that category before moving to the next category. The process is iterative until sufficient information is gathered to draw a conclusion—e.g., on a potential threat to public health.

Section 112 of the Clean Air Act mandates that EPA consider the hazards and possible regulation of 189 specified chemicals. Considering both the effort required to carry out complete risk assessments and the resources of the agency, it is unlikely that that can be accomplished within the time constraints of the act. Consequently, in the spirit of the act and in the interest of the public welfare, it is critical that EPA assign priorities to the chemicals listed. These priorities should be based first on their potential impact on human health and welfare.

Some of the 189 chemicals appear to present major problems because of their variety of sources, large exposures, or high potency. Other chemicals present simpler problems—e.g., some have relatively few sources, some have lower potential for human exposures, and some have very low potency. It is an inefficient use of resources to invest huge amounts of money and time in research and analysis to determine factors already known to be inconsequential for final risk assessment or to confirm credible estimates on which consensus can easily be obtained. Therefore, EPA should do preliminary analyses (screenings) on all listed compounds to ascertain which chemicals merit detailed risk-assessment efforts and which do not merit such work. These preliminary analyses should be reviewed by an independent board to ensure the validity of the resulting priorities for full-scale assessments. Priorities should be continually reevaluated and changed as appropriate in response to new data. The task of setting priorities and keeping them up to date is not trivial and should be specifically included, with adequate resources, in EPA's evolving program plan to implement CAAA-90. The iterative data-collection process can then help in setting priorities for ranking needed studies to avoid the accumulation of a surfeit of data, which would result in misuse of funds and waste of time.

DATA NEEDED FOR RISK ASSESSMENT

The following sections discuss the priority-setting and availability of data for each of the key data-processing steps in risk assessment: emissions, environmental fate and transport, exposure, and toxicity. The final section summarizes the data priorities in each of these areas, and indicates how this data can be used for overall priority-setting for data collection.

Emissions

Knowledge of emissions of a chemical into the air—specifically, the quantity emitted per unit of time (flux) from each place where it is made, stored, used, or disposed of plus its physical and chemical form—is fundamental to characterizing the magnitude of expected exposure to the chemical.

Priorities for Collecting Data

The specific methods for characterizing emissions are described and evaluated in Chapter 7. On the basis of this analysis, an iterative data-collecting process for emission characterization might proceed roughly as follows:

1. Plant-specific material balance
2. Industry-wide emission factors
3. Plant-specific emission factors
4. Facility measurements, including flux determinations.

Data quality is critical, because of the wide variety of emission-estimation techniques and the many types of facilities emitting hazardous air pollutants. EPA often uses whatever data are available at the time of decision-making and has not published guidelines or standards for the quality of emission data to be used in its risk assessments.

Because the emission-characterization database is extremely important for priority-setting, EPA should review the emission estimates submitted to ensure that they meet reasonable quality standards and that emission estimates from all sources within a site are submitted.

Data Availability

EPA plans to use emission information that is available in the Toxic Release Inventory (TRI) database as required by Title III of the Superfund Amendments and Recovery Act (SARA). The information available in this database is shown in the table provided by EPA to the committee in Appendix A. The TRI database includes information on annual emissions, facility location, and categorization of emissions as fugitive, point source, or both.

These data have two serious limitations for any use in risk assessment. First, the database does not include emissions from all operations at a facility; for example, transfer operations are not reported. Second, the database does not include emissions of less than 10 tons/year, nor does it have the locations of emission points or the frequency of emissions. Some information is available in emission inventory databases that are required by state implementation plans (SIPs) that states are required to submit to EPA to indicate how they plan to control emissions relative to CAAA-90, but that information is not necessarily well characterized. For example, emissions of volatile organic chemicals (VOCs) might be listed as a total, instead of as emissions of separate chemicals; but risk assessments should generally be done for separate chemicals, rather than for classes of chemicals.

A study by Amoco and EPA (1992) gives an example of the differences between estimated or calculated emissions (such as those listed in the TRI database) and emissions determined via direct measurement. This study found that the "existing estimates of environmental releases were not adequate for making a chemical-specific, multi-media, facility wide assessment." The report identified several specific problems in using the TRI database to conduct an in-depth evaluation of a facility:

- Lack of chemical characterization data.
- Difficulty of measuring and characterizing small sources.
- Use of estimated, rather than actual, data.
- Lack of identification of new sources leading to underestimation.
- Overestimation of some sources because of use of standardized industry-wide emission factors.
- No requirement that all chemicals be reported in the TRI database (e.g., only 9% of total hydrocarbons were required to be reported).
- Exclusion of some activities and emissions from record-keeping requirements (e.g., barge loading, which accounted for about 20% of benzene emissions).
- Lack of data in TRI on location of nearby populations and ecosystems.

EPA should develop a mechanism to gather the information just listed in a consistent fashion. This mechanism could include changes in Title III of SARA, which requires the TRI reporting requirement or development of information for Title I or V of CAAA-90. Although development of emission characterization databases for all of the 189 chemicals might initially seem to be a major task, CAAA-90 requires states to develop more detailed emission inventories by November 1992 and to update them. Most facilities are then required to estimate their emissions on a point basis to satisfy state requirements for emission inventories. Much of this information is also required for permit purposes.

Even simple changes, such as modifying the SARA Title III requirements to include all 189 hazardous air pollutants on the list, would help. Sixteen of the 189 compounds in CAAA-90 Title III are not on the TRI list (see Table 8-1). In addition, the TRI database includes only sources that have 10 or more full-time employees and that manufacture, process, or use specified chemicals above a certain production rate. That restriction excludes smaller sources within the manufacturing sector for which risk assessments must be conducted under the Title III requirements. Instituting an emission threshold relative to the Title III requirements (e.g., 10 tpy for single compound; 25 tpy for multiple compounds) might be more appropriate for gathering information for risk-assessment purposes.

TABLE 8-1 List of Section 112 Pollutants Not in Toxic Release Inventory Data Base

2,2,4-Trimethyl pentane
Acetophenone
Caprolactan
Dichlorodiphenyldichloroethylene (DDE)
Dimethyl formanide
Fine mineral fibers
1 texamethylene-t,t,-diisocyanate
Hexane
Isophorone
Phosphine
Polycylic organic matter
Sulfur dioxide, anhydrous
TCDD
Triethylamine

For evaluation of VOCs, many of which are on the list of 189 compounds under Title III, emission estimates developed for other regulatory purposes (such as the ozone provisions of CAAA-90) can be used. However, these data are frequently not speciated in terms of the chemical composition of the VOCs. In addition, the reporting of VOC emission information is required only in nonattainment areas, so this information may not always be available.

Environmental Fate and Transport

Emitted pollutants can move within and between environmental media and be converted to different forms. A thorough understanding of what happens to a chemical in the environment forms part of the basis for estimating human exposure and hence determining risk.

Priorities for Collecting Data

In the proposed iterative data-collection process described at the beginning of this chapter, data on environmental fate and transport would be acquired in roughly the following order:

1. Physical properties.
2. Physicochemical properties of environment.
3. Chemical properties or reactivity.
4. Rates of potential removal processes.

Once that information is available, a model calculation of expected concentrations in nearby air is relatively straightforward. If the information is not available, it must be obtained or assumed.

Data Availability

Data on emissions and physical properties are generally available or can be estimated (Lyman et al., 1982). For chemical properties and reactivity, they are available for some environmental reactions, but not all. In the case of physicochemical properties, the environment data are generally available at most locations in the United States. Information on the rates of potential removal processes are more difficult and costly to obtain.

Careful evaluation of data is necessary. For example, published vapor pressures of organic chemicals of moderate to low volatility determined under laboratory conditions can be seriously inaccurate and misleading. For all chemicals, vapor-phase reaction rate constants, when extrapolated from the laboratory to outdoor ambient air, can be seriously in error. The literature is not always for purposes of risk assessment.

Exposure

Accurate exposure data are crucial to valid risk assessment. For example, exposure data must match up temporally with the health end points of concern. Key issues in the evaluation of exposure are

- The end points of interest (e.g., acute vs. chronic toxicity).
- The populations at risk (i.e., the general population and defined subpopulations with potentially increased risks).
- The routes of exposure (e.g., air, diet, or skin).
- The duration (e.g., lifetime, annual, or instantaneous).
- The nature and degree of simultaneous toxicant exposures.

Rarely are all those issues resolved by the exposure data available for a risk assessment. Efforts to collect the data should focus on the minimum needed to meet the goals of the assessment in its risk-management context.

Priorities for Collecting Data

In the proposed iterative data-collection process, the order of data collection might be as follows:

1. *Ambient-air monitoring.* Most commonly, ambient-air monitoring produces interval concentrations in samples averaged over a fixed time, such as 8 hr or 24 hr at fixed sampling stations. The number of stations, their times of operation, and their locations relative to known emission sources and popula-

tions at risk must be known, as well as concentration averages, variances or ranges (to estimate uncertainty), and a description of the methods used, including potential error. The time interval of ambient-air monitoring should be commensurate with the time needed to elicit the physiological effects of concern.

2. *Targeted fixed-point monitoring data.* These data are often generated from samples placed near sources of high-volume emissions (i.e., "hot spots") or in response to some real or perceived public-health need. They should be accompanied by the same information as for ambient-air monitoring. Targeted monitoring is often more useful than monitoring at pre-existing sampling stations if it can focus on higher concentrations of a pollutant, a population at greater risk, or both.

3. *Peak-concentration data.* Either ambient-air or targeted monitoring can miss peak concentrations, because the sampling interval is so long as to "average out" all peaks and valleys in the sampled air mass. Sampling with instantaneous analyzers (e.g., spectrophotometers) or interval analyzers that can accept a sample of short duration is needed to define peaks. That might be of special importance for a toxicant released intermittently.

4. *Personal monitoring.* Concentration data from personal monitors are often more useful for risk assessment, because they show the exposure of individual subjects and can be used to relate activity patterns to exposure. If enough subjects are selected for monitoring, a population exposure can be constructed. Such information is not yet generally available, except for a few toxicants, because of the time and expense of a comprehensive study. This in turn is primarily due to a lack of low-cost, portable sampling devices for most chemicals. Active samplers may provide more information directly for risk assessment than passive samplers for personal monitoring, because pollutant concentrations (and thus the dose) can be estimated more directly with active sampling. Passive samplers do not provide specific concentrations; however, they are far less costly and bulky than active samplers. They are useful in screening (i.e., to determine whether exposure has occurred). Research to correlate the concentrations detected by passive samplers with exposure and dose would further enhance their potential.

5. *Biological markers.* If a toxicant produces a metabolite, enzyme alteration, or other signal that exposure has occurred and so leads to a high correlation between that marker and degree of exposure, such information can reduce the uncertainty in a predicted risk and could be useful for risk assessment. In one respect, this would be the best exposure information, because it would show that the toxicant has been absorbed and has already had some biological effect (NRC, 1987); but it makes single-source exposure assessment difficult, because it reveals total uptake across all routes of exposure. Unless biologic-marker data are checked against external exposure data, they cannot be used to determine dose. Validation of the correlation between an external concentration and the magnitude of a biological marker in experimental animals can be helpful, but

one is left with the difficulty of extrapolating to humans, who may not respond in the same quantitative way as experimental animals. In some cases, markers in humans can be established in occupational settings.

Data Availability

Some of the 189 chemicals on the Clean Air Act Amendments list have relatively abundant data on concentrations; some have virtually none. When concentration data are available, they are more likely to be from ambient-air monitoring or, at best, targeted fixed-point monitoring. For only some of the compounds are sufficient exposure data available for preliminary evaluation of relative priority for more detailed risk assessment (see Appendix A). That is a major problem that can be solved only by a much more extensive state or federal monitoring program. Some states, such as California, are moving rapidly in developing a hazardous air-pollutant monitoring program. Coordination between states and with federal agencies is necessary to keep scarce resources from being wasted in duplicative efforts.

Collection of new exposure data on humans is limited by current methods of monitoring individual exposures (which are often expensive, often of low accuracy or precision, and often nonquantitative or lacking in the ability to determine the source of exposure) and by methods of obtaining information on human behavior that might affect uptake or exposures. In addition, no reference database is available for comparing new data, that is, for determining whether new data represent exposure outside the general norm or are within the realm of acceptability defined by prior studies. Furthermore, when exposure data are gathered, they should be probability-based to allow inferences to the population and estimation of the tails of the distribution of exposures.

Toxicity

A full assessment of the inherent toxicity of an agent requires some combination of structure-activity analyses, in vitro or whole-animal short-term tests, chronic or long-term animal bioassays, human biomonitoring, clinical studies, and epidemiological investigations (NRC, 1984, 1991c,d). A complete hazard identification might entail review of information in all those categories before a determination that a quantitative risk assessment of the agent is warranted (Bailar et al., 1993).

Estimation of dose-effect relationships requires data on the effects of a wide range of doses, on factors that influence the dose delivered to critical target cells by given magnitudes and patterns of exposure (e.g., uptake, anatomic distribution, metabolism, and excretion) (NRC, 1987), on the shapes and slopes of pertinent dose-effect curves, on the relevant mechanisms of effects (NRC, 1991c), and on the extent to which the response to an agent can vary with species, sex, age, previous exposure, health status, exposure to extraneous agents, and other variables (NRC, 1988a).

Priorities for Collecting Data

Strategies to fill data gaps in toxicity assessment are best developed case by case, but the following priority-setting of the major types of toxicological data that may be used are listed below. In the suggested iterative data-collection process, the toxicity data listed in the first three categories below (i.e., generic and acute toxicity, acute mammalian lethality) should be collected on every chemical as a starting point, and other, more expensive, data should be collected only on chemicals that give cause for concern based on the data in those categories.

1. Generic toxicity data (structure-activity relationships and results of other correlational analyses).
2. Data on acute toxicity (on lethality in microorganisms or effects on mammalian cells in vitro).
3. Acute mammalian lethality data (usually rodent).
4. Toxicokinetics data, phase 1 (on uptake, distribution, retention, and excretion in rodents).
5. Genotoxicity data (results of short-term in vitro tests in microorganisms, *Drosophila*, and mammalian cells).
6. Data on subchronic toxicity (on 14-day or 28-day inhalation toxicity in rodents).
7. Toxicokinetic data, phase 2 (on metabolic pathways and metabolic fate in rodents and other mammalian species, with special attention given to exposure by inhalation).
8. Data on chronic toxicity (on carcinogenicity, neurobehavioral toxicity, reproductive and developmental toxicity, and immunotoxicity in two rodent species of both sexes, with special attention given to the exposure by inhalation).
9. Human toxicity data (clinical, biomonitoring, and epidemiological data).
10. Data on toxic mechanisms, dose-effect relationships, influence of modifying factors (age, sex, and other variables) on susceptibility, and interactive effects of mixtures of chemical and physical agents.

This prioritization is based on the cost and complexity of gathering such data (NRC, 1984). It is generally not possible to plan the collection of clinical and epidemiological data. Toxicological studies conducted clinically in humans are usually planned and implemented under experimental control, but very few are done, because of the attendant hazards. Epidemiological studies are relatively expensive and often produce data that are difficult to interpret as to effects of specific toxic agents. If one were to set data-collection priorities without concern for cost, ethical, or other considerations, the sequence of collection might be

1. Toxicological human data.
2. Clinical data.
3. Epidemiological data.

Data Availability

Availability of requisite data varies widely among the 189 chemicals. On the one hand, some preliminary toxicity data are available on some of the chemicals, or at least can be estimated from structure-activity correlations. On the other hand, the toxicity data are incomplete on almost all 189 chemicals.

The amount of data available is highly variable and depends largely on the existence of uncontrollable chance events. Generally, better data sets exist on individual chemicals that have been used over long periods (vinyl chloride, some solvents, etc.) and on chemicals of wide use (such as pesticides) than on chemicals rarely used or chemicals that are byproducts of other chemicals (e.g., chemicals in automobile exhaust and cigarette smoke). Additional information and analysis on the Integrated Risk Information System (IRIS) used by EPA is provided in Chapter 12. Some of the partial data needed to test models are discussed in Chapter 6.

Overall Priority Setting

The data needed for each step of risk assessment are summarized in rough order of increasing complexity (see Table 8-2). In an iterative data-collection process, if information in the top one or two items of each of the four columns in Table 8-2 does not indicate increased risk potential the priority for full risk assessment should be low. Various combinations of negative information in the first few items of any two of the first three lists (e.g., emissions, environmental fate and transport, exposure) with positive information in the third list might lead to a medium priority. Positive information in the early items of two, or perhaps three, of the lists would argue for a high priority. Data for the more complex items of each list would be developed when evidence of potential hazard exceeded an agreed-on "bright line" of concern, i.e., a decision point set either by regulation or programmatic procedures.

Although a full priority scheme probably should be on a continuous scale, several important points to develop a more detailed scheme might appear as follows:

TABLE 8-2 Types of Data Available for Risk Assessment

Emissions	Environmental Fate and Transport	Exposure	Toxicity
1. Material balance	1. Physical properties	1. Ambient fixed-point monitoring	1. Generic toxicity
2. Industry-wide emission factors	2. Physicochemical properties of environment	2. Targeted fixed-point monitoring	2. Acute toxicity (lethality for microorganisms or mammalian cells in vitro)
3. Plant-specific emission factors (EPA protocol)	3. Chemical properties or reactivity	3. Duration and frequency of peak concentrations for populations at risk	3. Acute mammalian lethality (rodent)
4. Facility measurements, including flux determinations	4. Rates of potential removal processes	4. Personnel monitoring for average and maximally exposed people	4. Toxicokinetics, phase 1
		5. Biologic markers	5. Genotoxicity (short-term in vitro tests in microorganisms, *Drosophila*, or mammalian cells)
			6. Subchronic (13-day or 28-day) inhalation toxicity (rodent)
			7. Toxicokinetics, phase 2
			8. Chronic toxicity: carcinogenicity, neurobehavioral toxicity, reproductive and developmental toxicity, or immunotoxicity
			9. Human toxicity (clinical, biomonitoring, epidemiologic)
			10. Toxic mechanisms and dose-effect relationships

Screening risk assessment

Emissions—Items 1 and 2
Environmental fate and transport—Items 1-3
Exposure—Items 1-3
Toxicity—Items 1-3

• If the information for all the above items (or items lower on the list, if available) indicates no potential health concerns, assign "low priority."
• If any information on exposure (emissions, environmental fate and transport, exposure) is positive, assign the chemical "medium priority."
• If any information on exposure is positive (i.e., emission, environmental fate and transport, or exposure measurement), *and* toxicity data are positive, then assign the chemical "high priority"and proceed to the full-scale risk assessment.

Full risk assessment

Emissions—Items 1-4
Environmental fate and transport—Items 1-5
Exposure—Items 1-5
Toxicity—Items 1-10

• If the information is not positive for the higher-order items in all four lists, assign the chemical to Action Level 2 (more extended time response).
• If the information is positive for the higher-order items in all four lists, assign the chemical to Action Level 1 (short time-frame response).

Reliable positive human evidence will always result in a high priority and the full risk evaluation. Any positive clinical, toxicologic, or epidemiological human data would override a priority based on exposure and animal toxicity data alone and move a given chemical to the stage of full risk assessment.

The detailed nature of the process used to set priorities for full risk assessment needs to be addressed in a coordinated way by federal and state agencies, to ensure the best use of limited resources for this programmatic step. There might be, for example, a numerical weighting or scoring approach based on data in the four categories of emissions, environmental fate and transport, exposure, and toxicological data. EPA should consider convening a panel of experts to develop a priority-setting process and the requisite accompanying iterative approach to data collection.

DATA MANAGEMENT

More attention needs to be paid to data management to ensure that vital data gaps are filled, that data used in risk assessments are of the best possible quality, and that relevant information (such as negative epidemiological information) is not overlooked. The lack of a consistent data-collection scheme makes data analysis, and thus effective risk assessment, inconsistent and unreliable for risk-management purposes.

For example, risk assessment often requires that the assessor decide whether to set aside information from old studies when newer, supposedly better information is available. The ultimate desire is for credibility; therefore, it is important to use information that is widely acknowledged as the best representation of reality. If the results of a new study contradict information from an old study and if there is only a small difference in the "bottom-line" estimate of human health risk, then both should be used, and the error bounds of the current risk assessment should be revised. However, if the studies lead to quite different conclusions, use of both might be feasible. For example, some animal evidence might show a major health hazard while there may also be weak, negative, or equivocal animal studies. Such conflicting data should be carefully reviewed in the risk-assessment document, with detailed study of possible reasons for the discrepancy. When no reconciliation of results seems feasible, the committee recommends that the voice of prudence be heard and that the risk assessment be either based on the higher ultimate risk estimate or delayed (as was done in part on formaldehyde) until additional studies can be completed.

FINDINGS AND RECOMMENDATIONS

The committee's findings and recommendations follow.

Insufficient Data for Risk Assessment

EPA does not have sufficient data to assess fully the health risks of the 189 chemicals in Title III within the time permitted by the Clean Air Act Amendments of 1990.

• EPA should screen the 189 chemicals for priorities for the assessment of health risks, identify the data gaps, and develop incentives to expedite generation of the needed data by other public agencies (such as the National Toxicology Program, the Agency for Toxic Substances and Disease Registry, and state agencies) and by other organizations (industry, academia, etc.).

Need for Data-Gathering Guidelines

EPA has not defined the guidelines or process to be used for determining the types, quantities, and quality of data that are needed for conducting risk assessments for facilities emitting one or more of the 189 chemicals.

• EPA should develop an iterative approach to gathering and evaluating data in the categories of emission, transport and fate, exposure, and toxicology for use in both screening and full risk assessment. The data-gathering and data-evaluation process should be set forth by EPA in guidelines for use by those who conduct data-gathering activities. To develop these guidelines, EPA should convene a panel of experts to develop a priority-setting scheme that uses a numerical weighting or scoring approach.

Inadequacy of Emission and Exposure Data

EPA has often relied on non-site-specific emission and exposure data. These data are often not sufficient to assess the risk to individuals and the affected population at large.

• EPA should expand its efforts to gather emission and exposure data to personal monitoring and site-specific monitoring.

Inadequacy of TRI Database as a Source of Emission Data for Risk-Assessment Purposes

The SARA 313 Toxic Release Inventory data and other readily available data used by EPA for emission characterization may be adequate for screening purposes but are not adequate for developing detailed risk assessments for specific facilities. Present processes of gathering emission data do not yield information appropriate for all risk-assessment purposes under the Clean Air Act Amendments.

• EPA should modify its data-gathering activities related to emissions to ensure that it has or will acquire the data needed to conduct screening and full risk assessments, especially of the 189 chemicals listed in CAAA-90.

Lack of Adequate Natural Background-Exposure Database

EPA does not have an adequate database on natural background exposures to the 189 air pollutants against which to evaluate total human exposure data from facilities producing or using these substances.

• EPA should develop an ambient-outdoor-exposure database on the 189 listed hazardous air pollutants.

Inadequate Explanation of Analytical Techniques

EPA does not always explain adequately the analytical and measurement methods it uses for estimating ambient outdoor exposures.

• EPA should collate and explain the analytical and measurement methods it uses for ambient outdoor exposures, including the errors, precision, accuracy, detection limits, etc., of all methods that it uses for risk-assessment purposes.

Need for System of Data Management for Risk Assessment

EPA needs more adequate mechanisms to compile and maintain databases for use in health-risk screening and assessment.

• EPA should review its data-management systems and improve them as needed to ensure that the quality and quantity of the data are routinely updated and that the data are sufficiently accessible for risk screening and risk assessment. Its responsibilities under CAAA-90 should be prominent in this review and revision.

9

Uncertainty

The need to confront uncertainty in risk assessment has changed little since the 1983 NRC report *Risk Assessment in the Federal Government*. That report found that:

> The dominant analytic difficulty [in decision-making based on risk assessments] is pervasive uncertainty. . . . there is often great uncertainty in estimates or the types, probability, and magnitude of health effects associated with a chemical agent of the economic effects of a proposed regulatory action, and of the extent of current and possible future human exposures. These problems have no immediate solutions, given the many gaps in our understanding of the causal mechanisms of carcinogenesis and other health effects and in our ability to ascertain the nature or extent of the effects associated with specific exposures.

Those gaps in our knowledge remain, and yield only with difficulty to new scientific findings. But a powerful solution exists to some of the difficulties caused by the gaps: the systematic analysis of the sources, nature, and implications of the uncertainties they create.

CONTEXT OF UNCERTAINTY ANALYSIS

EPA decision-makers have long recognized the usefulness of uncertainty analysis. As indicated by former EPA Administrator William Ruckelshaus (1984):

> First, we must insist on risk calculations being expressed as distributions of estimates and not as magic numbers that can be manipulated without regard to

what they really mean. We must try to display more realistic estimates of risk to show a range of probabilities. To help do this, we need new tools for quantifying and ordering sources of uncertainty and for putting them into perspective.

Ten years later, however, EPA has made little headway in replacing a risk-assessment "culture" based on "magic numbers" with one based on information about the range of risk values consistent with our current knowledge and lack thereof.

As we discuss in more depth in Chapter 5, EPA has been skeptical about the usefulness of uncertainty analysis. For example, in its guidance to those conducting risk assessments for Superfund sites (EPA, 1991f), the agency concludes that quantitative uncertainty assessment is usually not practical or necessary for site risk assessments. The same guidance questions the value and accuracy of assessments of the uncertainty, suggesting that such analyses are too data-intensive and "can lead one into a false sense of certainty."

In direct contrast, the committee believes that uncertainty analysis is the only way to combat the "false sense of certainty," which is *caused* by a refusal to acknowledge and (attempt to) quantify the uncertainty in risk predictions.

This chapter first discusses some of the tools that can be used to quantify uncertainty. The remaining sections discuss specific concerns about EPA's current practices, suggest alternatives, and present the committee's recommendations about how EPA should handle uncertainty analysis in the future.

NATURE OF UNCERTAINTY

Uncertainty can be defined as a lack of precise knowledge as to what the truth is, whether qualitative or quantitative. That lack of knowledge creates an intellectual problem—that we do not know what the "scientific truth" is; and a practical problem—we need to determine how to assess and deal with risk in light of that uncertainty. This chapter focuses on the practical problem, which the 1983 report did not shed much light on and which EPA has only recently begun to address in any specific way. This chapter takes the view that uncertainty is always with us and that it is crucial to learn how to conduct risk assessment in the face of it. Scientific truth is always somewhat uncertain and is subject to revision as new understanding develops, but the uncertainty in quantitative health risk assessment might be uniquely large, relative to other science-policy areas, and it requires special attention by risk analysts. These analysts need to allow questions such as: What should we do in the face of uncertainty? How should it be identified and managed in a risk assessment? How should an understanding of uncertainty be forwarded to risk managers, and to the public? EPA has recognized the need for more and better uncertainty assessment (see EPA memorandum in Appendix B), and other investigators have begun to make substantial progress with the difficult computations that are often required (Monte Carlo methods, etc.). However, it appears that these changes have not yet affected the day-to-day work of EPA.

Some scientists, mirroring the concerns expressed by EPA, are reluctant to quantify uncertainty. There is concern that uncertainty analysis could reduce confidence in a risk assessment. However, that attitude toward uncertainty may be misguided. The very heart of risk assessment is the responsibility to use whatever information is at hand or can be generated to produce a number, a range, a probability distribution—whatever expresses best the present state of knowledge about the effects of some hazard in some specified setting. Simply to ignore the uncertainty in any process is almost sure to leave critical parts of the process incompletely examined, and hence to increase the probability of generating a risk estimate that is incorrect, incomplete, or misleading.

For example, past analyses of the uncertainty about the carcinogenic potency of saccharin showed that potency estimates could vary by a factor as large as 10^{10}. However, this example is not representative of the ranges in potency estimates when appropriate models are compared. Potency estimates can vary by a factor of 10^{10} only if one allows the choice of some models that are generally recognized as having no biological plausibility and only if one uses those models for a very large extrapolation from high to low doses. The judicious application of concepts of plausibility and parsimony can eliminate some clearly inappropriate models and leave a large but perhaps a less daunting range of uncertainties. What is important, in this context of enormous uncertainty, is not the best estimate or even the ends of this 10^{10}-fold range, but the best-informed estimate of the likelihood that the true value is in a region where one rather than or another remedial action (or none) is appropriate. Is there a small chance that the true risk is as large as 10^{-2}, and what would be the risk-management implications of this very small probability of very large harm? Questions such as these are what uncertainty analysis is largely about. Improvements in the understanding of methods for uncertainty analysis—as well as advances in toxicology, pharmacokinetics, and exposure assessment—now allow uncertainty analysis to provide a much more accurate, and perhaps less daunting, picture of what we know and do not know than in the past.

Taxonomies

Before discussing the practical applications of uncertainty analysis, it may be best to step back and discuss it as an intellectual endeavor. The problem of uncertainty in risk assessment is large, complex, and nearly intractable, unless it is divided into smaller and more manageable topics. One way to do so, as seen in Table 9-1 (Bogen, 1990a), is to classify sources of uncertainty according to the step in the risk assessment process in which they occur. A more abstract and generalized approach preferred by some scientists is to partition all uncertainties into the three categories of bias, randomness, and true variability. This method

TABLE 9-1 Some Generic Sources of Uncertainty in Risk Assessment

I. HAZARD IDENTIFICATION

Unidentified hazards

Definition of incidence of an outcome in a given study (positive-negative association of incidence with exposure)

Different study results

Different study qualities
  —conduct
  —definition of control population
  —physical-chemical similarity of chemical studied to that of concern

Different study types
  —prospective, case-control, bioassay, in vivo screen, in vitro screen
  —test species, strain, sex, system
  —exposure route, duration

Extrapolation of available evidence to target human population

II. DOSE-RESPONSE ASSESSMENT

Extrapolation of tested doses to human doses

Definition of "positive responses" in a given study
  —independent vs. joint events
  —continuous vs. dichotomous input response data

Parameter estimation

Different dose-response sets
  —results
  —qualities
  —types

Model selection for low-dose risk extrapolation
  —low-dose functional behavior of dose-response relationship (threshold, sublinear, linear, supralinear, flexible)
  —role of time (dose frequency, rate, duration; age at exposure; fraction of lifetime exposed)
  —pharmacokinetic model of effective dose as a function of applied dose
  —impact of competing risks

continued on next page

TABLE 9-1 *Continued*

III. EXPOSURE ASSESSMENT

Contamination-scenario characterization (production, distribution, domestic and industrial storage and use, disposal, environmental transport, transformation and decay, geographic bounds, temporal bounds)
—environmental-fate model selection (structural error)
—parameter estimation error
—field measurement error

Exposure-scenario characterization
—exposure-route identification (dermal, respiratory, dietary)
—exposure-dynamics model (absorption, intake processes)

Target-population identification
—potentially exposed populations
—population stability over time

Integrated exposure profile

IV. RISK CHARACTERIZATION

Component uncertainties
—hazard identification
—dose-response assessment
—exposure assessment

SOURCE: Adapted from Bogen, 1990a.

of classifying uncertainty is used by some research methodologists, because it provides a complete partition of types of uncertainty, and it might be more productive intellectually: bias is almost entirely a product of study design and performance; randomness a problem of sample size and measurement imprecision; and variability a matter for study by risk assessors but for resolution in risk management (see Chapter 10).

However, a third approach to categorizing uncertainty may be more practical than this scheme, and yet less peculiar to environmental risk assessment than the taxonomy in Table 9-1.

This third approach, a version of which can be found in EPA's new exposure guidelines (EPA, 1992a) and in the general literature on risk assessment uncertainty (Finkel, 1990; Morgan and Henrion, 1990), is adopted here to facilitate communication and understanding in light of present EPA practice. Although the committee makes no formal recommendation on which taxonomy to use, EPA staff might want to consider the alternative classification above (bias, randomness, and variability) to supplement their current approach in future documents. Our preferred taxonomy consists of:

• *Parameter uncertainty*. Uncertainties in parameter estimates stem from a variety of sources. Some uncertainties arise from measurement errors; these in turn can involve random errors in analytic devices (e.g., the imprecision of continuous monitors that measure stack emissions) or systematic biases (e.g., measuring inhalation from indoor ambient air without considering the effect of volatilization of contaminants from hot water used in showering). A second type of parameter uncertainty arises when generic or surrogate data are used instead of analyzing the desired parameter directly (e.g., the use of standard emission factors for industrialized processes). Other potential sources of error in estimates of parameters are misclassification (e.g., incorrect assignment of exposures of subjects in historical epidemiological studies due to faulty or ambiguous information), random sampling error (e.g., estimation of risk to laboratory animals or exposed workers from outcomes observed in only a small sample), and nonrepresentativeness (e.g., developing emission factors for dry cleaners based on a sample that included predominantly "dirty" plants due to some quirk in the study design).[1]

• *Model uncertainty*. These uncertainties arise because of gaps in the scientific theory that is required to make predictions on the basis of causal inferences. For example, the central controversy over the validity of the linear, no-threshold model for carcinogen dose-response is an argument over model uncertainty. Common types of model uncertainties include relationship errors (e.g., incorrectly inferring the basis for correlations between chemical structure and biologic activity) and errors introduced by oversimplified representations of reality (e.g., representing a three-dimensional aquifer with a two-dimensional mathematical model). Moreover, any model can be incomplete if it excludes one or more relevant variables (e.g., relating asbestos to lung cancer without considering the effect of smoking on both those exposed to asbestos and those unexposed), uses surrogate variables for ones that cannot be measured (e.g., using wind speed at the nearest airport as a proxy for wind speed at the facility site), or fails to account for correlations that cause seemingly unrelated events to occur much more frequently than would be expected by chance (e.g., two separate components of a nuclear plant are both missing a particular washer because the same newly hired assembler put both of them together). Another example of model uncertainty concerns the extent of aggregation used in the model. For example, to fit data on the exhalation of volatile compounds adequately in physiologically based pharmacokinetic (PBPK) models, it is sometimes necessary to break up the fat compartment into separate compartments reflecting subcutaneous and abdominal fat (Fiserova-Bergerova, 1992). In the absence of enough data to indicate the inadequacy of using a single aggregated variable (total body fat), the modeler might construct an unreliable model. The uncertainty in risk that results from uncertainty about models might be as high as a factor of 1,000 or even greater, even if the same data are used to determine the results from each. This can occur, for example, when the analyst must choose between a linear multi-stage model and a threshold model for cancer dose-response relations.

PROBLEMS WITH EPA'S CURRENT
APPROACH TO UNCERTAINTY

EPA's current practice on uncertainty is described elsewhere in this report, especially in Chapter 5, as part of the risk-characterization process. Overall, EPA tends at best to take a qualitative approach to uncertainty analysis, and one that emphasizes model uncertainty rather than parameter uncertainties. The uncertainties in the models and the assumptions made are listed (or perhaps described in a narrative way) in each step of the process; these are then presented in a nonquantitative statement to the decision-maker.

Quantitative uncertainty analysis is not well explored at EPA. There is little internal guidance for EPA staff about how to evaluate and express uncertainty. One useful exception is the analysis conducted for the National Emission Standards for Hazardous Air Pollutants (NESHAPS) radionuclides document (described in Chapter 5), which provides a good initial example of how uncertainty analysis could be conducted for the exposure portion of risk assessment. Other EPA efforts, however, have been primarily qualitative, rather than quantitative. When uncertainty is analyzed at EPA, the analysis tends to be piecemeal and highly focused on the sensitivity of the assessment to the accuracy of a few specified assumptions, rather than a full exploration of the process from data collection to final risk assessment, and the results are not used in a systematic fashion to help decision-makers.

The major difficulty with EPA's current approach is that it does not supplant or supplement artificially precise single estimates of risk ("point estimates") with ranges of values or quantitative descriptions of uncertainty, and that it often lacks even qualitative statements of uncertainty. This obscures the uncertainties inherent in risk estimation (Paustenbach, 1989; Finkel, 1990), although the uncertainties themselves do not go away. Risk assessments that do not include sufficient attention to uncertainty are vulnerable to four common and potentially serious pitfalls (adapted from Finkel, 1990):

1. They do not allow for optimal weighing of the probabilities and consequences of error for policy-makers so that informed risk-management decisions can be made. An adequate risk characterization will clarify the extent of uncertainty in the estimates so that better-informed choices can be made.

2. They do not permit a reliable comparison of alternative decisions, so that appropriate priorities can be established by policy-makers comparing several different risks.

3. They fail to communicate to decision-makers and the public the range of control options that would be compatible with different assessments of the true state of nature. This makes informed dialogue between assessors and stakeholders less likely, and can cause erosion of credibility as stakeholders react to the overconfidence inherent in risk assessments that produce only point estimates.

4. They preclude the opportunity for identifying research initiatives that might reduce uncertainty and thereby reduce the probability or the impact of being caught by surprise.

Perhaps most fundamentally, without uncertainty analysis it can be quite difficult to determine the conservatism of an estimate. In an ideal risk assessment, a complete uncertainty analysis would provide a risk manager with the ability to estimate risk for each person in a given population in both actual and projected scenarios of exposures; it would also estimate the uncertainty in each prediction in quantitative, probabilistic terms. But even a less exhaustive treatment of uncertainty will serve a very important purpose: it can reveal whether the point estimate used to summarize the uncertain risk is "conservative," and if so, to what extent. Although the choice of the "level of conservatism" is a risk-management prerogative, managers might be operating in the dark about how "conservative" these choices are if the uncertainty (and hence the degree to which the risk estimate used may fall above or below the true value) is ignored or assumed, rather than calculated.

SOME ALTERNATIVES TO EPA'S APPROACH

A useful alternative to EPA's current approach is to set as a goal a quantitative assessment of uncertainty. Table 9-2, from Resources for the Future's Center for Risk Management, suggests a sequence of steps that the agency could follow to generate a quantitative uncertainty estimate. To determine the uncertainty in the estimate of risk associated with a source probably requires an understanding of the uncertainty in each of the elements shown in Table 9-3. The following pages describe more fully the development of probabilities and the method of using probabilities as inputs into uncertainty analysis models.

Probability Distributions

A probability density function (PDF) describes the uncertainty, encompassing objective or subjective probability, or both, over all possible values of risk. When the PDF is presented as a smooth curve, the area under the curve between any two points is the probability that the true value lies between the two points. A cumulative distribution function (CDF), which is the integral or sum of the PDF up to each point, shows the probability that a variable is equal to or less than each of the possible values it can take on. These distributions can some-

TABLE 9-2   Steps That Could Improve a Quantitative Uncertainty Estimate

1. *Determine the desired measure of risk* (e.g., mortality, life years lost, risk to the individual who is maximally exposed, number of persons at more than arbitrary "unacceptable" risk.) More than one measure will often be desired, but the remaining steps will need to be followed *de novo* for each method.

2. *Specify one or more "risk equations," mathematical relationships that express the risk measure in terms of its components.* For example, R = C x I x P (risk equals concentration times intake times potency) is a simple "risk equation" with three independent variables. Care must be taken to avoid both an excess and an insufficiency of detail.

3. *Generate an uncertainty distribution for each component.* This will generally involve the use of analogy, the use of statistical inference, of expert opinion, or a combination of these.

4. *Combine the individual distributions into a composite uncertainty distribution.* This step will often require Monte Carlo simulation (described later).

5. *"Recalibrate" the uncertainty distributions.* At this point, inferential analysis should enter or re-enter the process to corroborate or correct the outputs of step 4. In practice, it might involve altering the range of the distribution to account for dependence among the variables or truncating the distributions to exclude extreme values that are physically or logically impossible. Repeat steps 3, 4, and 5 as needed.

6. *Summarize the output, highlighting important implications for risk management.* Here the decision-maker and uncertainty analyst need to work together (or at least to understand each other's needs and limitations). In all written and oral presentations, the analyst should strive to ensure that the manager understands the following four aspects of the results:

• Their implications for supplanting any point estimate that might have been produced without consideration of uncertainty. In particular, presentations of uncertainty will help in advancing the debate over whether the standardized procedures used to generate point estimates of risk are too "conservative" in general or particular cases.

• Their insights regarding the balance between the costs of overestimating and underestimating risk (i.e., the shape and breadth of the uncertainty distribution informs the manager about how prudent various risk estimates might be).

• Their sensitivity to fundamentally unresolved scientific controversies.

• Their implications for research, identifying which uncertainties are most important and which uncertainties are amenable to reduction by directed research efforts. As part of this process, the analyst should attempt to quantify in absolute terms how much total effort might be put into reducing uncertainty before a control action is implemented (i.e., estimate the value of information using standard techniques).

SOURCE: Adapted from Finkel, 1990.

times be estimated empirically with statistical techniques that can analyze large sets of data adequately. Sometimes, especially when data are sparse, a normal or lognormal distribution is assumed and its mean and variance (or standard deviation) are estimated from available data. When data are in fact normally distributed over the whole range of possible values, the mean and variance completely characterize the distribution, including the PDF and CDF. Thus, with certain assumptions (such as normality), only a few points might be needed to estimate the whole distribution for a given variable, although more points will both im-

TABLE 9-3 Some Key Variables in Risk Assessment for Which Probability Distributions Might Be Needed

Model Component	Output Variable	Independent Parameter Variable
Transport	Air concentration	Chemical emission rate Stack exit temperature Stack exit velocity Mixing heights
Deposition	Deposition rate	Dry-deposition velocity Wet-deposition velocity Fraction of time with rain
Overland	Surface-water load	Fraction of chemical in overload runoff
Water	Surface-water concentration	River discharge Chemical decay coefficient in river
Soil	Surface-soil concentration	Surface-soil depth Exposure duration Exposure period Cation-exchange capacity Decay coefficient in soil
Food Chain	Plant concentration	Plant interception fraction Weathering elimination rate Crop density Soil-to-plant bioconcentration factor
	Fish concentration	Water-to-fish bioconcentration factor
Dose	Inhalation dose	Inhalation rate Body weight
	Ingestion dose	Plant ingestion rate Soil ingestion rate Body weight
	Dermal-absorption dose	Exposed skin surface area Soil absorption factor Exposure frequency Body weight
Risk	Total carcinogenic risk	Inhalation carcinogenic potency factor Ingestion carcinogenic potency factor Dermal-absorption carcinogenic potency factor

SOURCE: Adapted from Seigneur et al., 1992.

prove the representation of the uncertainty and allow examination of the normality assumption. However, the problem remains that apparently minor deviations in the extreme tails may have major implications for risk assessment (Finkel, 1990). Furthermore, it is important to note that the assumption of normality may be inappropriate.

When data are flawed or not available or when the scientific base is not understood well enough to quantify the probability distributions of all input variables, a surrogate estimate of one or more distributions can be based on analysis of the uncertainty in similar variables in similar situations. For example, one can approximate the uncertainty in the carcinogenic potency of an untested chemical by using the existing frequency distribution of potencies for chemicals already tested (Fiering et al., 1984).

Subjective Probability Distributions

A different method of probability assessment is based on expert opinion. In this method, the beliefs of selected experts are elicited and combined to provide a subjective probability distribution. This procedure can be used to estimate the uncertainty in a parameter (cf., the subjective assessment of the slope of the dose-response relationship for lead in Whitfield and Wallsten, 1989). However, subjective assessments are more often used for a risk assessment component for which the available inference options are logically or reasonably limited to a finite set of identifiable, plausible, and often mutually exclusive alternatives (i.e., for *model* uncertainty). In such an analysis, alternative scenarios or models are assigned subjective probability weights according to the best available data and scientific judgment; equal weights might be used in the absence of reliable data or theoretical justifications supporting any option over any other. For example, this approach could be used to determine how much the risk assessor should rely on relative surface area vs. body weight in conducting a dose-response assessment. The application of particular sets of subjective probability weights in particular inference contexts could be standardized, codified, and updated as part of EPA's implementation of uncertainty analysis guidelines (see below).

Objective probabilities might seem inherently more accurate than subjective probabilities, but this is not always true. Formal methods (Bayesian statistics)[2] exist to incorporate objective information into a subjective probability distribution that reflects other matters that might be relevant but difficult to quantify, such as knowledge about chemical structure, expectations of the effects of concurrent exposure (synergy), or the scope of plausible variations in exposure. The chief advantage of an objective probability distribution is, of course, its objectivity; right or wrong, it is less likely to be susceptible to major and perhaps undetectable bias on the part of the analyst; this has palpable benefits in defending a risk assessment and the decisions that follow. A second advantage is that objective probability distributions are usually far easier to determine. However, there can be no rule that objective probability estimates are always preferred to subjective estimates, or vice versa.

Model Uncertainty: "Unconditional" Versus "Conditional" PDFs

Regardless of whether objective or subjective methods are used to assess them, the distinction between parameter uncertainty and model uncertainty remains pivotal and has implications for implementing improved risk assessments that acknowledge uncertainty. The most important difference between parameter uncertainty and model uncertainty, especially in the context of risk assessment, concerns how to interpret the output of an objective or subjective probability assessment for each.

One can readily construct a probability distribution for risk, exposure, potency, or some other quantity that reflects the probabilities that various values, corresponding to fundamentally different scientific models, represent the true state. Such a depiction, which we will call an "unconditional PDF" because it tries to represent all the uncertainty surrounding the quantity, can be useful for some decisions that agencies must make. In particular, EPA's research offices might be able to make more efficient decisions about where resources should be channeled to study particular risks, if the uncertainty in each risk were presented unconditionally. For example, an unconditional distribution might be reported in this way: "the potency of chemical X is 10^{-2} per part per million of air (with an uncertainty of a factor of 5 due to parameter uncertainty surrounding this value), but only if the LMS model is correct; if instead the chemical has a threshold, the potency at any ambient concentration is effectively zero." It might even help to assign subjective weights to the current thinking about the probability that each model is correct, especially if research decisions have to be made for many risks.

In addition, some specified *regulatory* decisions—those involving the ranking of different risks for the purpose of allowing "tradeoffs" or "offsets"—can also suffer if model uncertainty is not quantified. For example, two chemicals (Y and Z) with the same potency—assuming that the LMS model is correct—might involve different degrees of confidence in the veracity of that model assumption. If we judged that chemical Y had a 90%, or even a 20%, chance of acting in a threshold fashion, it might be a mistake to treat it as having the same potency as a chemical Z that is virtually certain to have no threshold and then to allow increased emissions of Z in exchange for greater reductions in Y.

However, unconditional statements of uncertainty can be misleading if managers use them for standard-setting, residual-risk decisions, or risk communication, and especially if others then misinterpret these statements. Consider two situations, involving the same hypothetical chemical, in which the same amount of uncertainty can have different implications, depending on whether it stems from parameter uncertainty (Situation A) or ignorance about model choice (Situation B). In Situation A, suppose that the uncertainty is due entirely to parameter sampling error in a single available bioassay involving few test animals. If 3 of 30 mice tested in that bioassay developed tumors, then a reasonable central-tendency estimate of the risk to mice at the dose used would be 0.1 (3/30). However, because of sampling error, there is approximately a 5% probability that the true number of tumors might be as low as zero (leading to zero as the lower confidence limit, LCL, of risk) and about a 5% probability that the true number of tumors is 6 or higher (leading to 0.2 (6/30) as the upper confidence limit, UCL, of risk).

In Situation B, suppose instead that the uncertainty is due entirely to ambiguity over which model of biological effect is correct. In this hypothetical situation, there was one bioassay in which 200 of 1,000 mice developed tumors; the risk to mice at that dose would be 0.2 (with essentially no parameter uncertainty due to the very large sample size). But suppose scientists disagree about whether the effect in mice is at all relevant to humans, because of profound metabolic or other differences between the two species, but can agree to assign equal probabilities of 50% to each eventuality. In this case as well, the LCL of the risk to humans would be zero (if the "nonrelevance" theory were correct), and the UCL would be 0.2 (if the "relevance" theory were correct), and it would be tempting to report a "central estimate" of 0.1, corresponding to the expected value of the two possible outcomes, weighted by their assigned probabilities. In either situation A or B, it would be *mathematically* correct to say the following: "The expected value of the estimate of the number of annual excess cancer deaths nationwide caused by exposure to this substance is 1,000; the LCL of this estimate is zero deaths, and the UCL is 2,000 deaths." [3]

We contend that in such cases, which typify the two kinds of uncertainties that risk managers must deal with, it would be a mistake simply to report the confidence limits and expected value in Situation B as one might do more routinely in Situation A, especially if one then used these summary statistics to make a regulatory decision. The risk-communication problem in treating this dichotomous model uncertainty (Situation B) as though it were a continuous probability distribution is that it obscures important information about the scientific controversy that must be resolved. Risk managers and the public should be given the opportunity to understand the sources of the controversy, to appreciate why the subjective weights assigned to each model are at their given values, and to judge for themselves what action is appropriate when the two theories, *at least one of which must be incorrect*, predict such disparate outcomes.

More critically, the expected value in Situation B might have dramatically different properties as an estimate for decision-making from the one in Situation A. The estimate of 1,000 deaths in Situation B is a contrivance of multiplying subjective weights that corresponds to no possible true value of risk, although this is not itself a fatal flaw; indeed, it is possible that a strategy of deliberately inviting errors of both overprotection and underprotection at each decision will turn out to be optimal over a long-run set of similar decisions. The more fundamental problem is that any estimate of central tendency does not necessarily lead to optimal decision-making. This would be true even if society had no desire to make conservative risk management decisions.

Simply put, although classical decision theory does encourage the use of expected values that take account of all sources of uncertainty, it is not in the decision-maker's or society's best interest *to treat fundamentally different predictions as quantities that can be "averaged" without considering the effects of each prediction on the decision that it leads to*. It is possible that a coin-toss gamble between zero deaths and 2,000 deaths would lead a regulator rationally to act as though 1,000 deaths were the certain outcome. But this is only a shorthand description of the actual process of expected-value decision-making, which asks how the *decisions* that correspond to estimates of zero deaths, 1,000 deaths, and 2,000 deaths perform relative to each other, in light of the possibility that each estimate (and hence each decision) is wrong. In other words, the choice to use an unconditional PDF when there is the kind of model uncertainty shown in situation B is a choice between the *possibility* of overprotecting or underprotecting—if one model is accepted and the other rejected—and the *certainty* of erring in one direction or the other if the hybrid estimate of 1,000 is constructed. Because in this example the outcomes are numbers that can be manipulated mathematically, it is tempting to report the average, but this would surely be nonsensical if the outcomes were not numerical. If, for example, there were model uncertainty about where on the Gulf Coast a hurricane would hit, it would be sensible to elicit subjective judgment about the probability that a model predicting that the storm would hit in New Orleans was correct, versus the probability that an alternative model—say, one that predicted that the storm would hit in Tampa—was correct. It would also be sensible to assess the expected losses of lives and property if relief workers were irrevocably deployed in one location and the storm hit the other ("expected" losses in the sense of probability times magnitude). It would be foolish, however, to deploy workers irrevocably in Alabama on the grounds that it was the "expected value" of halfway between New Orleans and Tampa under the model uncertainty—and yet this is just the kind of reasoning invited by indiscriminate use of averages and percentiles from distributions dominated by model uncertainty.

Therefore, we recommend that analysts present *separate* assessments of the parameter uncertainty that remains for *each* independent choice of the underlying model(s) involved. This admonition is not inconsistent with our view that model uncertainty is important and that the ideal uncertainty analysis should consider and report all important uncertainties; we simply suspect that comprehension and decision-making might suffer if all uncertainties are lumped together indiscriminately. The subjective likelihood that each model (and hence each parameter uncertainty distribution) might be correct should still be elicited and reported, *but primarily to help the decision-maker gauge which depiction of risk and its associated parameter uncertainty is the correct one,* and not to construct a single hybrid distribution (except for particular purposes involving priority-setting, resource allocation, etc.). In the hypothetical Situation B, this would mean presenting both models, their predictions, and their subjective weights, rather than simple summary statistics, such as the unconditional mean and UCL.

The existence of default options for model uncertainty (as discussed in the introduction to Part II and in Chapter 6) also places an important curb on the need for and use of unconditional depictions of uncertainty. If, as we recommend, EPA develops explicit principles for choosing and modifying its default models, it will further codify the practice that for every risk assessment, a sequence of "preferred" model choices will exist, with only one model being the prevailing choice at each inference point where scientific controversy exists. Therefore, the "default risk characterization," including uncertainty, will be the uncertainty distribution (embodying the various sources of parameter and scenario uncertainty) that is conditional on the approved choices for dose-response, exposure, uptake, and other models made under EPA's guidelines and principles. For each risk assessment, this PDF, rather than the single point estimate currently in force, should serve as the quantitative-risk input to standard-setting and residual-risk decisions that EPA will make under the act.

Thus, given the current state of the art and the realities of decision-making, model uncertainty should play only a subsidiary role in risk assessment and characterization, although it might be important when decision-makers integrate all the information necessary to make regulatory decisions. We recognize the intellectual and practical reasons for presenting alternative risk estimates and PDFs corresponding to alternative models that are scientifically plausible, but that have not supplanted a default model chosen by EPA. However, we suggest that to create a single risk estimate or PDF out of various different models not only could undermine the entire notion of having default models that can be set aside for sufficient reason, but could lead to misleading and perhaps meaningless hybrid risk estimates. We have presented this discussion of the pitfalls of combining the results of incompatible models to support our view urging caution in applying these techniques in EPA's risk assessment. Such techniques should not be used for calculating unit risk estimates, because of the potential for misinterpretation of the quantitative risk characterization.[4] However, we encourage risk assessors and risk managers to work closely together to explore the implications of model uncertainty for risk management, and in this context explicit characterization of model uncertainty may be helpful. The characterization of model uncertainty may also be appropriate and useful for risk communication and for setting research priorities.

Finally, an uncertainty analysis that carefully keeps separate the influence of fundamental model uncertainties versus other types of uncertainty can reveal which controversies over model choice are actually important to risk management and which are "tempests in teapots." If, as might often be the case, the effect of all parameter uncertainties (and variabilities) is as large as or larger than that contributed by the controversy over model choice, then resolving the controversy over model choice would not be a high priority. In other words, if the "signal" to be discerned by a final answer as to which model or inference option is correct is not larger than the "noise" caused by parameter uncertainty in either (all) model(s), then effort should be focused on data collection to reduce the parameter uncertainties, rather than on basic research to resolve the modeling controversies.

SPECIFIC GUIDANCE ON UNCERTAINTY ANALYSIS

Generating Probability Distributions

The following examples indicate how probability distributions might be developed in practice and illustrate many of the principles and recommended procedures discussed earlier in the chapter.

• *Example 1.* Estimated emission rates can differ significantly from actual values. Experience might show that emission estimates based on emission factors, mass balances, or material balances have an inherent uncertainty of a factor of about 100, whereas those based on testing tend to be within a factor of about 10. Expert opinion and analysis of past studies of such emission estimates could provide more definitive bounds on the estimates and result in a probability distribution. For example, a lognormal distribution with the median at the calculated emission estimate and a geometric standard deviation[5] of 10 (i.e., the case of emission factors) or $\sqrt{10}$ (for emissions based on testing).

• *Example 2.* A standard animal carcinogenicity bioassay provides the raw material for three related features of a complete uncertainty analysis. First, there is the random sampling uncertainty due to the limitation on the number of animals that can be tested. Suppose that at a particular dose 10 of 50 mice develop leukemia. The most likely estimate of the risk to each mouse would be calculated as 0.2 (the observed risk to the group, 10/50). However, chance dictates that if different groups of 50 animals were exposed to a risk of 0.2, some number n other than 10 might develop leukemia at each replication of the experiment. According to the binomial theorem, which governs independent dichotomous chance events (such as a coin falling either "heads" or "tails"), between 4 and 16 animals would develop cancer 99% of the time if many groups of 50 animals were exposed to identical lifetime risks of 0.2. EPA's standard procedure of reporting only the "q_1^*" value for potency is equivalent to computing the 95th percentile of random uncertainty using the binomial theorem (e.g., assuming that if 10 tumors were observed, 14 tumors would be a "conservative" estimate), and then finding the slope of the straight line drawn between this hypothetical response and the control-group response. Such a point estimate is informative neither about the plausible slopes greater and less than this value nor about the relative probabilities of the different plausible values. A distribution for q_1 derived from the entire binomial probability distribution for n, on the other hand, would answer both of these concerns.

A second opportunity, which allows the analyst to draw out some of the model uncertainty in dose-response relationships, stems from the flexibility of the LMS model. Even though this model is often viewed as unduly restrictive (e.g., it does not allow for thresholds or for "superlinear" dose-response relations at low doses), it is inherently flexible enough to account for sublinear dose-response relations (e.g., a quadratic function) at low doses. EPA's point-estimation procedure forces the q_1^* value to be associated with a linear low-dose model, but there is no reason why EPA could not fit an unrestricted model through all the values on the binomial uncertainty distribution of tumor response, thereby generating a distribution for potency that might include some probability that the true dose-response function is of quadratic or higher order (see, for example, Guess et al., 1977; Finkel, 1988).

Finally, EPA could account for another source of parameter uncertainty if it made use of more than one data set for each carcinogen. Techniques of meta-analysis, more and more frequently used to generate composite *point estimates* by averaging together the results of different studies (e.g., a second mouse study that might have found 20 leukemic animals out of 50 at the same dose), can perhaps more profitably be used to generate a composite *uncertainty distribution*. This distribution could be broader than the binomial distribution that would arise from considering the sampling uncertainty in a single study, if the new study contradicted the first, or it could be narrower, if the results of each study were reinforcing (i.e., each result was well within the uncertainty range of the other).

• *Example 3.* The linearized multistage (LMS) model is often used to estimate dose-response relationships. Although many models could be used to estimate this relationship, two—the LMS and the biologically motivated (BM) models—seem to have the best biologic and mechanistic underpinning. Others, such as the probit and logit models, do not have a similar underpinning and are generic dose-response models. An additional possible advantage of BM models is their flexibility to accommodate the possibility of zero added response at low doses, even when there is a response at high doses. At present, there is rarely enough information to use BM models with great confidence, and a key issue is the plausibility of no increased hazard at low doses. If available information on such matters as biochemistry, genotoxicity, and induced cell replication suggests that low doses do not increase risk above background levels, then the question arises whether the subjective probability of risk at low doses should include both a positive probability that the risk is zero and a probability distribution for the degree of potency if it is not zero. In application, that might result in one of the following three decisions:

— If the data are sufficient to use the BM model, specify its parameters, and conclude scientifically (using whatever principles and evidentiary standards EPA sets forth in response to the committee's recommendation that it develop such principles) that this model is appropriate, the BM model could be used. Such occurrences are likely to be uncommon in the near term because of the need for extensive data of special types.

— If the data lead to a scientific conclusion that there is a substantial possibility that the low-dose potency is zero, the potency distributions from the BM and LMS models could be presented separately, perhaps with a narrative or quantitative statement of the probability weights to be assigned to each model.

— If the data do not suggest a substantial possibility of zero risk at low doses, the LMS model would continue to be used exclusively.

Statistical Analysis of Generated Probabilities

Once the needed subjective and objective probability distributions are estimated for each variable in the risk assessment, the estimates can be combined to determine their impact on the ultimate risk characterization. Joint distributions of input variables are often mathematically intractable, so an analyst must use approximating methods, such as numerical integration or Monte Carlo simulation. Such approximating methods can be made arbitrarily precise by appropriate computational methods. Numerical integration replaces the familiar operations of integral calculus by summarizing the values of the dependent variable(s) on a very fine (multivariate) grid of the independent variables. Monte Carlo methods are similar, but sum the variables calculated at random points on the grid; this is especially advantageous when the number or complexity of the input variables is so large that the costs of evaluating all points on a sufficiently fine grid would be prohibitive. (For example, if each of three variables is examined at 100 points in all possible combination, the grid would require evaluation at $100^3 = 1,000,000$ points, whereas a Monte Carlo simulation might provide results that are almost as accurate with only 1,000-10,000 randomly selected points.)

Barriers to Quantitative Uncertainty Analysis

The primary barriers to determining objective probabilities are lack of adequate scientific understanding and lack of needed data. Subjective probabilities are also not always available. For example, if the fundamental molecular-biologic bases of some hazards are not well understood, the associated scientific uncertainties cannot be reasonably characterized. In such a situation, it would be prudent public-health policy to adopt inference options from the conservative end of the spectrum of scientifically plausible available options. Quantitative dose-response assessment, with characterization of the uncertainty in the assessment, could then be conducted conditional on this set of inference options. Such a "conditional risk assessment" could then routinely be combined with an uncertainty analysis for exposure (which might not be subject to fundamental model uncertainty) to yield an estimate of risk and its associated uncertainty.

The committee recognizes the difficulties of using subjective probabilities in regulation. One is that someone would have to provide the probabilities to be used in a regulatory context. A "neutral" expert from within EPA or at a university or research center might not have the knowledge needed to provide a well-informed subjective probability distribution, whereas those who might have the most expertise might have or be perceived to have a conflict of interest, such as persons who work for the regulated source or for a public-interest group that has taken a stand on the matter. Allegations of conflict of interest or lack of knowledge regarding a chemical or issue might damage the credibility of the ultimate product of a subjective assessment. We note, however, that most of the same problems of real or perceived bias pervade EPA's current point-estimation approach.

At bottom, what matters is how risk managers and other end-users of risk assessments interpret the uncertainty in risk analysis. Correct interpretation is often difficult. For example, risks expressed on a logarithmic scale are commonly misinterpreted by assuming that an error of, say, a factor of 10 in one direction balances an error of a factor of 10 in the other. In fact, if a risk is expressed as 10^{-5} within a factor of 100 uncertainty in either direction, the average risk is approximately 1/2,000, rather than 1/100,000. In some senses, this is a problem of risk communication within the risk-assessment profession, rather than with the public.

Uncertainty Guidelines

Contrary to EPA's statement that the quantitative techniques suggested in this chapter "require definition of the distribution of all input parameters and knowledge of the degree of dependence (e.g., covariance) among parameters," (EPA, 1991f) complete knowledge is not necessary for a Monte Carlo or similar approach to uncertainty analysis. In fact, such a statement is a tautology: it is the uncertainty analysis that tells scientists how their lack of "complete knowledge" affects the confidence they can have in their estimate. Although it is always better to be able to be precise about how uncertain one is, an imprecise statement of uncertainty reflects how uncertain the situation is—it is far better to acknowledge this than to respond to the "lack of complete knowledge" by holding fast to a "magic number" that one knows to be wildly overconfident. Uncertainty analysis simply estimates the logical implications of the assumed model and whatever assumed or empirical inputs the analyst chooses to use.

The difficulty in documenting uncertainty can be reduced by the use of uncertainty guidelines that will provide a structure for how to determine uncertainty for each parameter and for each plausible model. In some cases, objective probabilities are available for use. In others, a subjective consensus about the uncertainty may be based on whatever data are available. Once these decisions are documented, many of the difficulties in determining uncertainty can be alleviated. However, it is important to note that consensus might not be achieved. If a "first-cut" characterization of uncertainty in a specific case is deemed to be inappropriate or superseded by new information, it can be changed by means of such procedures as those outlined in Chapter 12.

The development of uncertainty guidelines is important, because a lack of clear statements as to how to address uncertainty in risk assessment might otherwise lead to continuing inconsistency in the extent to which uncertainty is explicitly considered in assessments done by EPA and other parties, as well as to inconsistencies in how uncertainty is quantified. Developing guidelines to promote consistency in efforts to understand the uncertainty in risk assessment should improve regulatory and public confidence in risk assessment, because guidelines would reduce inappropriate inconsistencies in approach, and where inconsistencies remain, they could help to explain why different federal or state agencies come to different conclusions when they analyze the same data.

RISK MANAGEMENT AND UNCERTAINTY ANALYSIS

The most important goal of uncertainty analysis is to improve risk management. Although the process of characterizing the uncertainty in a risk analysis is also subject to debate, it can at a minimum make clear to decision-makers and the public the ramifications of the risk analysis in the context of other public decisions. Uncertainty analysis also allows society to evaluate judgments made by experts when they disagree, an especially important attribute in a democratic society. Furthermore, because problems are not always resolved and analyses often need to be repeated, identification and characterization of the uncertainties can make the repetition easier.

Single Estimates of Risk

Once EPA succeeds in supplanting single point estimates with quantitative descriptions of uncertainty, its risk assessors will still need to summarize these distributions for risk managers (who will continue to use numerical estimates of risk as inputs to decision-making and risk communication). It is therefore crucial to understand that uncertainty analysis is *not* about replacing "risk numbers" with risk distributions or any other less transparent method; it is about *con-*

sciously selecting the appropriate numerical estimate(s) from out of an understanding of the uncertainty.

Regardless of whether the applicable statute requires the manager to balance uncertain benefits and costs or to determine what level of risk is "acceptable," a bottom-line summary of the risk is a very important input, as it is critical to judging how confident the decision-maker can be that benefits exceed costs, that the residual risk is indeed "acceptable," or whatever other judgments must be made. Such summaries should include at least three types of information: (1) a fractile-based summary statistic, such as the median (the 50th percentile) or a 95th-percentile upper confidence limit, which denotes the probability that the uncertain quantity will fall an unspecified distance above or below some associated value; (2) an estimate of the mean and variance of the distribution, which along with the fractile-based statistic provides crucial information about how the probabilities and the absolute magnitudes of errors interrelate; and (3) a statement of the potential for errors and biases in these estimates of fractiles, mean, and variance, which can stem from ambiguity about the underlying models, approximations introduced to fit the distribution to a standard mathematical form, or both.

One important issue related to uncertainty is the extent to which a risk assessment that generates a point estimate, rather than a range of plausible values, is likely to be too "conservative" (that is, to excessively exaggerate the plausible magnitude of harm that might result from specified environmental exposures). As the two case studies that include uncertainty analysis (Appendixes F and G) illustrate, these investigations can show whether "conservatism" is in fact a problem, and if so, to what extent. Interestingly, the two studies reach opposite conclusions about "conservatism" in their specific risk-assessment situations; perhaps this suggests that facile conclusions about the "conservatism" of risk assessment in general might be off the mark. On the one hand, the study in Appendix G claims that EPA's estimate of MEI risk (approximately 10^{-1}) is in fact quite "conservative," given that the study calculates a "reasonable worst-case risk" to be only about 0.0015.[6] However, we note that this study essentially compared different and incompatible models for the cancer potency of butadiene, so it is impossible to discern what percentile of this unconditional uncertainty distribution any estimate might be assigned (see the discussion of model uncertainty above). On the other hand, the Monte Carlo analysis of parameter uncertainty in exposure and potency in Appendix F claims that EPA's point estimate of risk from the coal-fired power plant was only at the 83rd percentile of the relevant uncertainty distribution. In other words, a standard "conservative" estimate of risk (the 95th percentile) *exceeds* EPA's value, in this case by a factor of 2.5. It also appears from Figure 5-7 in Appendix F that there is about a 1% chance that EPA's estimate is too low by more than a factor of 10. Note that both case studies (Appendixes F and G) fail to distinguish sources of uncertainty from sources of interindividual variability, so the corresponding "uncertainty" distributions obtained cannot be used to properly characterize uncertainty either in predicted incidence or in predicted risk to some particular (e.g., average, highly exposed, or high-risk) individual (see Chapter 11 and Appendix I-3).

As discussed above, access to the entire PDF allows the decision-maker to assess the amount of "conservatism" implicit in any estimate chosen from the distribution. In cases where the risk manager asks the analyst to summarize the PDF via one or more summary statistics, the committee suggests that EPA might consider a particular kind of point estimate to summarize uncertain risks, in light of the two distinct kinds of "conservatism" discussed in Appendix N-1 (the "level of conservatism," the relative percentile at which the point estimate of risk is located, and the "amount of conservatism," the absolute difference between the point estimate and the mean). Although the specific choice of this estimate should be left to EPA risk managers, and may also need to be flexible enough to accommodate case-specific circumstances, estimates do exist that can account for both the percentile and the relationship to the mean in one single number. For example, EPA could choose to summarize uncertain risks for reporting the mean of the upper five percent of the distribution. It is a mathematical truism that (for right-skewed distributions commonly encountered in risk assessment) the larger the uncertainty, the greater the chance that the mean may exceed any arbitrary percentile of the distribution (see Table 9-4). Thus, the mean of the upper five percent is by definition "conservative" both with respect to the overall mean of the distribution and to its 95th percentile, whereas the 95th percentile may not be a "conservative" estimate of the mean. In most situations, the amount of "conservatism" inherent in this new estimator will not be as extreme as it would be if a very high percentile (e.g. the 99.9th) was chosen without reference to the mean.

Thus, the issue of uncertainty subsumes the issue of conservatism in point estimates. Point estimates chosen without regard to uncertainty provide only the barest beginnings of the story in risk assessment. Excessive or insufficient conservatism can arise out of inattention to uncertainty, rather than out of a particular way of responding to uncertainty. Actions taken solely to reduce or eliminate potential conservatism will not reduce and might increase the problem of excessive reliance on point estimates.

In summary, EPA's position on the issue of uncertainty analysis (as represented in the Superfund document) seems plausible at first glance, but it might be somewhat muddled. If we know that "all risk numbers are only good to within a factor of 10," why do *any* analyses? The reason is that both the variance and the conservatism (if any) are case-specific and can rarely be estimated with adequate precision until an honest attempt at uncertainty analysis is made.

Risk Communication

Inadequate scientific and technical communication about risk is sometimes a source of error and uncertainty, and guidance to risk assessors about what to

TABLE 9-4 Calculation Showing How Mean of Upper 5% of Lognormal Distribution (M_{95}) Relates to Other Distribution Statistics

σ_{lnx}	Uncertainty factor	M_{95}	Mean	M_{95}/Mean	X_{95}	M_{95}/X_{95}	Percentile location of mean	Percentile location of M_{95}
0.25	1.3	1.75	1.03	1.7	1.51	1.16	54	98.8
0.5	1.6	2.95	1.13	2.6	2.28	1.29	60	99.2
0.75	2.1	5.03	1.32	3.8	3.43	1.46	65	98.5
1	2.7	8.57	1.65	5.2	5.18	1.65	69	98.4
1.5	4.5	27.72	3.08	9	11.79	2.35	77	98.6
1.645	5.2	38.70	3.87	10	14.97	2.59	79	98.7
1.75	5.8	49.94	4.62	10.8	17.79	2.81	81	98.7
2	7.4	94.6	7.39	12.8	26.84	3.52	84	98.8
2.5	12.2	364.16	22.76	16	61.10	5.9	89	99
3	20.1	1647.3	90.02	18.3	139.07	11.84	93.3	99.2
4	54.6	59023	2981	19.8	720.54	81.92	97.7	99.7

include in a risk analysis should include guidance about how to present it. The risk assessor must strive to be understood (as well as to be accurate and complete), just as risk managers and other users must make themselves understood when they apply concepts that are sometimes difficult. This source of uncertainty in interprofessional communication seems to be almost untouched by EPA or any other official body (AIHC, 1992).

COMPARISON, RANKING, AND HARMONIZATION OF RISK ASSESSMENTS

As discussed in Chapter 6, EPA makes no attempt to apply a single set of methods to assess and compare default and alternative risk estimates with respect to parameter uncertainty. The same deficiency occurs in the comparison of risk estimates. When EPA ranks risks, it usually compares point estimates without considering the different uncertainties in each estimate. Even for less important regulatory decisions (when the financial and public-health impacts are deemed to be small), EPA should at least make sure that the point estimates of risk being compared are of the same type (e.g., that a 95% upper confidence bound for one risk is not compared with a median value for some other risk) and that each assessment has an informative (although perhaps sometimes brief) analysis of the uncertainty. For more important regulatory decisions, EPA should estimate the uncertainty in the *ratio* of the two risks and explicitly consider the probabilities and consequences of setting incorrect priorities. For any decisions involving risk-trading or priority-setting (e.g., for resource allocation or "offsets"), EPA should take into account information on the uncertainty in the quantities being ranked so as to ensure that such trades do not increase expected risk and that such priorities are directed at minimizing expected risk. When one or both risks are highly uncertain, EPA should also consider the probability and consequences of greatly erring in trading one risk for another, because in such cases one can lower the risk on average and yet introduce a small chance of greatly increasing risk.

Finally, EPA sometimes attempts to "harmonize" risk-assessment procedures between itself and other agencies, or among its own programs, by agreeing on a single common model assumption, even though the assumption chosen might have little more scientific plausibility than alternatives (e.g., replacing FDA's body-weight assumption and EPA's surface-area assumption with body weight to the 0.75 power). Such actions do not clarify or reduce the uncertainties in risk assessment. Rather than "harmonizing" risk assessments by picking one assumption over others when several assumptions are plausible and none of the assumptions is clearly preferable, EPA should use the preferred models for risk calculation and characterization, but present the results of the alternative models (with their associated parameter uncertainties) to further inform decision-makers and the public. However, "harmonization" does serve an important purpose in the context of uncertainty analysis—it will help, rather than hinder, risk assessment if agencies cooperate to choose and validate a common set of *uncertainty distributions* (e.g., a standard PDF for the uncertain exponent in the "body weight to the X power" equation or a standard method for developing a PDF from a set of bioassay data).

FINDINGS AND RECOMMENDATIONS

The committee strongly supports the inclusion of uncertainty analysis in risk assessments despite the potential difficulties and costs involved. Even for lower-tier risk assessments, the inherent problems of uncertainty need to be made explicit through an analysis (although perhaps brief) of whatever data are available, perhaps with a statement about whether further uncertainty analysis is justified. The committee believes that a more explicit treatment of uncertainty is critical to the credibility of risk assessments and to their utility in risk management.

The committee's findings and recommendations are summarized briefly below.

Single Point Estimates and Uncertainty

EPA often reports only a single point estimate of risk as a final output. In the past, EPA has only qualitatively acknowledged the uncertainty in its estimates, generally by referring to its risk estimates as "plausible upper bounds" with a plausible lower bound implied by the boilerplate statement that "the number could be as low as zero." In light of the inability to discern how "conservative" an estimate might be unless one does an uncertainty analysis, both statements might be misleading or untrue in particular cases.

• Use of a single point estimate suppresses information about sources of error that result from choices of model, data sets, and techniques for estimating values of parameters from data. EPA should not necessarily abandon the use of single-point estimates for decision-making, but such numbers must be the product of a consideration of both the estimate of risk and its uncertainties, not appear out of nowhere from a formulaic process. In other words, EPA should be free to choose a particular point estimate of risk to summarize the risk in light of its knowledge, uncertainty, and its desire to balance errors of overestimation and underestimation; but it should first derive that number from an uncertainty analysis of the risk estimate (e.g., using a summary statistic such as the "mean of the upper 5% of the distribution"). EPA should not simply state that its generic procedures yield the desired percentile. For example (although this is an analogous procedure to deal with variability, not uncertainty), EPA's current way of calculating the "high-end exposure estimate" (see Chapter 10) is ad hoc, rather than systematic, and should be changed.

• EPA should make uncertainties explicit and present them as accurately and fully as is feasible and needed for risk management decision-making. To the greatest extent feasible, EPA should present quantitative, as opposed to qualitative, representations of uncertainty. However, EPA should not necessarily quantify model uncertainty (via subjective weights or any other technique), but should try to quantify the parameter and other uncertainty that exists for each plausible choice of scientific model. In this way, EPA can give its default models the primacy they are due under its guidelines, while presenting useful, but distinct alternative estimates of risk and uncertainty. In the quantitative portions of their risk characterizations (which will serve as one important input to standard-setting and residual-risk decisions under the Act), EPA risk assessors should consider only the uncertainty conditional on the choice of the preferred models for dose-response relationships, exposure, uptake, etc.

• In addition, uncertainty analyses should be refined only so far as improvements in the understanding of risk and the implications for risk management justify the expenditure of the professional time and other resources that are required.

Uncertainty Guidelines

EPA committed itself in a 1992 internal memorandum (see Appendix B) to doing some kind of uncertainty analysis in the future, but the memorandum does not define when or how such analysis might be done. In addition, it does not distinguish between the different types of uncertainty or provide specific examples. Thus, it provides only the first, critical step toward uncertainty analysis.

• EPA should develop uncertainty analysis guidelines—both a general set and specific language added to its existing guidelines for each step in risk assessment (e.g., the exposure assessment guidance). The guidelines should consider in some depth all the types of uncertainty (model, parameter, etc.) in all the stages of risk assessment. The uncertainty guidelines should require that the uncertainties in models, data sets, and parameters and their relative contributions to total uncertainty in a risk assessment be reported in a written risk-assessment document.

Comparison of Risk Estimates

EPA makes no attempt to apply a consistent method to assess and compare default and alternative risk estimates with respect to parameter uncertainty. Presentations of numerical values in an incomplete form lead to inappropriate and possibly misleading comparisons among risk estimates.

• When an alternative model is plausible enough to be considered for use in risk communication, or for potentially supplanting the default model when sufficient evidence becomes available, EPA should analyze parameter uncertainty at a similar level of detail for the default and alternative models. For example, in comparing risk estimates derived from delivered-dose versus PBPK models, EPA should qualify uncertainty in the interspecies scaling factor (for the former case) and in the parameters used to optimize the PBPK equations (for the latter case). Such comparisons may reveal that given current parameter uncertainties, the risk estimate chosen would not be particularly sensitive to the judgment about which model is correct.

Harmonization of Risk Assessment Methods

EPA sometimes attempts to "harmonize" risk-assessment procedures between itself and other agencies or among its own programs by agreeing on a single common model assumption, even though the assumption chosen might have little more scientific plausibility than alternatives, (e.g., replacing FDA's body-weight assumption and EPA's surface-area assumption with body weight to the 0.75 power). Such actions do not clarify or reduce the uncertainties in risk assessment.

• Rather than "harmonizing" risk assessments by picking one assumption over others when several assumptions are plausible and none of the assumptions is clearly preferable, EPA should maintain its own default assumption for regulatory decisions but indicate that any of the methods might be accurate and present the results as an uncertainty in the risk estimate or present multiple estimates and state the uncertainty in each. However, "harmonization" does serve an important purpose in the context of uncertainty analysis—it will help, rather than hinder, risk assessment if agencies cooperate to choose and validate a common set of *uncertainty distributions* (e.g., a standard PDF for the uncertain exponent in the "body weight to the X power" equation or a standard method for developing a PDF from a set of bioassay data).

Ranking of Risk

When EPA ranks risks, it usually compares point estimates without considering the different uncertainties in each estimate.

• For any decisions involving risk-trading or priority-setting (e.g., for resource allocation or "offsets"), EPA should take into account information on uncertainty in quantities being ranked so as to ensure that such trades do not increase expected risk and such priorities are directed at minimizing expected risk. When one or both risks are highly uncertain, EPA should also consider the probability and consequences of greatly erring in trading one risk for another, because in such cases one can lower the risk on average and yet introduce a small chance of greatly increasing risk.

NOTES

1. Although variability in a risk-assessment parameter across different individuals is itself a type of uncertainty and is the subject of the following chapter, it is possible that new parameters might be incorporated into a risk assessment to model that variability (e.g., a parameter for the standard deviation of the amount of air that a random person breathes each day) and that those parameters themselves might be uncertain (see "uncertainty and variability" section in Chapter 11).

2. It is important to note that the distributions resulting from Bayesian models include various subjective judgments about models, data sets, etc. These are expressed as probability distributions but the probabilities should not be interpreted as probabilities of adverse effect but, rather, as expressions of strengths of conviction as to what models, data sets, etc. might be relevant to assessing risks of adverse effect. This is an important distinction which should be kept in mind when interpreting and using such distributions in risk management as a quantitative way of expressing uncertainty.

3. Assume that to convert from risk to the test animals to the predicted number of deaths in the human population, one must multiply by 10,000. Perhaps the laboratory dose is 10,000 times larger than the dose to humans, but 100 million humans are exposed. Thus, for example,

$$0.2\left(\frac{risk}{laboratory\ dose}\right) \times 10^{-4}\left(\frac{laboratory\ dose}{ambient\ dose}\right) \times 10^{8} = \left(\frac{deaths}{ambient\ dose}\right).$$

4. Note that characterizing risks considering only the parameter uncertainty under the preferred set of models might not be as restrictive as it appears at first glance, in that some of the model choices can be safely recast as parameter uncertainties. For example, the choice of a scaling factor between rodents and humans need not be classified as a model choice between body weight and surface area that calls for two separate "conditional PDFs," but instead can be treated as an uncertain parameter in the equation $R_{human} \alpha\ R_{rodent}\ BW^{a}$, where a might plausibly vary between 0.5 and 1.0 (see our discussion in Chapter 11). The only constraint in this case is that the scaling model is some power function of BW, the ratio of body weights.

5. It is not always clear what percent of the distribution someone is referring to by "correct to within a factor of X." If instead of assuming that the person means with 100% confidence, we assumed that the person means 98% confidence, then the factor of X would cover two standard deviations on either side of the median, so one geometric standard deviation would be equal to $\sqrt{X}$.

6. We arrive at this figure of 0.0015, or 1.5×10^{-3}, by noting that the "base case" for fenceline risk (Table 3-1 in Appendix G) is 5×10^{-4} and that "worst case estimates were two to three times higher than base case estimates."

10

Variability

INTRODUCTION AND BACKGROUND

It is always difficult to identify the true level of risk in an endeavor like health risk assessment, which combines measurement, modeling, and inference or educated guesswork. Uncertainty analysis, the subject of Chapter 9, enables one to come to grips with how far away from the desired answer one's best estimate of an unknown quantity might be. Before we can complete an assessment of the uncertainty in an answer, however, we must recognize that many of our questions in risk assessment have *more than one useful answer*. Variability—typically, either across space, in time, or among individuals—complicates the search for the desired value of many important risk-assessment quantities.

Chapter 11 and Appendix I-3 discuss the issue of how to aggregate uncertainties and interindividual differences in each of the components of risk assessment. This chapter describes the sources of variability[1] and appropriate ways to characterize these interindividual differences in quantities related to predicted risk.

Variability is a very well-known "fact of life" in many fields of science, but its sources, effects, and ramifications are not yet routinely appreciated in environmental health risk assessment and management. Accordingly, the first section of this chapter will step back and deal with the general phenomenon (using some examples relevant to risk assessment, but not exclusively), and then for the remainder of the chapter focus only on variability in quantities that directly influence calculations of individual and population risk.

When an important quantity is both uncertain and variable, opportunities are created to fundamentally misunderstand or misestimate the behavior of the quantity.

To draw an analogy, the exact distance between the earth and the moon is both difficult to measure precisely (at least it was until the very recent past) and changeable, because the moon's orbit is elliptical, rather than circular. Thus, as seen in Figure 10-1, uncertainty and variability can complement or confound each other. When only scattered measurements of the earth-moon distance were available, the variation among them might have led astronomers to conclude that their measurements were faulty (i.e., ascribing to uncertainty what was actually caused by variability) or that the moon's orbit was random (i.e., not allowing for uncertainty to shed light on seemingly unexplainable differences that are in fact variable *and* predictable). The most basic flaw of all would be to simply misestimate the true distance (the third diagram in Figure 10-1) by assuming that a few observations were sufficient (after correcting for measurement error, if applicable). This is probably the pitfall that is most relevant for health risk assessment: treating a highly variable quantity as if it was invariant or only uncertain, thereby yielding an estimate that is incorrect for some of the population (or some of the time, or over some locations), or even one that is also an inaccurate estimate of the average over the entire population.

In the risk-assessment paradigm, there are many sources of variability. Certainly, the regulation of air pollutants has long recognized that chemicals differ from each other in their physical and toxic properties and that sources differ from each other in their emission rates and characteristics; such variability is built into virtually any sensible question of risk assessment or control. However, even if we focus on a single substance emanating from a single stationary source, variability pervades each stage from emission to health or ecologic end point:

- *Emissions* vary temporally, both in flux and in release characteristics, such as temperature and pressure.
- The *transport and fate* of the pollutant vary with such well-understood factors as wind speed, wind direction, and exposure to sunlight (and such less-acknowledged factors as humidity and terrain), so its concentrations around its source vary spatially and temporally.
- Individual human *exposures* vary according to individual differences in breathing rates, food consumption, and activity (e.g., time spent in each micro-environment).
- The *dose-response* relationship (the "potency") varies for a single pollutant, because each human is uniquely susceptible to carcinogenic or other stimuli (and this inherent susceptibility might well vary during the lifetime of each person, or vary with such things as other illness or exposures to other agents).

Each of these variabilities is in turn often composed of several underlying variable phenomena. For example, the natural variability in human weight is due to the interaction of genetic, nutritional, and other environmental factors.

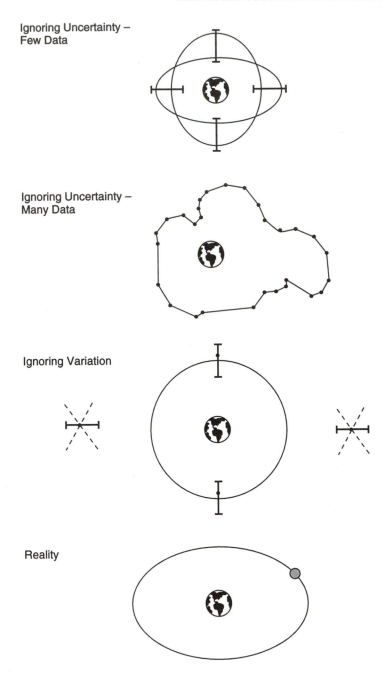

FIGURE 10-1   Effects of ignoring uncertainty versus ignoring variability in measuring the distance between the earth and the moon.

According to the central limit theorem, variability that arises from independent factors that act multiplicatively will generally lead to an approximately lognormal distribution across the population or spatial/temporal dimension (as is commonly observed when concentrations of air pollutants are plotted).

When there is more than one desired answer to a scientific question where the search for truth is the end in itself, only two responses are ultimately satisfactory: gather more data or rephrase the question. For example, the question "How far away is the moon from the earth?" cannot be answered both simply and correctly. Either enough data must be obtained to give an answer of the form "The distance ranges between 221,460 and 252,710 miles" or "The moon's orbit is approximately elliptical, with a minor axis of 442,920 miles, a major axis of 505,420 miles, and an eccentricity of 0.482," or the question must be reduced to one with a single right answer (e.g., "How far away is the moon from the earth *at its perigee?"*).

When the question is not purely scientific, but is intended to support a social decision, the decision-maker has a few more options, although each course of action will have repercussions that might foreclose other courses. Briefly, variability in the substance of a regulatory or science-policy question can be dealt with in four basic ways:

1. *Ignore the variability and hope for the best.* This strategy tends to be most successful when the variability is small and any estimate that ignores it will not be far from the truth. For example, the Environmental Protection Agency's (EPA's) practice of assuming that all adults weigh 70 kg is likely to be correct to within ±25% for most adults and probably valid to within a factor of 3 for virtually all adults. However, this approach may not be appropriate for children, where variability may be large (NRC, 1993e).

2. *Explicitly disaggregate the variability.* Where the quantity seems to change smoothly and predictably over some range, continuous mathematical models may be fitted to the data in place of a discrete step function. An example might be the fitting of sine waves to annual concentration cycles for a particular pollutant. In other cases, it is easier to disaggregate the data by considering all or the relevant subgroups or subpopulations. For interindividual variability, this involves dividing the population into as many subpopulations as deemed necessary. For example, one might perform a separate risk assessment for short-term exposure to high levels of ionizing radiation for each 10-year age interval in the population, to take account of age-related differences in susceptibility. For temporal variability, it involves modeling or measuring in a discrete, rather than a continuous, fashion, on an appropriate time scale. For example, a specific type of air-pollution monitor might collect air for 15 min of each hour and report the 15-min average concentration of some pollutant. Such values might then be further aggregated to produce summary values at an even coarser time scale. For spatial variability, it involves choosing an appropriate subregion, e.g., modeling

the extent of global warming or cooling for each 10-deg swath of latitude around the globe, rather than predicting a single value for the entire planet, which might mask substantial and important regional differences. In each case, the common thread appears: when variability is "large" over the entire data set, the variability within each subset can become sufficiently "small" ("small" in the sense of the body-weight example in the paragraph above), if the data are disaggregated into an appropriate number of qualitatively distinct subsets. The strategy tends to be most successful when the stakes are so high (or the data or estimates so easy to obtain) that the proliferation of separate assessments does not consume inordinate amounts of resources. In contrast, in studies of a phenomenon such as global climate change, where the stakes are quite high, the estimates may also be quite hard to obtain on a highly disaggregated basis.

In health risk assessment, the choice of the averaging time used to transform the variable quantity into a more manageable form is crucially important. In general, for the assessment of acute toxicity, estimates of the variability in exposure and/or uptake over relatively short periods (minutes, hours, days) are needed. For chronic effects such as cancer, one might model exposure and/or update over months or years without losing needed information, since short-term "peaks and valleys" would matter for cancer risk assessment only insofar as they affected the long-term or lifetime average exposure.[2] The longer-term variability will generally, though not always, be significantly less marked than the variation over the short-term (but see Note 3). Moreover, the shorter the averaging time, the more such periods will be contained in an individual's lifetime, and the more opportunity there will be for rare fluctuations in exposure or uptake to produce significant risks. This, for example, explains why regulators concerned with the health effects of tropospheric ozone consider the combination of peak short-term concentration *and* peak activity (e.g., the "exercising asthmatic"). In all cases, the exposure assessor needs to determine which time periods are relevant for which toxic effects, and then see whether available data measuring exposure, uptake, internal dose rates, etc., can provide estimates of both the average and the variability over the necessary averaging time.

3. *Use the average value of a quantity that varies.* This strategy is not the same as ignoring the variability; ideally, it follows from a decision that the average value can be *estimated reliably in light of the variability,* and that it is a good surrogate for the variable quantity. For example, EPA often uses 70 kg as the average body weight of an adult, presumably because although many adults weigh as little as 40 kg and as much as 100 kg, the average weight is almost as useful as (and less complicated than) three different "scenario" values or an entire distribution of weights. In the same vein, a layperson might be content in knowing the average value of the moon's distance from the earth, rather than the minimum, average, and maximum (let alone a complete mathematical description of its orbit)—whereas the average alone would be useless, or even dangerous, to the National Aeronautics and Space Administration in

planning an Apollo mission. Thus, this strategy tends to be most successful (and indeed might be the only sensible strategy) when the variability is small[3] or when the quantity is itself an input for a model or decision in which the average value of the end result (the combination of several quantities) is all that matters, either for scientific or policy reasons. An example of a scientific rationale for using the average value is the long-term average concentration of a carcinogen in air. If the dose-response function is linear (i.e., "potency" is a single number), the end result (risk) is proportional to the average concentration. If the concentration is, say, 10 ppm higher than the average in one week and 10 ppm lower than the average in another week, this variability will have no effect on an exposed person's lifetime risk, so it is biologically unimportant. An example of a policy rationale is the use of the expected number of cancer cases in a population exposed to varying concentrations of an airborne carcinogen. If it is determined for a particular policy rationale that the distribution of individual risks across the population does not matter, then the product of *average* concentration, potency and population size equals the expected incidence, and the spread of concentrations about the average concentration is similarly unimportant. The average value is also the summary statistic of choice for social decisions when there is an opportunity for errors of underestimation and overestimation (which lead to underregulation and overregulation) to even out over a large set of similar choices over the long run.

There are at least two reasons why large variabilities can lead to precarious decisions if the average value is used. The obvious problem is that individual characteristics of persons or situations far from the average are "averaged away" and can no longer be identified or reported. A less obvious pitfall occurs when the variability is dichotomous (or has several discrete values) and the average corresponds to a value that does not exist in nature. If men and women respond markedly differently to some exposure situation, for example, the decision that would be appropriate if there existed an "average person" (midway between man and woman) might be inappropriate for either category of real person (see Finkel, 1991).

4. *Use a maximum or minimum of a quantity that varies.* This is perhaps the most common way of dealing with variability in risk assessment—to focus attention on one period (e.g., the period of peak exposure), one spatial subregion (e.g., the location where the "maximally exposed individual" resides), or one subpopulation (e.g., exercising asthmatics or children who ingest pathologically large amounts of soil) and ignore the rest. This strategy tends to be most successful when the measures needed to protect or account for the person (or situation) with the extreme value will *also* suffice for the remainder of the distribution. It is also important to ensure that this strategy will not impose inordinate costs, compared with other approaches (such as using different controls for each subregion or population or simply controlling less stringently by using the average value instead of the extreme "tail").

The crucial point to bear in mind about all four of those strategies for dealing with variability is that unless someone measures, estimates, or at least roughly models the extent and nature of the variability, any strategy will be precarious. It stands to reason that strategy 1 ("hope for the best") hinges on the assumption that the variability is small—an assumption whose verification requires at least some attention to variability. Similarly, strategy 2 requires the definition of subregions or subpopulations in each of which the variability is small, so care must be taken to avoid the same conundrum that applies to strategy 1. (It is difficult to be sure that you can ignore variability until you think about the possible consequences of ignoring it.) Less obviously, one still needs to be somewhat confident that one has a handle on the variability in order to reduce the distribution to either an average (strategy 3) or a "tail" value (strategy 4). We *know* that 70 kg is an average adult body weight (and that virtually no adults are above or below 70 kg by more than a factor of 3), because weight is directly observable *and* because we know the mechanism by which people grow and the biologic limits of either extreme. Armed with our senses and this knowledge, we might need only a few observations to pin down roughly the minimum, the average, and the maximum. But what about a variable like "the rate at which human liver cells metabolize ethylene dibromide into its glutathione conjugate"? *Here a few direct measurements or a few extrapolations from animals may not be adequate,* because in the absence of any firm notion of the spread of this distribution within the human population (or the mechanisms by which the spread occurs), we cannot know how reliably our estimate of the average value reflects the true average, nor how well the observed minimum and maximum mirror the *true* extremes.

The distribution for an important variable such as metabolic rate should thus explicitly be considered in the risk assessment, and the reliability of the overall risk estimate should reflect knowledge about both the uncertainty and the variability in this characteristic. The importance of a more accurate risk estimate may motivate additional measurements of this variable, so that its distributions may be better defined with these additional data.

This chapter concentrates on how EPA treats variability in emissions, exposures, and dose-response relationships, to identify which of the four strategies it typically uses and to assess how adequately it has considered each choice and its consequences. The goals of this chapter are three: (1) to indicate how EPA can increase its sophistication in defining variability and handling its effects; (2) to provide information as to how to improve risk *communication*, so that Congress and the public understand at least which variabilities are and which are not accounted for, and how EPA's handling of variability affects the "conservatism" (or lack thereof) inherent in its risk numbers; and (3) to recommend specific research whose results could lead to useful changes in risk-assessment procedures.

In recent years, EPA has begun to increase its attention to variability. Moreover, the lack of attention in the past was due in part to a set of choices to erect a set of conservative default options (strategy 4 above) instead of dealing with variability explicitly. In theory at least, the question "How do you determine the extreme of a distribution without knowing the whole distribution?" can be answered by setting a highly conservative default and placing the burden of proof on those who wish to relax the default by showing that the extreme is unrealistic even as a "worst case." For example, the concept of the MEI (someone who breathes pollutants from the source for 70 years, 24 hours per day, at a specified location near a plant boundary) has been criticized as unrealistic, but most agree that as a summary of the population distribution of "number of hours spent at a given location during a lifetime" it might be a reasonable place to start from as a conservative short-cut for the entire distribution.

EPA has also tackled interindividual variability squarely in *Exposure Factors Handbook* (EPA, 1989c), which provides various percentiles (e.g., 5th, 25th, 50th, 75th, 95th) of the observed variability distributions for some components of exposure assessment, such as breathing rates, water ingestion, and consumption of particular foodstuffs. This document has not yet become a standard reference for many of EPA's offices, however. In addition, as we will discuss below, EPA has not dealt adequately with several other major sources of variability. *As a result, EPA's methods to manage variability in risk assessment rely on an ill-characterized mix of some questionable distributions, some verified and unverified point values intended to be "averages," some verified and unverified point values intended to be "worst cases," and some "missing defaults," that is, hidden assumptions that ignore important sources of variability.*

Moreover, several trends in risk assessment and risk management are now increasing the urgency of a broad and well-considered strategy to deal with variability. The three most important of these trends are the following:

• *The emergence of more sophisticated biological models for risk assessment.* As pharmacokinetic models replace the administered assumption and as cell-kinetics models (such as the Moolgavkar-Venzon-Knudson model) replace the linearized-multistage model, default models that ignored human variability or took conservative measures to sidestep it will be supplanted by models that explicitly contain values of biologic measures intended to represent the human population. If the latter models ignore variability or use unverified surrogates for presumed average or worst-case properties, risk assessment might take a step backwards, becoming either less or more conservative without anyone's knowledge.

• *The growing interest in detailed assessments of the actual exposures that people face, rather than hypothetical worst-case exposures.* To be trustworthy, both average and worst-case surrogates for variability require some knowledge of the rest of the distribution, as mentioned above. However, it is not well recognized that the average might be *more* sensitive to the extreme portions of the whole distribution than an upper percentile might be, such as the 95th. In addition, the use of such terms as *actual* and *best estimates* carries an expectation of precision that might apply to only *part* of the exposure assessment, dose-response relationship, or risk assessment. If, for example, we could precisely measure the airborne concentration of a pollutant in a community around a stationary source (i.e., understand the spatial variability), but did not know the population distribution of breathing rates, we could not predict *anyone's* "actual exposure." In fact, even if we knew *both* distributions but could not superimpose them (i.e., know which breathing rates went with which concentrations), the predictions would be as variable as either of the underlying distributions. These circumstances speak to the need for progress in many kinds of research and data collection at once, if we wish to improve the power and the realism of risk assessment.

• *The growing interest in risk-reduction measures that target people, rather than sources.* It should go without saying that if government or industry wishes to eliminate unacceptably high risks to particular persons by purchasing their homes, providing them with bottled water, restricting access to "hot spots" of risk, etc., it needs to know precisely who those persons are and where or when those hot spots are occurring. Even if such policies were not highly controversial and difficult to implement in an equitable and socially responsive way, merely identifying the prospective targets of such policies may well presuppose a command of variability beyond our current capabilities.

EXPOSURE VARIABILITY

Variability in human response to pollutants emitted from a particular source or set of sources can arise from differences in characteristics of exposure, uptake, and personal dose-response relationships (susceptibility). Exposure variability in turn depends on variability in all the factors that affect exposure, including emissions, atmospheric processes (transport and transformation), personal activity, and the pollutant concentration in the microenvironments where the exposures occur. Information on those variabilities is not routinely included in EPA's exposure assessments, probably because it has been difficult to specify the distributions that describe the variations.

Human exposure results from the contact of a person with a substance at some nonzero concentration. Thus, it is tied to personal activities that determine a person's location (e.g., outdoors vs. indoors, standing downwind of an industrial facility vs. riding in a car, in the kitchen vs. on a porch); the person's level of activity and breathing rate influences the uptake of airborne pollutants. Exposure is also tied to emission rates and atmospheric processes that affect pollutant concentrations in the microenvironment where the person is exposed. Such processes include infiltration of outside air indoors, atmospheric advection (i.e., transport by the prevailing wind), diffusion (i.e., transport by atmospheric turbulence), chemical and physical transformation, deposition, and re-entrainment—variability in each process tends to increase the overall variability in exposure. The variabilities in emissions atmospheric processes, characteristics of the microenvironment, and personal activity are not necessarily independent of each other; for example, personal activities and pollutant concentrations at a specific location might change in response to outdoor temperature; they might also differ between weekends and weekdays because the level of industrial activity changes.

Emissions Variability

There are basically four categories of emission variability that may need separate assessment methods, depending on the circumstances:

- Routine—this is the type most frequently covered by current approaches.
- Ordinary maintenance—special emissions may occur, for example, when the bag house is cleaned. In other cases certain emissions may only occur during maintenance, as when a specific volatile cleaner is routinely used to scour or wash out a reaction tank. These can be deliberately observed and monitored to obtain needed emissions information, if this mode is deemed likely to be significant.
- Upsets and breakdowns—unusual operating conditions that may recur within average periods of days, weeks, or months, depending on the facility/process. A combination of observations and modeling approaches may be needed here.
- Catastrophic failures—large explosions, ruptures of storage tanks, etc.

The last category is addressed in a separate section of the Clean Air Act and is not discussed in this report.

At least two major factors influence variability in emissions as it affects exposure assessment. First, a given source typically does not emit at a constant rate. It is subject to such things as load changes, upsets, fuel changes, process modifications, and environmental influences. Some sources are, by their nature, intermittent or cyclical. A second factor is that two similar sources (e.g., facilities in the same source category) can emit at different rates because of differences in such things as age, maintenance, or production details.

The automobile is an excellent example of both causes. Consider a single, well-characterized car with an effective control system. When it is started, the catalyst has not warmed up, and emissions can be high. Almost half the total automobile emissions in, say, Los Angeles can occur during the cold-start period. After the catalyst reaches its appropriate temperature range, it is extremely effective (>90%) at removing organic substances, such as benzene and formaldehyde, during most of the driving period. However, hard accelerations can overwhelm the system's capabilities and lead to high emissions. Those variations can lead to spatial and temporal distributions of emissions in a city (e.g., high emissions in areas with a large number of cold starts, particularly in the morning). The composition of the emissions, including the toxic content, differs between cold-start and driving periods. Emissions also differ between cars— often dramatically. Because of differences in control equipment, total emissions can vary, and emissions between cycles can vary between cars (e.g., cold-start vs. evaporative emissions). A final notable contribution to emission variability in automobiles is the presence of super-emitters, whose control systems have failed and may emit organic substances at a rate 10 times that of a comparable vehicle that is operating properly.

Thus, an exposure analysis based on source-category average emissions will miss the variability in sources within that category. And, exposure analyses that do not account for temporal changes in emissions from a particular source will miss an important factor, especially to the extent that emissions are linked to meteorologic conditions. In many cases, it is difficult or impossible to know a priori how emissions will vary, particularly because of upsets in processes that could lead to high exposures over short periods.

Atmospheric Process Variability

Meteorologic conditions greatly influence the dispersion, transformation, and deposition of pollutants. For example, ozone concentrations are highest during summer afternoons, whereas carbon monoxide and benzene concentrations peak in the morning (because of the combination of large emissions and little dilution) and during the winter. Formaldehyde can peak in the afternoon during the summer (because of photochemical production) and in the morning in the winter (because of rush-hour emissions and little dilution). Concentrations of primary (i.e., emitted) pollutants, such as benzene and carbon monoxide, are higher in the winter in urban areas, whereas those of many secondary pollutants (i.e., those resulting from atmospheric transformations of primary pollutants), such as ozone, are higher in the summer. Meteorologic conditions may also play a role in regional variations. Some areas experience long periods of stagnant air, which lead to very high concentrations of both primary and secondary pollutants. An extreme example is the London smog that led to high death rates before the mid-1950s. Wind velocity and mixing height also influence pollutant concentrations. (Mixing height is the height to which pollutants are rapidly mixed due to atmospheric turbulence; in effect, it is one dimension of the atmospheric volume in which pollutants are diluted.) They are usually correlated; the prevailing winds and velocities in the winter, when the mixing height is low, can be very different from those in the summer.

Some quantitative information is available about the impact of meteorologic variability on pollutant concentrations. Concentrations measured at one location over some period tend to follow a lognormal distribution. There are significant fluctuations in the concentrations about the medians (e.g., Seinfeld, 1986), which often vary by a factor of more than 10. The extreme concentrations are usually related to time and season. The relative magnitudes and frequencies of such fluctuations in concentration increase as distance from the source decreases. Pollutant transport over complex terrain (e.g., presence of hills or tall buildings), which is generally difficult to model, can further increase relative differences in extreme concentrations about the medians. Two examples of the influence of complex terrain are Donora, Pennsylvania (in a river valley), and the Meuse Valley in Belgium. In those areas, as in London, periods of extremely high pollutant concentrations led to a period of increased deaths. Estimates of concentration over flat terrain cannot capture such effects.

Empirical data on concentration variability are sparse, except for a few pollutants, notably the criteria pollutants (including carbon monoxide, ozone, sulfur dioxide, and particulate matter). Some information on variations in formaldehyde and benzene concentrations is also available. One interesting study that considered air-pollutant exposure during commuting in the Los Angeles area was conducted by the South Coast Air Quality Management District (SCAQMD, 1989). The authors looked at exposure dependence on seasonal, vehicular-age, and freeway-use variations. They found that drivers of older vehicles had greater exposure to benzene and that exposure to benzene, formaldehyde, ethylene, and chromium was greater in the winter, although exposure to ethylene dichloride was greater in the summer. They did not report the variability in exposure between similar vehicles or distributions of the exposures (e.g., probability density functions).

Microenvironmental and Personal-Activity Variability

Microenvironmental variability, particularly when compounded with differences in personal activity, can contribute to substantial variability in individual exposure. For example, the lifetime-exposed 70-year-old has been faulted as an extreme case, but it is instructive to consider this hypothetical person in the distribution of personal activity traits. Although it is unlikely, this 70-year lifetime exposure activity pattern is one end of the spectrum in the variability of personal activity and time spent in a specific microenvironment.

Concentrations in various microenvironments vary considerably and depend on a variety of factors, such as species, building type, ventilation system, locality of other sources, and street canyon width and depth. Both the Los Angeles study (SCAQMD, 1989) and a New Jersey study (Weisel et al., 1992) revealed that exposure can be increased during commuting, particularly if the automobile itself is defective. The primary sources of many air pollutants are indoors, so their highest concentrations are found there. Those concentrations can be 10-1,000 times the outdoor concentrations (or even greater). However, the difference between outdoor and indoor concentrations of pollutants is not nearly so great when the indoor location is ventilated. Concentrations of compounds that do not react rapidly with or settle on surfaces, such as carbon monoxide and many organic compounds might not decrease significantly when ventilated indoors. If there are additional sources of these compounds indoors, their concentrations might, in fact, increase. Concentrations of more reactive compounds, such as ozone, can decrease by a factor of 2 or more, depending on ventilation rate and the ventilation system used (Nazaroff and Cass, 1986). Particles can also be advected indoors (Nazaroff et al., 1990). One concern is that the ventilation of outdoor pollutants indoors can increase the formation of other pollutants (Nazaroff and Cass, 1986; Weschler et al., 1992). The lifetime-exposed person sitting on the porch outside his home may be at one extreme for exposure to emissions from an outdoor stationary source, but may be at the other extreme for net air-pollutant exposure; such a person may have effectively avoided "hot" microenvironments in both the home and the automobile.

Increased personal activity leads to a larger uptake, and this will add to variability by as much as a factor of about 2 or more. The activity-related component of variability depends on both the microenvironmental variability (e.g., outdoors vs. indoors) and personal characteristics (e.g., children vs. adults).

VARIABILITY IN HUMAN SUSCEPTIBILITY

Person-to-person differences in behavior, genetic makeup, and life history together confer on individual people unique susceptibilities to carcinogenesis (Harris, 1991). Such interindividual differences can be inherited or acquired. For example, inherited differences in susceptibility to physical or chemical carcinogens have been observed, including a substantially increased risk of sunlight-induced skin cancer in people with xeroderma pigmentosum, of bladder cancer in dyestuff workers whose genetic makeup results in the "poor acetylator" phenotype, and of bronchogenic carcinoma in tobacco smokers who have an "extensive debrisoquine hydroxylator" phenotype (both are described further in Appendix H). Similarly among different inbred and outbred strains of laboratory animals (and within particular outbred strains) exposed to carcinogenic initiators or tumor promoters there may be a factor of 40 variation in tumor response (Boutwell, 1964; Drinkwater and Bennett, 1991; Walker et al., 1992). Acquired differences that can significantly affect an individual's susceptibility to carcinogenesis include the presence of concurrent viral or other infectious diseases, nutritional factors such as alcohol and fiber intake, and temporal factors such as stress and aging.

Appendix H describes three classes of factors that can affect susceptibility: (1) those which are rare in the human population but which confer very large increases in susceptibility upon those affected; (2) those which are very common but only marginally increase susceptibility; and (3) those which may be neither rare nor of marginal importance to those affected. The Appendix provides particular detail on five of the determinants that fall into this third group. This material in Appendix H represents both a compilation of existing literature as well as some new syntheses of recent studies; we commend the reader's attention to this important information.

Overall Susceptibility

Taken together, the evidence regarding the individual mediators of susceptibility described in Appendix H supports the plausibility of a continuous distribution of susceptibility in the human population. Some of the individual determinants of susceptibility, such as concentrations of activating enzymes or of proteins that might become oncogenic, may themselves exist in continuous gradations across the human population. Even factors that have long been thought to be dichotomous are now being revealed as more complicated—e.g., the recent finding that a substantial fraction of the population is heterozygous for ataxia-telangiectasia and has a susceptibility midway between that of ataxia-telangiectasia homozygotes and that of "normal" people (Swift et al., 1991). Most important, the combination of a large number of genetic, environmental, and lifestyle influences, even if each were bimodally distributed, would likely generate an essentially continuous overall susceptibility distribution. As Reif (1981) has noted, "we would expect to find in [the outbred human population] what would be the equivalent result of outbreeding different strains of inbred mice: a spectrum of different genetic predispositions for any particular type of tumor."

A working definition of the breadth of the distribution of "interindividual variability in overall susceptibility to carcinogenesis" is as follows: If we identified persons of high susceptibility (say, we knew them to represent the 99th percentile of the population distribution) and low susceptibility (say, the 1st percentile), we could estimate the risks that each would face if subjected to the same exposure to a carcinogen. If the estimated risk to the first type of person were 10^{-2} and the estimated risk to the second type of person were 10^{-6}, we could say that "human susceptibility to this chemical varies by at least a factor of 10,000."[4]

There are two distinct but complementary approaches to estimating the form and breadth of the distribution of interindividual variability in overall susceptibility to carcinogenesis. The biologic approach is a "bottom-up" method that uses empirical data on the distribution of particular factors that mediate susceptibility to model the overall distribution. In the major quantitative biologic analysis of the possible extent of human variations in susceptibility to carcinogenesis, Hattis et al. (1986) reviewed 61 studies that contained individual human data on six characteristics that are probably involved causally in the carcinogenic process. The six were the half-life of particular biologically active substances in blood, metabolic activation of drugs (in vivo) and putative carcinogens (in vitro), enzymatic detoxification, DNA-adduct formation, the rate of DNA repair (as measured by the rate of unscheduled DNA synthesis induced by UV light), and the induction of sister-chromatid exchanges after exposure of lymphocytes to x-rays. They estimated the overall variability in each factor by fitting a lognormal distribution to the data and then propagated the variabilities by using Monte Carlo simulation and assuming that the factors interacted multiplicatively and were statistically independent. Their major conclusion was that the logarithmic standard deviation of the susceptibility distribution lies between 0.9 and 2.7 (90% confidence interval). That is, the difference in susceptibility between the most sensitive 1% of the population and the least sensitive 1% might be as small as a factor of 36 (if the logarithmic standard deviation was 0.9) or as large as a factor of 50,000 (if the logarithmic standard deviation was 2.7).[5]

The alternative approach is inferential or "top-down," and combines epidemiologic data with a demographic technique known as heterogeneity dynamics. Heterogeneity dynamics is an analytic method for describing the changing characteristics of a heterogeneous population as its members age. The power of the heterogeneity-dynamics approach to explain initially puzzling aspects of demographic data, as well as to challenge simplistic explanations of population behavior, stems from its emphasis on the divergence between forces that affect individuals and forces that affect populations (Vaupel and Yashin, 1983). The most fundamental concept of heterogeneity dynamics is that individuals change at rates different from those of the cohorts they belong to, because the passage of time affects the composition of the cohort as it affects the life prospects of each member. In a markedly heterogeneous population, the overall death rate can decline with age, even though every individual faces an ever-increasing risk of death, simply because the population as a whole grows increasingly more "resistant" to death as the more susceptible members are preferentially removed. Specifically with regard to cancer, heterogeneity dynamics can examine the progressive divergence of observed human age-incidence functions (for many tumor types) away from the function that is believed to apply to an individual's risk as a function of age—namely, the power function of age formalized in the 1950s by Armitage and Doll (which posits that risk increases proportionally with age raised to an integral exponent, probably 4, 5, or 6). In contrast with groups of inbred laboratory animals, which do exhibit age-incidence functions that generally obey the Armitage-Doll model, in humans the age-incidence curves for many tumor types begin to level off and plateau at higher ages.

Many of the pioneering studies that used heterogeneity dynamics to infer the amount of variation in human susceptibility to cancer used cross-sectional data, which might have been confounded by secular changes in exposures to carcinogenic stimuli (Sutherland and Bailar, 1984; Manton et al., 1986). One investigation that built on the previous body of work was that of Finkel (1987), who assembled longitudinal data on cancer mortality, including the age at death and cause of death of all males and females born in 1890, for both the United States and Norway. That study separately examined deaths due to lung cancer and colorectal cancer and tried to infer the amount of population heterogeneity that could have caused the observed age-mortality relationships to diverge from the Armitage-Doll (age^N) function that should apply to the population if all humans are of equal sensitivity. The study concluded that as a first approximation, the amount of variability (for either sex, either disease, and either country) could be roughly modeled by a lognormal distribution with a logarithmic standard deviation on the order of 2.0 (i.e., general agreement with the results of Hattis et al., 1986). That is, about 5% of the population might be about 25 times more susceptible than the average person (and a corresponding 5% about 25 times less susceptible); about 2.5% might be 50 times more (or less) susceptible than the average, and about 1% might be at least 100 times more (or less) susceptible.

A later analysis (Finkel, in press) showed that such a conclusion, if borne out, would have important implications not only for assessing risks to individuals, but for estimating population risk in practice. In a highly heterogeneous population, quantitative uncertainties about epidemiological inferences drawn from relatively small subpopulations (thousands or fewer), as well as the frequent application of animal-based risk estimates to similarly "small" subpopulations, will be increased by the possibility that the *average* susceptibility of small groups varies significantly from group to group.

The issue of susceptibility is an important one for acute toxicants as well as carcinogens. The NRC Committee on Evaluation of the Safety of Fishery Products addressed this issue in depth in their report entitled *Seafood Safety* (NRC, 1991b). Guidelines for the assessment of acute toxic effects in humans have recently been published by the NRC Committee on Toxicology (NRC, 1993d).

CONCLUSIONS

This section records the results of the committee's analysis of EPA's practice on variability.

Exposure Variability and the Maximally Exposed Individual

One of the contentious defaults that has been used in past air-pollutant exposure and risk assessments has been the maximally exposed individual (MEI), who was assumed to be the person at greatest risk and whose risk was calculated by assuming that the person resided outdoors at the plant boundary, continuously for 70 years. This is a worst-case scenario (for exposure to the particular source only) and does not account for a number of obvious factors (e.g., the person spends time indoors, going to work, etc.) and other likely events (e.g., changing residence) that would decrease exposure to the emissions from the specific source. This default also does not account for other, possibly countervailing factors involved in exposure variability discussed above. Suggestions to remedy this shortcoming have included decreasing the point estimate for residence time at the location to account for population mobility, and use of personal-activity models (see Chapters 3 and 6).

EPA's most recent exposure-assessment guidelines (EPA, 1992a) no longer use the MEI, instead coining the terms "high-end exposure estimates" (HEEE) and "theoretical upper-bounding exposure" (TUBE) (see Chapter 3). According to the new exposure guidelines (Section 5.3.5.1), a high-end risk "means risks above the 90th percentile of the population distribution, but not higher than the individual in the population who has the highest risk." The EPA Science Advisory Board had recommended that exposures or risks above the 99.9th percentile be regarded as "bounding estimates" (i.e., use of the 99.9th percentile as the HEEE) for large populations (assuming that unbounded distributions such as the lognormal are used as inputs for calculating the exposure or risk distribution). For smaller populations, the guidelines state that the choice of percentile should be based on the objective of the analysis. However, neither the HEEE nor the TUBE is explicitly related to the expected MEI.

The new exposure guidelines (Section 5.3.5.1) suggest four methods for arriving at an estimator of the HEEE. These are, in descending order of sophistication:

• "If sufficient data on the distribution of doses are available, take the value directly from the percentile(s) of interest within the high end;"

• "if . . . data on the parameters used to calculate the dose are available, a simulation (such as an exposure model or Monte Carlo simulation) can sometimes be made of the distribution. In this case, the assessor may take the estimate from the simulated distribution;"

• "if some information on the distribution of the variables making up the exposure or dose equation . . . is available, the assessor may estimate a value which falls into the high end. . . The assessor often constructs such an estimate by using maximum or near-maximum values for one or more of the most sensitive variables, leaving others at their mean values;"

• "if almost no data are available, [the assessor can] start with a bounding estimate and back off the limits used until the combination of parameter values is, in the judgment of the assessor, clearly in the distribution of exposure or dose . . . The availability of pertinent data will determine how easily and defensibly the high-end estimate can be developed by simply adjusting or backing off from the ultraconservative assumptions used in the bounding estimates."

The first two methods are much preferable to the last two and should be used whenever possible. Indeed, EPA should place a priority on collecting enough data (either case-specific or generic) that the latter two methods will not be needed in estimating variability in exposure. The distribution of exposures, developed from measurements or modeling results or both, should be used to estimate population exposure, as an input in calculating population risk. It can also be used to estimate the exposure of the maximally exposed person. For example, the most likely value of the exposure to the most exposed person is generally the $100[(N-1)/N]$th percentile of the cumulative probability distribution characterizing interindividual variability in exposures, where N is the number of persons used to construct the exposure distribution. This is a particularly convenient estimator to use because it is independent of the shape of the exposure distribution (see Appendix I-3). Other estimators of exposure to the highest, or jth highest for some $j<N$, person exposed are available (see Appendix I-3). The committee recommends that EPA explicitly and consistently use an estimator such as $100[(N-1)/N]$, because it, and not a vague estimate "somewhere above the 90th percentile," is responsive to the language in CAAA-90 calling for the calculation of risk to "the individual most exposed to emissions. . . ."

In recent times, EPA has begun incorporating into distributions of exposure assumptions that are based on a national average of years of residence in a home, as a replacement for its 70-year exposure assumption (e.g., an average lifetime). Proposals have been made for a similar "departure from default" for the time an individual spends at a residence each day, as a replacement for the 24 hours assumption. However, such analyses make the assumption that individuals move to a location of zero exposure when they change residences during their lifetime or leave the home each day. But, people moving from one place to another, whether it be changing the location of their residence or moving from the home to office, can vary greatly in their exposure to any one pollutant, from relatively high exposures to none. Furthermore, some exposures to different pollutants may be considered as interchangeable: moving from one place to another may yield exposures to different pollutants which, being interchangeable in their effects, can be taken as an aggregate, single "exposure." This assumption of interchangeability may or may not be realistic; however, because people moving from place to place can be seen as being exposed over time to a mixture of pollutants, some of them simultaneously and others at separate times, a simplistic analysis of residence times is not appropriate. The real problem is, in effect, a more complex problem of how to aggregate exposure to mixtures as well as one of multiple exposures of varying level of intensities to a single pollutant.

Thus, a simple distribution of residence times may not adequately account for the risks of movement from one region to another, especially for persons in hazardous occupations, such as agricultural workers exposed to pesticides, or persons of low socioeconomic status who change residences. Further, some subpopulations that might be more likely to reside in a high-exposure region might also be less mobile (e.g., owing to socioeconomic conditions). For these reasons, the default residency assumption for the calculation of the maximally exposed individual should remain at the mean of the current U.S. life expectancy, in the absence of supporting evidence otherwise. Such evidence could include population surveys of the affected area that demonstrate mobility outside regions of residence with similar exposures to similar pollutants. Personal activity (e.g., daily and seasonal activities) should be included.

If in a given case EPA determines that it must use the third method (combining various different "maximum," "near-maximum," and average values for inputs to the exposure equation) to arrive at the HEEE, the committee offers another caution: EPA has not demonstrated that these combinations of point estimates do in fact yield an output that reliably falls at the desired location within the overall distribution of exposure variability (that is, in the "conservative" portion of the distribution, but not above the confines of the entire distribution). Accordingly, EPA should validate (through generic simulation analyses and specific monitoring efforts) that its point-estimation methods do reasonably and reliably approximate what would be achieved via the more sophisticated direct-measurement or Monte Carlo methods (that is, a point estimate at approximately the $100[(N-1)/N]$th percentile of the distribution). The fourth method, it should go without saying, is highly arbitrary and should not be used unless the bounding estimate can be shown to be "ultraconservative" and the concept of "backing off" is better defined by EPA.

Susceptibility

Human beings vary substantially in their inherent susceptibility to carcinogenesis, both in general and in response to any specific stimulus or biologic mechanism. No point estimate of the carcinogenic potency of a substance will apply to all individuals in the human population. Variability affects each step in the carcinogenesis process (e.g., carcinogen uptake and metabolism, DNA damage, DNA repair and misrepair, cell proliferation, tumor progression, and metastasis). Moreover, the variability arises from many independent risk factors, some inborn and some environmental. On the basis of substantial theory and some observational evidence, it appears that some of the individual determinants of susceptibility are distributed bimodally (or perhaps trimodally) in the human population; in such cases, a class of hypersusceptible people (e.g., those with germ-line mutations in tumor-suppressor genes) might be at tens, hundreds, or thousands of times greater risk than the rest of the population. Other determinants seem to be distributed more or less continuously and unimodally, with either narrow or broad variances (e.g., the kinetics or activities of enzymes that activate or detoxify particular pollutants).

To the extent that those issues have been considered at all with respect to carcinogenesis, EPA and the research community have thought almost exclusively in terms of the bimodal type of variation, with a normal majority and a hypersusceptible minority (ILSI, 1992). That model might be appropriate for noncarcinogenic effects (e.g., normal versus asthmatic response to SO_2), but it ignores a major class of variability vis-à-vis cancer (the continuous, "silent" variety), and it fails to capture even some bimodal cases in which hypersusceptibility might be the rule, rather than the exception (e.g., the poor-acetylator phenotype).

The magnitude and extent of human variability due to particular acquired or inherited cancer-susceptibility factors should be determined through molecular epidemiologic and other studies sponsored by EPA, the National Institutes of Health, and other federal agencies. Two priorities for such research should be

• To explore and elucidate the relationships between variability in each measurable factor (e.g., DNA adduct formation) and variability in susceptibility to carcinogenesis.
• To provide guidance on how to construct appropriate samples of the population for epidemiologic studies and risk extrapolation, given the influence of susceptibility variation on uncertainty in population risk and the possible correlations between individual susceptibility and such factors as race, ethnicity, age, and sex.

Results of the research should be used to adjust and refine estimates of risks to individuals (identified, identifiable, or unidentifiable) and estimates of expected incidence in the general population.

The population distribution of interindividual variation in cancer suscepti- bility cannot now be estimated with much confidence. Preliminary studies of this question, both biologic (Hattis et al., 1986) and epidemiologic (Finkel, 1987) have concluded that the variation might be described as approximately lognor- mal, with about 10% of the population being different by a factor of 25-50 (either more or less susceptible) from the median individual (i.e., the logarithmic standard deviation of the distribution is approximately 2.0). While the estimated standard deviation of a susceptibility distribution suggested by these studies is uncertain, in light of the biochemical and epidemiological data reviewed earlier in this chapter it is currently not scientifically plausible that the U.S. population is strictly homogeneous in susceptibility to cancer induction by cancer-causing chemicals. EPA's guidelines are silent regarding person-to-person variations in susceptibility, thereby treating all humans as identical, despite substantial evi- dence and theory to the contrary. This is an important "missing default" in the guidelines. EPA does assume (although its language is not very clear in this regard) that the median human has susceptibility similar to that of the particular sex-strain combination of rodent that responds most sensitively of those tested in bioassays, or susceptibility identical with that of the particular persons observed in epidemiologic studies. These latter assumptions are reasonable as a starting point (Allen et al., 1988), but of course they could err substantially in either direction for a specific carcinogen or for carcinogens as a whole.

The missing default (variations in susceptibility among humans) and ques- tionable default (average susceptibility of humans) are related in a straightfor- ward manner. Any error of overestimation in rodent-to-human scaling (or in epidemiologic analysis) will tend to counteract the underestimation errors that must otherwise be introduced into some individual risk estimates by EPA's cur- rent practice of not distinguishing among different degrees of human susceptibility. Conversely, any error of underestimation in interspecies scaling will exacerbate the underestimation of individual risks for every person of above-average susceptibility. Therefore, EPA should increase its efforts to validate or improve the default assumption that the median human has similar susceptibility to that of the rodent strain used to compute potency, and should attempt to assess the plausible range of uncertainty surrounding the existing assumption. For further information, see the discussion in Chapter 11.

It can be argued, in addition, that EPA has a responsibility, insofar as it is practicable, to protect persons regardless of their individual susceptibility to carcinogenesis (we use *protect* here not in the absolute, zero-risk sense, but in the sense of ensuring that excess individual risk is within acceptable levels or below a de minimus level). It is unclear from the language in CAAA-90 Section 112(f)(2) whether the "individual most exposed to emissions" is intended to mean the person at highest risk when both exposure and susceptibility are taken into account, but this interpretation is both plausible and consistent with the fact that a major determinant of susceptibility is the degree of metabolism of inhaled or ingested pollutants and the resulting exposure of somatic and germ cells to carcinogenic compounds (i.e., two people of different susceptibilities will likely be "exposed" to a different extent even if they breathe or ingest identical ambient concentrations). Moreover, EPA has a record of attempting to protect people with a combination of high exposure and high sensitivity, as seen in the National Ambient Air Quality Standards (NAAQS) program for criteria air pollutants (e.g., SO_2, NO_x, ozone, etc.).

Therefore, EPA should adopt an explicit default assumption for susceptibility before it begins to implement those decisions called for in the Clean Air Act Amendments of 1990 that require the calculation of risks to individuals. EPA could choose to incorporate into its cancer risk estimates for individual risk (not for population risk) a "default susceptibility factor" greater than the implicit factor of 1 that results from treating all humans as identical. EPA should explicitly choose a default factor greater than 1 if it interprets the statutory language to apply to individuals with both high exposure and above-average susceptibility.[6] EPA could explicitly choose a default factor of 1 for this purpose, if it interprets the statutory language to apply to the person who is average (in terms of susceptibility) but has high exposure. Or, preferably, EPA could develop a "default distribution" of susceptibility, and then generate the joint distribution of exposure and cancer potency (in light of susceptibility), to find the upper 95th or 99th percentile of risk for use in a risk assessment. The distribution is the more desirable way of dealing with this problem, because it takes explicit account of the joint probability (which may be large or small) of a highly exposed individual who is also highly susceptible.

Many of the currently known individual determinants of susceptibility vary by factors of hundreds or thousands at the cellular level; however, many of these risk factors (see Appendix I-2) tend to confer excess risks of approximately a factor of 10 on predisposed people, compared with "normal" ones. Although the total effect of the many such factors may cause susceptibility to vary upwards by more than a factor of 10, some members of the committee suggest that a default factor of 10 might be a reasonable starting point, if EPA wished to apply the statutory risk criteria (see Chapter 2) to the more susceptible members of the human population. Conversely, other members of the committee do not consider an explicit factor of 10 to be justified at this time. A 10-fold adjustment might yield a reasonable best estimate of the high end of the susceptibility distribution for some pollutants when only a single predisposing factor divides the population into normal and hypersusceptible people.

If any susceptibility factor greater than 1 is applied, the short-term practical effect will be to increase all risk assessments for individual risk by the same factor, except for chemical-specific risk estimates where there is evidence that the variation in human susceptibility is larger or smaller for that chemical than for other substances. Such a general adjustment of either the default factor or default distribution might become appropriate when more information becomes available about the nature and extent of interindividual variations in susceptibility.

Individual risk assessments may depart from the new default when it can be shown either that humans are systematically either more or less sensitive than rodents to a particular chemical or that interindividual variation is markedly either more or less broad for this chemical than for the typical chemical. Therefore, in the spirit of our recommendations in Chapter 6 and Appendixes N-1 and N-2, the committee encourages EPA both to rethink the new default in general and to depart from it in specific cases when appropriately justified by general principles the agency should articulate.

Although it is known that there are susceptibility differences among people due to such factors as age, sex, race, and ethnicity, the nature and magnitude of these differences is not well known or understood; therefore, it is critical that additional research be pursued. As knowledge increases, science may be able to describe differences in the population at risk and recognize these differences with some type of default or distribution, although caution will be necessary to ensure that broad *correlations* between susceptibility and age, sex, etc., are not interpreted as deterministic *predictions*, valid for all individuals, or used in areas outside of risk assessment without proper respect for autonomy, privacy, and other social values.

In addition to adopting a default assumption for the effect of variations in susceptibility on individual risk, EPA should consider whether these variations might affect calculations of *population risk* as well. Estimates of population risk (i.e., the number of cases of disease or the number of deaths that might occur as a result of some exposure) are generally based on estimates of the average individual risk, which are then multiplied by the number of exposed persons to obtain a population risk estimate. The fact that individuals have unique susceptibilities should thus be irrelevant to calculating population risk, except if ignoring these variations *biases the estimate of average risk*. Some observers have pointed out a logical reason why EPA's current procedures might misestimate average risk. Even assuming that allometric or other interspecies scaling procedures correctly map the risk to test animals onto the "risk to the average human" (an assumption we encourage EPA to explore, validate, or refine), it is not clear *which* "average" is correctly estimated—the *median* (i.e., the risk to a person who has susceptibility at the 50th percentile of the population distribution) or the *expected value* (i.e., the average individual risk, taking into account *all* of the risks in the population and their frequency or likelihood of occurrence).

If person-to-person variation in susceptibility is small or symmetrically distributed (as in a normal distribution), the median and the average (or mean) are likely to be equivalent, or so similar that this distinction is of no practical importance. However, if variation is large and asymmetrically distributed (as in a lognormal distribution with logarithmic standard deviation on the order of 2.0 or higher—see earlier example), the mean may exceed the median by roughly an order of magnitude or more.[7]

The committee encourages EPA to explore whether extrapolations made from animal bioassay data (or from epidemiological studies) at high exposures are likely to be appropriate for the median or for the average human, and to explore what response is warranted for the estimation and communication of population risk if the median and average are believed to differ significantly. As an initial position, EPA might assume that animal tests and epidemiological studies in fact lead to risk estimates for the *median* of the exposed group. This position would be based on the logic that at high exposures and hence high risks (that is, on the order of 10^{-2} for most epidemiologic studies, and 10^{-1} for bioassays), the effect of any variations in susceptibility within the test population would be truncated or attenuated. In such cases, any test animal or human subject whose *susceptibility* was X-fold higher than the median would face *risks* (far) less than X-fold higher than the median risk, because in no case can risk exceed 1.0 (certainty), and thus the effect of these individuals on the population average would not be in proportion to their susceptibilities. On the other hand, when extrapolating to ambient exposures where the median risk is closer to 10^{-6}, the full divergence between median and average in the general population would presumably manifest itself.

If, therefore, current procedures correctly estimate the median risk, then estimates of population risk would have to be increased by a factor corresponding to the ratio of the average to the median.

Other Changes in Risk-Assessment Methods

(1) Children are a readily identifiable subpopulation with its own physiologic characteristics (e.g., body weight), uptake characteristics (e.g., food consumption patterns), and inherent susceptibilities. When excess lifetime risk is

the desired measure, EPA should compute an integrated lifetime risk, taking into account all relevant age-dependent variables, such as body weight, uptake, and average susceptibility (for one example of such a computation, see Appendix C of NRDC, 1989). If there is reason to believe that risk is not linearly related to biologically effective dose, and if the computed risks for children and adults are found to be significantly different, EPA should present separate risk assessments for children and adults.

(2) Although EPA has tried to take account of interindividual variability in susceptibility for non-cancer effects (e.g., in standards for criteria air pollutants such as ozone or SO_2), such efforts have neither seen exhaustive nor part of an overall focus on variability. In particular, the "10-fold safety factor" used to account for interindividual variability when extrapolating from animal toxicity data has not been validated, in the sense that EPA is generally not aware how much of the human population falls within an order of magnitude of the median susceptibility for any particular toxic stimulus.

Although this chapter has focused on susceptibility to carcinogens, because this subject has received even less attention than that of susceptibility to non-carcinogens, the committee urges EPA to continue to improve its treatment of variability in the latter area as well.

(3) EPA has not sufficiently accounted for interindividual variability in biologic characteristics when it has used various physiologic or biologically based risk-assessment models. The validity of many of these models and assumptions depends crucially on the accuracy and precision of the human biological characteristics that drive them. In a wide variety of cases, interindividual variation can swamp the simple measurement uncertainty or the uncertainty in modeling that is inherent in deriving estimates for the "average" person. For example, physiologically based pharmacokinetic (PBPK) models require information about partition coefficients and enzyme concentrations and activities; Moolgavkar-Venzon-Knudson and other cell-kinetics models require information about cell growth and death rates and the timing of differentiation; and specific alternative models positing dose-response thresholds for given chemicals require information about ligand-receptor kinetics or other cellular phenomena. EPA has begun to collect data to support the development of distributions for the key PBPK parameters (such as alveolar ventilation rates, blood flows, partition coefficients, and Michaelis-Menten metabolic parameters) in both rodents and humans (EPA, 1988f). However, this database is still sparse, especially with respect to the possible variability in human parameters. EPA has developed point estimates for human PBPK parameters for 72 volatile organic chemicals, only 26 of which are on the list of 189 hazardous air pollutants covered in CAAA-90. For only five chemicals (benzene, n-hexane, toluene, trichloroethylene, and n-xylene) does EPA have any information on the presumed average and range of the parameters in the human population. It is perhaps noteworthy that in the one major instance in which EPA has revised a unit risk factor for a hazardous air pollutant on the

basis of PBPK data (the case of methylene chloride), no information on the possible effect of human variability was used (EPA, 1987d; Portier and Kaplan, 1989).

Even when the alternative to the default model hinges on a qualitative, rather than a quantitative, distinction, such as the possible irrelevance to humans of the alpha-2μ-globulin mechanism involved in the initiation of some male rat kidney tumors, the new model must be checked against the possibility that some humans are qualitatively different from the norm. Any alternative assumption might be flawed, if it turns out to be biologically inappropriate for some fraction of the human population. Finally, although epidemiology is a powerful tool that can be used as a "reality check" on the validity of potency estimates derived from animal data, there must be a sufficient amount of human data for this purpose. The sample size needed for a study to have a given power level *increases* under the assumption that humans are not of identical susceptibility.

When EPA proposes to adopt an alternative risk-assessment assumption (such as use of a PBPK model, use of a cell-kinetics model, or the determination that a given animal response is "not relevant to humans"), it should consider human interindividual variability in estimating the model parameters or verifying the assumption of "irrelevance." If the data are not available that would enable EPA to take account of human variability, EPA should be free to make any reasonable inferences about its extent and impact (rather than having to collect or await such data), but should encourage other interested parties to collect and provide the necessary data. In general, EPA should ensure that a similar level of variability analysis is applied to both the default and the alternative risk assessment, so that it can compare estimates of equal conservatism from each procedure.

Risk Communication

EPA often does not adequately communicate to its own decision-makers, to Congress, or to the public the variabilities that are and are not accounted for in any risk assessment and the implications for the conservatism and representativeness of the resulting risk numbers. Each of EPA's reports of a risk assessment should state its particular assumptions about human behavior and biology and what these do and do not account for. For example, a poor risk characterization for a hazardous air pollutant might say "The risk number R is a plausible upper bound." A better characterization would say, "The risk number R applies to a person of reasonably high-end behavior living at the fenceline 8 hours a day for 35 years." EPA should, whenever possible, go further and state, for example, "The person we are modeling is assumed to be of average susceptibility, but eats F grams per day of food grown in his backyard; the latter assumption is quite conservative, compared with the average."

Risk-communication and risk-management decisions are more difficult when, as is usually the case, there are both uncertainty and variability in key risk-assessment inputs. It is important, whenever possible, to separate the two phenomena conceptually, perhaps by presenting multiple analyses. For its full (as opposed to screening-level) risk assessments, EPA should acknowledge that all its risk numbers are made up of three components: the estimated risk itself (X), the level of confidence (Y) that the risk is no higher than X, and the percent of the population (Z) that X is intended to apply to in a variable population. EPA should use its present practice of saying that "the plausible upper-bound risk is X" only when it believes that Y and Z are both close to 100%. Otherwise, it should use statements like, "We are Y% certain that the risk is no more than X to Z% of the population," or use an equivalent pictorial representation (see Figure 10-2).

As an alternative or supplement to estimating the value of Z, EPA can and should try to present multiple scenarios to explain variability. For example, EPA could present one risk number (or preferably, an uncertainty distribution—see Chapter 9) that explicitly applies to a "person selected at random from the population," one that applies to a person of reasonably high susceptibility but "average" behavior (mobility, breathing rate, food consumption, etc.), and one that applies to a person whose susceptibility and behavioral variables are both in the "reasonably high" portion of their distributions.

Identifiability and Risk Assessment

Not all the suggestions presented here, especially those regarding variation in susceptibility, might apply in every regulatory situation. The committee notes that in the past, whenever persons of high risk or susceptibility have been identified, society has tended to feel a far greater responsibility to inform and protect them. For such identifiable variability, the recommendations in this section are particularly salient. However, interindividual variability might be important even when the specific people with high and low values of the relevant characteristic cannot currently be identified.[8] Regardless of whether the variability is now identifiable (e.g., consumption rates of a given foodstuff), difficult to identify (e.g., presence of a mutant allele of a tumor-suppressor gene), or unidentifiable (e.g., a person's net susceptibility to carcinogenesis), the committee agrees that it is important to think about its potential magnitude and extent, to make it possible to assess whether existing procedures to estimate average risks and population incidence are biased or needlessly imprecise.

In contrast with issues involving average risk and incidence, however, some members of the committee consider the distribution of individual susceptibilities and the uncertainty as to where each person falls in that distribution to be irrelevant if the variation is and will remain unidentifiable. For example, some argue that people should be indifferent between a situation wherein their risk is determined to be precisely 10^{-5} or one wherein they have a 1% chance of being highly

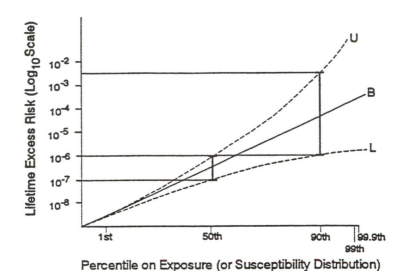

- *Curve B* presents the best estimate of the relationship between exposure (or susceptibility) and risk [expressed in a scale relative to the rest of the population, not in absolute exposure (or susceptibility) units.]

- *Curve L* presents the 5th (or other lower) percentile of this relationship.

- *Curve U* presents the 95th (or other upper) percentile of this relationship.

Thus, for this hypothetical example, the risk communicator could say:

"We are 90% certain that the risk to the person with median exposure is between 10^{-7} and 10^{-6}," AND/OR

"We are 90% certain that the risk to the person with high (90th percentile) exposure is between 10^{-6} and 3×10^{-3}."

FIGURE 10-2   Communicating risk, uncertainty, and variability graphically.

Note: To translate from percentile-relative exposure to absolute exposure, you could add a second x-axis scale based on a figure as follows:

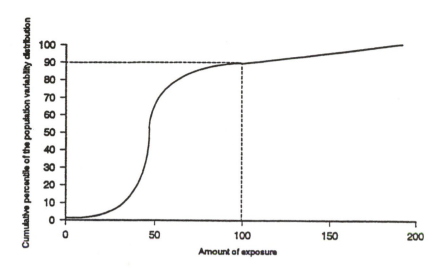

Thus, combining figures 1 and 2 gives you the 2 x-axes:

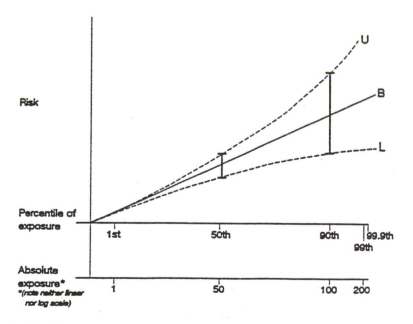

FIGURE 10-2 *Continued*

susceptible (with risk $= 10^{-3}$) and a 99% chance of being immune, with no way to know which applies to whom. In both cases, the expected value of individual risk is 10^{-5}, and it can be argued that the distribution of risks is the same, in that without the prospect of identifiability no one actually faces a risk of 10^{-3}, but just an equal chance of facing such a risk (Nichols and Zeckhauser, 1986).

Some of the members also argue that as we learn more about individual susceptibility, we will eventually reach a point where we will know that some individuals are at extremely high risk (i.e., carried to its extreme, an average individual risk of 10^{-6} may really represent cases where one person in each million is guaranteed to develop cancer while everyone else is immune). As we approach this point, they contend, society will have to face up to the fact that in order to guarantee that everyone in the population faces "acceptable" low levels of risk, we would have to reduce emissions to an impossibly low extent.

Other committee members reject or deem irrelevant the notion that risk is ultimately either zero or 1; they believe that, both for an individual's assessment of how foreboding or tolerable a risky situation is and for society's assessment of how just or unjust the distribution of risks is, the information about the unidentifiable variability must be reported—that it affects both judgments. To bolster their contentions, these members cite literature about the limitations of expected utility theory, which takes the view, contradicted by actual survey data, that the distribution of risky outcomes about their mean values should not affect the individual's evaluation of the situation (Schrader-Frechette, 1985; Machina, 1990), and empirical findings that the skewness of lotteries over risky outcomes matters to people even when the mean and variance are kept constant (Lopes, 1984). They also argue that EPA should maintain consistency in how it handles exposure variability, which it reports even when the precise persons at each exposure level cannot be identified; i.e., EPA reports the variation in air concentration and the maximal concentration from a source even when (as is usually the case) it cannot predict exactly where the maximum will occur. If susceptibility is in large part related to person-to-person differences in the amount of carcinogenic material that a person's cells are exposed to via metabolism, then it is essentially another form of exposure variability, and the parallel with ambient (outside-the-body) exposure is close. Finally, they claim that having agreed that issues of pure uncertainty are important, EPA (and the committee) must be consistent and regard unidentifiable variability as relevant (see Appendix I-3). Our recommendations in Chapter 9 reflect our view that uncertainty is important because individuals and decision-makers do regard values other than the mean as highly relevant. If susceptibility is unidentifiable, then to the individual it represents a source of uncertainty about his or her individual risk, and many members of the committee believe it must be communicated just as uncertainty should be.

Social-science research aimed at clarifying the extent to which people care about unidentifiable variability in risk, the costs of accounting for it in risk management, and the extent to which people want government to take such variation and costs into account in making regulatory decisions and in setting priorities might be helpful in resolving these issues.

FINDINGS AND RECOMMENDATIONS

The committees findings and recommendations are briefly summarized below.

Exposure

Historically, EPA has defined the maximally exposed individual (MEI) as the worst-case scenario—a continuous 70-year exposure to the maximal estimated long-term average concentration of a hazardous air pollutant. Departing from this practice, EPA has recently published methods for calculating bounding and "reasonably high-end" estimates of the highest actual or possible exposures using a real or default distribution of exposure within a population. The new exposure guidelines do not explicitly define a point on this distribution corresponding to the highest expected exposure level of an individual.

• The committee endorses the EPA's use of bounding estimates, but only in screening assessments to determine whether further levels of analysis are necessary. For further levels of analysis, the committee supports EPA's development of distributions of exposure values based on available measurements, modeling results, or both. These distributions can also be used to estimate the exposure of the maximally exposed person. For example, the most likely value of the exposure to the most exposed person is generally the $100[(N - 1)/N]$th percentile of the cumulative probability distribution characterizing interindividual variability in exposure, where N is the number of persons used to construct the exposure distribution. This is a particularly convenient estimator to use because it is independent of the shape of the exposure distribution. The committee recommends that EPA explicitly and consistently use an estimator such as $100[(N - 1)/N]$, because it, and not a vague estimate "somewhere above the 90th percentile," is responsive to the language in CAAA-90 calling for the calculation of risk to "the individual most exposed to emissions. . . ."

In recent times, EPA has begun incorporating into distributions of exposure assumptions that are based on a national average of years of residence in a home, as a replacement for its 70-year exposure assumption (e.g., an average lifetime). Proposals have been made for a similar "departure from defaults" for the time an individual spends at a residence each day, as a replacement for the 24 hours assumption. However, such analyses make the assumption that individuals move to a location of zero exposure when they change residences during their lifetime or leave the home each day. But, people moving from one place to another, whether it be changing the location of their residence or moving from the home to office, may vary greatly in their exposure to any one pollutant, from relatively high exposures to none. Further, some exposures to different pollutants may be considered as interchangeable: moving from one place to another may yield exposures to different pollutants which, being interchangeable in their effects, can be taken as an aggregate, single "exposure." This assumption of interchangeability may or may not be realistic; however, because people moving from place to place can be seen as being exposed, over time to a mixture of pollutants, some of them simultaneously and others at separate times, a simplistic analysis of residence times is not appropriate. The real problem is, in effect, a more complex problem of how to aggregate exposure to mixtures as well as one of multiple exposures of varying level of intensities to a single pollutant. Thus, a simplistic analysis based on a simple distribution of residence times is not appropriate.

• EPA should use the mean of current life expectancy as the assumption for the duration of individual residence time in a high-exposure area, or a distribution of residence times which accounts for the likelihood that changing residences might not result in significantly lower exposure. Similarly, EPA should use a conservative estimate for the number of hours a day an individual is exposed, or develop a distribution of the number of hours per day an individual spends in different exposure situations. Such information can be gathered through neighborhood surveys, etc. in these high-exposure areas. Note that the distribution would correctly be used only for individual risk calculations, as total population risk is unaffected by the number of persons whose exposures sum to a given total value (if risk is linearly related to exposure rate).

EPA has not provided sufficient documentation in its exposure-assessment guidelines to ensure that its point-estimation techniques used to determine the "high-end exposure estimate" (HEEE) when data are sparse reliably yield an estimate at the desired location within the overall distribution of exposure (which, according to these guidelines, lies above the 90th percentile but not beyond the confines of the entire distribution).

• EPA should provide a clear method and rationale for determining *when* point estimators for the HEEE can or should be used instead of a full Monte Carlo (or similar) approach to choosing the desired percentile explicitly. The rationale should more clearly indicate how such estimators are to be generated, should offer more documentation that such point-estimation methods do yield reasonably consistent representations of the desired percentile, and should justify the choice of such a percentile if it differs from that which corresponds to the expected value of exposure to the "person most exposed to emissions".

Potency

EPA has dealt little with the issue of human variability in susceptibility; the limited efforts to date have focused exclusively on variability relative to noncarcinogenic effects (e.g., normal versus asthmatic response to SO_2). The appropriate response to variability for noncancer end points (i.e., identify the characteristics of "normal" and "hypersusceptible" individuals, and then decide whether or not to protect both groups) might not be appropriate for carcinogenesis, in which variability might well be continuous and unimodal, rather than either-or.

• EPA, NIH, and other federal agencies should sponsor molecular epidemiologic and other research on the extent of interindividual variability in various factors that affect susceptibility and cancer, on the relationships between variability in each factor and in the health end point, and on the possible correlations between susceptibility and such covariates as age, race, ethnicity, and sex. Results of the research should be used to adjust and refine estimates of risks to individuals (identified, identifiable, or unidentifiable) and estimates of expected incidence in the general population. As this research progresses, the natural science and social science community should collaborate to explore the implications of any susceptibility factors that can be tested for or that strongly correlate with other genetic traits, so as to ensure that any findings are not misinterpreted or used outside of the environmental risk assessment arena without proper care.

Susceptibility

EPA does not account for person-to-person variations in susceptibility to cancer; it thereby treats all humans as identical in this respect in its risk calculations.

• EPA should adopt a default assumption for susceptibility before it begins to implement those decisions called for in the Clean Air Act that require the calculation of risks to individuals. EPA could choose to incorporate into its cancer risk estimates for individual risk a "default susceptibility factor" greater than the implicit factor of 1 that results from treating all humans as identical. EPA should explicitly choose a default factor greater than 1 if it interprets the statutory language to apply to an individual with high exposure and above-average susceptibility. EPA could explicitly choose a default factor of 1 for this purpose, if it interprets the statutory language to apply to an individual with high exposure but average susceptibility. Preferably, EPA could develop a "default distribution" of susceptibility, and then generate the joint distribution of exposure and cancer potency (in light of susceptibility) to find the upper 95th percentile (or 99th percentile) of risk for each risk assessment.

EPA makes its potency calculations on the assumption that, on average, humans have susceptibility similar to that of the particular sex-strain combination of rodent that responds most sensitively of those tested in bioassays or susceptibility identical with that of the particular groups of persons observed in epidemiologic studies.

- EPA should continue and increase its efforts to validate or improve the default assumption that, on average, humans to be protected at the risk-management stage have susceptibility similar to that of humans included in relevant epidemiological studies, the most-sensitive rodents tested, or both.

It is possible that ignoring variations in human susceptibility may cause significant underestimation of population risk, if both of two conditions hold: (1) current procedures to extrapolate results of laboratory bioassays or epidemiologic studies to the general population correctly map the observed risk in the test population to the human with *median* susceptibility, not to the expected value averaged over the entire general population; and (2) there is sufficient skewed variability in susceptibility in the general population to cause the expected value to exceed the median to a significant extent.

- In addition to continuing to explore the assumption that interspecies scaling (or epidemiologic extrapolation) correctly predicts average human susceptibility, EPA should investigate whether the average that is predicted corresponds to the median or the expected value. If there is reason to suspect the former is true, EPA should consider whether it needs to adjust its estimates of population risk to account for this discrepancy.

Children are a readily identifiable subpopulation with its own physiologic characteristics (e.g., body weight), uptake characteristics (e.g., food consumption patterns), and inherent susceptibilities.

- If there is reason to believe that risk of adverse biological effects per unit dose depends on age, EPA should present separate risk estimates for adults and children. When excess lifetime risk is the desired measure, EPA should compute an integrated lifetime risk, taking into account all relevant age-dependent variables.

EPA does not usually explore or consider interindividual variability in key biologic parameters when it uses or evaluates various physiologic or biologically based risk-assessment models (or else evaluates some data but does not report on this in its final public documents). In some other cases, EPA does gather or review data that bear on human variability, but tends to accept them at face value without ensuring that they are representative of the entire population. As a general rule, the larger the number of characteristics with an important effect on risk or the more variable those characteristics are, the larger the sample of the human population needed to establish confidently the mean and range of each of those characteristics.

- When EPA proposes to adopt an alternative risk-assessment assumption (such as use of a PBPK model, use of a cell-kinetics model, or the determination that a given animal response is "not relevant to humans"), it should consider human interindividual variability in estimating the model parameters or verifying the assumption of "irrelevance." If the data are not available to take account of human variability, EPA should be free to make any reasonable inferences about its extent and impact (rather than having to collect or await such data), but should encourage other interested parties to collect and provide the necessary data. In general, in parallel to recommendation UAR4, EPA should ensure that a similar level of variability analysis is applied to both the default and the alternative risk assessment, so that it can compare equivalently conservative estimates from each procedure.

Risk Communication

EPA does not adequately communicate to its own decision-makers, to Congress, or to the public the variabilities that are and are not accounted for in any risk assessment and the implications for the conservatism and representativeness of the resulting risk numbers.

• EPA should carefully state in each risk assessment what its particular assumptions about human behavior and biology do and do not account for.

For its full (as opposed to screening-level) risk assessments, EPA makes risk-communication and risk-management decisions more difficult when, as is usually the case, both uncertainty and variability are important.

• Whenever possible, EPA should separate uncertainty and variability conceptually, perhaps by presenting multiple analyses. EPA should acknowledge that all its risk numbers are made up of three components: the estimated risk itself (X), the level of confidence (Y) that the risk is no higher than X, and the percent of the population (Z) that X is intended to apply to in a variable population. In addition, rather than reporting both Y and Z, EPA can and should try to present multiple scenarios to explore and explain the variability dimension.

NOTES

1. Some specialists in different fields often use the term "variability" to refer to a dispersion of possible or actual values associated with a particular quantity, often with reference to random variability associated with any estimate of an unknown (i.e., *uncertain*) quantity. This report, unless stated otherwise, will use the terms *interindividual variability*, *variability*, and *interindividual heterogeneity* all to refer to individual-to-individual differences in quantities associated with predicted risk, such as in measures of or parameters used to model ambient concentration, uptake or exposure per unit ambient concentration, biologically effective dose per unit exposure, and increased risk per unit effective dose.

2. This assumes that risk is linear in long-term average dose, which is one of the bases of the classical models of carcinogenesis (e.g., the LMS dose-response model using administered dose). However, when one moves to more sophisticated models of the dose-exposure (i.e., PBPK) and exposure-response (i.e., biologically motivated or cell-kinetics models) relationships, shorter averaging times become important even though the health endpoint may manifest itself over the long-term. For example, the cancer risk from a chemical that is both metabolically activated and detoxified *in*

vivo may not be a function of total exposure, but only of those periods of exposure during which detoxification pathways cannot keep pace with activating ones. In such cases, data on average long-term concentrations (and interindividual variability therein) may completely miss the only toxicologically relevant exposure periods.

3. As discussed above, in many cases variability that exists over a short averaging time may grow less and less important as the averaging time increases. For example, if on average, adults breathe $20m^3$ of air per day, then over any random 1-minute period, in a group of 1,000 adults there would probably be some (those involved in heavy exertion) breathing much more than the average value of 0.014 (m^3/min), and other (those asleep) breathing much less. Over the course of a year, however, the variation around the average value of 7300 m^3/yr would be much smaller, as periods of heavy exercise, sleep, and average activity "average out." On the other hand, some varying human characteristics do not substantially converge over longer averaging periods. For example, the daily variation in the amount of apple juice people drink probably mirrors the monthly and yearly variation as well—those individuals who drink no apple juice on a random day are probably those who rarely or never drink it, while those at the other "tail" of the distribution (drinking perhaps three glasses per day) probably tend to repeat this pattern day after day (in other words, the distribution of "glasses drunk per year" probably extends all the way from zero to 365×3, rather than varying narrowly around the midpoint of this range).

4. Similarly, the two persons might face equal cancer risks at exposures that were 10,000-fold different. However, an alternative definition, which would be more applicable for threshold effects, would be to call the difference in susceptibility the ratio of doses needed to produce the same effect in two different individuals.

5. The logarithmic standard deviation is equivalent to the standard deviation of the normal distribution corresponding to the particular lognormal distribution. If one takes the antilog of the logarithmic standard deviation, one obtains the "geometric standard deviation", or GSD, which has a more intuitively appealing definition: N standard deviations away from the median corresponds to multiplying or dividing the median by the GSD raised to the power N.

6. Moreover, existing studies of overall variations in susceptibility suggest that a factor of 10 probably subsumes one or perhaps 1.5 standard deviations above the median for the normal human population. That is, assuming (as EPA does via its explicit default) that the median human and the rodent strain used to estimate potency are of similar susceptibility, an additional factor of 10 would equate the rodent response to approximately the 85th or 90th percentiles of human response. That would be a protective, but not a *highly* conservative, safety factor, inasmuch as perhaps 10 percent or more of the population would be (much) more susceptible than this new reference point.

Inclusion of a default factor of 10 could bring cancer risk assessment partway into line with the prevailing practice in noncancer risk assessment, wherein one of the factors of 10 that are often added is meant to account for person-to-person variations in sensitivity.

However, if EPA decides to use a factor of 10, it should emphasize that this is a default procedure that tries to account for some of the interindividual variation in dose-response relationships, but that in specific cases may be too high or too low to provide the optimum degree of "protection" (or to reduce risks to "acceptable" levels) for persons of truly unusual susceptibility. Nor does it ensure that (in combination with exposure estimates that might actually correspond to a maximally exposed or reasonably high-end person) risk estimates are predictive or conservative for the actual "maximally-at-risk" person. In contrast, some persons of extremely high susceptibility might, as a consequence of their susceptibility, not face high exposures. It might also be the case that some risk factors for carcinogenesis also predispose those affected to other diseases from which it might be impossible to protect them.

7. For example, suppose the median income in a country was $10,000, but 5 percent of the population earned 25 times less or more than the median and an additional 1 percent earned 100 times less or more. Then the average income would be $[(0.05)(400) + (0.05)(250,000) + (0.01)(100) + (0.01)(1,000,000) + (0.88)(10,000)] = \$31,321$, or more than three times the median income.

8. "Currently" is an important qualifier given the rapid increases in our understanding of the molecular mechanisms of carcinogenesis. During the next several decades, science will doubtless become more adept at identifying individuals with greater susceptibility than average, and perhaps even pinpoint specific substances to which such individuals are particularly susceptible.

11

Aggregation

INTRODUCTION

A recurring issue in quantitative risk assessment and quantitative risk characterization is the aggregation (and disaggregation) of separate but related causes and effects of risk. Questions about the aggregation of causes or agents differ somewhat from questions about the aggregation effects or end points, but the similarities are great enough for us to treat them together in this chapter. For example, people may be exposed to mixtures of compounds from a single stack, and each compound may be associated with an increase in the degree or probability of occurrence of one or more toxic end points; the situation can be further complicated by questions about synergy. In contrast, dose-response data are often available only on single end points in response to doses of single agents. How should we characterize and estimate the potential aggregate toxicity posed by exposure to a mixture of toxic agents?

The aggregation problem is simplified when all end points of concern are believed to have dose-response thresholds or no-adverse-effect levels. Under this restriction, "acceptable," "allowable," or "reference" doses are typically calculated by dividing empirically determined threshold estimates (such as no-observed-adverse-effect levels, NOAELs) by appropriate safety or uncertainty factors (Dourson and Stara, 1983; Layton et al., 1987; Barnes and Dourson, 1988; Lu, 1988; Shoaf, 1991). The risk-management goal for mixed exposures is generally to avoid exposures that exceed any of the relevant thresholds, while taking into account the possible joint effects of multiple agents. One strategy that has been implemented in environmental and occupational settings is to ensure that the sum of all the ratios of incurred dose to acceptable dose relevant to a given end point total less than 1 (NRC, 1972a, 1989; OSHA, 1983; ACGIH, 1977, 1988; EPA, 1987a, 1988g; Calabrese, 1991; Pierson et al., 1991). That approach is based on an assumption that doses of different agents can be treated as roughly additive with regard to inducing the end point; this assumption is reasonably consistent with much of the experimental evidence on the joint actions of chemicals in mixtures.

Among the key problems associated with the general strategy is that the procedures currently used for defining acceptable exposures to systemic toxicants are rather crude. Proposals to incorporate more quantitative treatment of data and to focus on risk prediction without reference to thresholds (e.g., Crump, 1984; Dourson et al., 1985; Dourson, 1986) have not been widely adopted. The additivity assumption for systemic toxicants further complicates the crude approaches taken to identifying safe intakes of the components of a complex mixture. As an Environmental Protection Agency (EPA) technical support document (EPA, 1988g) comments, this use of the additivity assumption implies that,

> as the acceptable level is approached or exceeded, the level of concern increases linearly . . . and in the same manner for all mixtures [which is incorrect, because the estimates used to derive such recommended acceptable levels] do not have equal accuracy or precision, and are not based on the same severity of toxic effect. Moreover, slopes of dose-response curves in excess of [such levels] in theory are expected to differ widely. The determinations of accuracy, precision or slope are exceedingly difficult because of the general lack of toxicity data.

Despite its drawbacks, the crude additivity approach to the problem of aggregation of potential threshold effects has had relatively straightforward and uncontroversial regulatory applications.

Much more debate has focused on quantitative risk-assessment methods for end points assumed *not* to have threshold dose-response relationships, such as cancer. Particularly with regard to environmental exposures to multiple chemicals, risk-management decisions (e.g., cleanup criteria) tend to be driven by the estimated low-dose risk associated with exposure to materials that lead to assumed nonthreshold end points. This chapter focuses on aggregation of different risks and different types of risk attributable to integrated, multiroute exposure to multiple chemicals that are assumed to have nonthreshold effects.

EXPOSURE ROUTES

Any comprehensive assessment of health risk associated with environmental exposure to any particular compound must consider all possible routes by which people might be exposed to that compound, even if expected applications in risk management are limited to some particular medium, such as air, or particular source generator or category, such as a coke-oven facility. That is because compounds present in one environmental medium might be transferred to another at any time before exposure. The major routes of exposure are inhalation, ingestion, and dermal absorption. In the context of environmental exposures, *inhalation* pertains to uptake of compounds present in respired air during rest or activity both indoors and outdoors; *ingestion* refers to gastrointestinal absorption of compounds that are intentionally or unintentionally present in any ingested material, including water, liquid foods, mother's milk, solid foods (including crops and game), and soil; and *dermal absorption* refers to percutaneous uptake of compounds deposited on skin, including those present in water during showering, bathing, or recreational swimming. Assessments of exposure to a substance from a given source must account for all potentially important routes by which the substance might come into contact with people (or environmental biota, if an ecological impact assessment is being undertaken). For example, mercury emitted into air from an industrial smoke stack might be inhaled by nearby residents, but might pose an even greater health risk by the ingestion of bioconcentrated mercury in fish that are caught locally after mercury from the stack plume has been deposited onto lake water.

EPA has given the issue of integrated multiroute exposure considerable attention in the context of risk-assessment guidance for Superfund-related regulatory compliance (EPA, 1989a). For example, EPA suggested that assessment of the environmental fate and transport of compounds in ambient air address a range of issues as diverse as volatilization and occurrence in wild game (EPA, 1988h, 1989a,c,d,e). Additional information on multimedia transport and multiroute exposure assessments is available (Neely, 1980; Neely and Blau, 1985; Cohen, 1986; McKone and Layton, 1986; Allen et al., 1989; Cohen et al., 1990; McKone and Daniels, 1991; McKone, 1991, 1992).

RISK-INDUCING AGENTS

Quantitative environmental risk assessment is often needed for exposure to multiple toxic agents, for example, in the context of hazardous-waste, drinking-water, and air-pollution control. The 1990 Amendments to the Clean Air Act in particular list 189 airborne pollutants of immediate regulatory concern that can be emitted singly or in combination from a variety of specified emission-source categories.

Over the last 2 decades, environmental remediation involving complex chemical mixtures has required general reviews of issues and cases of potential toxicity associated with concurrent exposure to multiple chemical agents (e.g., NRC, 1972a, 1980a,b, 1988a, 1989; EPA, 1988i; Goldstein et al., 1990; Calabrese, 1991). The earlier reviews supported the concept that toxicity predicted by dose additivity or concentration additivity was reasonably consistent with data on the joint action of acute toxicants (NRC, 1972a, 1980a,b; ACGIH, 1977;

EPA, 1987a). Although some cases of supra-additivity for acute toxicants are known, such as the synergistic interaction of organophosphate pesticide combinations in which one compound inhibits the detoxification of another compound, additivity has nevertheless been viewed as a reasonable expectation at the low doses at which detoxification enzymes are not expected to be saturated (NRC, 1988b; Calabrese, 1991).

The EPA Database on Toxic Interactions, as of 1988, covered 331 studies involving roughly 600 chemicals (EPA, 1988g). Most of the studies focused on the effects of two-compound mixtures on acute lethality; fewer than 10% examined chronic or lifetime toxicity. Less than 3% of all the studies reported clear evidence of a synergistic interaction—i.e., a "response to a mixture of toxic chemicals that is greater than that suggested by the component toxicities" (EPA, 1988g). However, EPA also concluded that in only one of 32 studies chosen as a 10% random sample of the 331 studies was the design and use of statistics "appropriate with the conclusion justified" (EPA, 1988g). As a consequence, EPA has asserted that

> given the quality and quantity of the available data on chemical interactions, few generalizations can be made concerning the likelihood, nature or magnitude of interactions. Most interactions that have been quantified are within a factor of 10 of the expected activity based on the assumption of dose addition (EPA, 1988g).

Results of the few detailed comparative studies in which *Salmonella*-mutation assays were applied to complex mixtures (kerosene-combustion particles, coal-hydrogenation material, and heterocyclic amines from cooked food) are also generally consistent with approximate additivity of mutagenic potencies of constituents within complex mutagenic mixtures (Thilly et al., 1983; Felton et al., 1984; Schoeny et al., 1986).

Epidemiological evidence concerning the synergistic potential of human carcinogens (usually involving long-term cigarette-smoking) has been extensively reviewed (Saracci, 1977; Steenland and Thun, 1986; EPA, 1988g; NRC, 1988a,b; Kaldor and L'Abbé, 1990; Pershagen, 1990; Calabrese, 1991). Although no single mathematical expression is likely to give an accurate representation of joint effects, especially given the heterogeneity of human responses, the discussion here has often focused on whether responses are more clearly additive or multiplicative. The best-studied interactions (such as in joint exposure to tobacco and radon or tobacco and asbestos) suggest that a strictly additive model within the dose ranges studies may underestimate the true joint effects by a factor of 3-10. Results of epidemiological studies of joint exposure to radon progeny and cigarette smoke, for example, have been interpreted as showing an additive or possibly multiplicative interaction of the two agents with respect to the number of cancers induced and a synergistic decrease in latency period for tumor induction (NCRP, 1984; NRC, 1988a). The NRC (1988b) BEIR IV committee concluded that results of epidemiological studies of smoking and non-smoking uranium miners exposed to radon gas, particularly the large study by Whittemore and McMillan (1983), were consistent with a multiplicative effect of the combined agents.

The effects of asbestos exposure among workers who have a history of cigarette-smoking have been described (NRC, 1988a) as "one of the most current and well-recognized examples [based on epidemiological data] of how two distinct agents administered together can produce an increased incidence of [lung] cancer that is greater than that predicted from the administration of either agent alone [and that] is considered multiplicative by most investigators who have studied the problem." A study not cited by NRC of more than 1,600 British asbestos workers suggests an additive, rather than multiplicative, increase in relative risk after joint tobacco and asbestos exposure (Berry et al., 1985). Other investigators have also concluded that the overall evidence of multiplicative interaction of these agents is questionable (Saracci, 1977; Steenland and Thun, 1986).

Epidemiological detection of possible multiplicative action among human carcinogens is not surprising, given the large amount of experimental data on the action of cancer promoters in animals, including clear examples of supra-additive interaction (EPA, 1988g; Calabrese, 1991; Krewski and Thomas, 1992). Highly nonlinear, supra-additive synergistic interaction of some types of non-genotoxic cancer promoters with genotoxic agents is predicted by "biomechanistic" multistage models of carcinogenesis. In those models, increased cell replication can play a pivotal role either by directly increasing the rates of production of premalignant or malignant lesions, by amplifying the incidence of malignant lesions through stimulated growth of spontaneously occurring premalignant lesions, or both (Armitage and Doll, 1957; Moolgavkar and Knudson, 1981; Moolgavkar, 1983; Bogen, 1989; Cohen and Ellwein, 1990a,b; 1991; Ames and Gold, 1990a,b; Preston-Martin et al., 1990). From that mechanistic perspective, several nongenotoxic compounds are now thought to be capable of promoting carcinogenesis, both spontaneous and experimentally chemically induced, solely by increasing target-cell replication, a phenomenon that might have a threshold-like dose-response relation (Weisburger and Williams, 1983; Weisburger, 1988; Butterworth, 1989, 1990; Bogen, 1990b; IARC, 1991; Flamm and Lehman-McKeeman, 1991). EPA is considering formal recognition of such threshold carcinogens from the mechanistic perspective (e.g., EPA 1988g, 1991d), although these cases remain awkward to accommodate within EPA's currently-used 1986 general scheme for classifying potential chemical carcinogenicity (EPA, 1987a).

In general, both biological and statistical considerations make it difficult to rule out a nonthreshold mutation-related component of chemically induced carcinogenesis, and this effect might be dominant at low environmental exposures (Portier, 1987; Portier and Edler, 1990; Kopp-Schneider and Portier, 1991; Weinstein, 1991). For example, an increase in target-cell replication induced by some nongenotoxic chemicals might have a low-dose, linear, nonthreshold dose-response relation. Alternatively, a broad distribution of thresholds within a highly heterogeneous human population might give rise to practical quasilinearity or superlinearity for low-dose promotional effects. Therefore, low-dose linearity has been recommended as a reasonable default assumption, even for agents known to increase cancer risk through nongenotoxic promotional mechanisms, in the absence of data establishing a pertinent, clearly defined, generally applicable threshold dose-response relation (Lutz, 1990; Perera, 1991). Under this default assumption, the mechanistic type of cancer-risk model and the classical multistage cancer-risk model both predict that small amounts of increased risk will be approximately linearly proportional to the risk associated with small combined doses of genotoxic or nongenotoxic carcinogens, or both, and that their joint action will be approximately additive (Gibb and Chen, 1986; NRC, 1988a; Brown and Chu, 1989; Krewski et al., 1989; Kodell et al., 1991b).

The general assumption of low-dose linearity for a presumed nonthreshold quantal end point (i.e., an end point observed only as present or absent), such as cancer occurrence before age 70, is equivalent to assuming $P = p + qD$, where P is the risk of such occurrence after a lifetime exposure at dose rate D, p is the background cancer risk by age 70, and q is the potency (increased risk per unit dose) for small values of D. Of interest is the aggregate increased probability P of cancer occurrence due to exposure to a low-dose environmental mixture of nonthreshold toxic agents. If the linear model is assumed for each of two such agents, and if an additional independent-action assumption is made that the agents act through statistically independent events to increase risk R, it follows that $P \approx q_1 D_1 + q_2 D_2$ for very small D_1 and D_2 (NRC, 1980b, 1988b; Berenbaum, 1989). A more general sum of potency-dose products has been used by EPA for approximating P in cases of exposure to a mixture of carcinogens (EPA, 1987a, 1988g). Appendix I-1 shows that the same general assumptions imply that a similar sum-of-products relation may be used to approximate the risk associated with mixtures of agents, each having one or more different end-point-specific effective dose rates. Multiple nonthreshold end points can be of interest in quantitative risk assessment, as discussed in more detail below.

TYPES OF NONTHRESHOLD RISK

Quantitative risk assessment can involve multiple toxic end points, as well as multiple toxic agents. In particular, toxic end points other than cancer might at some point also be assumed to have nonthreshold dose-response relations for public-health regulatory purposes. Furthermore, cancer is not a single disease, but a variety of neoplastic disorders with different characteristics that occur in different tissues of animals and humans at different times in the life history. Aggregate human cancer risk is often estimated from animal bioassay data that indicate statistically significant increases in dose-related risk of more than a single tumor type (e.g., cancer of the lung and cancer of the kidney). Similarly, genetic, reproductive, and developmental risks can arise in multiple forms that are measured separately in toxicity assays (e.g., reduced fertility and incomplete ossification of some bone). The issues of aggregating risk of both multiple end points and multiple types of a given end point are discussed below. Both these aggregation problems can be addressed simultaneously by using Expression 6 in Appendix I-1, if independent actions and effects are assumed.

Cancer

The issue of how to use bioassay data that indicate dose-related effects for multiple tumor types is addressed by the EPA (1987a) cancer-risk guidelines as follows:

> To obtain a total estimate of carcinogenic risk, animals with one or more [histo-logically distinct] tumor sites or types showing significantly elevated . . . incidence should be pooled and used for [risk] extrapolation. The pooled estimates will generally be used in preference to risk estimates based on single types or sites.

If different tumor types observed to have increased incidences are known to occur in a statistically independent fashion within and among the bioassay animals tested, this EPA-recommended procedure leads to inconsistently biased estimates of aggregate potency or risk because, under the independence assumption, the pooled tumor-incidence data may randomly exclude relevant information (Bogen, 1990a). For potency estimates based on classical multistage models, that statistical problem is avoided if aggregate potency is estimated as the sum of tumor-type-specific potencies (Bogen, 1990a). If the latter approach is used, then the aggregate increased risk P of incurring one or more tumor types at a very low dose can be estimated from Expression 7 in Appendix I-1 (for one carcinogen). The type-specific potencies are uncertain quantities (one reason is that they are generally estimated from bioassay data), so appropriate procedures must be used for summation.

This alternative (Expression 7 in Appendix I-1) to EPA's procedure for estimating aggregate cancer potency depends on the validity of the assumption that different tumor types occur independently within individual bioassay animals. If substantial interanimal heterogeneity exists in susceptibility to cancer, or if tumor types are positively correlated, the occurrence of multiple tumor types would be expected to cluster in the more susceptible individuals. Although some significant tumor-type associations have been identified in some species, they have tended to involve a relatively small number of tumor types (see Appendix I-2).

Appendix I-2 summarizes an investigation of independence in interanimal tumor-type occurrence in a subset of the National Toxicology Program (NTP) 2-

year cancer-bioassay data, which has been used by EPA as the basis for quantifying the potency of most chemical carcinogens. Separate analyses were conducted for four sex-species combinations (male and female mice, male and female rats) by using control-animal data from 61 rat studies and 62 mouse studies and treated-animal data from a subset of studies in which there were significant increases in multiple tumor types. Correlations in the occurrence of pairs of tumor types in individual animals were evaluated. Little evidence was found of tumor-type correlation for most of the tumor-type pairs in control and treated mice and rats. Some tumor-type pairs were statistically significantly (and generally negatively) correlated, but in no case was the correlation large. These findings indicate that a general assumption of statistical independence of tumor-type occurrences within animals is not likely to introduce substantial error in assessing carcinogenic potency from NTP rodent-bioassay data.

Other Nonthreshold End Points

Two major categories of possible nonthreshold toxicity other than cancer that may often be relevant in quantitative risk assessment are genetic mutation (which might be caused by material that reaches and damages gonadal DNA) and developmental and reproductive toxicity (such as developmental neurotoxicity of lead). In general, however, if both dose-response linearity at low doses and independent dose induction of these effects are assumed, then they may also be incorporated with cancer into the general additive strategy already discussed. The extent to which those assumptions might apply to genetic toxicity and reproductive and developmental toxicity is considered below.

Genetic Effects

Mutagenic agents can cause detrimental inherited effects with an important genetic component, such as clinically autosomal dominant and recessive mutations, X-linked mutations, congenital birth defects, chromosomal anomalies, and multifactorial disorders of complex origin. Inherited genetic effects other than complex multifactorial effects have been found to occur spontaneously in roughly 2% of all liveborn people, appearing either at birth or thereafter; about 40-80% often involve chromosomal anomalies or dominant or X-linked mutations ("CADXMs") (Mohrenweiser, 1991). In addition, more than 25% of all spontaneous abortions are thought to be due to genetic defects, the majority involving CADXMs (Mohrenweiser, 1991). Rates of those genetic effects are known to be increased in animals by exposure to environmental agents, such as ionizing radiation (which also causes cancer); furthermore, the risks of both genetic and cancer end points associated with low doses of ionizing radiation are currently modeled as being increased above background in linear proportion to dose (NRC, 1972b, 1980c, 1990b; NCRP, 1989; Favor, 1989; Sobels, 1989; Vogel, 1992).

Exposure of experimental animals to mutagenic chemicals can also cause some of these genetic effects, although specific characteristics of chemically induced genetic damage appear to differ in some ways from those induced by irradiation, e.g., in the fraction of dominant versus recessive specific-locus effects (Ehling and Neuhauser, 1979; Lyon, 1985; Favor, 1989; Rhomberg et al., 1990).

Experimental data are not all consistent with a linear nonthreshold dose-response relation for genetic end points induced by either chemicals or ionizing radiation (ICPEMC, 1983a; Sobels, 1989). Chemical mutagenesis, in particular, involves many potentially nonlinear and threshold processes, such as transport of reactants, metabolic activation and deactivation, DNA repair, and chemically induced functional change and lethality (ICPEMC, 1983a). However, it is difficult (if not impossible) to show experimentally that a complex, inherently statistical biological response does not differ from background (ICPEMC, 1983a). In light of such complexities, several National Research Council committees (NRC, 1975, 1977, 1983b) have concluded that the linear nonthreshold dose-response assumption used for ionizing radiation is also a reasonable default hypothesis for mutagenic chemicals. That conclusion reflects the fact that "if an effect can be caused by a single hit, a single molecule, or a single unit of exposure, then the effect in question cannot have a threshold in the dose-response relationship, no matter how unlikely it is that the single hit or event will produce the effect." It has been similarly concluded that a linear nonthreshold dose-response relation is a reasonable default assumption for chemical mutagens (Ehling and Neuhauser, 1979; ICPEMC, 1983a,b; Lyon, 1985; Ehling, 1988; Favor, 1989; Sobels, 1989; Rhomberg et al., 1990).

Such support of a default assumption of nonthreshold linearity in induced genetic risk has highlighted the uncertainty that exists in quantitative assessment of the total genetic risk to humans associated with exposure to ionizing radiation or genotoxic chemicals. That uncertainty, due particularly to problems in estimating possible increases in rates of human genetic disease, has led some to conclude that realistic assessment of total genetic risk associated with environmental exposure will not soon be possible (NRC, 1990b; Mohrenweiser, 1991; Vogel, 1992). The degree of uncertainty varies greatly among different end points, but dose-response data for mutations in mice, supplemented by corresponding estimates of human spontaneous incidence rates, appear to provide a basis for reasonable quantitative risk assessment for some genetically simple and straightforward end points, such as those involving CADXMs (NRC, 1990b; Mohrenweiser, 1991; Vogel, 1992).

In 1986, EPA adopted guidelines for mutagenicity risk assessment that do not specifically endorse a linear nonthreshold default assumption. Rather, they state that EPA "will strive to use the most appropriate extrapolation models for risk analysis" and "will consider all relevant models for gene and chromosomal mutations in performing low-dose extrapolations and will chose the most appropriate model" (EPA, 1987a). The 1986 guidelines committed EPA to "assess risks associated with all genetic end points" to the greatest extent possible when data are available, with risk to be "expressed in terms of the estimated increase of genetic disease per generation, or the fractional increase in the assumed background spontaneous mutation rate of humans." In pursuit of methods to implement the goals of its guidelines, EPA sponsored a major effort concerning genetic-risk assessment for the direct-acting mutagen ethylene oxide (Dellarco and Farland, 1990; Dellarco et al., 1990; Rhomberg et al., 1990). But EPA does not now routinely perform quantitative assessments of genetic risk posed by chemical mutagens in the environment as part of any of its regulatory programs.

EPA's 1986 guidelines are nonspecific not only regarding particular methods to be used by the agency for estimating mutagenic risk, but also regarding how such risk might be aggregated with risks estimated for other end points, such as cancer. The suggested measures of genetic risk in the guidelines cannot readily be aggregated with EPA's commonly used measures of increased cancer risk to individuals or populations. However, individual genetic risk could be expressed as increased lifetime risk of expression of a serious inherited genetic end point in a person whose parents were both exposed from birth to a given relevant compound at a given effective dose rate. And addition of such a predicted risk to a corresponding magnitude of predicted somatic (cancer) risk would be appropriate under assumptions of low-dose linearity and independence as discussed above and in Appendix I.

Risk assessments of ionizing radiation provide precedents for the simple addition of quantitative estimates of genetic and cancer risk (e.g., Anspaugh and Robison, 1968; ICRP, 1977a,b, 1984, 1985). However, EPA has made no systematic effort to consider the combination of mutagenic and cancer risks. In the context of setting radiological National Emission Standards for Hazardous Air Pollutants (NESHAPs), the agency's Office of Radiation Programs made a substantial effort to describe quantitative risk estimates for both cancer and genetic end points (EPA, 1989b). However, the genetic risk factors were not used later in EPA's corresponding quantitative radiologic-risk assessments for radioactive air contaminants (EPA, 1989b), nor are they considered in current EPA guidance on how to calculate preliminary Superfund remediation goals for radionuclides at hazardous-waste sites (EPA, 1991f).

The importance of considering a quantitative combination of genetic and cancer end points depends on the ratio of genetic-to-cancer potency of any given chemical. If the ratio is much less than 1, genetic-risk assessment of the chemical is probably unwarranted, because it is likely to have little impact on regulatory action. For example, the upper-bound estimate of the potency of ethylene oxide (ETO) to produce heritable translocations (HTs) in children of exposed men was recently estimated to be equivalent to 0.00066 per part of ETO per million parts of air continuously inhaled. This estimate was based on an EPA analysis that applied a linearized multistage extrapolation model to dose-response data on HT induction in mice; a 21-day critical exposure period was assumed to be potentially damaging to human males (Rhomberg et al., 1990). In contrast, EPA had previously estimated ETO's cancer potency to be 0.19 per part of ETO per million part of air continuously inhaled over a lifetime—a value almost 290 times its estimated HT potency (EPA, 1985c). The genetic risk associated with ETO could not therefore constitute a substantial fraction of the genetic-plus-cancer risk unless HT represented a very small fraction (e.g., less than 1/290) of all reasonably quantifiable ETO-induced genetic end points. This appears to be unlikely, given that HTs constitute between about 5% and 10% of CADXMs (ICPEMC, 1983b).

Reproductive/Developmental Risks

There are continuing concerns about the adequacy of current approaches (threshold, linear, nonlinear, BD, etc., described in Chapter 4) to characterize the risks associated with potential reproductive and developmental hazards (Barnes and Dourson, 1988; Mattison, 1991). Particular questions remain regarding thresholds. Although threshold mechanisms might seem plausible, the estimation of an upper limit to ensure that doses are safe depends heavily on available methods of study and measurement and our knowledge of organ- and tissue-specific repair mechanisms. The issue merits continued consideration. This issue is also discussed in the NRC report entitled *Seafood Safety* (NRC, 1991b).

The current and proposed EPA guidelines concerning reproductive- and developmental-toxicity risk assessment are based on the controversial assumption that chemical induction of reproductive or developmental toxicity generally has a true or practical threshold dose-response relationship. As noted by EPA (1991a), such thresholds might differ among exposed people, and EPA has traditionally accommodated such interindividual variability by using an extra uncertainty factor or safety factor of 10, whose adequacy remains to be established.

MEASURES AND CHARACTERISTICS OF RISK

Overall Characterization Goals

An essential component of risk characterization is the aggregation of different measures and characteristics of risk; the risk assessor must communicate measures and characteristics of predicted risk in ways that are useful in risk management. The technical aspects of risk aggregation and characterization cannot and should not be separated from the design of useful, politically responsible, and legally tenable criteria of risk acceptability, because such criteria must generally be based on risk characterizations that follow some standard format, and the format must accommodate the criteria. As new, more sophisticated approaches to risk assessment and characterization are proposed—such as the incorporation of integrated uncertainty and variability analysis—the corresponding more complicated criteria for risk acceptability have not been agreed on. It is therefore appropriate to establish as an interim goal of risk characterization the adoption of a format that includes a summary of predicted risk that is accurate, comprehensive, easily understood, and responsive to a wide array of public concerns about risk. The format should include the magnitude and uncertainty of estimated population risk (that is, predicted incidence) as well as individual risk, the uncertainty of estimates of costs and competing risks inherent in alternative risk-management options, the degree to which estimated risks might vary among exposed individuals, and the time frame of risks imposed.

Consistency in Characterization:
Example of Aggregation of Uncertainty

To the extent that a given aggregated characteristic of a risk assessment, such as uncertainty, is addressed in an overall characterization of predicted risk, it should be determined with a consistent approach to estimates of the magnitudes of the components considered (e.g., ambient concentration, uptake, and potency). In the case of uncertainty aggregation, such consistency will come about through a rigorous, fully quantitative approach (see Chapter 9). But such a fully quantitative approach might be deemed impractical; for example, quantification of subjective probability judgments in the assessment might be considered difficult or misleading. A screening-level alternative to a fully quantitative approach to uncertainty aggregation is to use a qualitative or categorical approach that describes, in narrative or tabular form, the impact of each component of the analysis on each aspect of predicted risk. However, an exclusively qualitative, categorical approach is generally impractical because it fails to communicate effectively the fundamental quantitative conclusions of the risk analysis in terms that are of direct use to risk managers.

Thus, the approach to uncertainty aggregation most often used has been a semiquantitative approach incorporating specific key assumptions whose merits and impact are discussed verbally. The difficulty with this approach lies in ensuring that resulting semiquantitative characteristics are properly interpreted and communicated. For example, it would be illogical and potentially misleading to characterize a final risk estimate as a "plausible upper bound" on risk, if it were derived by aggregating component-specific point estimates that represent a mixture of best estimates and statistical upper confidence limits. That is particularly true if the components for which *best* estimates are used are also the components known to be the most uncertain among those considered. When, for example, risk is modeled as a simple product of estimated quantities (such as concentration, potency, etc.) a great deal of conservatism is lost whenever a best estimate is used in place of a far larger corresponding upper-bound value (and little conservatism is gained by using an upper-bound value if it is close to the corresponding best estimate). Thus, if a semiquantitative approach is to be used, the only way to obtain a meaningful "upper-bound" point estimate of risk from component-specific point estimates would be to base the "upper-bound" point estimate *entirely* on "upper-bound" estimates of *all* the component quantities. This point is illustrated by the following example involving EPA's cancer-risk guidelines.

The EPA guidelines for cancer-risk characterize the estimate produced by following the guidelines as a "plausible upper bound" on increased cancer risk. Such a risk estimation will generally involve a pertinent set of animal bioassay data, an animal-cancer potency estimate, and an interspecies dose-scaling factor. According to the 1986 guidelines, the risk assessment is to be based on the data showing the most sensitive response (i.e., that give the highest estimated potency value or set of related values), and the animal-cancer potency value used is a statistical upper confidence limit of potency estimated from the animal-bioassay data set selected. The guidelines specify a dose-scaling factor—based on what was intended by EPA to be a deliberately conservative assumption that carcinogenic doses are equivalent between species if they are expressed as daily mass per unit of body surface area. Recently, EPA (1992e) proposed adopting a new scaling factor that is somewhat less conservative because this new factor appears to be close to a "best" estimate of what the factor might actually be. However, EPA (1992e) noted that

> Although scaling doses by [the newly proposed factor] characterizes the trend [relating epidemiologically based human-cancer potencies with corresponding experimentally determined ones for animals] fairly well, individual chemicals may deviate from this overall pattern by two orders of magnitude or more in either direction....The proposed scaling [approach] . . . represents a best guess . . . surrounded by an envelope of considerable uncertainty. . . . [It] is intended to be...an unbiased projection; i.e., it is to be thought of as a "best" estimate rather than one with some conservatism built in . . . [such] as a "safety factor" or other intentional bias designed to "err on the side of safety."

A similarly large degree of uncertainty associated with interspecies dose scaling was also indicated in a recent reassessment of uncertainty pertaining to interspecies extrapolation of acute toxicity (Watanabe et al., 1992). Other studies (Raabe et al., 1983; Kaldor et al., 1988; Dedrick and Morrison, 1992) provide evidence that a milligram-per-kilogram-per-lifetime dose metric may be roughly equivalent across species. These studies compare human carcinogenicity and animal carcinogenicity for alkylating or radioactive agents (administered for therapeutic purposes in the case of humans). Dose-scaling uncertainty may thus be substantially far greater than that associated with parameter-estimation error for cancer potency in bioassay animals and be at least as great as that associated with the selection of a bioassay data set for analysis. EPA's proposed dose-scaling policy would therefore be an exception to its reasonably consistent practice of using component-specific upper bounds when semiquantitative aggregation of uncertainty is used to derive a "plausible upper bound" on increased risk. The most straightforward way to obtain such an upper-bound dose-scaling factor would be to calculate it directly from the best available relevant empirical data that relate epidemiologically based human-cancer potencies to corresponding experimentally determined animal-cancer potencies (e.g., Raabe et al., 1983; Allen et al., 1988; Kaldor et al., 1988; Dedrick and Morrison, 1992). An uncertainty distribution for the scaling factor could also readily be developed from these data, and an appropriate summary statistic chosen explicitly from this distribution, rather than by fiat and without reference to uncertainty (see, for example, Watanabe et al., 1992).

Uncertainty and Variability

We have deliberately treated these two concepts separately up to this point in the report, because we view them as conceptually quite different even though they share much of the same terminology (e.g., "upper confidence limit," "standard deviation"). Indeed, as emphasized in Chapters 9 and 10, the realms of uncertainty and variability have fundamentally different ramifications for science and judgment: uncertainty forces decision-makers to judge how *probable* it is that risks will be overestimated or underestimated for every member of the exposed population, whereas variability forces them to cope with the *certainty* that different individuals will be subjected to risks both above and below any reference point one chooses.[1]

Thus, any criticism that EPA has assessed or managed a risk too "conservatively" needs to consider and explain which type of conservatism is being decried. The use of a plausible but highly conservative scientific model, if it imposes large costs on society or the regulated community, can throw into question whether it is wise to be "better safe than sorry." The attempt to provide protection to persons at the "conservative" end of a distribution of exposure or risk, in contrast, determines who ends up with what degree of safety and thus requires a different decision calculus. In particular situations, either uncertainty or variability (or perhaps both) might be handled "conservatively." For example, society might in one case determine that the marginal costs of protecting individuals with truly unusual hypersusceptibility were too large relative to the costs of protecting only the majority, but might still choose to assess the risk to each group in a highly conservative manner. In another case, society might view the central tendency of an uncertain risk as an appropriate summary statistic, yet deem it important to extend protection to individuals whose risks are far above the central tendency with respect to the varied risks across the population.

On the other hand, this risk management distinction between uncertainty and variability should not blind people to a central fact of environmental health risk assessment: that in general, risks are both uncertain and variable simultaneously. In the prototypical hazardous air pollutant risk assessment case, one can think of the source exposing each nearby resident to a different ambient concentration of each emitted pollutant; each of these concentration values is made still more variable by the unique activity patterns, uptake parameters, and susceptibility of each individual. Simultaneously, each of these "individualized" parameters is either hard to measure or impossible to model with certainty (or both), and all of the "generalized" parameters (such as the inherent carcinogenic potency of each substance) are also surrounded by uncertainty. In sum, the source does not impose "a risk"—it imposes a *spectrum* of individual risks, *each* of which can only be completely described as a probability distribution rather than a single number.

Elsewhere in the report, we have commented on two aspects of the challenge of assessing variable and uncertain risks: communicating them correctly and comprehensively (see the findings and recommendations for this chapter), and describing how to relate variability to uncertainty in order to explicitly target risk management to the desired members of the population (average, "high-end," maximally at risk, etc.) in light of the uncertainty (again, see the findings and recommendations for this chapter).

Here, we briefly mention two additional complications that arise because uncertainty and variability work in tandem. We make no specific recommendations regarding either issue, because we feel EPA analysts and other risk assessors need flexibility to account for these technical problems as they gradually improve their treatment of the separate phenomena of uncertainty and variability. Nevertheless, it is important to keep in mind two other relationships between these phenomena:

(1) *Variability in one quantity can contribute to uncertainty in another.* The most relevant example of this general phenomenon involves the influence of variability in a quantity on the uncertainty in its mean. As mentioned in the introduction to Chapter 10, one way to deal with interindividual variability is to substitute the average value of the varying quantity, although this does preclude conducting analyses that are meaningful at the individual level. However, even this short-cut is not without additional complications, because the new parameter (the population average value of the variable quantity) may be rather uncertain if the variability is substantial. Although the central limit theorem states that the uncertainty in the mean is inversely proportional to the number of observations made, when the quantity varies by orders of magnitude, even "large" data sets (tens or even hundreds of observations) may not be sufficient to pin down the mean with the precision desired. A group of 1000 workers observed in an epidemiologic study, for example, may have an *average* susceptibility to cancer significantly greater or less than the true mean of the entire population, if by chance (or due to a systematic bias) the occupational group has slightly more or slightly fewer outliers (particularly those of extremely high susceptibility) than the overall population. In such cases, estimates of potency or population incidence drawn from the worker study may be overly "conservative" (or insufficiently so).

(2) *The amount of variability is generally itself an uncertain parameter.* There are at least three factors that work to complicate the estimation of variability. Thus, risk assessment parameters that attempt to summarize variability (either as inputs to other calculations or as the output for risk management or communication) should be regarded as uncertain unless these three factors are deemed unimportant: (1) "double-counting" and overestimation of variability may occur when error-prone measurements are made—these errors will tend to make the extremes in the population seem more divergent than they truly are; (2) even when measurements are perfect, the amount of variability cannot be perfectly determined from any single data set—random parameter uncertainty introduces the possibility that by chance, the population observed might be inherently less or more variable than the entire population; and (3) there may be "model uncertainty" in deciding what kind of probability distribution to fit to variable observations, and hence statistics such as the standard deviation or the upper confidence limit might be in error if they apply to a distribution that does not precisely describe the actual variability.

In sum, EPA should realize that estimates of variability themselves may be too large or too small—if "conservatism" is crucial, it may make sense to take account of this impreciseness of variability as well as taking account of variability itself (e.g., if fish consumption is deemed to be lognormal with a standard deviation somewhere between $(x - \delta)$ and $(x + \delta)$, it might be appropriate to use an upper confidence limit for fish consumption that is in turn based on the larger of the two estimates of variability, $x + \delta$).

Aggregation of Uncertainty and Variability

To the extent that both uncertainty and interindividual variability (that is, heterogeneity or differences among people at risk) are addressed quantitatively with separate input components (e.g., ambient concentration, uptake, and potency) for aggregation into an assessment of risk, the distinction between uncertainty and variability ought to be maintained rigorously throughout the analytic process, so that uncertainty and variability can be distinctly reflected in calculated risk. If no distinction were made between uncertainty-related and heterogeneity-related distributions associated with inputs to a given risk calculation, then whatever distribution might be obtained as a characteristic of risk would necessarily reflect risk to an individual selected at random from the exposed population (Bogen and Spear, 1987). This restricted result would render such analyses less useful for environmental regulatory purposes, in light of the tendency to focus substantial regulatory attention on increased risk to highly sensitive or highly exposed members of the population.

Another advantage of distinguishing between uncertainty and variability is that it permits one to estimate the uncertainty in the risk to the individual who is "average" with respect to all characteristics that are heterogeneous among individuals at risk, and the latter risk may be used to estimate uncertainty in predicted population risk or number of cases (Bogen and Spear, 1987). Technical issues that arise in aggregating uncertainty and interindividual variability for the purpose of calculating estimated individual and population risk are described in Appendix I-3.

FINDINGS AND RECOMMENDATIONS

Multiple Routes of Exposure

Although the Clean Air Act Amendments of 1990 do not specifically refer to multiple exposure pathways, EPA has routinely considered multiple exposure routes in regulatory contexts, such as Superfund, that logically concern source-specific pollutants that might transfer to other media before human exposure.

• Health-risk assessments should generally consider all possible routes by which people at risk might be exposed, and this should be done universally for compounds regulated by EPA under the Clean Air Act Amendments of 1990. The agency's risk-assessment guidance for Superfund-related regulatory compliance (EPA, 1989a) can serve as a guide in this regard, but EPA should take advantage of new developments and approaches to the analysis of multimedia fate and transport data. This will facilitate systematic consideration of multi-route exposures in designing and measuring compliance with Clean Air Act requirements.

Multiple Compounds and End Points

When aggregating cancer risk associated with exposures to multiple compounds, EPA adds the risk related to each compound in developing its risk estimate. That is appropriate when the only risk characterization desired is a point estimate used for screening-level analysis. However, if a quantitative uncertainty characterization is desired, simple addition of upper confidence limits may not be appropriate.

• EPA should consider using appropriate statistical (e.g., Monte Carlo) procedures to aggregate cancer risks from exposure to multiple compounds if a quantitative uncertainty characterization is desired.

EPA currently uses a specific procedure when analyzing animal bioassay data involving the occurrence of multiple tumor types (e.g., lung, stomach, etc.) to estimate the total cancer risk associated with exposure to a single compound. In this procedure EPA adds the numbers of animals with tumor types that are significantly increased above control levels, such that an animal with multiple tumor types counts the same as one with a single tumor type. This procedure does not allow full use of the data available and can overestimate or underestimate total cancer risk.

• When analyzing animal bioassay data involving the occurrence of multiple tumor types, EPA should use the following default procedure. Cancer potencies should first be separately estimated for each tumor type involved with the procedure normally used in the case of bioassays involving a single tumor type. The type-specific potencies should then be added as upper bounds or using appropriate statistical (e.g., Monte Carlo) methods. This procedure should be used unless specific data indicate that occurrence of the different tumor types within individual animals are significantly correlated.

Genetic Effects

Current EPA guidelines do not clearly state a default option of nonthreshold low-dose linearity for genetic effects that can be reasonably estimated for quantitative risk assessment.

• EPA's guidelines should clearly state a default option of nonthreshold low-dose linearity for genetic effects on which adequate data (e.g., data on chromosomal aberrations or dominant or X-linked mutations) might exist. This default option allows a reasonable quantitative estimate of, for example, first-generation genetic risk due to environmental chemical exposure.

Reproductive and Developmental Toxicants

While EPA is increasing its use of the benchmark dose, it still uses a threshold model in its proposal for regulation for reproductive and developmental toxicants. Although the threshold model is generally accepted for these toxicants, it is not known how accurately it predicts human risk. Current evidence on some toxicants, most notably lead and alcohol, does not unequivocally demonstrate any "safe" threshold and thus has raised concerns that the threshold model might only reflect the limits of current scientific knowledge, rather than the limits of safety.

• EPA should continue to collect and use the data needed to evaluate the validity of the threshold assumption, and it should make any needed revisions in the proposed model so that human risks, particularly those of individuals with above-average sensitivity or susceptibility, are accurately estimated.

"Upper-Bound Estimates" versus "Best Estimates"

In a screening-level or semiquantitative risk characterization, component uncertainties associated with predicted cancer risk are not generally aggregated in a rigorous quantitative fashion. In such cases, it is practical to calculate an "upper-bound" point estimate of risk by combining similarly "upper-bound" (and not "best") point estimates of the component quantities involved, particularly for quantities (such as the dose-scaling factor) that are highly uncertain. For screening-level analyses, the EPA (1992d) proposal to adopt a new interspecies dose-equivalence factor is inconsistent with the 1986 guideline stipulation that risk estimated under the guidelines represents a "plausible upper bound" on increased cancer risk, and it is inconsistent with the corresponding stipulation that "upper-bound" or health-conservative assumptions are to be used at each point in cancer-potency assessment that involves substantial scientific uncertainty.

• For a screening-level or semiquantitative approach in which component uncertainties associated with predicted upper-bound cancer risk are not aggregated in a rigorous quantitative fashion, the EPA guidelines, to determine upper-bound cancer risk, should require the use of an upper-bound (i.e., reasonably health-conservative), rather than a "best," interspecies dose-scaling factor consistent with the best available scientific information.

Uncertainty versus Variability

A distinction between uncertainty (i.e., degree of potential error) and interindividual variability (i.e., population heterogeneity) is generally required if the resulting quantitative risk characterization is to be optimally useful for regulatory purposes, particularly insofar as risk characterizations are treated quantitatively.

• The distinction between uncertainty and individual variability ought to be maintained rigorously at the level of separate risk-assessment components (e.g., ambient concentration, uptake, and potency) as well as at the level of an integrated risk characterization.

NOTE

1. For example, in the 1980s the Consumer Products Safety Commission (CPSC) had to issue a standard regarding how close together manufacturers had to place the vertical slats in cribs used by infants, with the aim of minimizing the number of accidental strangulations nationwide. Presumably, there was virtually no uncertainty about the diameter of an average infant's head, but there was significant variability in distinguishing different infants from each other. CPSC thus had to make a decision about which estimation of head size to peg the standard to—an "average" estimate, a "reasonable worst case," the smallest (i.e., most conservative) plausible value, etc. We suggest that it is not apropos to use the phrase "better safe than sorry" to apply to this kind of reasoning, because uncertainty is not at work here. Rather, deciding whether to be conservative in the face of variability rests on a policy judgment about how far to extend the attempt to provide safety.

PART
III

Implementation of Findings

The committee believes that a major portion of its charge is to consider how its findings and recommendations should be implemented in light of the comprehensive rewriting of Section 112 by Title III of the 1990 amendments to the Clean Air Act. Many of the common problems in health risk assessment might have arisen because of the two most salient features of EPA's implementation of the Red Book paradigm over the last 10 years: the emphasis on single outputs of each step, which are then processed into single numbers for risk; and the separation of the research and analysis functions into discrete, sequential stages.

A tiered system of priority-setting would be an important positive development in the practice of standard-setting and risk analysis. Currently, standards (goals for achieving health and safety) are set in accordance with a Congressional mandate to provide "an adequate margin of safety." Where data do not exist (particularly with respect to responses to low doses and mechanisms of toxicity), EPA has generally chosen default options that, in addition to being in keeping with current scientific knowledge, are intended to be conservative (i.e., health protective) in the outcomes to which they lead. This protective approach provides the basis for developing a stepwise, tiered system for assigning priorities to chemicals to be examined for potential regulation. As a first tier—usually in the absence of data—computations can be made (with the appropriate default assumptions) that lead to a possible regulatory standard. If this standard is readily achievable, no further analysis is called for. If the standard is not achievable, data will be sought to replace the possibly too-conservative default assumptions. Substituting more chemical-specific information for default assumptions will usually lead to less rigid and thus more easily attainable standards (or higher "safe" doses). (The rare situations in which this relaxation of the standards does not occur would imply that the default assumptions were not sufficiently health protective and so needed to be re-examined.)

A stepwise process that replaces default assumptions with specific data can be expected to yield more and more firmly established standards (regulatory doses); i.e., uncertainty should be reduced as a consequence of having more information. The tiered process for setting standards thus reflects the philosophical process of proceeding from conjecture ("it is reasonable that . . .") through information to (one hopes) wisdom.

The issue of implementation is discussed in Chapter 12, the final chapter, from two points of view. First, technical guidance is provided on EPA's implementation of the recommendations in a regulatory context. Second, the committee discusses institutional issues in risk assessment and risk management.

12

Implementation

Health risk assessment is one element of most environmental decision-making—a component of decisions about whether, how, and to what degree the assessed risk requires reduction. The factors that may be considered by decision-makers depend on the requirements of applicable statutes, precedents established within the responsible government agencies, and good public policy. This chapter discusses how the risk-assessment recommendations in this report could be implemented in the context of Section 112 of the Clean Air Act (as amended in 1990), and it discusses several institutional issues in risk assessment and risk management.

PRIORITY-SETTING AND SECTION 112

As we explained in Chapter 2, Section 112 calls for EPA to regulate hazardous air pollutants in two stages. In the first, sources will be required to do what is feasible to reduce emissions. In the second, EPA must set "residual-risk" standards to protect public health with an ample margin of safety if it concludes that implementation of the first stage of standards does not provide such a margin of safety. This second stage will require use of risk assessment.

Neither the resources nor the scientific data exist to perform a full-scale risk assessment on each of the 189 chemicals listed as hazardous air pollutants by Section 112. Nor, as we noted in Part II, is such an assessment needed in many cases.

We therefore urge an iterative approach to risk assessment. Such an approach would start with relatively inexpensive screening techniques and move to more resource-intensive levels of data-gathering, model construction, and model application as the particular situation warranted. To guard against the possibility of underestimating risk, screening techniques must be constructed so as to err on the side of caution when there is uncertainty. The results of these techniques should be used to set priorities for the gathering of further data and the application of successively more complex techniques. These techniques should then be used to the extent necessary to make a judgment. The result would be a process that supports the risk-management decisions required by the Clean Air Act and that provides incentives for further research, without the need for costly case-by-case evaluations of individual chemicals.

Under an iterative approach, a screening analysis is followed by increases in the refinement of the estimate, as appropriate. In effect, each iteration amounts to a more detailed screen. As we have explained in Chapter 6, screening analyses need to incorporate conservative assumptions to preclude the possibility that a pollutant that poses dangers to health or welfare will not receive full scrutiny.

Considering the effort required to carry out a "full-scale" risk assessment of 189 potentially hazardous substances and the current resources of the agency, it is unlikely that this task can be accomplished within the time permitted by the act if full-scale risk assessments must be conducted by EPA itself. This committee recommends a priority-setting scheme (as described in the following sections) based on initial assessments of each chemical's possible impact on human health and welfare. But Congress should recognize that the resources now available to EPA probably will not support a full-scale risk analysis for each source or even each source category within the time permitted, even with priority-setting. Thus, EPA will need alternatives to full-scale risk assessment, and attention should be given to setting priorities for the allocation of resources. In addition, a full statement of resource requirements should be developed and presented to Congress for its use in decisions about budget and for its understanding and guidance with regard to reducing the task.

ITERATIVE RISK ASSESSMENT

To implement Section 112, the committee generally supports the tiered, iterative risk-assessment process proposed by EPA in its draft document as shown in Appendix J. As stated by EPA, this process is based on the concept that as the comprehensiveness of a risk assessment increases, the uncertainty in the assessment decreases.

In the absence of sufficient data or resources to characterize each risk-assessment parameter accurately, EPA deliberately uses default options that are intended to yield health-protective risk estimates. Lower-tier risk assessments that are used for preliminary screening rely heavily on default options, and their results should be health-protective. If a lower-tier risk assessment indicates that an unacceptable health risk could be associated with a particular exposure and a regulated party believes that the risk has been overestimated, a higher-tier risk assessment can be performed. The higher-tier risk assessment would be based on more precise (and less uncertain) exposure and health information instead of relying on the default options. Conversely, if EPA believes that a lower-tier risk assessment has underestimated the health risk associated with a particular exposure, a higher-tier risk assessment might yield a more reliable estimate.

The following sections evaluate each step in the health-risk assessment process with reference to how EPA plans to implement its tiered approach.

Exposure Assessment

EPA (1992f) has proposed a tiered scheme for using health risk assessments to delist source categories and eliminate residual risk. EPA asserts that this scheme provides health-protective estimates of risk by assuming maximal exposure levels, except for cases related to complex terrains (for which an alternative dispersion model should be selected from the complex-terrain models available to EPA to estimate maximal concentrations of chemicals in air and hence maximal exposure levels).

In the initial step of the tiered approach (see Table 12-1), the emission rate for a facility is multiplied by a dispersion value obtained from a table and chosen on the basis of two site-specific parameters: stack height and the approximate distance to the site boundary line. A generic "worst-case" meteorology applicable to all noncomplex terrain is used to obtain the dispersion factors for a simple Gaussian-plume model with worst-case plant parameters (e.g., zero-buoyancy plume and zero exit velocity).

The second tier uses a simple, single Gaussian-plume model that incorporates site-specific data on the site boundary distance; the stack height, exit velocity, temperature, and diameter; the urban-rural classification; and the building dimensions. Again, a generic worst-case meteorology is used in the calculation.

In the third tier, the modeling would include multiple-point release, local meteorologic characteristics, and the choice of specific local receptor-site locations. The maximal exposure is calculated by multiplying the estimated concentration by residence time. EPA is debating the extent to which it will use less than lifetime residence (i.e., alter the 70-year-lifetime assumption).

In a presentation made to the committee by EPA staff, a fourth tier was described that would incorporate time-activity modeling, as in the Human Exposure Model II (HEM II). HEM-II uses an approach similar to that of the National Ambient Air Quality Standards (NAAQS) Exposure Model (NEM), which has been used in exposure assessments for criteria pollutants (tropospheric ozone, sulfur dioxide, etc.). However, the NEM has not been fully evaluated and validated (NRC, 1991a).

There are a number of difficulties associated with the method in the initial tiers. First, EPA does not specify that a conservative emission rate should be

TABLE 12-1   Summary of EPA's Draft Tiered Risk-Assessment Approach as
Presented to the Committee

Tier 1: Lookup Tables
* Two tables: short- and long-term (based on EPA's SCREEN model)
* Inputs: emissions rate, release height, fenceline distance
* Outputs:
  Maximum offsite concentration (focus on maximum exposed individual, MEI)
  Maximum offsite cancer risk (based on unit risk estimate, URE)
  Chronic noncancer hazard index (based on chronic health thresholds)
  Acute noncancer hazard index (based on acute health thresholds)

Tier 2: Screening Dispersion
* Based on EPA's SCREEN model (uses conversion factor for long-term)
* Inputs: Tier 1 + stack diameter, exit velocity and temperature, rural/urban classification, and building dimensions
* Outputs:
  Maximum offsite concentration and downwind distance (focus on MEI)
  Cancer risk and/or noncancer hazard index

Tier 3: Site-Specific Dispersion Model
* Based on EPA's TOXLT, TOXST models (uses the ISC dispersion model)
* Inputs: Tier 1 + Tier 2 + local meteorology, release point and fenceline layout, terrain features, release frequency, and duration
* Outputs:
  Long-term - receptor-specific risk, chronic noncancer hazard index (MEI)
  Short-term - receptor-specific hazard index exceedance rate (MEI)
* Ambient monitoring used to enhance modeling or as alternative on case-by-case basis for difficult modeling applications

Tier 4: Site-Specific Dispersion and Exposure Model
* Based on EPA's HEM II model
* Inputs: Tier 3 + population model
* Outputs: Maximum offsite concentration (MEI), exposure distribution, and population risk (incidence) with optional characterization of uncertainties
* Personal monitoring used as alterative on case-by-case basis for difficult modeling applications

NOTES: (1) Approach considers flat or rolling terrain only;
       (2) Complex terrain alternatives used on case-by-case basis;
       (3) Analysis considers only direct inhalation exposure;
       (4) Tiers proceed from most conservative and least data intensive (Tier 1) to least conservative and most data intensive (Tier 4).

SOURCE: Guinnup, 1992 (see Appendix J).

used; it will use the emission rate for normal operation of a plant at full capacity. In addition, none of EPA's current emission estimation methods accounts for "upset" situations with higher than normal emissions or for the emission-estimate uncertainty. Therefore, current emission estimates cannot be relied on as necessarily conservative.

Second, the committee reiterates its earlier concern (see Chapter 7) about the use of the Gaussian-plume model beyond the lower-tier screening level. Even there, complex terrain can create substantial problems. The EPA complex-terrain models have focused on emissions released from tall stacks toward the side of a hill or valley, and not on poor dispersion of material from a point or area source within a valley. Models for complex terrain have been developed and evaluated by the atmospheric-research community. The committee does not recommend any specific model, but suggests that EPA look beyond its set of existing models to find the best possible ones for the dispersion of hazardous air pollutants in the particular type of complex terrain that applies in each case. In addition, models should be considered that account for the possibility of a negative buoyancy plume (i.e., gas heavier than air).

For the conditions under which hazardous air pollutants are emitted from many emission points within a plant, EPA has not demonstrated that the simple, single Gaussian-plume approach (choosing dispersion values from a table generated on the basis of a generic worst-case meteorology and worst-case plant-dispersion characteristics) will be appropriate for all the situations to which it might be applied. The Gaussian-plume models have been tested for the dispersion of criteria pollutants from point sources that typically have good dispersive characteristics (e.g., tall stacks, high thermal buoyancy, and high exit velocity). However, it has not been demonstrated that this generic worst-case meteorology is fully representative of any location, such as cities with substantial local perturbations in the dispersion characteristics (surface roughness, street canyon, heat-island effects, etc.). The committee recommends that, until the evaluations can be completed, exposure assessment for source delisting and evaluating residual risk begin at EPA's current Tier 3, where the industrial source complex (ISC) model with local meteorology and local receptor-site choices will provide better estimates of the worst-case possibilities. If Tiers 1 and 2 can be shown definitively to estimate exposure conservatively, they could be incorporated into the delisting, priority-setting, and residual-risk process.

In accordance with the discussion in Chapter 7, the committee recommends that distributions of pollutant concentration values be estimated with available evaluated stochastic dispersion models that provide more realistic descriptions of the atmospheric dispersion process and that incorporate variability and uncertainty in their estimates. If the screening process suggests that a source cannot be excluded from further review, exposure estimation should be more comprehensive and incorporate more advanced methods of emission characterization, stochastic modeling of dispersion, and time-activity patterns, as discussed in

Chapter 7. Exposure assessment can be improved as necessary by incorporating more explicit local topographic, meteorologic, and other site-specific characteristics. However, if the regulated sources find it acceptable to be regulated on the basis of a (truly conservative) screening analysis, then there should be no obligation to go further. If they are not content, then the sources should bear the burden of doing the higher-tier analysis, subject to EPA guidelines and review.

Assessment of Toxicity

In EPA's proposed approach, four metrics will be used to determine whether the predicted impact of a source should warrant concern: lifetime cancer risk, chronic noncancer hazard index, acute noncancer hazard index, and frequency with which acute hazard index is exceeded. The toxicity data needed to evaluate these metrics, such as weight-of-evidence characterizations and cancer potencies for carcinogenicity and reference concentrations (RfCs) for noncancer end points, can be found by referring to the Integrated Risk Information System (IRIS) on-line database (Appendix K). This database is maintained by EPA's Environmental Criteria and Assessment Office within the Office of Research and Development for use by EPA's various program offices, by state air-quality and health agencies, and by other parties that look to EPA to provide current information on chemical toxicity.

The IRIS database will be the primary source of toxicity data for the tiered risk-assessment approach described here. The committee believes that it is appropriate for EPA to use IRIS as its preferred data source for toxicity information, rather than duplicate the effort needed to assemble and maintain such information for those of the 189 chemicals specified in Section 112 that are in IRIS. For chemicals that require a higher-tier risk assessment, EPA could supplement the information in IRIS with additional data, probability distributions, and modeling approaches. For Section 112 chemicals not yet in IRIS, EPA must collect and enter data on carcinogenic and noncarcinogenic effects.

For many of the 189 chemicals now on the Section 112 list, there are no IRIS entries, or the existing entries do not include cancer potencies for suspected carcinogens or RfCs for chemicals suspected of causing acute or chronic noncarcinogenic health effects. In these cases, it will be appropriate for EPA to develop crude screening estimates of cancer potencies and RfCs for use in research planning; if the screening values are entered in IRIS, they should be clearly identified as screening values. These estimates should be combined with exposure estimates to calculate potential cancer risks and the likelihood of acute and chronic noncancer health effects. Such estimates may be based, for example, on in vitro tests for carcinogenicity, expert judgment on structure-activity relationships, and other available information and judgment on the toxicity of the chemical in question. These crude estimates should not be used as a basis for regulatory decisions when the supporting data are not adequate for such use. However, an entry can and should summarize current information on the extent to which a chemical might be a potentially important threat to public health. If a bioassay of the chemical is under way through the National Toxicology Program or elsewhere, the estimated date of availability of results should be stated in IRIS.

A review of IRIS by the EPA Science Advisory Board (SAB) noted the importance of IRIS for both EPA and non-EPA users (Appendix K). If IRIS entries are to be used for risk assessments that lead to major risk-management decisions, then EPA must ensure their quality and keep them up to date. It is EPA's standard practice that IRIS files must be assessed in their entirety so that cancer potencies and RfCs are not distributed without an accompanying narrative description of their scientific basis; IRIS is intended not only as a source of numerical data, but also as an important source of qualitative risk-assessment information. The appropriate caveats and explanations of numerical values are important for keeping risk managers and other IRIS users fully informed about health-risk information.

The SAB noted that chemical-specific risk assessments, such as Health Assessment Documents (HADs) and SAB reviews of HADs, should be referenced and summarized in IRIS. Where different risk assessments have yielded different cancer potencies or RfCs, the file should include an explanation that relates these differences to variations in data, assumptions, or modeling approaches. Data deficiencies and weaknesses in a risk assessment that might be remedied through further data collection and research should be described in the file. In this way, IRIS can evolve into a high-quality information support system for the needs of EPA and other users relative to the Section 112 chemicals, providing not just one set of numbers for dose-response assessment, but also a summary of alternative approaches, their strengths and weaknesses, and opportunities for further research that could improve risk estimates.

Summary

The committee supports EPA's general concept of tiered risk assessment with two modifications. First, the tiered approach requires a conservative first level of analysis. EPA asserts that its approach provides a conservative risk estimate, except in the case of complex terrain. But EPA has not yet demonstrated that this assertion is valid. Second, rather than stopping a risk assessment at a particular point, EPA should encourage and support an iterative risk-assessment process wherein improvements in the accuracy of the risk estimate will replace the initial screening estimate. This process will continue until one of three possible conclusions is reached: (1) the risk, assessed conservatively, is found to be lower than the applicable decision level (e.g., 1 in a million excess lifetime risk of cancer); (2) further improvements in the model or data would not significantly change the risk estimate; or (3) the source or source category determines that the cost of reducing emissions of this pollutant are not high enough for it to justify the investment in research required for further improvements in the accuracy and precision of analysis. This procedure provides private parties with the opportunity to improve the models and data used in the analysis.

EPA must avoid interminable analysis. At some point, the risk-assessment portion of a decision should end, and a decision should be made. Reasonable limits on time (consistent with statutory time limits) and resources must be set for this effort, and they should be based on a combination of the regulatory constraints and the benefits gained from additional scientific analysis. It is not necessary to determine or measure every variable to high accuracy in the risk-assessment process. Rather, the uncertainties that have the most influence on a risk assessment should be the ones that the risk assessor most seeks to quantify and then reduce.

EPA PRACTICES: POINTS TO CONSIDER

The committee throughout this report has noted differences between the methods EPA is currently using and practices the committee considers useful in the risk-assessment process. The committee's recommendations (summarized below) highlight differences that should be considered in the process of EPA's undertaking its proposed tiered risk assessment approach.

• Select and validate an appropriate emission and exposure-assessment model for each given implementation in the risk-assessment process.
• Use a carcinogen-classification scheme that reflects the strength and relevance of evidence as a supplement to the proposed narrative description.
• Screen the 189 chemicals for programmatic priorities for the assessment of health risks, identify gaps in the data on the 189 chemicals, develop incentives to expedite generation of the needed data, and evaluate the quality of data before their use.
• Clarify defaults and the rationales for them, including defaults now "hidden," and develop criteria for selecting and departing from the defaults.
• Clarify the sources and magnitudes of uncertainties in risk assessment.
• Develop a default factor or procedure to account for the differences in susceptibility among humans.
• Use a specific conservative mathematical estimation technique to determine exposure variability.
• Conduct pediatric risk assessments whenever children might be at greater risk than adults.
• Evaluate all routes of exposure to address multimedia issues.
• Use an upper-bound interspecies dose-scaling factor for screening-level estimates.
• Fully communicate to the public each risk estimate, the uncertainty in the risk estimates, and the degree of protection.

Implications for Priority-Setting for Title III Activities

With a large number of hazardous air pollutants, hundreds of source catego-ries, and perhaps hundreds of sources within many of those categories, and with strains on personnel and financial resources, EPA will need to set priorities on its actions under Section 112. In addition, Title IX of the Amendments requires EPA to perform health assessments at a rate sufficient to make them available when needed for the residual risk assessments under Title III (approximately 15 per year). To respond to these requirements, EPA will have to determine data needs, the level of analysis needed, and the criteria for determining priorities under the Clean Air Act, as well as seek sufficient funds for conducting these analyses.

It is important that EPA establish priorities for its risk assessment activities. In the past, EPA has often appeared to base its priorities on the ease of obtaining data on a particular chemical. Rather, EPA should acknowledge the relevance and strength of the existing data on each of the 189 chemicals (and mixtures) on the list, identify the gaps in scientific knowledge, and set priorities for filling the gaps so that research that is likely to contribute the most relevant information in the most time- and cost-effective manner will be conducted first.

At a minimum, an inventory of the relevant chemical, toxicologic, clinical, and epidemiologic literature should be compiled for each of the 189 chemicals (or mixtures). For each chemical without animal test data, a structure-activity evaluation should be conducted; and for each mixture, results of available short-term toxicity tests should be analyzed. If the evidence from this step or from reviews of the clinical, epidemiologic, or toxicologic literature suggests potential human health concerns, aggregate emission data and estimates of potentially exposed populations should be reviewed. The completed preliminary analyses, including a description of the assessment process used and the findings, should be placed in the public domain (e.g., IRIS or another mechanism readily accessi-ble to the public). The inclusion of exposure data would represent a departure from past practices, and the database might need to be restructured to accommo-date this new information.

For any chemical (or mixture) for which preliminary results suggest a poten-tial health concern, it is appropriate to use more accurate emission data (includ-ing existing source-specific data), information on the environmental fate and transport of the chemical (or mixture), and more accurate characterizations (e.g., types and estimated numbers) of the populations that may be at risk of exposure, including potentially sensitive subpopulations such as children and pregnant women. In addition, a more intensive review of the relevance and strength of the available animal and human evidence (including toxicologic, clinical, and epide-miologic) data should be developed to refine insights into the probable human-health end points. If the evidence on a chemical (or mixture) and exposure still suggests potential human health effects, the agency should conduct a comprehensive risk assessment. This assessment should be conducted and communicated in accord with the recommendations elsewhere in this report, and the limitations of the data and the related assumptions, limitations, uncertainties, and variability should be appropriately stated with the final output of the assessment.

In summary, this iterative approach to gathering and evaluating the existing evidence is intended to produce a risk assessment for each of the 189 chemicals (or mixtures) that is appropriate to the quality and quantity of available evidence, the estimated size of the problem, and the most realistic scientific judgment of potential human-health risks based on that evidence. The committee believes that the process will result in a time- and cost-efficient mechanism that will effectively set priorities among the 189 chemicals (or mixtures) that fit the probable public-health concerns about them.

Model Evaluation and Data Quality

Data should not be used unless they are explicitly judged to be of sufficiently high quality for use in an activity as sensitive as risk analysis. No data should be incorporated into the risk-assessment process unless the method used to generate them has been peer-reviewed before its use. Table 12-2 indicates some steps that EPA could take to substantiate and validate its models and assumptions before use.

EPA should take additional steps to ensure that methods used to generate data for risk assessments are scientifically valid, perhaps through the use of its Science Advisory Board or other advisory mechanisms. A process for public review and comment, with a requirement for EPA to respond, should be available so that industry, environmental groups, or the general public may raise questions regarding the scientific basis of a decision made by EPA on the basis of its risk-assessment process.

Default Options

We have noted in previous chapters that EPA should articulate more explicit criteria by which it will decide whether it is appropriate to use an alternative to a default in risk assessment. Such criteria may be expressed either in the form of a general standard or in terms of specific types of evidence that the agency considers acceptable.

Critics of EPA's use of defaults have characterized the issue of their scientific validity in binary terms: either they are supported by science, in which case they are deemed legitimate, or they are contradicted by new knowledge, in which case they might be too conservative or not sufficiently protective. The reality that EPA confronts is more complex than that dichotomy. New scientific knowledge is rarely conclusive at its first appearance and rarely gains acceptance overnight. Rather, evidence accumulates, and its validity and weight are gradually

TABLE 12-2 Example of Procedure for Methods, Data, and Model Evaluation

Database Evaluation and Validation

1. Develop data-quality guidelines that require all data submitted to agency to meet minimal quality level relative to their intended use before use in given risk-assessment tier.
2. Conduct critical review of data-gathering and data-management systems to ensure that quality and quantity of data are sufficient to meet EPA's risk-assessment responsibilities under act.
3. Document procedures used to develop data, including why particular analytic or measurement method was chosen and its limitations (e.g., sources of error, precision, accuracy, and detection limits).
4. Characterize and document data quality by indicating overall robustness, spatial and temporal representativeness, and degree of quality control implemented; define and display accuracy and precision of measurements; indicate how missing information is treated; identify outliers in data.
5. Account for uncertainty and variability in collection and analysis of data.

Model Evaluation and Validation

1. Develop model-validation guidelines that indicate minimal quality of model that can be used for given risk-assessment purpose.
2. Conduct critical review of each model used in risk-assessment process to ensure that quality and quantity of output of each model are sufficient to meet EPA's risk-assessment responsibilities under act.
3. Assess database and establish and document its appropriateness for model selected.
4. Conduct sensitivity testing to identify important input-controlling parameters.
5. Assess accuracy and predictive power of model.

established through a transition period. The challenge for EPA is to decide when in the course of this evolutionary development the evidence has become strong enough to justify overriding or supplementing an existing default assumption.

Management considerations can appropriately be permitted to influence science-policy decisions related to deviations from established default positions. The committee emphasizes the desirability of well-articulated criteria for deviation from defaults. If new scientific evidence suggests that a supposedly conservative default option is not as conservative as previously believed, a new default option might be substituted. EPA needs a procedural mechanism that will allow departure from existing default models and assumptions. A more formal process should be developed.

Uncertainty Analysis

Not characterizing the uncertainty in an analysis can lead to inappropriate decisions. In addition, attempting to incorporate default assumptions of unknown conservatism into each step of a risk assessment can lead to an insufficiently or too conservative analysis.

The committee believes that the uncertainty on a risk (i.e., risk characterization) can be handled in three ways:

1. Conduct a conservative screening analysis.
2. Conduct a generic uncertainty analysis.
3. Conduct testing or analysis to develop plant-specific and chemical-specific probability distributions.

A possible uncertainty-analysis process is described in Table 12-3. As stated earlier, a key factor in deciding to increase the scope and depth of uncertainty analysis should be the extent to which expected costs and risks might alter decisions.

For parameter uncertainty, enough objective probability data are available in some cases to permit estimation of the probability distribution. In other cases, subjective probabilities might be needed. For example, a committee might conclude on the basis of engineering judgment that emission estimates calculated with emission factors are likely to be correct to within a factor of 100 (see discussion in Chapter 7) and be approximately lognormally distributed. Thus, the median of the estimated distribution would be set equal to the observed or modeled emission estimate, and the geometric standard deviation would be taken as approximately 10. If making such a generic-uncertainty assumption and then picking a conservative estimator from the distribution leads to an estimate that is above the relevant decision-making threshold, that should govern the decision unless affected parties wish to devote more resources to improving the risk characterization. If the risk characterization is sufficient for decision-making purposes, then it will not be necessary to improve it.

INSTITUTIONAL ISSUES IN RISK ASSESSMENT AND MANAGEMENT

EPA's conduct of risk assessment has been evaluated in previous chapters largely from a technical perspective, with the aim of increasing the scientific reliability and credibility of the process. But EPA operates in a decision-making context that imposes pressures on the conduct of risk assessment, and these contextual pressures have led to recurrent problems of scientific credibility, the most important of which were noted in Chapter 2.

Criticisms of EPA's risk assessments take a variety of forms, but many of them focus on three basic decision-making structural and functional problems: unjustified conservatism, often manifested as unwillingness to accept new data or abandon default options; undue reliance on point estimates generated by risk assessment; and a lack of conservatism due to failure to accommodate such issues as synergism, human variability, unusual exposure conditions, and ad hoc departures from established procedures. Although some of those criticisms might have been overstated (and we provide evidence in earlier chapters that they

TABLE 12-3 Example of Procedure for Uncertainty Analysis

Preliminary Steps

1. Conduct generic review of each parameter in each step of risk-assessment process and determine default distribution on the basis of objective probabilities, if possible, or subjective probabilities, if sufficient information is not available. If subjective probabilities cannot be used because of lack of consensus, assume either continuous-uniform or discontinuous-dichotomous distribution between reasonable lower and upper bounds or if unimodality is reasonable assumption, triangular distribution (between reasonable upper and lower bounds) or lognormal distribution (with reasonable estimates of geometric mean and standard deviation).

Improving Generic Uncertainty Analysis

2. If it is decided that default uncertainty analysis should be improved, conduct review of default probability distribution for each parameter to determine whether default distribution is reasonable. If this distribution is *not* reasonable, conduct step 3 by either of two methods:.

3a. Conduct sensitivity analysis by replacing each component distribution with its corresponding mean.	3a. Conduct sensitivity analysis by identifying most influential parameters for each model component (sensitivity index = change in model result per unit change in value of parameter)
3b. Select probability distribution for most-sensitive parameters on basis of experience, judgment, and available information from existing data samples, parameter-value ranges, most likely values, or range of most likely values.	3b. Determine uncertainty in each parameter—e.g., uncertainty index = ratio of standard deviation of parameter × to mean value of parameter ×, or number of orders of magnitude of uncertainty = log(ratio of upper-bound order statistic corresponding lower-bound order statistic).
	3c. Determine model sensitivity-uncertainty index by multiplying sensitivity index for each parameter by its uncertainty index.
	3d. Select probability distribution for *only* influential parameters on basis of experience, judgment, and available information from existing data samples, parameter-value ranges, most likely values, or range of most likely values.

Uncertainty Analysis Flowchart

4. Conduct a Monte Carlo analysis by using probability distributions of parameters as input for simplified versions of each model (e.g., emissions, exposure, and dose-response relationship) to generate a set of synthetic (Monte Carlo) probability distributions for output from each model. Approaches other than Monte Carlo might be equally feasible.

continued on next page

TABLE 12-3 *Continued*

Uncertainty Analysis Flowchart—*continued*

5. For each plausible scientific model (i.e., the default plus any plausible alternatives), conduct a numerical analysis (e.g., Monte Carlo) to determine the probability-distribution function of risk due to uncertainty in the parameters selected in Step 4. Present each distribution separately or combine them into a single representation, clearly indicating which portions of the distribution are derived from fundamental controversy about which model might be correct.

6. Conduct a "reality check" to ensure that resulting risk-estimate distribution makes scientific sense; if not, adjustments may be made. Objective of this analysis is to improve representation of uncertainty. Clearly state that uncertainty representation does not characterize all uncertainty associated with estimated risk.

7. Repeat analysis for each type of risk measurement (e.g., individual risk, population risk, and years of life lost) needed for decision-making.

Risk Management

8. Judge what probability provides sufficient level of confidence relative to regulatory decision needed. For example, risk manager might judge mean of upper 5% of distribution to be point estimate that is appropriately "conservative" within context of regulatory decision. This is simple way to guarantee modest but tangible amount of conservatism with respect to *both* average and upper tail of uncertainty distribution.

---

might have been), it is important for EPA to understand the features of its internal organization, decision-making practices, and interactions with other federal agencies that lead to these criticisms of its performance. The agency's prevailing assumptions concerning the appropriate role of risk assessment and its relationship to risk management also should be re-examined.

Stability and Change

Like any other complex organization, EPA is subject to many competing institutional pressures that affect the quality and credibility of its decisions. The agency is expected to use the best possible science in risk assessment; yet assessments must often be carried out under conditions that preclude deliberation or continued study. Problems of intra-agency coordination that have persisted throughout EPA's history create communication gaps between risk assessors and managers. The firefighting mode in which the agency all too often operates hinders the design of effective long-range research programs and even the formulation of the right questions for science to answer. As in all bureaucracies, it often seems safest to take refuge in established approaches, even if these have begun to appear scientifically outdated. External pressures, such as the demands of state agencies for precise guidance, strengthen this tendency.

These overarching managerial problems are faced by any regulatory body that is responsible for rendering consistent decisions based on changing scientific knowledge. Uncertainty, variability, and imperfections in knowledge make it difficult to control environmental risks. To remain accountable to the public under these circumstances, regulatory agencies like EPA must assess uncertain science in accordance with principles that are fully and openly articulated and applied in a predictable and consistent manner from case to case. Risk-assessment guidelines and default assumptions were designed to accomplish those objectives, and they have succeeded to a large extent in making EPA's decisions both transparent and predictable.

But an unintended side effect of such explicit decision-making rules is that they can run the risk of becoming rigid over time to the detriment of scientific credibility. Science-policy rules might ensure a valuable degree of consistency from one case to another, but they do so in part by sometimes failing to stay abreast of changing consensus in the scientific community. Some have criticized EPA for allowing bureaucratic considerations of consistency to override good scientific judgment. In trying to ensure that like cases are treated alike, the agency might fail to acknowledge, or even recognize, the scientific reasons why a new case is substantially unlike others in ostensibly the same category. In short, risk-assessment guidelines can be applied in practice like unchangeable rules. That is unfortunate, as articulated earlier in the discussion on guidelines versus requirements.

Since the mid-1970s, numerous reports and proposals have addressed the generic problem of enlisting the best possible science for EPA's decision-making. We note, for example, a January 1992 report, *Safeguarding the Future* (EPA, 1992f), submitted to the EPA administrator and containing detailed recommendations for strengthening EPA's scientific capabilities. Such reports have stressed the need for high-quality scientific advice, expanded peer review, and adequate incentives for staff scientists—clearly important issues that have attracted attention at the highest levels of EPA's administration, but have not been effectively implemented. The agency's decision-making practices have evolved since the mid-1970s, defining a positive, although gradual, learning curve. There can be little doubt that EPA is aware, at a conceptual level, of steps that can be taken to improve both its in-house scientific capabilities and its collaboration with the independent scientific community.

Management As Guide To Assessment

A more subtle and less widely recognized impediment to good decision-making on risk arises from a rigid adherence to the principle of separating risk assessment from risk management. The call to keep these two functions distinct was originally articulated in response to a widespread perception that EPA was making judgments on the risk posed by a particular substance not on the basis of science, but rather on the basis of its willingness to regulate the substance. The purpose of separation, however, was not to prevent any exercise of policy judgment at all when evaluating science or to prevent risk managers from influencing the type of information that assessors would collect, analyze, or present. Indeed, the Red Book made it clear that judgment (also referred to as risk-assessment policy or science policy) would be required even during the phase of risk assessment. The present committee concludes further that the science-policy judgments that EPA makes in the course of risk assessment would be improved if they were more clearly informed by the agency's priorities and goals in risk management. Protecting the integrity of the risk assessment, while building more productive linkages to make risk assessment more accurate and relevant to risk management, will be essential as the agency proceeds to regulate the residual risks of hazardous air pollutants.

Risk assessment should be an adjunct to the Clean Air Act's primary goal of safeguarding public health, not an end in itself. A legitimate desire for accuracy and objectivity in representing risk can induce such an obsession with numbers that too much energy is expended on representing the results of risk assessment in precise numerical form. Thus, new research might be commissioned because there is insufficient notice of how marginal the results would be in a given case or without consideration of new, less resource-intensive methods of providing essential inputs.

Moreover, there might be a vast difference between having "the truth" and having enough information to enable a risk manager to choose the best course of action from the options available. The latter criterion is more applicable in a world with resource and time constraints. Determining whether "enough information" exists to decide in turn implies the need to evaluate a full range of decisions. Thus, further improvement of a risk-assessment estimate might or might not be the most desirable course in a given situation, especially if the refinement is not likely to change the decision or if disproportionate resources have been directed to studying the risk at the expense of creating a full set of decision options from which to choose.

Comparisons of Risk

It can be questioned whether risk assessment is sufficiently developed for the particular class of decisions regarding "offsets" or other tradable actions. In general, because of the substantial and varied degrees of model and parameter uncertainties in risk estimates, it is almost impossible to rank relative risks accurately unless the uncertainty in each risk is quantified or otherwise accounted for in the comparison. If the regulatory need for comparison of risks is imperative, one might attempt to compute the uncertainty distribution of the ratio of the two risks and choose from it one or more appropriate summary statistics. For example, one might determine in a given case that there is a 90% chance that chemical A is riskier than chemical B and a 50% chance that it is at least 10 times as risky. Also, if EPA decides to undertake the proposed iterative approach to risk assessment, it will not be possible to apply this kind of ratio comparison to estimates derived from different tiers of analysis. That is because the analyses at each level will be conducted differently and will produce risk estimates of differing accuracy and conservatism. The same might be true of aggregation of risks associated with different exposures.

Even more difficult is the issue of the relative degrees of reliability in the risk figures being compared. Is it appropriate, for example, to compare actuarial risks with modeled risks? Those and other difficulties suggest that EPA should pay more attention than it now does to the appropriateness of various procedures for risk comparison. A scientifically sound way to do this would be to modify risk-assessment procedures to characterize more specifically the uncertainties in each comparison of risks—some larger, some smaller than the uncertainties in individual risk assessments—and this could be done across tiers.

Risk Management and Research

Improved cooperation between EPA's Office of Air Quality Planning and Standards (OAQPS), which conducts the regulatory work of the air program, and its Office of Research and Development (ORD), which conducts research and revises the risk-assessment guidelines, would be helpful in ensuring that research needs of the risk-management side were met by the research side. For example, the two groups might jointly publish a research agenda on hazardous air pollutants, submit the agenda for public comment and SAB review, publish a final agenda based on these comments, and then report annually on how much progress has been made on the agenda. EPA should have a review and research-management system that catalogs risk-assessment weaknesses as identified by the SAB and other peer-review activities and that helps to direct research within EPA (and to guide strategies in other federal and state agencies and in the private sector) to remedy the weaknesses when the importance of a risk assessment justifies the expenditure of research funds. In many cases, the regulated parties may be willing to fund research that will enable health-protective default options in risk assessment to be replaced by more complex and less conservative alternatives. EPA will need to maintain its own substantial research capability to understand and evaluate advances in risk assessment. In some cases, EPA will want to support targeted risk-assessment research and data collection on specific chemicals that could lead to revisions in risk assessments of such chemicals. Situations might be discovered where current risk-assessment practice is underestimating health risk or where the information base for a chemical is not sufficient to allow regulation to proceed.

Present EPA practice is to remove IRIS listings while cancer potencies or RfCs are under review. This practice is frustrating to non-EPA users, not only because the information becomes inaccessible, but also because EPA has been reluctant to state when such information will be returned to the system. The committee believes that a better practice would be for EPA to retain listings in the database, inform users that it is conducting a review, and perhaps include alternatives that can be used in the interim as the basis for calculated cancer potencies and RfCs. The narrative supporting the information on each chemical in IRIS should inform users about the assumptions underlying each calculation, about sources of data and judgments about uncertainty and variability, and about research under way to improve risk assessment on the chemical in support of future regulatory decisions.

Risk Assessment as a Policy Guide

Allocations of public-health resources reflect, among other things, some estimate of the potential benefits from health improvements achieved, and risk assessment is an important tool for understanding potential public-health impacts. Seen from this perspective, risk assessment should be a principal component of public-health and regulatory programs. Risk-management approaches will differ, perhaps greatly, depending on political choices. But establishing the relative impacts of various resource-allocation for achieving risk reduction, by continuing pursuit of comprehensive assessments of risks, should always be an objective.

For example, the committee is concerned that neither Section 112 nor other legislation provides for appropriate control of toxic emissions from mobile and indoor sources. There is strong evidence that public exposure to chemicals (and radiation) in these settings can give rise to higher public-health risk in many cases than outdoor exposure due to stationary-source emissions.

Focusing regulation on the source, rather than on the overall reduction of the pollutant (and its potential risk to public health), is unlikely to be very cost-effective in reducing disease, although it might effectively reduce high individual risks and reduce public concern over involuntary exposures. Given limited funds for both the analysis and control of environmental problems, some believe that EPA should focus on environmental toxicants that pose the greatest public-health threat.

Social and Cultural Factors

Although the principle of maximal risk reduction is of central importance, some social and cultural factors that might introduce different risk-management priorities also need to be considered.

First, it is apparent from many studies that people's perceptions of relative risk do not always match those of technical experts. When it comes to comparing risks, most people evaluate not only the mathematical probability that an adverse outcome will occur—the principal concern of the technical expert—but also other less tangible features of the risk context, most of which are not generally considered by the risk assessor.   These other concerns should be expressed and reflected at the stage of risk management.

For example, people generally feel greater anxiety about relatively low probability events with catastrophic outcomes (such as an airplane crash) than about higher-probability activities that take only a few lives at a time (such as an automobile collision). People are reluctant to accept risks, no matter how small, unless they feel that the risky activity or exposure provides some personal benefit. Risks believed to be imposed by others are less well tolerated than those voluntarily assumed. In a related vein, risks perceived as being of natural origin are less threatening than risks created by other human beings. Risks that scientists do not understand well, and over which they publicly disagree, are more feared than those about which scientific consensus is strong. Buttressing these observations is additional research that helps us to understand why people, and their governments, seem at times much more anxious about, and willing to act against, the risks associated with industrial chemicals than risks that scientists believe are more important from a public-health perspective (Slovic, 1987). We know, for example, that public perceptions of the need for regulation are influenced by such concerns as people's trust in government, their experience with experts' reassurances, and their views about social justice. When public opinion appears to be exaggerating the risks associated with industrial products, their fear might in fact be founded on an understandable mistrust of the institutional context in which those risks are produced, assessed, and eventually controlled.

SUMMARY

Apart from its specific findings and recommendations, the committee's report is dominated by a number of central themes:

1. EPA should retain its conservative, default-based approach to risk assessment for the purposes of screening analysis for standard-setting; however, a number of corrective actions are necessary for this approach to work properly.

2. EPA should rely more on scientific judgment and less on rigid procedures by taking an iterative approach to its work. Such judgment demands more understanding of the relationship between risk assessment and risk management and a creative but disciplined blending of the two.

3. The iterative approach proposed by the committee provides the ability to make improvements in both the models and data used in its analysis. However, in order for this approach to work properly, EPA needs to provide justification for its current defaults and set up a procedure such as that proposed in the report that permits departures from the default options.

4. When reporting estimates of risk to decision-makers and the public, EPA should report not only point estimates of risk but also the sources and magnitudes of uncertainty associated with these estimates.

FINDINGS AND RECOMMENDATIONS

General findings and recommendations regarding implementation and risk management are presented below.

Tiered vs. Iterative Risk Assessment

EPA proposes to adopt a tiered risk-assessment approach that will begin with a "lookup" table and move to deeper analysis with the amount of conservatism generally decreasing as estimated uncertainty decreases.

• Rather than a tiered risk-assessment process, EPA should develop the ability to conduct iterative risk assessments, allowing improvements in the process until the risk, assessed conservatively, is below the applicable decision-making level (e.g., 1×10^{-6}, etc.); until further improvements would not significantly change the risk estimate; or until EPA, the source, or the public determines that the stakes are not high enough to warrant further analysis.

Verification of Amount of Risk-Assessment Conservatism

In its tiered approach, EPA plans to use exposure models developed and validated for criteria pollutants, but not fully evaluated for the broader group of situations including hazardous air pollutants. In particular, it has not shown that analysis conducted with a simple, single Gaussian-plume approach with the generic worst-case conditions will necessarily be conservative over all situations in which it would be applied.

• Until the accuracy and conservatism of the proposed models can be evaluated, EPA should consider beginning at Tier 3, where site-specific data will provide better estimates needed for such key decisions as delisting, priority-setting, and residual-risk decisions.

Full Set of Exposure Models

Even at Tier 3, EPA plans to use a Gaussian-plume model that does not hold over complex terrain. EPA's complex-terrain models focus on tall stacks, rather than the effects of hills or valleys, and emissions from a low point or area source disperse poorly in these models.

• The committee recommends no specific model, but EPA should look beyond the set of models it now uses to find the best possible models of dispersion of hazardous air pollutants in complex terrain.

IRIS Data Quality

EPA plans to use IRIS as the database for as many as possible of the 189 Section 112 chemicals. The IRIS database has quality problems and is not fully referenced.

• EPA should enhance and expand the references in the data files on each chemical and include information on risk-assessment weaknesses for each chemical and the research needed to remedy such weaknesses. In addition, EPA should expand its efforts to ensure that IRIS maintains a high level of data quality. The chemical-specific files in IRIS should include references and brief summaries of EPA health-assessment documents and other major risk assessments of the chemicals carried out by the agency, reviews of these risk assessments by the EPA Science Advisory Board, and the agency's responses to the SAB reviews. Important risk assessments carried out by other government agencies or private parties should also be referenced and summarized.

Toxicity Data Development

Some of the 189 chemicals lack cancer potencies or RfCs.

• If IRIS does not contain a cancer potency or RfC, EPA should develop a procedure for making crude screening estimates. These estimates should generally not be used for regulation, but only as a means of setting research priorities for carrying out the animal studies from which cancer potencies and RfCs could be calculated with EPA standard default methods. EPA should develop a summary of health-risk research needs from a review of the IRIS files on the 189 chemicals. EPA should determine which research is most important, how much of it is likely to be carried out by other parties, and what research should be carried out by EPA and other federal agencies under their mandates to protect public health.

Full Data Set for Priority-Setting

EPA often appears to base priorities on the simple availability of data on a particular chemical.

• At a minimum, EPA should compile for each of the 189 chemicals an inventory of the existing and relevant chemical, toxicologic, clinical, and epidemiologic literature. For each specific chemical, EPA should have at a minimum a structure-activity evaluation; and for each important mixture, it should complete an analysis of available short-term toxicity tests (such as the Ames test). If review of toxicity information suggests a possible need for regulation to protect human health, it should develop aggregate emission data and estimates of populations potentially exposed.

Iterative Priority-Setting

EPA sometimes appears to base its priorities on a one-time analysis of incomplete and preliminary data.

• EPA should take an iterative approach to gathering and evaluating existing evidence to use in a level of risk assessment for each of the 189 chemicals that is appropriate for the quality and quantity of available evidence and the most realistic scientific judgment of potential human health risks. On the basis of that evidence, EPA should further maintain a continuing oversight of new scientific results so that it can identify needs to re-examine chemicals that it has already assessed.

Full and Complete Documentation of Priority-Setting

EPA does not always clearly communicate the methods and data on which it bases its priority-setting analysis. In addition, emission, exposure, and toxicity information is not often collected in the same database.

• Once EPA's preliminary priority-setting analyses are completed for a chemical on the list, a description of the assessment process used, the findings, and the emission, exposure, and toxicity information should be placed in one location in the public domain (e.g., in IRIS).

Guidelines vs. Requirements

EPA and others often interpret the term *risk assessment* as a specific methodologic approach to extrapolating from sets of human and animal carcinogenicity data, often obtained in intense exposures, to quantitative estimates of carcinogenic risk associated with the (typically) much lower exposures experienced by human populations.

• EPA should recognize that the conduct of risk assessment does not require any specific methodologic approach and that it is best seen not as a number or even a document, but as a way to organize knowledge regarding potentially hazardous activities or substances and to facilitate the systematic analysis of the risks that those activities or substances might pose under specified conditions. The limitations of risk assessment thus broadly conceived will be clearly seen as resulting from limitations in our current state of scientific understanding. Therefore, risk-assessment guidelines should be just that—guidelines, not requirements. EPA should give specific long-term attention to ways to improve this process, including changes in guidelines.

Process for Public Review and Comment

EPA does not always provide a method by which industry, environmental groups, or the general public can raise questions regarding the scientific basis of a decision made by EPA during the risk-assessment process.

• EPA should provide a process for public review and comment with a requirement that it respond, so that outside parties can be assured that the methods used in risk assessments are scientifically justifiable.

Petitions for Departure from Default Options

EPA does not have a procedural mechanism that allows those outside EPA to petition for departures from default options.

• EPA should develop a formal process to allow those outside the agency to petition for departures.

Iterative Uncertainty Analysis

Because EPA often fails to characterize fully the uncertainty in risk assessments, inappropriate decisions and insufficiently or excessively conservative analyses can result.

• The committee believes that the uncertainty in a risk estimate can be handled through an iterative process with the following parts: conduct a conservative screening analysis, conduct a default-uncertainty analysis, and conduct testing or analysis to develop site-specific probability distributions for each important input. The key factor in deciding to increase the intensiveness of uncertainty analysis should be the extent to which changes in estimates of costs and risks could affect risk-management decisions.

Risk Assessment vs. Risk Management

The principle of separation of risk assessment from risk management has led to systematic downplaying of the science-policy judgments embedded in risk assessment. Risk assessment accordingly is sometimes mistakenly perceived as a search for "truth" independent of management concerns.

• EPA should increase institutional and intellectual linkages between risk assessment and risk management so as to create better harmony between the science-policy components of risk assessment and the broader policy objectives of risk management. This must be done in a way that fully protects the accuracy, objectivity, and integrity of its risk assessments—but the committee does not see these two aims as incompatible. Interagency and public understanding would be served by the preparation and release of a report on the science-policy issues and decisions that affect EPA's risk-assessment and risk-management practices.

Comparisons of Risk

EPA often does not elucidate all relevant considerations of technical accuracy when it compares and ranks risks.

• EPA should further develop its methods for risk comparison, taking account of such factors as differing degrees of uncertainty and of conservatism in different categories of risk assessment.

Policy Focus on Stationary Sources

Title III focuses primarily on outdoor stationary sources of hazardous air pollutants and does not consider indoor or mobile sources of those pollutants.

• EPA should clearly communicate to Congress that emissions and exposure, and thus the aggregate risk to the public, related to indoor and mobile sources might well be higher than those related to stationary sources.

Risk Management and Research

EPA does not appear to use risk assessment adequately as a guide to research and might abandon some important risk-assessment and regulatory efforts prematurely because of data inadequacies.

• The conduct of risk assessment reveals major scientific uncertainties in a highly systematic way, so it is an excellent guide to the development of research programs to improve knowledge of risk. EPA should, therefore, not abandon risk assessments when data are inadequate, but should seek to explore the implications for research. Risk-assessment uncertainties can also help to determine the urgency with which such research should be developed. In particular, improved cooperation between EPA's Office of Air Quality Planning and Standards (OAQPS) and its Office of Research and Development (ORD) through such actions as joint publication of a research agenda on hazardous air pollutants would be most helpful.

References

Abelson, P. 1993. Health risk assessment. Regul. Toxicol. Pharmacol. 17(2 Pt. 1):219-223.

ACGIH (American Conference of Governmental and Industrial Hygienists). 1977. TLVs: Threshold Limit Values for Chemical Substances and Physical Agents in the Workroom Environment with Intended Changes for 1977. Cincinnati, Ohio: American Conference of Governmental and Industrial Hygienists.

ACGIH (American Conference of Governmental and Industrial Hygienists). 1988. TLVs: Threshold Limit Values for Chemical Substances and Physical Agents in the Workroom Environment with Intended Changes for 1988. Cincinnati, Ohio: American Conference of Governmental and Industrial Hygienists.

AIHC (American Industrial Health Council). 1992. Improving Risk Characterization. Summary report of a workshop, Sept. 26-27, 1991, Washington, D.C., sponsored by the AIHC, Center for Risk Management, Resources for the Future, and EPA. Washington, D.C.: American Industrial Health Council.

Akland, G.G., T.D. Hartwell, T.R. Johnson, and R.W. Whitmore. 1985. Measuring human exposure to carbon monoxide in Washington, D.C., and Denver, Colorado, during the winter of 1982-83. Environ. Sci. Technol. 19:911-918.

Albert, R.E., R.E. Train, and E. Anderson. 1977. Rationale developed by the Environmental Protection Agency for the assessment of carcinogenic risks. J. Natl. Cancer Inst. 58:1537-1541.

Allen, B.C., K.S. Crump, and A.M. Shipp. 1988. Correlation between carcinogenic potency of chemicals in animals and humans. Risk Anal. 8:531-544.

Allen, D.T., Y. Cohen, and I.R. Kaplan, eds. 1989. Intermedia Pollutant Transport: Modeling and Field Measurements. New York: Plenum Press.

Ames, B.N., and L.S. Gold. 1990a. Too many rodent carcinogens: Mitogenesis increases mutagenesis. Science 249:970-971.

Ames, B.N., and L.S. Gold. 1990b. Chemical carcinogenesis: Too many rodent carcinogens. Proc. Natl. Acad. Sci. USA 87:7772-7776.

Amoco/EPA (U.S. Environmental Protection Agency). 1992. Amoco-U.S. EPA Pollution Prevention Project. Project Summary, Yorktown, Va., and Amoco Corp., Chicago, Ill.

Andersen, M.E. 1991. Quantitative risk assessment and chemical carcinogens in occupational environments. Appl. Ind. Hyg. 3:267-273.

Andersen, M.E., H.J. Clewell III, M.L. Gargas, F.A. Smith, and R.H. Reitz. 1987. Physiologically based pharmacokinetics and the risk assessment process for methylene chloride. Toxicol. Appl. Pharmacol. 87:185-205.

Andersen, M.E., H.J. Clewell III, M.L. Gargas, M.G. MacNaughton, R.H. Reitz, R.J. Nolan, and M.J. McKenna. 1991. Physiologically based pharmacokinetic modeling with dichloromethane, its metabolite, carbon monoxide, and blood carboxyhemoglobin in rats and humans. Toxicol. Appl. Pharmacol. 108:14-27.

Anspaugh, L.R., and W.L. Robison. 1968. Quantitative Evaluation of the Biological Hazards of Radiation Associated with Project Ketch. UCID-15325. Biomedical Division, Lawrence Livermore Radiation Laboratory, University of California, Livermore, Calif.

Armitage, P., and R. Doll. 1957. A two-stage theory of carcinogenesis in relation to the age distribution of human cancer. Br. J. Cancer 11:161-169.

Bacci, E., D. Calamari, C. Gaggi, and M. Vighi. 1990. Bioconcentration of organic chemical vapors in plant leaves: Experimental measurements and correlation. Environ. Sci. Technol. 24:885-889.

Bailar, J.C., III, E.A. Crouch, R. Shaikh, and D. Spiegelman. 1988. One-hit models of carcinogenesis: Conservative or not? Risk Anal. 8:485-497.

Bailar, J.C., III, J. Needleman, B.L. Berney, J.M. McGinnis, and J. Michael. 1993. Determining Risks to Health: Methodological Approaches. Westport, Conn.: Auburn House Publishing.

Barbacid, M. 1986. Mutagens, oncogenes and cancer. Trends Genet. 2:188-192.

Barnes, D.G., and M. Dourson. 1988. Reference dose (RfD): Description and use in health risk assessments. Regul. Toxicol. Pharmacol. 8:471-486.

Berenbaum, M.C. 1989. What is synergy? Pharmacol. Rev. 41:93-141.

Berry, G., M. Newhouse, and P. Antonis. 1985. Combined effect of asbestos and smoking on mortality from lung cancer and mesothelioma in factory workers. Br. J. Ind. Med. 42:12-18.

Bishop, J.M. 1987. The molecular genetics of cancer. Science 235:305-311.

Bogen, K.T. 1989. Cell proliferation kinetics and multistage cancer risk models. J. Natl. Cancer Inst. 81:267-277.

Bogen, K.T. 1990a. Uncertainty in Environmental Health Risk Assessment. New York: Garland Publishing.

Bogen, K.T. 1990b. Risk extrapolation for chlorinated methanes as promoters vs. initiators of multistage carcinogenesis. Fundam. Appl. Toxicol. 15:536-557.

Bogen, K.T., and R.C. Spear. 1987. Integrating uncertainty and interindividual variability in environmental risk assessment. Risk Anal. 7:427-436.

Boughton, B.A., J.M. Delaurentis, and W.E. Dunn. 1987. A stochastic model of particle dispersion in the atmosphere. Boundary-Layer Meteorol. 40:147-163.

Boutwell, R.K. 1964. Some biological aspects of skin carcinogenesis. Prog. Exp. Tumor Res. 4:207-250.

Bowne, N.E., and R.J. Londergan. 1983. Overview, Results, and Conclusions for the EPRI Plume Model Validation and Development Project: Plains Site. EPRI Report No. EA-3074. Prepared by TRC Environmental Consultants, Inc., East Hartford, Conn. Palo Alto, Calif.: Electric Power Research Institute.

Brown, C.C., and K.C. Chu. 1989. Additive and multiplicative models and multistage carcinogenesis theory. Risk Anal. 9:99-105.

Brown, K.G., and L.S. Erdreich. 1989. Statistical uncertainty in the no-observed-adverse-effect level. Fundam. Appl. Toxicol. 13:235-244.

Butterworth, B.E. 1989. Nongenotoxic carcinogens in the regulatory environment. Regul. Toxicol. Pharmacol. 9:244-256.

Butterworth, B.E. 1990. Consideration of both genotoxic and nongenotoxic mechanisms in predicting carcinogenic potential. Mutat. Res. 239:117-132.

Calabrese, E.J. 1991. Multiple Chemical Interactions. Chelsea, Mich.: Lewis Publishers.

CDHS (California Department of Health Services). 1985. Guidelines for Chemical Carcinogen Risk Assessments and Their Scientific Rationale. Sacramento, Calif.: California Department of Health Services, Health and Welfare Agency.

Chen, J.J., and R.L. Kodell. 1989. Quantitative risk assessment for teratological effects. J. Am. Statist. Assoc. 84:966-971.

Cleverly, D.H., G.E. Rice, and C.C. Travis. 1992. The Analysis of Indirect Exposures to Toxic Air Pollutants Emitted from Stationary Combustion Sources: A Case Study. Paper No. 92-149.07. U.S. Environmental Protection Agency and Oak Ridge National Laboratory. Paper presented at the 85th Annual Meeting of the Air and Waste Management Association, June 21-26, 1992, Kansas City, Mo. Pittsburgh, Pa.: Air and Waste Management Association.

CMA (Chemical Manufacturing Association). 1989. Improving Air Quality: Guidance for Estimating Fugitive Emissions from Equipment, Vol. 2. Washington, D.C.: Chemical Manufacturing Association.

Cohen, Y., ed. 1986. Pollutants in a Multimedia Environment. New York: Plenum Press.

Cohen, S.M., and L.B. Ellwein. 1990a. Cell proliferation in carcinogenesis. Science 249:1007-1011.

Cohen, S.M., and L.B. Ellwein. 1990b. Proliferation and genotoxic cellular effects in 2-acetylaminofluorene bladder and liver carcinogenesis: Biological modeling of the ED_{01} study. Toxicol. Appl. Pharmacol. 104:79-93.

Cohen, S.M., and L.B. Ellwein. 1991. Genetic errors, cell proliferation, and carcinogenesis. Cancer Res. 51:6493-6505.

Cohen, Y., W. Tsai, S.L. Chetty, and G.J. Mayer. 1990. Dynamic partitioning of organic chemicals in regional environments: A multimedia screening-level modeling approach. Environ. Sci. Technol. 24:1549-1558.

Cohen, S.M., E.M. Garland, and L.B. Ellwein. 1992. Cancer enhancement by cell proliferation. Prog. Clin. Biol. Res. 374:213-229.

Conolly, R.B., K.T. Morgan, M.H. Andersen, T.M. Monticello, and H.J. Clewell III. 1992. A biologically-based risk assessment strategy for inhaled formaldehyde. Comments Toxicol. 4:269-288.

Crump, K.S. 1984. A new method for determining allowable daily intakes. Risk Anal. 4:854-871.

Crump, K.S. 1989. Correlation of carcinogenic potency in animals and humans. Cell Biol. Toxicol. 5:393-403.

Dedrick, R.L., and P.F. Morrison. 1992. Carcinogenic potency of alkylating agents in rodents and humans. Cancer Res. 52:2464-2467.

Dellarco, V.L., and W.H. Farland. 1990. Introduction to the U.S. Environmental Protection Agency's genetic risk assessment on ethylene oxide. Environ. Mol. Mutagen. 16:83-84.

Dellarco, V.L., L. Rhomberg, and M.D. Shelby. 1990. Perspectives and future directions for genetic risk assessment. Environ. Mol. Mutagen. 16:132-134.

Doll, R., and R. Peto. 1981. The causes of cancer. Quantitative estimates of avoidable risks of cancer in the United States today. J. Natl. Cancer Inst. 66:1191-1308.

Dourson, M.L. 1986. New approaches in the derivation of acceptable daily intake (ADI). Comments Toxicol. 1:35-48.

Dourson, M.L., and J.F. Stara. 1983. Regulatory history and experimental support of uncertainty (safety) factors. Regul. Toxicol. Pharmacol. 3:224-238.

Dourson, M.L., R.C. Hertzberg, R. Hartung, and K. Blackburn. 1985. Novel methods for the estimation of acceptable daily intake. Toxicol. Ind. Health 1:23-33.

Drinkwater, N.R., and L.M. Bennett. 1991. Genetic control of carcinogenesis in experimental animals. Prog. Exp. Tumor Res. 33:1-20.

Duan, N. 1981. Micro-Environment Types: A Model for Human Exposure to Air Pollution. SIMS Tech. Rep. No. 47. Department of Statistics, Stanford University, Stanford, Calif.

Duan, N. 1987. Cartesianized Sample Mean: Imposing Known Independence Structures on Observed Data. WD-3602-SIMS/RC. Santa Monica, Calif.: RAND Corp.

Duan, N. 1988. Estimating microenvironment concentration distributions using integrated exposure measurements. Pp. 15-114 in Proceedings of the Research Planning Conference on Human Activity Patterns, T.H. Starks, ed. EPA-600/4-89/004. Washington, D.C.: U.S. Environmental Protection Agency, Office of Research and Development.

Duan, N. 1991. Stochastic microenvironment models for air pollution exposure. J. Expos. Anal. Care Environ. Epidemiol. 1:235-257.

Duan, N., H. Sauls, and D. Holland. 1985. Application of the Microenvironment Monitoring Approach to Assess Human Exposure to Carbon Monoxide. Research Triangle Park, N.C.: U.S. Environmental Protection Agency, Environmental Systems Laboratory.

Ehling, U.H. 1988. Quantification of the genetic risk of environmental mutagens. Risk Anal. 8:45-57.

Ehling, U.H., and A. Neuhauser. 1979. Procarbazine-induced specific-locus mutations in male mice. Mutat. Res. 59:245-256.

Ellwein, L.B., and S.M. Cohen. 1992. Simulation modeling of carcinogenesis. Toxicol. Appl. Pharmacol. 113:98-108.

Ennever, F., T. Noonan, and H. Rosenkranz. 1987. The predictivity of animal bioassays and short-term genotoxicity tests for carcinogenicity and non-carcinogenicity to humans. Mutagenesis 2:73-78.

EPA (U.S. Environmental Protection Agency). 1981a. Evaluation of Maintenance for Fugitive Emissions Control. EPA-600/52-81-080. Research Triangle Park, N.C.: U.S. Environmental Protection Agency, Office of Air Quality Planning and Standards.

EPA (U.S. Environmental Protection Agency). 1981b. Health Assessment Document for Cadmium. EPA-600/8-81-023. Washington, D.C.: U.S. Environmental Protection Agency, Office of Research and Development.

EPA (U.S. Environmental Protection Agency). 1984a. Review of Health Assessment Document on Trichloroethylene by Science Advisory Board. Washington, D.C.: U.S. Environmental Protection Agency.

EPA (U.S. Environmental Protection Agency). 1984b. Letter report from the Science Advisory Board to the EPA administrator, Dec. 5. Washington, D.C.: U.S. Environmental Protection Agency.

EPA (U.S. Environmental Protection Agency). 1985a. Inorganic Arsenic NESHAPS: Response to Public Comments on Health, Risk Assessment, and Risk Management. EPA-450/5-85-001. Research Triangle Park, N.C.: U.S. Environmental Protection Agency, Office of Air Quality Planning and Standards.

EPA (U.S. Environmental Protection Agency). 1985b. Compilation of Air Pollutant Emission Factors, Vol. I, Stationary Point and Area Sources, 4th ed. Research Triangle Park, N.C.: U.S. Environmental Protection Agency, Office of Air Quality Planning and Standards.

EPA (U.S. Environmental Protection Agency). 1985c. Health Risk Assessment Document for Ethylene Oxide. EPA-600/8-84/009F. Washington, D.C.: U.S. Environmental Protection Agency, Office of Health and Environmental Assessment.

EPA (U.S. Environmental Protection Agency). 1985d. Health Assessment Document for Trichloroethylene. Washington, D.C.: U.S. Environmental Protection Agency, Office of Research and Development.

EPA (U.S. Environmental Protection Agency). 1985e. Updated Mutagenicity and Carcinogenicity Assessment of Cadmium. Final Report. EPA-600/8-83-025F. Washington, D.C.: U.S. Environmental Protection Agency, Office of Health and Environmental Assessment.

EPA (U.S. Environmental Protection Agency). 1985f. Letter from Lee Thomas, EPA administrator, to Norton Nelson, chair of Executive Committee of the Science Advisory Board, Aug. 14. Washington, D.C.: U.S. Environmental Protection Agency.

EPA (U.S. Environmental Protection Agency). 1985g. Health Assessment Document for Polychlorinated Dibenzo-*p*-dioxins. Final Report EPA-600/8-84-014F. Washington, D.C.: U.S. Environmental Protection Agency, Office of Health and Environmental Assessment.

EPA (U.S. Environmental Protection Agency). 1985h. Letter report to the EPA administrator from the Science Advisory Board, April 26. Washington, D.C.: U.S. Environmental Protection Agency.

EPA (U.S. Environmental Protection Agency). 1986a. Guidelines for carcinogen risk assessment. Fed. Regist. 51:33992-34003.

EPA (U.S. Environmental Protection Agency). 1986b. Health Assessment Document for Nickel and Nickel Compounds. EPA-600/8-83-012F. Washington, D.C.: U.S. Environmental Protection Agency, Office of Health and Environmental Assessment.

EPA (U.S. Environmental Protection Agency). 1986c. Letter report to the EPA administrator from the Science Advisory Board on nickel and nickel compounds. Washington, D.C.: U.S. Environmental Protection Agency.

EPA (U.S. Environmental Protection Agency). 1986d. Interim Procedures for Estimating Risks Associated with Exposures to Mixtures of Chlorinated Dibenzo-*p*-Dioxins and Dibenzofurans (CDDs and CDFs). Risk Assessment Forum. Washington, D.C.: U.S. Environmental Protection Agency.

EPA (U.S. Environmental Protection Agency). 1986e. Letter report to EPA administrator, Lee Thomas, from the Science Advisory Board, Nov. 4. SAB-EC-87-008. Washington, D.C.: U.S. Environmental Protection Agency.

EPA (U.S. Environmental Protection Agency). 1987a. Risk Assessment Guidelines of 1986. EPA-600/8-87/045. (Guidelines for Carcinogen Risk Assessment; Guidelines for Mutagenicity Risk Assessment; Guidelines for the Health Risk Assessment of Chemical Mixtures; Guidelines for the Health Assessment of Suspect Developmental Toxicants; Guidelines for Estimating Exposures). Washington, D.C.: U.S. Environmental Protection Agency, Office of Health and Environmental Assessment.

EPA (U.S. Environmental Protection Agency). 1987b. Unfinished Business: A Comparative Assessment of Environmental Problems. Overview Report. Washington, D.C.: U.S. Environmental Protection, Office of Policy, Planning and Evaluation.

EPA (U.S. Environmental Protection Agency). 1987c. Assessment of Health Risks to Garment Workers and Certain Home Residents from Exposure to Formaldehyde. Washington, D.C.: U.S. Environmental Protection, Office of Pesticides and Toxic Substances.

EPA (U.S. Environmental Protection Agency). 1987d. Update to the Health Assessment Document and Addendum for Dichloromethane (Methylene Chloride): Pharmacokinetics, Mechanism of Action, and Epidemiology (Draft Document). EPA-600/8-87/030A. Washington, D.C.: U.S. Environmental Protection Agency, Office of Research and Development.

EPA (U.S. Environmental Protection Agency). 1987e. Addendum to Health Assessment Document on Trichloroethylene. Washington, D.C.: U.S. Environmental Protection Agency.

EPA (U.S. Environmental Protection Agency). 1988a. Proposed guidelines for assessing female reproductive risk. Notice. Fed. Regist. 53(June 30): 24834-24847.

EPA (U.S. Environmental Protection Agency). 1988b. Proposed guidelines for assessing male reproductive risk and request for comments. Fed. Regist. 53 (June 30):24850-24869.

EPA (U.S. Environmental Agency). 1988c. Proposed guidelines for exposure-related measurements and request for comments. Notice. Fed. Regist. 53(Dec. 2):48830-48853.

EPA (U.S. Environmental Protection Agency). 1988d. Protocols for Generating Unit-Specific Emission Estimates for Equipment Leaks of VOC and VHAP. EPA 450/3-88/010. Research Triangle Park, N.C.: U.S. Environmental Protection Agency, Office of Air Quality Planning and Standards.

EPA (U.S. Environmental Protection Agency). 1988e. A Compilation of Emission Factors, Vol. 1. AP-42. Washington, D.C.: U.S. Environmental Protection Agency.

EPA (U.S. Environmental Protection Agency). 1988f. Reference Physiological Parameters in Pharmacokinetic Modeling. EPA-600/8-88/004. Washington, D.C.: U.S. Environmental Protection Agency.

EPA (U.S. Environmental Protection Agency). 1988g. Technical Support Document on Risk Assessment of Chemical Mixtures. EPA-600/8-90/064. Washington, D.C.: U.S. Environmental Protection Agency, Office of Research and Development.

EPA (U.S. Environmental Protection Agency). 1988h. Superfund Exposure Assessment Manual. EPA-540/1-88/001. Washington, D.C.: U.S. Environmental Protection Agency, Office of Emergency and Remedial Response.

EPA (U.S. Environmental Protection Agency). 1988i. Request for comment on the EPA guidelines for carcinogen risk assessment. Fed. Regist. 53:32656-32658.

EPA (U.S. Environmental Protection Agency). 1988j. Review of Addendum to Health Assessment Document on Trichloroethylene by the Science Advisory Board. Washington, D.C.: U.S. Environmental Protection Agency.

EPA (U.S. Environmental Protection Agency). 1988k. Letter report from the Science Advisory Board to the EPA administrator, March 9. SAB-EHC-88-011. Washington, D.C.: U.S. Environmental Protection Agency.

EPA (U.S. Environmental Protection Agency). 1988*l*. Letter report from the Science Advisory Board to the EPA administrator, March 9. SAB-EHC-88-012. Washington, D.C.: U.S. Environmental Protection Agency.

EPA (U.S. Environmental Protection Agency). 1989a. Risk Assessment Guidance for Superfund, Volume I, Human Health Evaluation Manual (Part A). Interim Final. EPA-540/1-89/002. Washington, D.C.: U.S. Environmental Protection Agency, Office of Emergency and Remedial Response.

EPA (U.S. Environmental Protection Agency). 1989b. Risk Assessments Methodology. Environmental Impact Statement, NESHAPS for Radionuclides, Background Information Documents, Vols. 1, 2. EPA-520/1-89/005 and EPA-520/1-89/006-1. Washington, D.C.: U.S. Environmental Protection Agency, Office of Radiation Programs.

EPA (U.S. Environmental Protection Agency). 1989c. Exposure Factors Handbook. EPA-600/8-89/043. Washington, D.C.: U.S. Environmental Protection Agency, Office of Research and Development.

EPA (U.S. Environmental Protection Agency). 1989d. Exposure Assessment Methods Handbook, Revised Draft Report. Washington, D.C.: U.S. Environmental Protection Agency, Office of Research and Development.

EPA (U.S. Environmental Protection Agency). 1989e. Guidance Manual for Assessing Human Health Risks from Chemically Contaminated Fish and Shellfish. EPA-503/8-89/002. Washington, D.C.: U.S. Environmental Protection Agency, Office of Marine and Estuarine Protection.

EPA (U.S. Environmental Protection Agency). 1989f. Letter report to EPA administrator, William Reilly, from the Science Advisory Board, Nov. 28. SAB-EC-90-003. Washington, D.C.: U.S. Environmental Protection Agency.

EPA (U.S. Environmental Protection Agency). 1991a. Guidelines for developmental toxicity risk assessment. Fed. Regist. 56(Dec. 5):63798-63826.

EPA (U.S. Environmental Protection Agency). 1991b. Health Effects Assessment Summary Tables. Annual FY-1991. Washington, D.C.: U.S. Environmental Protection Agency.

EPA (U.S. Environmental Protection Agency). 1991c. Procedures for Establishing Emissions for Early Reduction Compliance Extensions, Vol. 1. EPA-450/3-91-012a. Washington, D.C.: U.S. Environmental Protection Agency.

EPA (U.S. Environmental Protection Agency). 1991d. Alpha-2u-globulin: Association With Chemically-Induced Renal Toxicity and Neoplasia in the Male Rat. EPA-625/ 3-91/019F. Washington, D. C.: U.S. Environmental Protection Agency.

EPA (U.S. Environmental Protection Agency). 1991e. Formaldehyde Risk Assessment Update. Washington, D.C.: U.S. Environmental Protection Agency, Office of Pesticides and Toxic Substances.

EPA (U.S. Environmental Protection Agency). 1991f. Risk Assessment Guidance for Superfund, Vol. 1. Human Health Evaluation Manual (Part B, Development of Risk-based Preliminary Remediation Goals) Interim Publ. 9285.7-01B. Washington, D.C.: U.S. Environmental Protection Agency, Office of Emergency and Remedial Response.

EPA (U.S. Environmental Protection Agency). 1992a. Guidelines for exposure assessment. Fed. Regist. 57(May 29):22888-22938.

EPA (U.S. Environmental Protection Agency). 1992b. Health Effects Assessment Summary Tables. Annual FY-1992. Washington, D.C.: U.S. Environmental Protection Agency.

EPA (U.S. Environmental Protection Agency). 1992c. Guidance on Risk Characterization for Risk Managers. Internal Memorandum by F. Henry Habicht II. Feb. 26, 1992. Washington, D.C.: U.S. Environmental Protection Agency, Office of the Administrator.

EPA (U.S. Environmental Protection Agency). 1992d. Methylene Chloride. Integrated Risk Information System (IRIS). IRIS Record No. 68, as of August 1992. Washington, D.C.: U.S. Environmental Protection Agency.

EPA (U.S. Environmental Protection Agency). 1992e. Draft report: A cross-species scaling factor for carcinogen risk assessment based on equivalence of mg/kg3/4/day. Fed. Regist. 57(June 5):24152-24172.

EPA (U.S. Environmental Protection Agency). 1992f. Safeguarding the Future: Credible Science, Credible Decisions. EPA-600/9-91/050. Washington, D.C.: U.S. Environmental Protection Agency.

FAO/WHO (Food and Agriculture Organization/World Health Organization). 1965. Evaluation of the Toxicity of Pesticide Residues in Food: Report of the Second Joint Meeting of the FAO Committee on Pesticides in Agriculture and the WHO Expert Committee on Pesticide Residues. FAO Meeting Report No. PL/1965/10; WHO/ Food Add./26.65. Geneva: World Health Organization.

FAO/WHO (Food and Agriculture Organization/World Health Organization). 1982. Evaluation of Certain Food Additives and Contaminants. 26th Report of the Joint FAO/WHO Expert Committee on Food Additives (WHO Tech. Rep. Ser. No. 683). Geneva: World Health Organization.

Farrar, D., B. Allen, K. Crump, and A. Shipp. 1989. Evaluation of uncertainty in input parameters to pharmacokinetic models and the resulting uncertainty in output. Toxicol. Lett. 49:371-385.

Favor, J. 1989. Risk estimation based on germ-cell mutations in animals. Genome 31:844-852.

FDA (U.S. Food and Drug Administration, Advisory Committee on Protocols for Safety Evaluation). 1971. Panel on carcinogenesis report on cancer testing in the safety evaluation of food additives and pesticides. Toxicol. Appl. Pharmacol. 20:419-438.

Felton, J.S., M. Knize, C. Wood, B. Wuebbles, S.K. Healy, D.H. Steurmer, L.F. Bjeldannes, B.J. Kimble, and F.T. Hatch. 1984. Isolation and characterization of new mutagens from fried ground beef. Carcinogenesis 5:95-102.

Fiering, M., R. Wilson, E. Kleiman, and L. Zeise. 1984. Statistical distributions of health risks. J. Civil Engin. Syst. 1:129-138.

Finkel, A. 1987. Uncertainty, Variability, and the Value of Information in Cancer Risk Assessment. D.Sc. Dissertation. Harvard School of Public Health, Harvard University, Cambridge, Mass.

Finkel, A. 1988. Computing Uncertainty in Carcinogenic Potency: A Bootstrap Approach Incorporating Bayesian Prior Information. Report to the Office of Policy Planning and Evaluation, U.S. Environmental Protection Agency, Washington, D.C.

Finkel, A. 1990. Confronting Uncertainty in Risk Management: A Guide for Decision Makers. Washington, D.C.: Center for Risk Management, Resources for the Future.

Finkel, A. 1991. Edifying presentation of risk estimates: Not as easy as it seems. J. Policy Anal. Manage. 10:296-303.

Finkel, A. In press. A Quantitative Estimate of the Extent of Human Susceptibility to Cancer and Its Implications for Risk Management. Washington, D.C.: International Life Sciences Institute, Risk Science Institute.

Fiserova-Bergerova, V. 1992. Inhalation anesthesia using physiologically based pharmacokinetic models. Drug Metab. Rev. 24:531-557.

Flamm, W.G., and L.D. Lehman-McKeeman. 1991. The human relevance of the renal tumor-inducing potential of d-Limonene in male rats: Implications for risk assessment. Regul. Toxicol. Pharmacol. 13:70-86.

Gaylor, D.W. 1989. Quantitative risk analysis for quantal reproductive and developmental effects. Environ. Health Perspect. 79:243-246.

Gibb, H.J., and C.W. Chen. 1986. Multistage model interpretation of additive and multiplicative carcinogenic effects. Risk Anal. 6:167-170.

Gold, L.S., T.H. Slone, B.R. Stern, N.B. Manley, and B.N. Ames. 1992. Rodent carcinogens: Setting priorities. Science 258:261-265.

Goldstein, R.S., W.R. Hewitt, and J.B. Hook, eds. 1990. Toxic Interactions. San Diego, Calif.: Academic Press.

Goodman, G., and R. Wilson. 1991. Quantitative prediction of human cancer risk from rodent carcinogenic potencies: A closer look at the epidemiological evidence for some chemicals not definitively carcinogenic in humans. Regul. Toxicol. Pharmacol. 14:118-146.

Guess, H., K. Crump, and R. Peto. 1977. Uncertainty estimates for low-dose-rate extrapolation of animal carcinogenicity data. Cancer Res. 37:3475-3483.

Guinnup, D.E. 1992. A Tiered Modeling Approach for Assessing the Risks Due to Sources of Hazardous Air Pollutants. EPA 450/4-92-001. Research Triangle Park, N.C.: U.S. Environmental Protection Agency, Office of Air Quality Planning and Standards.

Harris, C.C. 1991. Chemical and physical carcinogenesis: Advances and perspectives for the 1990s. Cancer Res. 51:5023s-5044s.

Hattis, D., L. Erdreich, and T. DiMauro. 1986. Human Variability in Parameters That Are Potentially Related to Susceptibility to Carcinogenesis—I. Preliminary Observations. Report No. CTPID 86-4. Center for Technology, Policy, and Industrial Development, Massachusetts Institute of Technology, Cambridge, Mass.

Hattis, D., P. White, L. Marmorstein, and P. Koch. 1990. Uncertainties in pharmacokinetic modeling for perchloroethylene. I. Comparison of model structure, parameters, and preductions for low-dose metabolism rates for models derived by different authors. Risk Anal. 10:449-458.

Heck, H., M. Casanova, and T.B. Starr. 1990. Formaldehyde toxicity—New understanding. Crit. Rev. Toxicol. 20:397-426.

IARC (International Agency for Research on Cancer). 1972. Some Inorganic Substances, Chlorinated Hydrocarbons, Aromatic Amines, N-nitroso Compounds, and Natural Products. IARC Monographs on the Evaluation of Carcinogenic Risk of Chemicals to Man, Vol. 1. Lyon, France: World Health Organization International Agency for Research on Cancer.

IARC (International Agency for Research on Cancer). 1982. General principles for evaluating the carcinogenic risk of chemicals. In Chemicals, Industrial Processes and Industries Associated with Cancer in Humans. IARC Monographs on the Evaluation of the Carcinogenic Risk of Chemicals to Humans, Suppl. 4. Lyon, France: World Health Organization International Agency for Research on Cancer.

IARC (International Agency for Research on Cancer). 1987. Overall evaluations of carcinogenicity. In Overall Evaluations of Carcinogenicity: An Updating of IARC Monographs, Vols. 1-42. IARC Monographs on the Evaluation of Carcinogenic Risk to Humans, Suppl. 7. Lyon, France: World Health Organization International Agency for Research on Cancer.

IARC (International Agency for Research on Cancer). 1991. Mechanisms of Carcinogenesis in Risk Identification: A Consensus Report of an IARC Monographs Working Group, June 11-18, 1991. IARC Internal Tech. Rep. No. 91/002. Lyon, France: World Health Organization International Agency for Research on Cancer.

ICPEMC (International Commission for Protection Against Environmental Mutagens and Carcinogens). 1983a. ICPEMC Publ. No. 10. Review of the evidence for the presence or absence of thresholds in the induction of genetic effects by genotoxic chemicals. Mutat. Res. 123:281-341.

ICPEMC (International Commission for Protection Against Environmental Mutagens and Carcinogens). 1983b. Committee 4 Final Report: Estimation of genetic risks and increased incidence of genetic disease due to environmental mutagens. Mutat. Res. 115:255-291.

ICRP (International Commission on Radiological Protection). 1977a. Recommendations of the International Commission on Radiological Protection. ICRP Publ. No. 26. Annals of the ICRP, Vol. 1, No. 3. Oxford, U.K.: Pergamon Press.

ICRP (International Commission on Radiological Protection). 1977b. Problems Involved in Developing an Index of Harm. ICRP Publ. No. 27. Annals of the ICRP, Vol. 1, No. 4. Oxford, U.K.: Pergamon Press.

ICRP (International Commission on Radiological Protection). 1984. Protection of the Public in the Event of Major Radiation Accidents: Principles for Planning. ICRP Publ. No. 40. Annals of the ICRP, Vol. 14, No. 2. Oxford, U.K.: Pergamon Press.

ICRP (International Commission on Radiological Protection) 1985. Quantitative Bases for Developing a Unified Index of Harm. ICRP Publ. No. 45. Annals of the ICRP, Vol. 15, No 3. Oxford, U.K.: Pergamon Press.

ILSI (International Life Sciences Institute). 1992. Summary Report of the Fourth Workshop on Mouse Liver Tumors, Nov. 9-10. Washington, D.C.: International Life Sciences Institute.

IRLG (Interagency Regulatory Liaison Group, Work Group on Risk Assessment). 1979. Scientific bases for identification of potential carcinogens and estimation of risks. J. Natl. Cancer Inst. 63:241-268.

Jenkins, P., T.J. Philips, E.J. Mulberg and S.P. Hui. 1992. Activity patterns of Californians: Use of and proximity to indoor pollutant sources. Atmos. Environ. 26A:2141-2148.

Johnson, T.R., and R.A. Paul. 1981. The NAAQS Model (NEM) and Its Application to Particulate Matter. Draft report prepared for the U.S. Environmental Protection Agency by PEDCo Environmental, Inc., Durham, N.C. Research Triangle Park, N.C.: U.S. Environmental Protection Agency, Office of Air Quality Planning and Standards.

Johnson, T.R., and R.A. Paul. 1983. The NAAQS Exposure Model (NEM) Applied to Carbon Monoxide. EPA-450/5-83-003. Report prepared for the U.S. Environmental Protection Agency by PEDCo Environmental, Inc., Durham, N.C. Research Triangle Park, N.C.: U.S. Environmental Protection Agency, Office of Air Quality Planning and Standards.

Johnson, T.R., and R.A. Paul. 1984. The NAAQS Exposure Model (NEM) Applied to Nitrogen Dioxide. Draft report prepared for the U.S. Environmental Protection Agency by PEDCo Environmental, Inc., Durham, N.C. Research Triangle Park, N.C.: U.S. Environmental Protection Agency, Office of Air Quality and Planning and Standards.

Kaldor, J., and K.A. L'Abbé. 1990. Interaction between human carcinogens. Pp. 35-43 in Complex Mixtures and Cancer Risk, H. Vainio, M. Sorsa, and A.J. McMichael, eds. IARC Scientific Publ. No. 104. Lyon, France: International Agency for Research on Cancer.

Kaldor, J.M., N.E. Day, and K. Hemminki. 1988. Quantifying the carcinogenicity of antineoplastic drugs. Eur. J. Cancer Clin. Oncol. 24:703-711.

Kerns, W.D., K.L. Pavkov, D.J. Donofrio, E.J. Gralla, and J.A. Swenberg. 1983. Carcinogenicity of formaldehyde in rats and mice after long-term inhalation exposure. Cancer Res. 43:4382-4392.

Kimmel, C.A., and D.W. Gaylor. 1988. Issues in qualitative and quantitative risk analysis for developmental toxicology. Risk Anal. 8:15-20.

Kodell, R.L., R.B. Howe, J.J. Chen, and D.W. Gaylor. 1991a. Mathematical modeling of reproductive and developmental toxic effects for quantitative risk assessment. Risk Anal. 11:583-590.

Kodell, R.L., D. Krewski, and J.M. Zielinski. 1991b. Additive and multiplicative relative risk in the two-stage clonal expansion model of carcinogenesis. Risk Anal. 11:483-490.

Kopp-Schneider, A., and C.J. Portier. 1991. Distinguishing between models of carcinogenesis: The role of clonal expansion. Fundam. Appl. Toxicol. 17:601-613.

Krewski, D., and R.D. Thomas. 1992. Carcinogenic mixtures. Risk Anal. 12:105-113.

Krewski, D., T. Thorslund, and J. Withey. 1989. Carcinogenic risk assessment of complex mixtures. Toxicol. Ind. Health 5:851-867.

Layton, D.W., B.J. Mallon, D.H. Rosenblatt, and M.J. Small. 1987. Deriving allowable daily intakes for systemic toxicants lacking chronic toxicity data. Regul. Toxicol. Pharmacol. 7:96-112.

Liljegren, J. 1989. A New Stochastic Model of Turbulent Dispersion in the Convective Planetary Boundary Layer and the Results of the Atterbury 87 Field Study. Ph.D. Thesis, Department of Mechanical Engineering, University of Illinois, Urbana, Ill.

Lopes, L.L. 1984. Risk and distributional inequality. J. Exp. Psychol. 10:465-485.

Lu, F.C. 1988. Acceptable daily intake: Inception, evolution, and application. Regul. Toxicol. Pharmacol. 8:45-60.

Lutz, W.K. 1990. Dose-response relationship and low dose extrapolation in chemical carcinogenesis. Carcinogenesis 11:1243-1247.

Lyman, W.J., W.F. Reehl, and D.H. Rosenblatt. 1982. Handbook of Chemical Property Estimation Methods: Environmental Behavior of Organic Compounds. New York: McGraw-Hill.

Lyon, M.F. 1985. Attempts to estimate genetic risks caused by mutagens to later generations. Pp. 151-160 in Banbury Report 19: Risk Quantitation and Regulatory Policy, D.G. Hoel, R.A. Merrill, and F.P. Perera, eds. Cold Spring Harbor, N.Y.: Cold Spring Harbor Laboratory Press.

Machina, M.J. 1990. Choice under uncertainty: Problems solved and unsolved. Pp. 134-188 in Valuing Health Risks, Costs, and Benefits for Environmental Decision Making: Report of a Conference, P.B. Hammond and R. Coppock, eds. Washington. D.C.: National Academy Press.

Manton, K.G., E. Stallard, and J.W. Vaupel. 1986. Alternative models for the heterogeneity of mortality risks among the aged. J. Am. Stat. Assoc. 81:635-644.

Mattison, D.R. 1991. An overview on biological markers in reproductive and developmental toxicology: Concepts, definitions and use in risk assessment. Biomed. Environ. Sci. 4:8-34.

McKone, T.E. 1991. Human exposure to chemicals from multiple media and through multiple pathways: Research overview and comments. Risk Anal. 11:5-10.

McKone, T.E. 1992. CalTOX, the California Multimedia-Transport and Multiple-Pathway-Exposure Model, Part III: Multimedia- and Multiple-Pathway-Exposure Model. Prepared for the Office of the Scientific Advisor, Department of Toxic Substances Control, California Environmental Protection Agency. UCRL-CR-111456PtIII. Livermore, Calif.: Lawrence Livermore National Laboratory.

McKone, T.E., and D.W. Layton. 1986. Screening the potential risks of toxic substances using a multimedia compartmental model: Estimation of human exposure. Regul. Toxicol. Pharmacol. 6:359-380.

McKone, T.E., and J.I. Daniels. 1991. Estimating human exposure through multiple pathways from air, water, and soil. Regul. Toxicol. Pharmacol. 13:36-61.

Melnick, R.L. 1993. An alternative hypothesis on the role of chemically induced protein droplet ($\alpha 2u$-globulin) nephropathy in renal carcinogenesis. Regul. Toxicol. Pharmacol. 16:111-125.

Mohrenweiser, H.W. 1991. Germinal mutation and human genetic disease. Pp. 67-92 in Genetic Toxicology, A.P. Li and R.H. Heflich, eds. Boca Raton, Fla.: CRC Press.

Moolgavkar, S.H. 1983. Model for human carcinogenesis: Action of environmental agents. Environ. Health Perspect. 50:285-291.

Moolgavkar, S.H. 1988. Biologically motivated two-stage model for cancer risk assessment. Toxicol. Lett. 43:139-150.

Moolgavkar, S.H., and D.J. Venzon. 1979. Two-event models for carcinogenesis: Incidence curves for childhood and adult tumors. Math. Biosci. 47:55-77.

Moolgavkar, S.H, and A.G. Knudson. 1981. Mutation and cancer: A model for human carcinogenesis. J. Natl. Cancer Inst. 66:1037-1052.

Moolgavkar, S.H., and G. Luebeck. 1990. Two-event model for carcinogenesis: Biological, mathematical, and statistical considerations. Risk Anal. 10:323-341.

Moolgavkar, S.H., A. Dewanji, and D.J. Venzon. 1988. A stochastic two-stage model for cancer risk assessment. 1. The hazard function and the probability of tumors. Risk Anal. 8:383-392.

Morgan, M.G., and M. Henrion. 1990. Uncertainty: A Guide to Dealing with Uncertainty in Quantitative Risk and Policy Analysis. New York: Cambridge University Press. 332 pp.

Nazaroff, W.W., and G.R. Cass. 1986. Mathematical modeling of chemically reactive pollutants in indoor air. Environ. Sci. Technol. 20:924-934.

Nazaroff, W.W., L. Salmon, and G.R. Cass. 1990. Concentration and fate of airborne particles in museums. Environ. Sci. Technol. 24:66-67.

NCRP (National Council on Radiation Protection and Measurements). 1984. Evaluation of Occupational and Environmental Exposures to Radon and Radon Daughters in the United States. NCRP Rep. No. 78. Bethesda, Md.: National Council on Radiation Protection and Measurements.

NCRP (National Council on Radiation Protection and Measurements). 1989. Comparative Carcinogenicity of Ionizing Radiation and Chemicals. NCRP Rep. No. 96. Bethesda, Md.: National Council on Radiation Protection and Measurements.

Neely, W.B. 1980. Chemicals in the Environment: Distribution, Transport, Fate, Analysis. New York: Marcel Dekker.

Neely, W.B., and G.E. Blau, eds. 1985. Environmental Exposure from Chemicals, Vols. I and II. Boca Raton, Fla.: CRC Press.

New York Department of Health. 1987. Biological Monitoring of School-Children in Middleport, N.Y., for Arsenic and Lead. New York Department of Health, New York.

Nichols, A.L., and J.J. Zeckhauser. 1986. The dangers of caution: Conservatism in the assessment and the mismanagement of risk. Pp. 55-82 in Advances in Applied Micro-Economics: Risk, Uncertainty, and the Valuation of Benefits and Costs, Vol. 4, V.K. Smith, ed. Greenwich, Conn.: JAI Press.

NRC (National Research Council). 1970. Evaluating the Safety of Food Chemicals: Report of the Food Protection Committee. Washington, D.C.: National Academy Press.

NRC (National Research Council). 1972a. Water Quality Criteria 1972. EPA Ecological Research Series, EPA-R3-73-033. Washington, D.C.: U.S. Environmental Protection Agency.

NRC (National Research Council). 1972b. The Effects on Populations of Exposure to Low Levels of Ionizing Radiation (BEIR I). Washington, D.C.: National Academy Press.

NRC (National Research Council). 1975. Principles for Evaluating Chemicals in the Environment. Washington, D.C.: National Academy Press.

NRC (National Research Council). 1977. Drinking Water and Health, Vol. 1. Washington, D.C.: National Academy Press.

NRC (National Research Council). 1980a. Problems in risk estimation. Pp. 25-65 in Drinking Water and Health, Vol. 3. Washington, D.C.: National Academy Press.

NRC (National Research Council). 1980b. Principles of Toxicological Interactions Associated with Multiple Chemical Exposures. Washington, D.C.: National Academy Press.

NRC (National Research Council). 1980c. The Effects on Populations of Exposure to Low Levels of Ionizing Radiation (BEIR III). Washington, D.C.: National Academy Press.

NRC (National Research Council). 1983a. Risk Assessment in the Federal Government: Managing the Process. Washington, D.C.: National Academy Press.

NRC (National Research Council). 1983b. Identifying and Estimating the Genetic Impact of Chemical Mutagens. Washington, D.C.: National Academy Press.

NRC (National Research Council). 1984. Toxicity Testing, Strategies to Determine Needs and Priorities. Washington, D.C.: National Academy Press.

NRC (National Research Council). 1986. Drinking Water and Health, Vol. 6. Washington, D.C.: National Academy Press.

NRC (National Research Council). 1987. Pharmacokinetics in Risk Assessment: Drinking Water and Health, Vol. 8. Washington, D.C.: National Academy Press.

NRC (National Research Council). 1988a. Complex Mixtures: Methods for In Vivo Toxicity Testing. Washington, D.C.: National Academy Press.

NRC (National Research Council). 1988b. Health Risks of Radon and Other Internally Deposited Alpha-Emitters (BEIR IV). Washington, D.C.: National Academy Press.

NRC (National Research Council). 1989. Part II. Mixtures. Pp. 93-181 in Drinking Water and Health. Selected Issues in Risk Assessment, Vol. 9. Washington, D.C.: National Academy Press.

NRC (National Research Council). 1990a. Tracking Toxic Substances at Industrial Facilities: Engineering Mass Balance Versus Material Accounting. Washington, D.C.: National Academy Press.

NRC (National Research Council). 1990b. Health Effects of Exposure to Low Levels of Ionizing Radiation (BEIR V). Washington, D.C.: National Academy Press.

NRC (National Research Council). 1991a. Human Exposure Assessment for Airborne Pollutants: Advances and Opportunities. Washington, D.C.: National Academy Press.

NRC (National Research Council). 1991b. Seafood Safety. Washington, D.C.: National Academy Press.

NRC (National Research Council). 1991c. Environmental Epidemiology. Public Health and Hazardous Wastes, Vol. 1. Washington, D.C.: National Academy Press.

NRC (National Research Council). 1991d. Opportunities in Applied Environmental Research and Development. Washington, D.C.: National Academy Press.

NRC (National Research Council). 1993a. Issues in Risk Assessment: III. A Paradigm for Ecological Risk Assessment. Washington, D.C.: National Academy Press.

NRC (National Research Council). 1993b. Issues in Risk Assessment: I. Use of the Maximum Tolerated Dose in Animal Bioassays for Carcinogenicity. Washington, D.C.: National Academy Press.

NRC (National Research Council). 1993c. Issues in Risk Assessment: II. The Two-Stage Model of Carcinogenesis. Washington, D.C.: National Academy Press.

NRC (National Research Council). 1993d. Guidelines for Developing Community Emergency Exposure Levels for Hazardous Substances. Washington, D.C.: National Academy Press.

NRC (National Research Council). 1993e. Pesticides in the Diets of Infants and Children. Washington, D.C.: National Academy Press.

NRDC (Natural Resources Defense Council). 1989. Intolerable Risk: Pesticides in Our Children's Food. Washington, D.C.: Natural Resources Defense Council.

OSHA (U.S. Occupational Safety and Health Administration). 1982. Identification, classification, and regulation of potential occupational carcinogens. Fed. Regist. 47(8):187-522.

OSHA (U.S. Occupational Safety and Health Administration). 1983. General Industry Standards: Subpart 2, Toxic and Hazardous Substances. 40 CFR 1910.1000(d)(2)(i).

OSTP (Office of Science and Technology Policy, Executive Office of the President). 1985. Chemical carcinogens: A review of the science and its associated principles. Fed. Regist. 50(March 14):10371-10442.

OTA (Office of Technology Assessment). 1993. Researching Health Risks. Washington, D.C.: Office of Technology Assessment.

Ott, W. 1981. Computer Simulation of Human Air Pollution Exposures to Carbon Monoxide. Paper 81-57.6 presented at the 74th Annual Meeting of the Air Pollution Control Association, Philadelphia, Pa.

Ott, W. 1984. Exposure estimates based on computer generated activity patterns. J. Toxicol.-Clin. Toxicol. 21:97-128.

Ott, W., J. Thomas, B. Mage, and L. Wallace. 1988. Validation of the Simulation of Human Activity and Pollution Exposure (SHAPE) model using paired days from the Denver, Colorado carbon monoxide field study. Atmos. Environ. 22:2101-2113.

Paustenbach, D. 1989. Health risk assessments: Opportunities and pitfalls. Columbia J. Environ. Law 14:379-410.

Perera, F.P. 1991. Perspectives on the risk assessment for nongenotoxic carcinogens and tumor promoters. Environ. Health Perspect. 94:231-235.

Perera, F.P., and P. Bofetta. 1988. Perspectives on comparing risks of environmental carcinogens. J. Natl. Cancer Inst. 80:1282-1293.

Pershagen, G. 1990. Air pollution and cancer. Pp. 240-251 in Complex Mixtures and Cancer Risk, H. Vainio, M. Sorsa, and A.J. McMichael, eds. IARC Scientific Publ. No. 104. Lyon, France: International Agency for Research on Cancer.

Pierson, T.K., R.G. Hetes, and D.F. Naugle. 1991. Risk characterization framework for noncancer end points. Environ. Health Perspect. 95:121-129.

Portier, C.J. 1987. Statistical properties of a two-stage model of carcinogenesis. Environ. Health Perspect. 76:125-131.

Portier, C.J., and N.L. Kaplan. 1989. The variability of safe dose estimates when using complicated models of the carcinogenic process. A case study: Methylene chloride. Fundam. Appl. Toxicol. 13:533-544.

Portier, C.J., and L. Edler. 1990. Two-stage models of carcinogenesis, classification of agents, and design of experiments. Fundam. Appl. Toxicol. 14:444-460.

Preston-Martin, S., M.C. Pike, R.K. Ross, P.A. Jones, and B.E. Henderson. 1990. Increased cell division as a cause of human cancer. Cancer Res. 50:7415-7421.

Raabe, O.G., S.A. Brook, and N.J. Parks. 1983. Lifetime bone cancer dose-response relationships in beagles and people from skeletal burdens of ^{226}Ra and ^{90}Sr. Health Phys. 44(Suppl.1):33-48.

Reif, A.E. 1981. Effect of cigarette smoking on susceptibility to lung cancer. Oncology 38:76-85.

Reitz, R.H., A.L. Mendrala, and F.P. Guengerich. 1989. In vitro metabolism of methylene chloride in human and animal tissues: Use in physiologically based pharmacokinetic models. Toxicol. Appl. Pharmacol. 97:230-246.

Rhomberg, L., V.L. Dellarco, C. Siegel-Scott, K.L. Dearfield, and D. Jacobson-Kram. 1990. Quantitative estimation of the genetic risk associated with the induction of heritable translocations at low-dose exposure: Ethylene oxide as an example. Environ. Mol. Mutagen. 16:104-125.

Roberts, L. 1991. Dioxin risks revisited. Science 251:624-626.

Ruckelshaus, W.D. 1984. Managing risk in a free society. Princeton Alumni Weekly, March 7, pp. 18-23.

Russell, A.G. 1988. Mathematical modeling of the effect of emission sources on atmospheric pollutant concentrations. Pp. 161-205 in Air Pollution, the Automobile, and Public Health, A.Y. Watson, R.R. Bates, and D. Kennedy, eds. Washington, D.C.: National Academy Press.

Saracci, R. 1977. Asbestos and lung cancer: An analysis of the epidemiological evidence on the asbestos-smoking interaction. Int. J. Cancer 20:323-331.

SCAQMD (South Coast Air Quality Management District). 1989. In-vehicle Characterization Study in the South Coast Air Basin. South Coast Air Quality Management District, El Monte, Calif.

Schoeny, R., D. Warshawsky, and G. Moore. 1986. Non-additive mutagenic responses by components of coal-derived materials. Environ. Mutagen 8:73-74.

Schrader-Frechette, K.S. 1985. Risk Analysis and Scientific Method: Methodological and Ethical Problems with Evaluating Societal Hazards. Dordrecht, The Netherlands: D. Reidel Publishing.

Seigneur, C., Constantinou, E., and T. Permutt. 1992. Uncertainty Analysis of Health Risk Estimates. Doc. No. 2460-009-510. Paper prepared for the Electric Power Research Institute, Palo Alto, Calif.

Seinfeld, J.H. 1986. Atmospheric Chemistry and Physics of Air Pollution. New York: Wiley-Interscience.

Sexton, K., and P.B. Ryan. 1988. Assessment of human exposure to air pollution: Methods, measurements, and models. Pp. 207-238 in Air Pollution, the Automobile, and Public Health, A.Y. Watson, R.R. Bates, and D. Kennedy, eds. Washington, D.C.: National Academy Press.

Shoaf, C.R. 1991. Current assessment practices for noncancer end points. Environ. Health Perspect. 95:111-119.

Shubik, P. 1977. General criteria for assessing the evidence for carcinogenicity of chemical substances: Report of the Subcommittee on Environmental Carcinogenesis, National Cancer Advisory Board. J. Natl. Cancer Inst. 58:461-465.

Slovic, P. 1987. Perception of risk. Science 236:280-285.

Sobels, F.H. 1989. Models and assumptions underlying genetic risk assessment. Mutat. Res. 212:77-89.

Steenland, K., and M. Thun. 1986. Interaction between tobacco smoking and occupational exposures in the causation of lung cancer. J. Occup. Med. 28:110-118.

Stevens, J.B. 1991. Disposition of toxic metals in the agricultural food chain. 1. Steady-state bovine milk biotransfer factors. Environ. Sci. Technol. 25:1289-1294.

Sutherland, J.V., and J.C. Bailar III. 1984. The multihit model of carcinogenesis: Etiologic implications for colon cancer. J. Chronic Dis. 37:465-480.

Swift, M., D. Morrell, R.B. Massey, and C.L. Chase. 1991. Incidence of cancer in 161 families affected by ataxia-telangiectasia. N. Engl. J. Med. 325:1831-1836.

Switzer, P. 1988. Developing Empirical Concentration Autocorrelation Functions and Average Time Models. Paper presented at Workshop on Modeling Commuter Exposure, Aug. 17-19, Research Triangle Park, N.C.

Thilly, W.G., J. Longweel, and B.M. Andon. 1983. General approach to biological analysis of complex mixtures. Environ. Health Perspect. 48:129-136.

Thomas, J. 1988. Validating SHAPE in a Second City. Paper presented at Workshop on Modeling Commuter Exposure, Aug. 17-19, Research Triangle Park, N.C.

Trapp, S., M. Matthies, I. Scheunert, and E.M. Topp. 1990. Modeling the bioconcentration of organic chemicals in plants. Environ. Sci. Technol. 24:1246-1252.

Travis, C.C., and H.A. Hattemer-Frey. 1988. Uptake of organics by aerial plant parts: A call for research. Chemosphere 17:277-284.

Vaupel, J.W., and A.I. Yashin. 1983. The Deviant Dynamics of Death in Heterogeneous Populations. No. RR-83-1. Laxenburg, Austria: International Institute for Applied Systems Analysis.

Vogel, F. 1992. Risk calculations for hereditary effects of ionizing radiation in humans. Hum. Genet. 89:127-146.

Walker, C., T.L. Goldsworthy, D.C. Wolf, and J. Everitt. 1992. Predisposition to renal cell carcinoma due to alteration of a cancer susceptibility gene. Science 255:1693-1695.

Wallace, L.A. 1987. The Total Exposure Assessment Methodology (TEAM) Study: Summary and Analysis, Vol. 1. EPA-600-87/002a. Washington, D.C.: U.S. Environmental Protection Agency, Office of Research and Development.

Wallace, L.A. 1989. Major sources of benzene exposure. Environ. Health Perspect. 82:165-169.

Warren, A.J., and S. Weinstock. 1988. Age and preexisting disease. Pp. 253-268 in Variations in Susceptibility to Inhaled Pollutants: Identification, Mechanisms, and Policy Implications, J.D. Brain, B.D. Beck, A.J. Warren, and R.A. Shaikh, eds. Baltimore, Md.: The Johns Hopkins University Press.

Watanabe, K., F.Y. Bois, and L. Zeise. 1992. Interspecies extrapolation: A reexamination of acute toxicity data. Risk Anal. 12:301-310.

Weinstein, I.B. 1991. Mitogenesis is only one factor in carcinogenesis. Science 2251:387-388.

Weisburger, J.H. 1988. Cancer risk assessment strategies based on mechanisms of action. J. Am. Coll. Toxicol. 7:417-425.

Weisburger, J.H., and G.M. Williams. 1983. The distinct health risk analyses required for genotoxic carcinogens and promoting agents. Environ. Health Perspect. 50:233-245.

Weisel, C.P., N.J. Lawryk, and P.J. Lioy. 1992. Exposure to emissions from gasoline within automobile cabins. J. Expos. Anal. Care Environ. Epidemiol. 2:79-96.

Weschler, C.J., M. Brauer, and P. Koutrakis. 1992. Indoor ozone and nitrogen dioxide: A potential pathway to the generation of nitrate radicals, dinitrogen pentoxide and nitric acid indoors. Environ. Sci. Technol. 26:179-184.

Whitfield, R.G., and T.S. Wallsten. 1989. A risk assessment for selected lead-induced health effects: An example of a general methodology. Risk Anal. 9:197-207.

Whittemore, A.S., and A. McMillan. 1983. Lung cancer mortality among U.S. uranium miners: A reappraisal. J. Natl. Cancer Inst. 71:489-499.

Zannetti, P. 1990. Air Pollution Modeling: Theories, Computational Methods, and Available Software. New York: Van Nostrand Reinhold.

APPENDIXES

A

Risk Assessment Methodologies: EPA's Responses to Questions from the National Academy of Sciences

DISCLAIMER

This document was prepared primarily by the staff of the Pollutant Assessment Branch within the Office of Air Quality Planning and Standards. Some of the responses that describe future risk assessment procedures and policies represent the opinions of the authors within the Office of Air Quality Planning and Standards and do not necessarily represent the U.S. Environmental Protection Agency policy.

Table of Contents

Page

LIST OF FIGURES

Figure   **Page**

LIST OF TABLES

**Question 1: What does EPA consider to be the risk assessment require-
ments needed to implement the Clean Air Act of 1990?**

I.A Introduction

Implementation of Title III of the Clean Air Act (CAA) requires the development and consideration of risk and hazard assessment in several provisions. The
extent of assessment appropriate for each implementation activity is dependent
on various factors. These include, but are not limited to, the purpose of the specific
provision, the statutory timing and relationship to other provisions, and the availability of data and analytical methods. The next sections describe the regulatory
flow and timing of Title III implementation, identify the levels of assessment
and review, and describe the provisions with risk-related requirements.

I.B Regulatory Flow and Chronology of Title III Implementation

Regulation under Title III is comprised of two major steps: the application
of technology-based emission standards to categories of major stationary industrial sources, followed by the evaluation of residual risks and the development of
further standards, as necessary, to insure that public health is being protected
with an ample margin of safety. Affected source categories are identified based
on emissions of listed pollutants. The list of source categories and agenda for
regulation are required to be published. Extensions from compliance with the
technology-based standards are available with demonstration of voluntary emissions reductions, documented problems with the installation of controls, or recently installed controls. Following compliance with the technology-based standards (maximum achievable control technology or MACT), EPA is required to
evaluate residual risks and promulgate further standards, if necessary. Compliance and enforcement of the regulations is implemented through an operating
permit program at the State level. The flow of the regulatory program under
Title III is summarized in Figure 1.

In addition to the regulatory requirements, there are a number of studies in
Title III that require reports to Congress on various schedules. The timing of
these studies and the principal regulatory milestones are illustrated in Figure 2.

I.C Levels of Risk Assessment

Table 1 presents a brief overview of those Title III provisions which contain
elements of risk assessment. Included is a categorization of the level of analysis
associated with each activity and the level of review. These are briefly described
below. Their use, as exemplified in the past and present or future efforts is
presented in the response to Question 2.

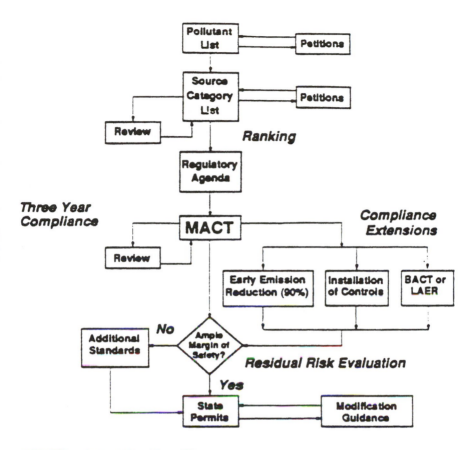

FIGURE 1   Title III Regulatory Flow.

a. <u>Problem Definition</u>: Problem definition activities generally include scoping studies to broadly assess the potential magnitude of the air toxics problem.

b. <u>Hazard Assessment</u>: A hazard assessment is the evaluation of the potential of a substance to cause human health or environmental effects. It would include an assessment of the available effects data and additional information such as environmental fate, potential for bioaccumulation, and identification of sensitive sub-populations.

c. <u>Hazard Ranking</u>: A hazard ranking is the relative comparison of information identified in individual pollutant hazard assessments. The purpose of this type of analysis is to rank or group pollutants that pose similar hazards to public health or the environment.

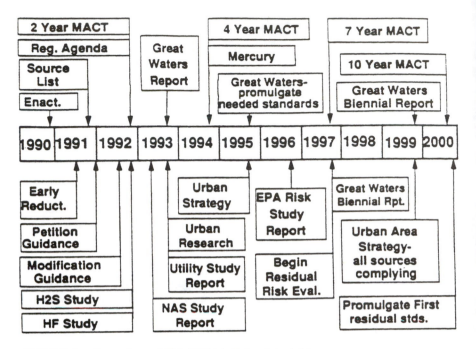

FIGURE 2   Chronology of Title III Risk-Related Activities.

d. Risk Ranking: A risk ranking is the comparative ranking that considers both emissions or exposure information and health effects data. The data may vary in quality depending upon the needs of the specific project.

e. Quantitative Risk Assessment: Quantitative risk assessment is the quantitative characterization of individual and population risk. It is typically conducted for individual sources, but the results may also be aggregated across an industrial source category. This level of analysis requires the most extensive collection of data and analytical resources.

I.D Risk Assessment Review Requirements

The assessments and methods used to implement various aspects of the air toxics program undergo a series of internal and external review procedures. The level of the review varies but will generally fall into one or more of the categories. The levels of review intended for each implementation activity under Title III are indicated in Table 1 and are broadly described below. It should be noted that individual components of a risk assessment may have a formal peer review.

For example, hazard assessment documents always undergo external peer review.

a. Internal Review: This generally consists of review by EPA technical and scientific staff, supervisors, and senior management. It may also include review by Agency-wide committees such as the Risk Assessment Forum (RAF) or the Risk Assessment Council (RAC). Internal review is included in all phases of regulatory and methods development.

b. External Review by Individuals: This review is conducted by individuals outside the Agency who are selected for their expertise in a specific area.

c. External Review by Panels: Such review is the result of a workshop or meeting of experts and representatives of interested groups or affected organizations.

d. Public Review: This consists of review by the public of all supporting documentation as part of the formal rulemaking process, and follows publication of a proposed rule in the Federal Register.

e. Formal External Review: This is review by established advisory committees (e.g., EPA's Science Advisory Board (SAB), National Academy of Sciences (NAS), National Air Pollution Control Techniques Advisory Committee (NAPC-TAC)).[1]

I.E Title III Risk-Related Provisions

Several provisions of Title III contain requirements for risk or hazard assessment. Beginning on the following page, Table 1 summarizes these provisions. The levels of analysis and review identified on the Table correspond to the levels discussed above. The codes used in the Table are explained in notes on the last page of the Table.

---

[1] The NAPCTAC is a committee composed of representatives of industry, environmental groups, and State and local agencies. It was established pursuant to Section 117 of the CAA. The primary focus of NAPCTAC is the review of control technology alternatives considered in the development of emission standards. The role has expanded to include other areas relevant to Title III implementation.

Table 1: Overview of CAA Title III Risk-Related Requirements

Statutory Cite/Project Description	Statutory Deadline	Data Considered for Finding	Level of Analysis[1]	Level of Review[2]
Section 112(a) Lesser Quantity Emission Rates - modification of major source definition based on potency, persistence, bioaccumulation potential, other characteristics of a pollutant, or other relevant factors - EPA currently considering implementation options	- discretionary activity	- sufficient data to identify critical health effect(s) - estimate of risk based on generic exposure modeling - consideration of bioaccumulation	RR	a,c,d
Section 112(b) Pollutant Listing and Delisting - modification of hazardous air pollutant (HAP) list by EPA on own accord or following petition - listing of HAP requires demonstration that emissions, ambient concentrations, bioaccumulation, or deposition of substance is known or reasonably anticipated to cause an adverse human health or environmental effect - delisting of HAP requires opposite demonstration - EPA currently drafting guidance for petitioners	- periodic review of list - response to petition within 18 months of receipt	- identification of known or reasonably anticipated adverse effect(s) Hazard data: sufficient to identify critical effect(s) Exposure data: sufficient to identify potential exposures	RR	a,d,e

Table 1: Overview of CAA Title III Risk-Related Requirements

Statutory Cite/Project Description	Statutory Deadline	Data Considered for Finding	Level of Analysis[1]	Level of Review[2]
Section 112(c) Source Category List (Delisting) - modification of source category list by EPA on own accord or following petition - max. individual cancer risk from any 1 source cannot >10⁻⁶ - noncancer risks: must protect public with "ample margin of safety" - consideration of environmental effects included - EPA currently drafting guidance for petitioners	- development of list within 12 months; list proposed 6/91 - response to petition within 12 months of receipt	- identification of effects and exposure as listed above - potential exposures may be assessed using tiered approach with increasing data requirements	QRA	a,d
Section 112(e) Source Category List (Schedule) - identification of categories to be covered by standards promulgated within 2, 4, 7, or 10 years - priorities to consider known or anticipated health and environmental effects, quantity and location of emissions, and potential for grouping categories	- publish schedule, Nov '92	- screening assessment using available effects and exposure data to assess relative ranking of source categories	RR	a,d

Table 1: Overview of CAA Title III Risk-Related Requirements

Statutory Cite/Project Description	Statutory Deadline	Data Considered for Finding	Level of Analysis[1]	Level of Review[2]
<u>Section 112(f) Residual Risk</u> - evaluation of risks associated with emissions following implementation of control technology standards promulgated under Section 112(d) - EPA/Surgeon General report to evaluate methods for evaluating health risks, significance of residual risks, additional control options and associated costs, uncertainties associated with analysis, and recommended legislative changes - default, if Congress does not act on recommended legislative changes, is to use previous methods (e.g., benzene decision)	- EPA/Surgeon General report, Nov '96 - promulgation of additional standards within 8 years of Section 112(d) standards	- characterization of effect(s) with quantification of exposure and dose-response - consideration of individual and population risks	QRA	a,d,e
<u>Section 112(g) Modifications</u> - identification of relative ranking of HAP's based upon potential to elicit health effects considering threshold and nonthreshold effects - ranking to be considered in evaluating emission offsets - establishment of de minimis levels	- guidance, May '92	- ranking using available effects data to assess relative toxicity of the HAPs and establish de minimis levels	IIR	a,c,d,e

Table 1: Overview of CAA Title III Risk-Related Requirements

Statutory Cite/Project Description	Statutory Deadline	Data Considered for Finding	Level of Analysis[1]	Level of Review[2]
Section 112(i) Early Reduction Program - extension of compliance with control technology standards allowed - special consideration of "highly toxic" pollutants	- proposed rule, Jun '91 - final rule, Spring '92	- establishment of highly toxic pollutants similar to Lesser Quantity Emission Rates program identified above	IIR	a,c,d,e
Section 112(k) Urban Area Source Program - development of research program on the sources of HAPs in urban areas - identification of ≥ 30 HAPs and associated source categories that pose the greatest threat to public health in urban areas - development of national strategy, accounting for ≥ 90% of the identified HAP emissions, resulting in ≥ 75% reduction in cancer incidence	- report to Congress on research activities, Nov '93 - report to Congress on national strategy, Nov '95 - implement strategy such that sources are in compliance, '99	- analysis of urban exposures using ambient monitoring data and modeling of estimated emissions - sufficient data to identify critical effect(s) and dose-response relationships	RR, QRA	a,b,c,e (d, as needed)

Table 1: Overview of CAA Title III Risk-Related Requirements

Statutory Cite/Project Description	Statutory Deadline	Data Considered for Finding	Level of Analysis[1]	Level of Review[2]
Section 112(m) Great Waters Study - investigation of the contribution of atmospheric deposition of HAPs to the Great Lakes, Chesapeake Bay, Lake Champlain, and coastal waters - development of regulations as necessary to prevent adverse health and environmental effects	- establish Great Lakes monitoring network, Dec '91 - report to Congress, Nov '93 - promulgation of additional regulatory measures, Nov '95	- sufficient data required to identify critical HAPs, potential exposures and critical effects - key parameters to consider include: environmental persistence, bioaccumulation potential, delineation of contribution of air emissions vs. other sources of pollutants	RR, QRA	a,b,c (d, as necessary)
Section 112(n) Electric Utility Study - evaluation of public health risks associated with HAPs emitted from electric steam generating units following implementation of Title IV (acid rain) regulations - report on alternative control strategies, if needed	- report to Congress, Nov '93	- sufficient data to identify critical HAPs, their exposures, potential and initial effects - assess individual and population risk	QRA	a,b,c

Table 1: Overview of CAA Title III Risk-Related Requirements

Statutory Cite/Project Description	Statutory Deadline	Data Considered for Finding	Level of Analysis[1]	Level of Review[2]
Section 112(n) NIEHS Study of Mercury - establishment of threshold concentrations of mercury	- report to Congress, Nov '93	- sufficient hazard and dose-response data relevant to establishing a threshold concentration for mercury	IIA	b,c (possible)
Section 112(n) Mercury Study - evaluation of mercury emissions from utilities, municipal waste combustors, and other sources	- report to Congress, Nov '94	- identify health and environmental effects	RR	a,b
Section 112(n) Hydrogen Sulfide Study - assessment of public health and environmental hazards associated with emissions from the extraction of oil and natural gas - development and implementation of control strategy, as needed	- report to Congress, Nov '92	- sufficient evidence on health and environmental effects - adequate emissions data from the oil and natural gas industry	RR	a,b,(d, as needed)

Table 1: Overview of CAA Title III Risk-Related Requirements

Statutory Cite/Project Description	Statutory Deadline	Data Considered for Finding	Level of Analysis[1]	Level of Review[2]
Section 112(n) Hydrofluoric Acid Study - assessment of the potential health and environmental hazards of HF emission releases including worst case accidental releases - recommendation for reducing hazards, if appropriate	- report to Congress, Nov '92	- sufficient evidence on health and environmental effects - adequate emissions data and probability of accidental release events	RR	a,b,(d, as needed)
Section 112(o) National Academy of Sciences Study - review of EPA risk assessment methods - exploration of opportunities to improve current methods	- report to Congress, May '93	- all data considered relevant by the committee		c

Table 1: Overview of CAA Title III Risk-Related Requirements

Statutory Cite/Project Description	Statutory Deadline	Data Considered for Finding	Level of Analysis[1]	Level of Review[2]
<u>Section 112(r) Accidental Release Program</u> - identification of principal pollutants ($\geq$ 100) and associated threshold quantities - pollutant petition process included - establishment of Chemical Safety Board - development of regulatory program	- report to Congress on use of hazard assessments, May '92 - promulgation of pollutant list and threshold quantities, Nov '92 - report to Congress on regulatory recommendations, Nov '92	- identification of potential for death, injury, serious adverse effect(s)	QRA	a,d,e

[1] Levels of Anaylsis: HA, Hazard Assessment; HR, Hazard Ranking; RR, Risk Ranking; QRA, Quantitative Risk Assessment, as described in Section I.C.
[2] Levels of Review as described in Section I.D

II. Question 2: What has EPA done in the past toward those or similar risk assessment requirements, and why did EPA take the specific actions it did?

II.A Introduction

The following sections describe the framework for risk assessment presented by specific activity. The first section describes these activities generically, and subsequent sections provide examples of past and current or planned assessments.

II.B Generic Discussion

The approach EPA follows in conducting risk assessments follows the framework proposed by the National Research Council (NRC) of the National Academy of Sciences in 1983. This process was described in a book entitled "Risk Assessment in the Federal Government: Managing the Process" and identified risk assessments as containing one or more of the following four components: hazard identification, dose-response assessment, exposure assessment, and risk characterization.

In response to the NRC proposal, EPA issued several risk assessment guidelines addressing such areas as carcinogenicity, developmental toxicity, chemical mixture assessment, reproductive toxicity, exposure assessment, and mutagenicity. The EPA is continuing to develop guidelines to address various issues including risk assessment methods for evaluating noncancer effects, e.g, guidelines discussing immunotoxicity and respiratory toxicity.

The sections that follow generally discuss the process of developing risk assessment guidelines and provide examples of the efforts undertaken by the Agency to address the four components of the risk assessment process. For instance, the hazard identification and dose-response assessment steps are incorporated into the development of the hazard assessment documents.

II.B.1 Risk Assessment Guideline Development

The EPA has published guidelines addressing various aspects of risk assessment to direct the Agency in the consistent evaluation of environmental pollutants. The process of developing Agency-wide risk assessment guidelines is a multi-year procedure incorporating the state-of-the science with both internal and external expertise. This process is illustrated in Figure 3. The guidelines serve two purposes: (1) to guide EPA scientists in conducting Agency risk assessments and (2) to inform EPA decision makers and the public about these procedures. The principles set forth in the EPA risk assessment guidelines apply across all risk-based decisions considered by the Agency.

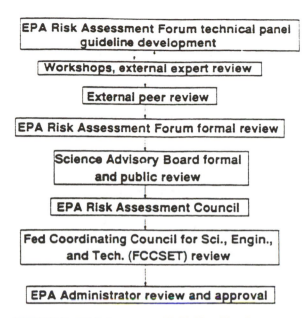

FIGURE 3   Risk Assessment Guidelines Development Process.

The emphasis of these guidelines is that the risk assessments should be conducted on a case-by-case basis considering all relevant information. The information considered includes: the level of analysis required to meet the needs of the risk manager, the availability of data, and the existing methods for appropriately interpreting the scientific data. The guidelines also stress the need to clearly articulate the scientific basis and rationale for each assessment along with its associated strengths and weaknesses. Included must be a description of the uncertainties, assumptions, and limitations of the risk assessment conducted.

II.B.2   Hazard Assessment Document Development

Hazard Assessment Documents (HADs) were commissioned at the request of EPA's Office of Air Quality Planning and Standards (OAQPS) to provide health information on the 30+ substances that were being considered for listing under section 112 of the Clean Air Act in the 1980's. In 1982, an EPA Office of Research and Development (ORD) committee was convened for the purposes of developing a plan for producing hazard assessment documents. They were specifically charged with determining the scope and content of the documents, procedures for production and peer review, and the schedules and resources necessary for production within the anticipated deadlines.

The most immediate purpose of the documents was to meet the needs of OAQPS by providing critical evaluations of all the pertinent health literature and data to determine whether or not significant human health effects were associated with exposure to chemicals at ambient air concentrations. The committee agreed these should focus on air-related health concerns, but attempts would be made to identify other EPA program offices as potential users, requiring the structure of the documents to consider multi-media assessments. The contents of each document would consider:

- physical and chemical characteristics
- man-made and natural sources and emissions
- environmental distribution and measurement, including measurement techniques, transport and fate, environmental concentrations and exposures (multi-media)
- ecological effects
- biological disposition, metabolism, and pharmacokinetics
- toxicological overview of health effects
- specific health effects, i.e., mutagenicity, carcinogenicity, and other non-cancer health effects
- synergism and antagonism
- health risk information

A multi-tiered assessment approach was employed, with sucessively more detailed and extensive assessments conducted as warranted by preceding outcomes. The results of each level would be reviewed by the program office (OAQPS) and considered along with exposure assessment information developed by OAQPS in order to determine the neccesity for further, more detailed assessment. The process is diagrammed below in Figure 4.

II.B.3 Exposure Methodology

The first systematic exposure assessments of hazardous air pollutants (HAPs) began as a result of provisions in the Clean Air Act of 1970 requiring the identification and listing of HAPs, as well as the promulgation of emissions standards for those listed HAPs. To assist in these assessments, the Human Exposure Model (HEM) was developed by OAQPS for use as a screening model in the identification and national assessment of candidate HAPs. This role expanded in the early 1980's to include more detailed quantitative evaluation of health risks (principally cancer) associated with stationary emission sources of HAPs.

In 1986, EPA published guidelines on conducting exposure assessments. The guidelines were developed to assist future assessment activities and encourage improvement in those EPA programs that require, or could benefit from, the use of exposure assessments. The authors of the guidelines also attempted to

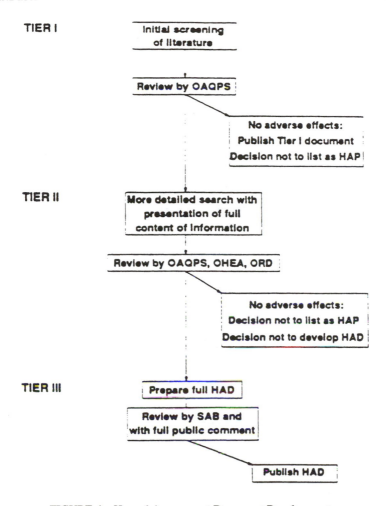

TIER I

Initial screening
of literature

Review by OAQPS

No adverse effects:
Publish Tier I document
Decision not to list as HAP

TIER II

More detailed search with
presentation of full
content of information

Review by OAQPS, OHEA, ORD

No adverse effects:
Decision not to list as HAP
Decision not to develop HAD

TIER III

Prepare full HAD

Review by SAB and
with full public comment

Publish HAD

FIGURE 4   Hazard Assessment Document Development.

promote consistency among various exposure assessment activities that are carried out by the Agency. The guidelines recognized that the main objective of an
exposure assessment is to provide reliable exposure data or estimates for a risk
characterization. Since a risk characterization requires coupling exposure information and toxicity or effects information, the exposure assessment process
should be coordinated with the effects assessment. The OAQPS has interpreted
this important consideration to mean a balancing of uncertainties in the exposure
assessment with the uncertainties in the effects assessment, i.e., quality toxicity
assessments are supported with quality exposure assessments. In 1991, EPA
revised the exposure assessment guidelines to substantially update the earlier

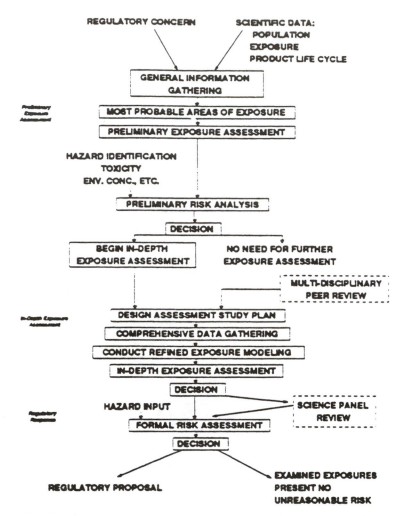

FIGURE 5   Integration of Exposure Assessment into Risk Assessment Process.

guidelines. The new guidelines incorporate developments in the exposure assessment field since 1986, both including the previous work and adding several
topics not covered previously. The EPA will be examining the exposure assessment process for HAPs to ensure consistency with the new guidelines. Figure 5
presents a diagram of the process.

II.B.4 Risk Characterization and Treatment of Uncertainty

One of the issues which EPA continues to address has been the characterization and communication of estimated risks and their uncertainties to a variety of

audiences, including Agency risk managers, State and local air pollution control agencies, the public, the affected industries, environmental groups, and other interested parties. The OAQPS traditionally conducted risk characterizations nationally by source category, rather than presenting risks posed by each emission point or facility. In the early 1980's risk estimates were used largely to rank source categories by their estimated potential risks. As experience was gained with risk assessments and the perceived need of risk managers to have more information to make more informed decisions increased, the national source category approach evolved into plant-by-plant and, in some cases, emission point-by-emission point analyses.

The process of risk characterization combines the results of the hazard identification, dose-response assessment, and exposure assessment. In evaluating HAPs, EPA reviews the available information and determines the most appropriate level of risk estimation that may be conducted using these data. The data generally can be categorized into four areas: (1) source and emissions; (2) transport of the pollutant from the source to the target population; (3) exposure of the target population; and (4) adverse effects resulting from the exposure. Depending upon the quantity and quality of the data, the risk assessment may be qualitative and/or quantitative in nature.

Qualitative risk assessments include an analysis of the existing data base and the potential for the pollutant to elicit an effect in a population. This assessment may involve the classification of the data into weight-of-evidence categories and would include a consideration of the severity of the effect anticipated in the exposed population.

Quantitative cancer risk assessments have frequently included the presentation of information in three ways: (1) estimated population risk, expressed as average annual incidence; (2) maximum individual lifetime risk; and (3) distribution of individual risk across the exposed population, i.e., the number of individuals at risk in various risk intervals (e.g., 10^{-4}, 10^{-5}, 10^{-6}).

The evaluation of potential noncancer risks has frequently involved the comparison of estimated ambient levels with a reference level. For example, the risk for developmental toxicity may be inferred by comparing the reference dose for this effect (RfD_{DT}) and the human exposure estimate or by calculating the "margin of exposure" (MOE). The RfD_{DT}, derived by applying uncertainty factors to the no observed adverse effect level (NOAEL) (or the lowest observed adverse effect level, LOAEL), differs from the RfD because the former is based on a short duration of exposure rather than chronic exposure situations. The MOE is the ratio of the NOAEL from the most appropriate or sensitive species to the estimated human exposure level, and is presented along with a discussion of the weight-of-evidence (WOE) classification. The WOE incorporates information from all relevant studies and represents a judgment based on the collective database as to the likelihood that exposure to a specific substance may pose a risk to humans. Placing an agent in a particular WOE category such as "adequate evidence for human developmental toxicity" does not mean that it will be a developmental toxicant at every dose, since the Agency assumes the existence of a threshold for the effect. Appendix A presents additional information on EPA's risk assessment guidelines for developmental toxicity.

As tools develop in the area of noncancer risk assessment along with expansion of existing data bases, quantitative presentation of risk assessments similar to analyses conducted for potential carcinogenic risks may be possible. It should be noted that the presentation of either a qualitative or quantitative risk assessment must always be accompanied by a description of the limitations associated with the analysis including attendant assumptions and uncertainties.

As risk managers seek to derive the maximum information possible for decision-making, greater emphasis has been placed on the characterization of uncertainty. The key uncertainties associated with the overall risk assessment process can be divided into three areas: uncertainties in the quantification of health effects; uncertainties in modeling the atmospheric dispersion of emitted HAPs; and uncertainties in the assessment of population exposure. Uncertainties associated with health effect quantification arise from use of the linear multistage model for estimating cancer potency, extrapolation from high-dose to low-dose, extrapolation from animal to sensitive human populations and extrapolation across various routes of exposure. A critical need is to expand our understanding about the relevant underlying physical, chemical, and biological mechanisms that affect the validity of extrapolation assumptions. Uncertainties associated with atmospheric dispersion modeling stem from uncertainties in emission rates, meteorological and terrain information, and relation of assumed stack parameters and locations to actual values. Uncertainties in the assessment of population exposure arise from uncertainties in the location and activity patterns of exposed populations, duration of actual exposures in each microenvironment, and extrapolations across exposure conditions. Appendix P includes a chart which illustrates the expected magnitude of uncertainties surrounding several exposure parameters evaluated in the assessment of benzene emissions. Activities within EPA to reduce the uncertainties in each of these areas are described later in the section on evolution of exposure and risk assessment methodologies (II.D.6).

II.B.5 Some Differences Between Past and Present Risk Assessment

The new CAA expands the scope of air toxics regulations. Consequently, expectations at each level of assessment have increased. For example, hazard assessment and hazard ranking currently place greater focus on the relative hazard and potency of the effects. Exposure information and emission data are also subject to this increased level of need. For example, the Toxic Release Inventory (TRI) data base, established under Section 313 of the Superfund Amendments and Reauthorization Act (SARA), contains emission data on many HAPs, but the data are generally not sufficient to use in quantitative risk assessments. The data base is limited in that it only covers certain industrial types, and is required only for relatively large plants. While this type of information may have been useful for defining a problem or to derive crude estimates of exposure in the past, it is not anticipated to be sufficient for quantitative risk assessments.

The questions and needs to be addressed under the CAA go beyond the data issue. The assessment procedures of the past will have to be reexamined in light of the new legislation. Requirements associated with the residual risk determinations bring about additional concerns for the quantitative risk assessment process. Some of these concerns are:

• assessing residual risk from multiple pollutants rather than individual pollutants within a source category
• determining the approach appropriate to evaluating risk to the most exposed individual
• assessing noncancer health risks
• determining the risks from less than chronic exposure, especially acute exposures
• factoring population mobility and activity patterns into the risk assessment process
• identifying sensitive populations
• assessing ecological risks

While these may not be new concerns, the CAA of 1990 has focused greater attention on these issues.

II.C Examples of Past Assessments

II.C.1 Problem Definition

Exposure to HAP emissions may result in a variety of adverse health effects considering both cancer and noncancer endpoints. In an effort to better understand the "big picture" of hazardous air pollutant exposures, EPA undertook broad, screening studies in the 1980's to evaluate the releases of these pollutants and the relative implication of the resulting exposures to human health.

One study, entitled "Cancer Risks from Outdoor Exposure to Air Toxics"

(Appendix B), assessed the magnitude and nature of potential cancer risks associated with exposure to hazardous air pollutants. Originally conducted in 1985 and updated in 1990, the work broadly assessed long-term exposures to HAPs and estimated potential cancer risks associated with these pollutants. The results of the updated analysis estimated an increase of cancer cases to be between 1700 and 2700 per year as a result of HAP exposure. Approximately 40 percent of these cases were associated with emissions from stationary sources versus mobile sources. In addition, maximum individual cancer risks were estimated to be in excess of 1 in 1,000 at several locations.

II.C.2 Hazard Assessment

A hazard assessment as defined by EPA guidelines is an evaluation of a chemical's toxicity and potential to cause adverse health and environmental effects. At minimum, it entails a search of the scientific literature and an assessment of the amount and quality of the data including the availability of dose-response data.

A qualitative assessment of data includes evaluation of available human, animal, and in vitro evidence in determining how likely a chemical is to elicit an adverse effect in humans or other exposed populations of interest. This type of information is generally examined within the framework of a weight-of-evidence classification scheme.

If sufficient quantitative data are available, a dose-response assessment may be conducted. For carcinogens, the Agency has traditionally developed unit risk estimates (UREs) to express the relationship between dose and carcinogenic response. An URE, under assumption of low-dose linearity, is an estimate of the excess, lifetime risk due to continuous exposure to one unit of concentration (e.g., ug/m^3 for inhalation). For noncarcinogens, limited data and risk assessment methods allowed only the identification of effect levels rather than a quantitative expression of the data.

In addition to toxicity data, other information that is typically included in a hazard assessment include data on a chemical's environmental fate, transport, or persistence in the environment. If the data are sufficient, a hazard assessment presents a profile of a chemical's toxicity, potential health and environmental risk, and related chemical characteristics. In practice, this is best exemplified by HADs (see discussion in Section II.B.2). The HADs incorporate all of the information listed above. These documents also undergo a peer-review by EPA's Science Advisory Board (refer back to Figure 4). This type of assessment formed the principal basis for decisions to list chemicals as HAPs under the previous Section 112.

II.C.3 Hazard Ranking

There are no past examples of hazard ranking. Rankings that were done used emission data to rank rather than toxicity data which, for the most part, lacked sufficient potency data to do adequate ranking.

II.C.4 Risk Ranking

Figure 6 illustrates the process used to identify HAPs prior to passage of the Clean Air Act Amendments of 1990. During the mid-1980's, the Agency modified this process to add in the "Intent-to-List" procedure prior to actually listing a chemical under Section 112 of the CAA. Table 2 identifies the pollutants that EPA formally evaluated during this time frame and the resulting decision to continue analysis (intent-to-list) or discontinue analysis (not-to-regulate). Examples of the notices published in the *Federal Register* are included in Appendix C.

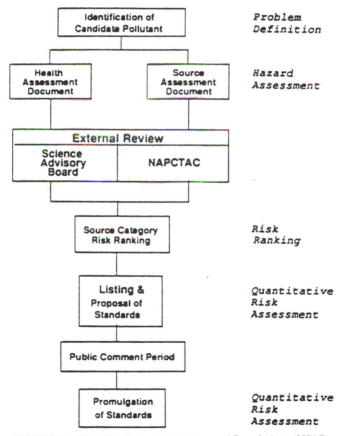

FIGURE 6   Identification, Assessment, and Regulation of HAPs.

TABLE 2 1984-1987 Hazardous Air Pollutant Decisions

POLLUTANT	ACTION	CITATION
Acrylonitrile	State Referral	50FR24319; June 10, 1985
1,3-Butadiene	Intent-to-List	50FR41466; October 10, 1985
Cadmium	Intent-to-List	50FR42000; October 16, 1985
Carbon Tetrachloride	Intent-to-List	50FR32621; August 13, 1985
Chlorofluorocarbon 113	Not-to-Regulate	50FR24313; June 10, 1985
Chlorinated Benzenes	Not-to-Regulate	50FR32628; August 13, 1985
Chloroform	Intent-to-List	50FR39626; September 27, 1985
Chloroprene	Not-to-Regulate	50FR39632; September 27, 1985
Chromium	Intent-to-List	50FR24317; June 10, 1985
Coke Oven Emissions	Listing Notice	49FR36560; September 18, 1984
Copper	Not-to-Regulate	52FR5496; February 23, 1987
Epichlorohydrin	Not-to-Regulate	50FR24575; June 11, 1985
Ethylene Dichloride	Intent-to-List	50FR41994; October 16, 1985
Ethylene Oxide	Intent-to-List	50FR40286; October 2, 1985
Hexachlorocyclopentadiene	Not-to-Regulate	50FR40154; October 1, 1985
Manganese	Not-to-Regulate	50FR32627; August 13, 1985
Methyl Chloroform	Not-to-Regulate	50FR24314; June 10, 1985
Methylene Chloride	Intent-to-List	50FR42037; October 17, 1985
Municipal Waste Combustion Emissions	Advance Notice of Proposed Rule-making	52FR25399; July 7, 1987
Naphthalene	Not-to-Regulate	53FR9138; March 1, 1988
Nickel	Not-to-Regulate	51FR34135; September 25, 1986
Perchloroethylene	Intent-to-List	50FR52880; December 26, 1985 *51FR7719; March 5, 1986
Phenol	Not-to-Regulate	51FR22854; June 23, 1986
Polycyclic Organic Matter	Not-to-Regulate	49FR31680; August 8, 1984
Toluene	Not-to-Regulate	49FR22195; May 25, 1984
Trichloroethylene	Intent-to-List	50FR52422; December 23, 1985 *51FR7714; March 5, 1986
Vinylidene Chloride	Not-to-Regulate	50FR32632; August 13, 1985
Zinc/Zinc Oxide	Not-to-Regulate	52FR32597; August 28, 1987

*Clarification Notice

II.C.5 Quantitative Risk Assessment: The Regulation of Benzene: 1977-1989

> Note: The following sections present an overview of the evolution of risk-based decision-making under the old Section 112, using the regulation of benzene as the principal example. The text is supplemented by several appendices that provide examples of decision documentation and briefing materials from these periods.

Introduction

In June of 1977, EPA added benzene to the list of HAPs under Section 112. For the next twelve years, under a succession of 6 Administrators, the air program wrestled with the regulation of a known human carcinogen for which a health effect threshold could not be established, under an authority requiring the protection of public health with an ample margin of safety. During this period, benzene became the test case for a series of procedural interpretations and re-interpretations of the statutory language, culminating in the 1987 vinyl chloride opinion by the D.C. Court of Appeals (NRDC v. U.S. EPA, July 28, 1987) and the revision of the statute in the 1990 amendments to the Clean Air Act.

The regulation of benzene also spans a period during which the methods for quantitatively estimating risks from exposure to airborne carcinogens evolved, and the appropriate role of such estimates in the decision process was hotly debated within, as well as outside, the EPA. For these reasons, benzene represents an interesting and illuminating case study of quantitative risk assessment and its use in determining the appropriate level of control under Section 112. A chronology of EPA's regulatory policy under Section 112 is summarized in Figure 7.

Benzene and the Airborne Carcinogen Policy (1977-1983)

The EPA listed benzene as a HAP in 1977 based on growing evidence of a link between occupational exposure and an increase in the incidence of acute myelogenous leukemia (Appendix D - Benzene Listing). Prior to the listing of benzene, EPA had regulated four pollutants under Section 112: asbestos, beryllium, and mercury in 1971; and vinyl chloride in 1974-75. In the absence of procedures for estimating cancer risk, the original asbestos standard was based on "no visible emissions". Beryllium had not been identified as a carcinogen (berylliosis was the effect of concern) and the toxic effects of mercury were addressed with an ambient air guideline, taking into consideration exposure by other routes (e.g., ingestion).

Era		Approach
1971	Asbestos	"No Visible Emissions"
	Beryllium	Best Technology
	Mercury	Ambient Guideline
1974-75	Vinyl Chloride	Best Available Technology (BAT)
1977-81	Benzene/Carcinogen Policy	BAT/Beyond BAT
1983-84	Risk Management	Weigh All Factors
1987	Vinyl Chloride Opinion	"Safe"/Ample Margin Of Safety
1988	Benzene Proposal	"Framing the Debate"
1989	Benzene Promulgation	"Fuzzy Bright Line"
1990	CAA Amendments	MACT Now/Residual Risk Later

FIGURE 7   Chronology of Section 112 Regulatory Policy Development.

By the listing of vinyl chloride in 1974, quantitative techniques were under development within the EPA. In conjunction with the promulgation of the vinyl chloride emission standards, rough estimates of projected incidence of angiosarcoma were made, but were considered too uncertain to be used in the determination. The vinyl chloride standards were principally based on the application of the best available control technology (BAT).

In May of 1976, EPA issued the first carcinogenicity guidelines (Appendix E - "Health Risk and Economic Impact Assessments of Suspected Carcinogens"). In the benzene listing notice the following year, EPA announced the conduct of a benzene health risk assessment and indicated that the "relative risk to the public" would be considered in judging "the degree of control which can and should be required". The risk assessment, containing the original unit risk estimate for benzene, was subsequently published in January 1979 (Appendix F - Benzene Population Risk).

The advent of a quantitative methodology and external pressure for a more aggressive program under Section 112 led to the development of EPA's airborne carcinogen policy. The policy was published in October 1979, as a proposed interpretive rule outlining procedures for the identification, assessment, and regulation of airborne carcinogens emitted from stationary sources (Appendix G - Airborne Carcinogen Policy). The policy reflected a technology-based approach to emission standard development with a limited role for quantitative risk assessment in establishing priorities and ensuring that the residual risks following the application of BAT were not unreasonable. The first round of benzene standards, beginning with the regulation of maleic anhydride plants in 1980, followed the proposed procedures, the shorthand for which became "BAT/Beyond BAT". Although a final version of the proposed policy was prepared in 1981, incorporating public comments, the policy was never promulgated. The procedures were informally followed, however, up to the introduction of the "risk management" approach in 1983.

Also in 1979, the development of the Human Exposure Model (HEM) (Appendix H - HEM Description) provided a means of estimating and summing ambient exposures across the populations living in the vicinity of emitting sources. These estimates were then combined with the unit risk estimate to yield cancer risk estimates. In the first benzene standards, estimates of maximum individual lifetime risk and annual incidence were calculated. The risk estimates were sometimes displayed as small ranges, incorporating some of the quantifiable sources of uncertainty. Other uncertainties were usually presented as tabular footnotes (Appendix I - Proposed Maleic Anhydride Standards) .

The Risk Management Era (1983-1985)

The change of Administrations in 1981 brought an increasing emphasis on the cost-effectiveness of regulation and regulatory reform. In this light, the presumption expressed in the proposed carcinogen policy - that, given the uncertainty in risk estimation, significant source categories of airborne carcinogens should be regulated, at a minimum, to a level of control constituting BAT - was called into question. The re-examination of this presumption resulted in a revised policy which held that risk information, as well as other relevant factors, should be considered in determining the appropriate level of control, including finding that control was unwarranted. One result of this change was to place greater weight on the risk assessment in the decision process.

In 1984, after "weighing all factors", EPA made several changes to the proposed benzene rules, including withdrawal of the maleic anhydride proposal, arguing that the risks were "too small to warrant Federal regulatory action" (Appendix J - Withdrawal of Proposed Standards). These decisions were promptly challenged by the NRDC, arguing the uncertainties in the risk estimates and the inappropriate consideration of cost in regulatory decisions made under Section 112. The issues raised were similar to litigation already pending on amendments to the original vinyl chloride standards.

Also during 1984, work was begun to revise the benzene unit risk estimate, based on new human and animal data and an improved methodology. A revised estimate was transmitted to the air program by the Office of Research and Development in early 1985 (Appendix K - "Interim Quantitative Unit Risk Estimates").

The Vinyl Chloride Opinion (1987)

On July 28, 1987, Judge Robert Bork, writing for the D.C. Circuit Court of Appeals, remanded the vinyl chloride amendments to EPA, finding that the Agency had placed too great an emphasis on technical feasibility and cost rather than the provision of an "ample margin of safety" as required by the statute (Appendix L - Vinyl Chloride Opinion). The opinion also laid out a process for making decisions, consistent with the requirements of the law.

The Bork opinion held that, in setting standards under Section 112, EPA must first determine a "safe" or "acceptable" level, and that this level must be established considering only the potential health impacts of the pollutant. Once an acceptable level was identified, the level could be reduced further, as appropriate and in consideration of other factors, including cost, technical feasibility, affordability, etc., to provide the required ample margin of safety. The Court also held, however, that "safe" did not require a finding of "risk-free" and that EPA should recognize that activities such as "driving a car or breathing city air" may not be considered "unsafe".

Benzene Proposal (1988)

The EPA accepted voluntary remand of the 1984-85 standards and issued a new proposal in July 1988, consistent with the vinyl chloride opinion. Given the requirement for a determination of "safe", the importance of the quantitative risk assessment took on even greater emphasis. This is evident in the senior management briefings on the proposal (Appendix M - Briefing for the Administrator). The determination of a "safe" or "acceptable risk" level continued to be problematic, however, in part due to the diversity of opinion within, and external to, the Agency on what constituted an "acceptable risk" but, also to the dicta of the legal opinion itself. The decision appeared to accept "driving a car or breathing city air" as examples of activities judged to be safe by society. This raises the issue of whether society's judgment to drive or live in cities is founded solely on the possible health impacts of these activities, rather than a consideration of all factors, which would be prohibited in the EPA framework.

Several options for the determination of "acceptable" risk were considered in the months preceding proposal. The preferred option, a case-by-case consideration of all of the relevant health information was described in a memorandum by the Administrator (Appendix N - "Proposed Benzene NESHAP Decisions").

Ultimately, however, EPA proposed four options for the determination of "safe", "framing the debate" for public comment (Appendix O - Proposed Benzene NESHAP). With the exception of the case-by-case alternative, the options represented "bright line" risk targets, either individual or population risk. All factors were to be considered in the determination of the ample margin of safety.

Benzene Promulgation (1989)

The EPA received a large volume of comments on the proposed rules. Again, the risk methodology and estimates, and the proposed acceptability criteria were extensively discussed. The appended briefing for the Assistant Administrator (Appendix P - "Consideration of Comments") illustrates the emphasis on the risk methods and underlying uncertainties. During this time, there was also increased interest in not only the estimates of maximum individual and population risk, but also the distribution of individual risk across the exposed population.

In September of 1989, EPA promulgated emission standards for several categories of benzene sources (Appendix Q - Final Benzene Rules). The decision criteria adopted represented a blend of several of the proposed options. The EPA argued for the consideration of all relevant health information and established "presumptive benchmarks" for risks that would be deemed "acceptable". The goal, which came to be known as the "fuzzy bright line", held that risks would be deemed acceptable if few, if any, individuals were exposed above a 1 in 10,000 lifetime cancer risk, and, as much of the exposed population as possible was below a lifetime risk of 1 in 1,000,000.

The selection of even "fuzzy" risk targets placed greater emphasis on the development and communication of risk characterization results. For the final benzene rules, this was evident in the decision briefings as well as the development of question and answer materials (Appendix R - Benzene Questions and Answers) and the decision to provide advance briefings for the news media (Appendix S - Background Information for the Media).

The Clean Air Act Amendments (1990)

The amendments to Section 112 require the application of technology-based standards to major and designated area source categories as a first step. Following compliance with the maximum achievable control technology (MACT) standards, EPA is required to evaluate residual risks, applying the decision criteria used in the final benzene rules, to determine whether the technology-based rules provide an ample margin of safety to protect public health. Risk assessment will continue to play an important role in the implementation of this and other provisions of Section 112 and the importance of appropriate methodologies and characterization of uncertainties cannot be understated.

II.D Examples of Present Assessments

II.D.1 Problem Definition

Sections 112(c) and (k) of Title III prescribe an *Urban Area Source* program that includes the development of a national strategy requiring 75% or more reduction in cancer incidence associated with emissions of 30 or more HAPs that "present the greatest threat to public health in the largest number of urban areas". For this national strategy to be implemented, many issues need to be defined and addressed including:

- types of sources covered
- selection of the urban areas covered
- selection of the 30 or more HAPs to be regulated on a variety of end-points and characterization of their ambient levels
- characterization of the emission release parameters
- establishment of an emissions inventory system to help demonstrate that the goals of the strategy are being met
- role of atmospheric transformation

This program requires policy decisions as well as research decisions to be made so that the goal of listing the sources in 1995 and promulgating the subsequent standards for affected sources can be met.

The *Great Waters Study* (Section 112[m]), requires that EPA, in cooperation with the National Oceanic and Atmospheric Administration, identify and assess the extent of atmospheric deposition of HAPs to the Great Lakes, Chesapeake Bay, Lake Champlain, and coastal waters. A report to Congress is due within 3 years of enactment and biennially thereafter. A plan is being developed to evaluate the information available, the information needed, and how to acquire that additional information. The Report to Congress requires the following information:

- contribution of atmospheric deposition to total pollution loading
- environmental and public health effects
- sources of the pollutants
- contribution of HAPs to water quality violations

To accomplish this, it will be necessary to:

- conduct atmospheric deposition monitoring for source identification and model validation
- conduct atmospheric transport and deposition modeling to include direct and indirect pathways
- develop emission inventories as input to models
- evaluate adverse effects of air toxics on public health and the environment

The Great Waters work shall also support data sharing and the development of remedial action plans (RAPs) and lakewide management plans (LaMPs). The final results of this study may be the promulgation of further emission standards or control measures as may be necessary and appropriate to prevent the adverse effects from occurring.

II.D.2 Hazard Assessment

With passage of the new CAA, the emphasis on hazard assessment changed from the generation of HADs to the generation of dose-response or potency based estimates where the data supported such an analysis. In selecting appropriated toxicity information, the data required for the statutory findings and mandated deadlines were considered including information on cancer and noncancer effects. For carcinogenic risks, emphasis to date has focused on existing quantitative assessments including unit risk estimates (UREs) (see discussion in Section II.C.2) and ED_{10}s.

The assessment of ambient concentrations of HAPs in relation to their potential to elicit adverse noncancer effects presents several challenges. Considerations must include: evaluation of short-term as well as long-term exposures, incorporation of severity-of-effect data, and consideration of reversible versus irreversible effects. The endpoints that may be of most concern could include respiratory effects, developmental/reproductive toxicity, and neurotoxicity.

The quantification of noncancer risks from exposure to inhaled hazardous air pollutants currently focuses on the derivation of inhalation reference concentrations (RfCs). The RfC is defined as an estimate (with uncertainty) of the concentration that is likely to be without appreciable risk of deleterious effects to the exposed population after continuous, lifetime exposure. The RfC focus is on the most sensitive members of the population who may be exposed and the respiratory system as the portal of entry. An experimental exposure level representing the highest level tested at which no adverse effect was observed (NOAEL) is selected from a given study and converted to a human equivalent concentration ($NOAEL_{HEC}$). The critical toxic effect used is the one generally characterized by the lowest NOAEL. This approach is based on the assumption that if the critical toxic effect is prevented, then all toxic effects are prevented. The RfC is derived from the ($NOAEL_{HEC}$) by the application of uncertainty factors to account for extrapolations that may be made. These estimates along with UREs are reviewed within the Agency before incorporation onto EPA's Integrated Risk Information System (IRIS).

Under Title IX of the CAA, EPA is required to develop environmental health assessments for the HAPs. In addition to hazard assessment information, these profiles are to identify data gaps and, where appropriate, identify the additional activities needed to better characterize the "types or levels of exposure which may present significant risk of adverse effects in humans."

Note: Information concerning revisions to the current EPA's cancer risk assessment guidelines will be provided under separate cover.

II.D.3 Hazard Ranking

A further step in the assessment process is the ranking of HAPs based on their relative hazard to human health. The data needs to be collected in a form which allows the comparison of chemical hazards, e.g. comparing similar endpoints of concern. Ideally, the ranking would rely on Agency-reviewed, benchmark risk values such as UREs or RfCs. In reality, due to the lack of health data, the ranking of chemicals may have to rely on less rigorously reviewed values and many assumptions or defaults. The ranking of HAPs for the purpose of offsets under the modifications section of the CAA (Section 112(g)) provides an example of one approach EPA has taken. This section of the CAA requires that EPA issue guidance which includes the ranking of threshold and non-threshold pollutants. Without sufficient data to the contrary, the EPA currently considers all noncarcinogens as threshold pollutants and carcinogens as non-threshold pollutants. As data become available, this general categorization may change for specific pollutants.

The ranking methodology currently being considered under Section 112(g) uses methods already in place, i.e. for establishing Reportable Quantities under the Comprehensive Emergency Response and Compensation Liability Act (CERCLA). Non-threshold pollutants (carcinogens) are ranked by comparing potency estimates ($1/ED_{10}$) and weight of evidence classification. The ED_{10}s are defined as the estimated dose associated with a lifetime increased cancer risk of 10%. Threshold pollutants are ranked by either their composite scores (CS) which reflect chronic toxicity, or their level of concern which reflects acute toxicity. Composite scores consider dose-response and severity of effect. The magnitude of the CS determines the ranking position of the chemical (pollutants with large composite scores elicit severe effects at low doses). Under section 112(g), increases in emissions of non-threshold pollutants cannot be offset by decreases in emissions of threshold pollutants, but the reverse is true. The ranking must provide a comparison of the relative hazard within categories of non-threshold and threshold compounds. It is also known that certain pollutants may cause severe effects resulting from acute exposures, therefore the guidance also provides a category for "high concern" threshold pollutants. These pollutants are considered (for the purposes of this section) more hazardous than threshold pollutants but no comparison can be made between these and non-threshold pollutants. If pollutants do not have adequate data to be ranked as a threshold, non-threshold, or "high concern" pollutant, then that pollutant is considered not tradeable under this section. The general methodology that has been developed to date was reviewed by the Science Advisory Board and National Air Pollution Control Techniques Advisory Committee (NAPCTAC) while the application underwent rounds of internal review and public comment.

II.D.4 Risk Ranking

A number of ongoing activities in Title III are associated with risk ranking, or have risk ranking as one of their components. Under the *Source Category Schedule* development program, the schedule for regulation of the listed source categories due to be published for comment this year, has been primarily based on a risk ranking of the various source categories included on the Section 112(c) list. This ranking process uses the Source Category Ranking System (SCRS), a methodology developed within OAQPS. The SCRS process uses health information, available or estimated emissions data, and population data to develop a numerical score for each category on the list. The scores are then ranked to develop a prioritized list. In general, the SCRS first develops a health score for each pollutant emitted by a source category. The health score for each pollutant is based on available data regarding carcinogenicity, reproductive toxicity, acute lethality, and other toxicity. The SCRS then develops an exposure score for each pollutant emitted by that source category. The exposure score is based on concentration approximations for each pollutant from each facility in the category combined with estimates of the numbers of people exposed to these concentration estimates. General assumptions concerning plant stack parameters, plant boundaries, population densities, and meteorological conditions are made on a category-wide basis to simplify the ranking process. Default assumptions and mass balance emission estimates are used where data are unavailable.

The end result of the SCRS process is not an estimate of risk, but rather a score which indicates the relative magnitude of risks between source categories. This score, along with other factors such as efficiency of grouping like sources for a particular regulation, availability of control technology information, and the specific nature of adverse health effects associated with a source category, is then used to assist in the scheduling of regulations.

The *Lesser Quantity Emission Rate* project is an example of a risk ranking assessment because of the use of exposure assessment and data on health effects. Title III (Section 112(a)) allows the Administrator to establish emission rates for less than 10 and 25 tons/year for HAPs based on their potency, persistence, potential for bioaccumulation, or other relevant factors. The HAPs with UREs and classified as a known, probable, or possible human carcinogen were initially selected. Added to these were chemicals of high concern under CERCLA. Non-carcinogens were selected on the basis of their inhalation RfC, RfD, LC50, or LOEL. Using standard parameters, a generic exposure modeling was done, including consideration of likely exposure duration. This modeling analysis yielded an estimated ambient concentration at a distance selected to represent the nearest residence. This ambient concentration was compared to cancer UREs or noncancer benchmarks, and HAPs of concern were identified. Lesser quantity emission rates (LQER) were assigned to selected carcinogens based upon order-of-magnitude changes in their potencies. The range of LQERs that resulted was .0001 to 1 ton/year. Selected noncarcinogens were assigned LQERs based on a comparison of the benchmark concentrations with the estimated ambient concentration. The major consequence of this analysis would be a redefinition of some sources as major sources if their emission rates of HAPs exceed the assigned LQER.

II.D.5 Quantitative Risk Assessment

With regard to quantitative risk assessment activities, two current CAA-related activities address the use of refined modeling techniques with site-specific data to quantify risks associated with both long- and short-term exposure to hazardous air pollutants from stationary sources.

Source Category Deletion Petition Process

Under Section 112(c), a source category may be deleted from the list of source categories subject to regulation via a petition process if a petition demonstrates, for the case of carcinogenic pollutants, that "no source in the category ... emits (carcinogenic) air pollutants in quantities which may cause a lifetime risk of cancer greater than one in one million to the individual in the population who is most exposed to emissions of such pollutants from the source," and, for the case of noncarcinogenic yet toxic pollutants, that "emissions from no source in the category ... exceed a level which is adequate to protect public health with an ample margin of safety and no adverse environmental effect will result from emissions from any source."

In support of the petition process, EPA is developing guidance for petitioners which suggests acceptable methodologies for assessing cancer and noncancer risk associated with sources of HAPs. This guidance references a document describing a tiered modeling approach for the estimation of maximum risks (see Appendix T - Tiered Modeling Approach). The tiered approach begins with a screening methodology which is used to identify facilities within a source category that do not present risks significant enough to warrant more refined analysis. The screening methodology uses minimal site-specific data (pollutant emission rates, stack heights, and minimum fenceline distances) in this assessment, and, as such, the results are very conservative. Facilities not screened out in this first tier are subjected to a more refined "Tier 2" assessment requiring additional site-specific information (stack diameters, exit velocities, exit temperatures, rural/urban classification, nearest building dimensions) concerning each modeled facility. The third modeling tier requires the most site-specific data (release point and fenceline locations, local meteorological data, release durations and annual frequencies) to provide the most refined estimate of risks due to each modeled source.

The analyses described above focus on the maximum risks presented by a facility outside its plant boundary, regardless of how many people are subjected to those risk levels. To the extent that population location and distribution data are available, they may be incorporated in the analysis on a case-by-case basis, to provide a more accurate estimate of the risk to the maximum exposed individual.

Residual Risk Evaluation

Under Section 112(f) of the CAA, EPA is required to assess the risks associated with a regulated source category within 8 years of the MACT standard promulgated for that category. The Agency is currently evaluating options for implementing this provision. In investigating various alternatives, many questions have been raised. The EPA is currently exploring many technical and policy issues. These issues must be addressed prior to establishing an implementation strategy for evaluating residual risks.

In fully characterizing the potential risks associated with emissions of HAPs following compliance with the MACT standards, EPA is evaluating the capabilities of current risk assessment methods. Presently, due to limited availability of data and methods, it is difficult to quantitatively characterize specific risks (e.g., noncancr risks). The EPA is evaluating various methods to collect additional effects data (see response to Question 4) as well as exploring the development of new methods and the modification of existing methods to improve the ability to quantify risks. Specific areas that are being explored include: evaluation of less-than-lifetime exposures, incorporation of severity-of-effect data, incorporation of data on reversibility (or irreversibility) of effects, and development of physiologically based pharmacokinetic and biologically based dose-response models.

Currently, it is envisioned that a tiered modeling approach (such as described above in the discussion of the source category deletion process) may be the basis for dispersion modeling associated with residual risk analyses of source categories. The EPA envisions that site-specific emission estimates may "drive"

TABLE 3 Data Sources for Exposure Assessment

DATA TYPES	SCREENING ANALYSES	SITE-SPECIFIC ASSESSMENTS
EMISSIONS AND RELEASE PARAMETERS (STACK HEIGHT, STACK GAS EXIT VELOCITIES)	Engineering estimates. Assume all emissions are emitted from plant center. Use of model plants (e.g., stack heights that typify the emission source). Use of emission factors (AP-42).	Data from industry via Section 114 CAA. Data from source testing, trade associations. Use site plans to place emission where they occur on plant property. Data from permits and plant site visits.
PLANT LOCATIONS (LATITUDE, LONGITUDE)	Some info from EPA data bases (AIRS/NEDS), other sources. Some plants sited by land use or at random.	From industry, topographic maps, site visits.
METEOROLOGY	National Climatic Center data from nearest airport (multiple years averaged to yield annual data sets).	Data collected on site. National Climatic Center data from nearest airport if similar to site.
POPULATION	Latest U.S. Census data. Block Group/Enumeration Districts, 300,000 data values.	Same. 1990 data will have 6,000,000 data values.
WHERE AMBIENT AIR BEGINS	At 200 meters from assumed plant center (approximately 30 acres).	Use actual plant size or some approximation.
EXPOSURE DURATION	Equal to that of the health information. If effects occur at one hour exposure, predict one hour concentrations/exposures. For cancer, estimate annual average values that are assumed to persist 70 years (the averaging time of unit risk factor).	Same.
EXPOSURE	Assume indoor concentrations same as outdoor.	Same.

the risk assessment process. Thus, EPA is planning its efforts to expand the available emission measurement methods and validation procedures (validated measurement methods currently exist for only about 15-20% of the listed HAPs). In addition, EPA believes that efforts should be extended to continue to improve available emission calculation methods (emission factors, surface impoundment emission estimation methods, etc.). To assist the process of obtaining sufficient site-specific data for quantitative risk analyses, EPA is investigating options for developing a user-access data entry system. Such a system would necessarily be designed to ease the burden of providing up-to-date data to EPA and to protect against the unauthorized access of proprietary information. Logistics, reporting requirements, and quality assurance associated with such a system are problems with no adequate answers at this point.

The EPA is also looking into improving risk assessments by factoring in more realistic approaches to exposure assessments including consideration of population mobility, population sensitivity, activity patterns, and indoor/outdoor exposures. Because of the intensive data requirements for addressing these factors adequately, sensitivity studies are being considered to assess the ranges of uncertainty induced by each of these factors on the predicted exposures and risks. The results from such studies would hopefully allow a more representative characterization of the distribution of risks among the exposed population in the future.

II.D.6 Evolution of Exposure and Risk Assessment

As previously mentioned, the role and scope of exposure assessments in the air toxics program is changing. Exposure estimates were conducted for two main purposes: 1) to estimate high end and population exposure to a candidate hazardous air pollutant, and 2) to evaluate the effectiveness of various air pollution control alternatives for reducing potential exposure and risk. Table 3 presents data sources and assumptions that were generally used in previous exposure assessments. The source category deletion and residual risk evaluation provisions in Title III place a much greater focus on source and individual exposures associated with an often complex mixture of source types and pollutants.

Procedures that the Agency develops for addressing residual risk will be designed to meet several criteria. State and local air pollution control agencies, affected industries, and private individuals may require access to and familiarity with available models. In addition, the procedure should be able to evaluate present and future control options as interested parties may wish to evaluate residual risk before air pollution control equipment are ordered.

As noted above, OAQPS is currently examining and developing improved techniques for conducting exposure assessments. Although these improvements will continue to chiefly rely on predictive methods (modeling), measured data, available from monitored levels or reconstructed from measurement of biological fluids and tissues, will remain an important source of information for validation and characterization purposes. The Agency will focus the improvements in three main areas:

1) Developing user-friendly models to enable diverse, interested parties to understand and operate the models if they choose. Data input and selection of specific models will be accomplished by menu screens that contain data checks.

2) Addition of Monte Carlo techniques to permit the representation of those parameters that greatly affect the exposure/risk estimates by distributions rather than point estimates (see Appendix U, Monte Carlo Approach).

3) A geographical information system (GIS) will be integrated with the models to improve the predicted ambient concentrations by incorporation of topography and land use information to aid in selection of appropriate meteorological data and the location of area source categories. In addition, GIS will allow OAQPS to more accurately locate areas where people may reside than is currently possible using U.S. Bureau of Census data alone (See Appendix V, GIS - Application to Exposure Assessment).

The HEM input parameters that can presently be described by distributions include:

- emission rates
- microenvironment concentrations
- time spent in each microenvironment
- information on the length of time people are expected to reside in their primary residences
- the ability to vary the location of the predicted ambient concentrations.

The EPA/OAQPS is also developing a separate model (the Hazardous Air Pollutant Exposure Model [HAPEM]) that examines the impact on exposure of population mobility (e.g., commuting) (see Appendix W, HAPEM - Mobility Considerations).

Since the process of conducting residual risk analyses for all regulated sources of HAPs is anticipated to be a resource-intensive process, the analytical methodology has evolved into a tiered approach, as mentioned above. This differs from most risk assessments performed in the past in that it allows for the incorporation of site-specific data where possible to refine the estimates of population exposure and risk. Since it will likely be difficult for EPA to require all regulated facilities to provide all of the necessary data for such site-specific analyses, EPA has plans to develop a voluntary data storage and retrieval system, whereby such facilities may provide site-specific data to EPA to facilitate the more rigorous risk assessment process. This will not only help EPA to perform residual risk analyses in a more efficient manner, but it will reduce the level of "unnecessary" conservatism associated with the risk assessment process. In situations where EPA does not have site-specific modeling parameters, the risk assessments will be performed at the Tier 1 level, consistent with risk assessments of the past. In situations where additional data have been provided by the facilities being analyzed, risk assessments will be more realistic, and risk estimates will generally be lower (sometimes by orders of magnitude).

Table 4 below summarizes the major differences between the 3 modeling tiers discussed above by briefly listing the input requirements, the major output parameters, and the assumptions associated with each tier. This table may be used to quickly determine whether a given scenario may be modeled at any particular tier based on available site-specific data. Within each tier, cancer unit risk estimates, chronic noncancer concentration thresholds, and acute concentration thresholds are required to convert concentration predictions into cancer risks, chronic noncancer risks, and acute noncancer risks, respectively.

In general, to perform a site-specific exposure assessment, Tiers 1 and 2 could be used to screen facilities with low risk estimates from further analysis at a higher Tier. In cases where facility-specific data are lacking, emissions estimates could be made using a model plant approach with emission factors or process engineering estimates of emissions. In such cases, all known or estimated emissions could be assumed to emanate from a single, typical stack at the plant center, and the plant could be assumed to have a circular boundary, 200 meters from the plant center. It is anticipated that plant location data (latitude and longitude) will be obtained from EPA permits, and this would allow predicted ambient concentration levels to be compared to potentially-exposed populations through the use of U.S. Census Bureau data. It is also anticipated that more rigorous analyses to provide the distribution of risks among exposed population would be performed where site data are sufficient to support such analyses.

The major influence of the guidelines on exposure assessments is in the quantification of uncertainty. The HEM is being redesigned to explicitly address uncertainty quantitatively where possible. A discussion of risk characterization and attempts to describe and communicate uncertainty was presented previously in Section II.B.4.

TABLE 4 Exposure Modeling Parameters

Modeling Tier	Input Requirements	Output Parameters	Major Assumptions
Tier 1	emission rate, stack height, minimum distance to fenceline	maximum off-site concentrations, worst-case cancer risk or worst-case noncancer hazard index (short- and long-term)	worst-case meteorology, worst-case downwash, worst-case stack parameters, short-term releases occur simultaneously, maximum impacts co-located, cancer risks additive, noncancer risks additive
Tier 2	emission rate, stack height, minimum distance to fenceline, stack velocity, stack temperature, stack diameter, rural/urban site classification, building dimensions for downwash calculation	maximum offsite concentrations, worst-case cancer risk and/or worst-case noncancer hazard index (short- and long-term)	worst-case meteorology, short-term releases occur simultaneously, maximum impacts co-located, cancer risks additive, noncancer risks additive
Tier 3	emission rate, stack height, actual fenceline and release point locations, stack velocity, stack temperature, stack diameter, rural/urban site classification, local meteorological data, receptor locations for concentration predictions, frequency and duration of short-term (intermittent) releases	concentrations at each receptor point, long-term cancer risk estimates, chronic noncancer hazard index estimates at each receptor point, annual hazard index exceedance rate at each receptor	cancer risks additive, noncancer risks additive

III. Question 3: What HAPs data are available now to implement the current risk assessment methodology?

III.A Introduction

The EPA has compiled currently available data on the hazardous air pollutants (HAPs) in developing strategies for implementing various provisions contained in Title III of the Clean Air Act. These data include: information on the schedule for control technology-based standards, recent annual air emissions data, preliminary estimates of the number of facilities that emit HAPs, and health effects information.

III.B Summaries of Available Data

Table 5 is an summary of the currently available health data (this is an updated version of the Table previously provided), taken from Table 6.

TABLE 5   Summary of Health Effects Data (November 1, 1993)[1]

Status	Cancer	Noncancer
Verified RfC		
On IRIS		40
Not on IRIS		2
Reviewed, not verifiable	58	
WOE and IUR	39	
WOE and OUR	14	
WOE Only	35	
Under review[2]	11	23
No status	87	63
Total HAPs	186	186

[1] Does not include lead, radionuclides, or glycol ethers

[2] Under review by Environmental Criteria and Assessment Office or Human Health Assessment Group for derivation of RfC or URE followed by verification review by RfC/RfD and CRAVE work groups before entering data onto IRIS

RfC: Inhalation reference concentration
WOE: Weight-of-evidence, includes A to D class.
IUR: Inhalation unit risk estimate
OUR: Oral unit risk estimate

TABLE 6 Current Data on the HAPs

CAS #	Chemical Name	M Y 2	A 4	C 7	T 10 (1)	1991 Emis (T/yr) 2	1990 Emis (T/yr) 2	1989 Emis (T/yr) 2	1988 Emis (T/yr) 2
	Column Number								
		#SC							
79345	1,1,2,2-Tetrachloroethane	2	X		X	32.1	22.3	17.7	22.9
79005	1,1,2-Trichloroethane	2	X		X	263.9	299.4	398.4	870.
57147	1,1-Dimethyl hydrazine	1	X			0.2	0.2	0.4	2.2
120821	1,2,4-Trichlorobenzene	1	X			204.8	188.4	575.7	760.
96128	1,2-Dibromo-3-chloropropane					0.1			
122667	1,2-Diphenylhydrazine	1	X						
106887	1,2-Epoxybutane (1,2-Butylene oxide)					29.9	39.7	59.8	54.1
75558	1,2-Propylenimine (2-Methyl aziridine)					0.2	0.3	0.3	0.3
106990	1,3-Butadiene	19	X	X	X	X	1975.2	2518.8	2768.7	326
542756	1,3-Dichloropropene (Telone II)	1	X			10.2	29.7	25.5	26.2
1120714	1,3-Propane sultone								
106467	1,4-Dichlorobenzene(p)	1	X			168.1	409.1	793.6	904.
123911	1,4-Dioxane (1,4-Diethyleneoxide)	1	X			359.3	299.2	390.1	270.
540841	2,2,4-Trimethylpentane	3	X	X	X				
1746016	2,3,7,8-Tetrachlorodibenzo-p-dioxin		X	X					
95954	2,4,5-Trichlorophenol	1	X					0.1	
88062	2,4,6-Trichlorophenol					0.0	0.1	0.1	
94757	2,4-D, salts,esters								
	(Dichlorophenoxyacetic acid)	1			X	8.1	3.9	3.6	3.5
51285	2,4-Dinitrophenol	1	X			12.1	12.3	6.8	10.4
121142	2,4-Dinitrotoluene	1	X			2.7	28.8	43.6	46.1
95807	2,4-Toluene diamine	1	X			1.9	2.0	2.2	1.5
584849	2,4-Toluene diisocyanate	2	X	X		661.9	28.7	61.3	113
53963	2-Acetylaminofluorene								
532274	2-Chloroacetophenone	1	X						
79469	2-Nitropropane	1	X			52.9	42.1	112.6	418
119904	3,3'-Dimethoxybenzidine						0.0	0.3	
119937	3,3'-Dimethyl benzidine	1	X						
91941	3,3-Dichlorobenzidine					0.0	0.1	0.1	
101779	4,4'-Methylenedianiline	1	X			6.6	9.8	23.9	75.
101144	4,4-Methylene bis(2-chloroaniline)	1	X			0.7	1.4	0.6	0.4
534521	4,6-Dinitro-o-cresol, and salts					0.0	0.1		
92671	4-Aminobiphenyl								
92933	4-Nitrobiphenyl								
100027	4-Nitrophenol	1	X			4.8	3.8	3.9	3.9
75070	Acetaldehyde	8	X	X	X	X	3540.5	3440.3	3762.1	333
60355	Acetamide	1	X				0.0		
75058	Acetonitrile	1	X			683.9	831.8	693.4	102
98862	Acetophenone	1	X						
107028	Acrolein	7	X		X	14.2	11.0	2.2	16.
79061	Acrylamide	2	X		X	32.1	25.0	12.5	13.
79107	Acrylic acid	2	X	X		205.3	213.5	178.7	399
107131	Acrylonitrile	7	X	X	X	X	1094.4	1574.0	2191.9	211
107051	Allyl chloride	1	X			90.1	103.0	87.8	73.
62533	Aniline	2	X		X	313.5	237.4	252.1	357
0	Antimony Compounds	3		X	X	43.1	73.0	79.3	78.
0	Arsenic Compounds								
	(inorganic including arsine)	11		X	X	95.2	82.9	87.9	156

	IUR per ug/m3 (5)	OUR per ug/L (6)	EPA WOE (7)	RfC mg/m3 Stat (8)	IARC WOE (9)	Exp. Asses. (10)	Genetic MVV So (11)	Genetic G	Toxicity MVT M (12)	Toxicity C	Data NM S (13)	Data E	Repr/Dev Data (14)
45	5.8E-5	5.8E-6	C		3						+		X
05	1.6E-5	1.6E-6	C	NV	3						-		
47			UR	NV	2B								
821			D	2.0E-1		+					-		X
28		4.0E-5	B2	2.0E-4	2B		-	+		+	+	+	X*
667	2.2E-4	2.2E-5	B2	NV							+	-	
887				2.0E-2					+	+	+		
58				NV	2B								
990	2.8E-4		B2		2B	B	+				+		X
756			B2	2.0E-2	2B						+		X
0714					2B								
467				8.0E-1	2B						-		
911		3.1E-7	B2		1B					-	-		
841				NV									
6016					2B	-			+		-		X*
54				NV							-		
62	3.1E-6	3.1E-7	B2	NV	2B				+	-	-		
57					2B		-	-	+	+	-		X*
85				NV							-	+	
142		1.9E-5	B2	NV			-	-	-	+	+	-	X*
07				NV	2B								
649				V	2B						-		
63											+	+	
274				3.0E-5									
69				2.0E-2	2B								
904					2B						+		
937				NV	2B								
41		1.3E-5	B2	NV	2B						+		
779				UR	2B						+		
144				NV	2A								
521				NV									
71					1			+	+		+	+	
33					3								
027				NV									X
70	2.2E-6		B2	9.0E-3	2B			+		+	+	+	X
55					2B								
58				UR									X
62			D	NV									
028			C	2.0E-5	3			-		+	+		X
61	1.3E-3	1.3E-4	B2	NV	2B	-		+			-		
07				3.0E-4	3								X
131	6.8E-5	1.5E-5	B1	2.0E-3	2A	B	-	-	+	+-	+	+	X
051			C	1.0E-3	3						+		X
33		1.6E-7	B2	1.0E-3	3	-			+	+	-	-	
	4.3E-3		A	5.0E-5	1	A	+	-	-	+	-	-	

CAS #	Chemical Name	#SC	M / Y / 2	A / E / 4	C / A / 7	T / R / 10	1991 Emis (T/yr)	1990 Emis (T/yr)	1989 Emis (T/yr)	1988 Emis (T/y)
	Column Number					i	2	2	2	2
1332214	Asbestos			A*			6.3	8.7	18.7	23.8
71432	Benzene	28	X	X	X	X	8737.2	12203.4	12341.5	141
92875	Benzidine	1			X					
98077	Benzotrichloride	1	X				3.9	4.2	12.6	12.5
100447	Benzyl chloride	3	X		X	X	13.4	16.8	13.6	21.7
0	Beryllium Compounds	3				X	0.1	0.1	1.7	2.3
57578	beta-Propiolactone	2	X		X					
92524	Biphenyl	3	X		X	X	430.2	560.5	544.1	604
111444	Bis(2-chloroethyl)ether (Dichloroethyl ether)	1	X				1.8	1.9	2.4	2.5
117817	Bis(2-ethylhexyl)phthalate (DEHP)						521.7	672.3	539.4	563
542881	Bis(chloromethyl)ether	1	X				0.3	0.0		
75252	Bromoform	1	X				0.1	24.1		
0	Cadmium Compounds	16		X	X	X	34.7	45.0	59.9	64.8
156627	Calcium cyanamide						6.3	6.3	6.3	6.3
105602	Caprolactam	4	X		X	X				
133062	Captan						3.6	9.6	12.6	7.9
63252	Carbaryl	1	X				3.4	4.6	5.1	3.7
75150	Carbon disulfide	5	X		X	X	44669.6	49111.3	49897.7	625
56235	Carbon tetrachloride	11	X	X	X	X	773.4	835.5	1683.6	188
463581	Carbonyl sulfide	3			X	X	8362.6	9317.4	9842.5	899
120809	Catechol						2.9	13.9	2.1	1.8
133904	Chloramben							0.0		
57749	Chlordane						0.7	2.2	1.9	0.3
7782505	Chlorine	11			X	X	38804.7	52458.9	66174.1	667
79118	Chloroacetic acid	3	X		X	X	256.8	12.7	12.4	13.1
108907	Chlorobenzene	3	X		X	X	1198.1	2023.4	2025.4	196
510156	Chlorobenzilate									
67663	Chloroform	6	X	X	X	X	9541.4	10881.2	12134.1	112
107302	Chloromethyl methyl ether						1.7	0.1	0.1	0.1
126998	Chloroprene (2-chloro-1,3-butadiene)	2	X	X			735.3	780.5	503.2	609
0	Chromium Compounds (+6 FOR IRIS)	31		X	X	X	278.2	384.8	1119.2	603
0	Cobalt Compounds	3			X		16.9	25.9	60.8	43.1
0	Coke Oven Emissions	2	X		X					
1319773	Cresols/Cresylic acid (isomers and mixture)	4	X	X		X	370.7	366.3	446.9	333
98828	Cumene (Isopropylbenzene)	3	X	X		X	1638.8	2051.6	2197.5	235
0	Cyanide Compounds	7	X	X	X		385.1	569.1	274.5	317
72559	DDE (p,p'-Dichlorodiphenyl-dichloroethylene)									
334883	Diazomethane									
132649	Dibenzofuran						20.1	15.1	31.9	35.4
84742	Dibutylphthalate	3		X	X	X	75.1	54.1	116.1	107
62737	Dichlorvos						0.3	0.7	0.7	0.5
111422	Diethanolamine	1	X				135.5	191.8	242.1	314
64675	Diethyl sulfate	1	X				2.1	2.7	4.4	3.1
60117	Dimethyl aminoazobenzene									
79447	Dimethyl carbamoyl chloride									
68122	Dimethyl formamide	2	X		X					
131113	Dimethyl phthalate	2	X		X		32.9	166.8	181.7	110
77781	Dimethyl sulfate	1	X				5.1	4.9	8.2	5.4

	IUR per ug/m3	OUR per ug/L	EPA WOE	RfC mg/m3 Stat	IARC WOE	Exp. Asses.	Genetic MVV So	Genetic MVV G	Toxicity MVT M	Toxicity MVT C	Data NM S	Data NM E	Repr/ Dev Data
	5	6	7	8	9	10	11		12		13		14
-													
32214	2.3E-1	Fib/ML	A		1	A	-		-+	+-	-	-	
432	8.3E-6	8.3E-7	A	UR	1	A	+	-	-+	-	-	-	X*
875	6.7E-2	6.7E-3	A	NV	1		+	-	+-	+	+	-	
077		3.6E-4	B2								+	+	
0447		4.9E-6	B2	NV			-		+-	+-	+	+	
	2.4E-3	1.2E-4	B2	UR	2A	A			+	+	-	+	
578				NV	2B								
524			D	NV									
	2.5	3.3E-4	B2	3.3E-5 NV	3		-						+
1444	3.3E-4	3.3E-5	B2	NV	3			-					X
7817		4.0E-7	B2		2		-	+	-	-			X
2881	6.2E-2	6.2E-3	A	NV	1						+		X
6252	1.1E-6	2.3E-7	B2							-	-		
	1.8E-3		B1	UR	2A	B	-	-	+	+-		+	
6627													
05602					4				-	-+	-		X
83062			UR	NV	3								
8252				NV	3								
6150				UR			+	+			-		X*
235	1.5E-5	3.7E-6	B2		2B	B	-				-	+	X
63581				NV									
0809					3								
3904													
749	3.7E-4	3.7E-5	B2	UR	3			-	-+		-		
82505			D										
118													
8907			D	UR		B	+		+	-	-		X
0156				NV	3								
663	2.3E-5	1.7E-7	B2	UR	2B	B	-		-	-	-	-	X
7302			A	NV	1						+		
6998				7.0E-3	3	B	+	+	-		+		X*
	1.2E-2		A	UR	1	B	+	+	+	+	+	+	
	6.2E-4		A		1	A							
19773				NV									
828				UR									
			D	V									
559			B2	9.7E-6					+	+	-	-	
84883				NV	3								
82649			D	NV									
742			D	NV						+			X
737		8.3E-6	C	5.0E-4	3								X
1422													
4675				NV	2A			+		+	+	+	+
0117													
447				NV	2A		+		+	-+	+		
122				3.0E-2			-		-+	-	-		X
31113			D	NV							-		
7781			B2	NV	2A		+	+	+	+	+	+	

CAS #	Chemical Name	M Y 2	A 4	C 7	T R 10 1	1991 Emis (T/yr) 2	1990 Emis (T/yr) 2	1989 Emis (T/yr) 2	1988 Emis (T/yr) 2
	Column Number								
		#SC							
106898	Epichlorohydrin (1-chloro-2,3-epoxypropane)	4	X	X		229.6	213.7	234.1	195.
140885	Ethyl acrylate	3	X	X	X	115.9	102.1	85.6	125.
100414	Ethyl benzene	23	X	X	X	X	4320.5	4308.8	4270.2	3358
51796	Ethyl carbamate (Urethane)					9.9	2.0	1.7	72.7
75003	Ethyl chloride (Chloroethane)	4	X		X	X	1431.6	1971.0	2394.1	2310
106934	Ethylene dibromide (Dibromoethane)	1	X			19.1	29.0	29.6	31.7
107062	Ethylene dichloride (1,2-Dichloroethane)	4	X	X	X	X	1997.7	2798.0	2055.1	2383
151564	Ethylene imine (Aziridine)							0.3	0.3
75218	Ethylene oxide	4	X	X	X	896.5	1223.7	1514.5	2300
96457	Ethylene thiourea					0.3	0.0	0.4	0.3
75343	Ethylidene dichloride (1,1-Dichloroethane)	6	X		X	X				
107211	Et-hylene glycol	3	X	X		5330.1	4694.4	6446.1	6640
0	Fine mineral fibers	1			X				
50000	Formaldehyde	22	X	X	X	X	5109.2	6383.0	6281.1	6199
0	Glycol ethers	2	X	X		21957.1	24429.1	24238.7	2420
76448	Heptachlor						0.4	1.7	24.5
118741	Hexachlorobenzene	1	X			0.4	0.7	2.3	2.5
87683	Hexachlorobutadiene	1	X			1.7	2.5	1.8	1.3
77474	Hexachlorocyclopentadiene					12.7	42.3	44.6	7.4
67721	Hexachloroethane	1	X			11.3	4.0	8.5	9.6
822060	Hexamethylene-1,6-diisocyanate								
680319	Hexamethylphosphoramide								
110543	Hexane	20	X	X	X	X				
302012	Hydrazine	1			X	14.2	13.9	15.1	13.9
7647010	Hydrochloric acid	7		X	X	X	41460.7	36723.5	30371.2	3696
7664393	Hydrogen fluoride (Hydrofluoric acid)	7	X	X	X	X	4590.6	4273.0	4990.1	6054
123319	Hydroquinone	4	X		X	X	5.4	5.7	6.4	5.1
78591	Isophorone	1	X						
0	Lead Compounds	22		X	X	X	703.8	812.2	1224.9	1339
58899	Lindane (gamma-hexachlorocyclohexane)					0.3	0.8	0.4	0.1
108316	Maleic anhydride	3	X		X	X	229.5	246.5	225.3	382.
0	Manganese Compounds	19		X	X	X	623.2	1126.4	2215.4	1582
0	Mercury Compounds (IRIS-INORGANIC)	11		X	X	X	1.4	0.6	14.6	12.9
67561	Methanol	6	X	X	X	99841.5	100706.2	105913.5	5108
72435	Methoxychlor					0.3	0.8	0.3	136.
74839	Methyl bromide (Bromomethane)	1	X			1222.8	1102.9	1289.8	595.
74873	Methyl chloride (Chloromethane)	10	X	X	X	X	2849.4	3821.9	4437.4	4693
71556	Methyl chloroform (1,1,1-Trichloroethane)	4	X	X	X	X	68753.1	80699.8	84309.1	8354
78933	Methyl ethyl ketone (2-Butanone)	20	X	X	X	X	51710.9	60663.5	63815.9	6276
60344	Methyl hydrazine	1	X						
74884	Methyl iodide (Iodomethane)					12.7	14.9	12.7	4.5
108101	Methyl isobutyl ketone (Hexone)	14	X	X	X	X	13599.3	13655.7	15341.4	1552
624839	Methyl isocyanate	1	X			3.9	7.2	7.5	5.1
80626	Methyl methacrylate	5	X	X	X	1278.7	1058.1	1571.9	1700
1634044	Methyl tert butyl ether	1	X			1519.1	1392.3	1495.1	1384
75092	Methylene chloride (Dichloromethane)	9	X	X	X	39669.2	46248.6	54636.1	606
101688	Methylene diphenyl diisocyanate (MDI)	1	X			313.2	338.2	312.8	143.

	IUR per ug/m3	OUR per ug/L	EPA WOE	RfC mg/m3 Stat	IARC WOE	Exp. Asses.	Genetic MVV So	Genetic MVV G	Toxicity MVT M	Toxicity MVT C	Data NM S	Data NM E	Repr/ Dev Data
	5	6	7	8	9	10	11		12		13		14
6898	1.2E-6	2.8E-7	B2	1.0E-3	2A	B	-+	-	+-	+	+	+	X
0885			UR		2B		+			+	-		X
0414			D	1.0E+0			-		+		-		X
796				NV	2B								
003				1.3E+1									X
6934	2.2E-4	2.5E-3	B2	2.0E-4	2A		-	-	+	+	+	+	X*
07062	2.6E-5	2.6E-6	B2		2B	B	-	-	+		+	-	X
51564				NV	3								
5218					2A		+	+	+	+	+	+	X
5457					2B	B	-	-	-	-	+	+	
5343			C	UR									
07211													
0000	1.3E-5		B1		2A		-+	-	+	+	+	+	X*
448	1.3E-3	1.3E-4	B2				-		-		+		
8741	4.6E-4	4.6E-5	B2	NV	2B		-		-		-		
683	2.2E-5	2.2E-6	C		3		-		-		-		X
474			D			B							
721	4.0E-6	4.0E-7	C	NV	3						-		X
2060				1.0E-5	2B								X
0319				7.0E-6	2B								
0543			UR	2.0E-1									X
02012	4.9E-3	8.5E-5	B2		2B		-		-		+	+	X
47010				7.0E-3									X
64393				UR									
3319				NV	3								
591		2.7E-8	C	NV							-		X
			B2		2B		+-		-+		-	-	
899				NV	2B								X*
8316													
			D	4.0E-4	B						+	+	
			D	3.0E-4									
561				UR									X
435			D	NV	3								
839			D	5.0E-3	3		+		+		+	+	X
873				UR	3	B		+	+		+		X
556			D	UR	3	B							X
933			D	1.0E+0									X
344				7.0E-5									
884				1.0E-2	3		+		+		-		
8101				UR									X
4839				NV									X*
626					3								X
34044				5.0E-1	3								X
092	4.7E-7	2.1E-7	B2	UR	2B		-		-	+	+		X
1688				5.0E-5									

CAS #	Chemical Name	#SC	M Y 2	A E 4	C A 7	T R 10 1	1991 Emis (T/yr) 2	1990 Emis (T/yr) 2	1989 Emis (T/yr) 2	1988 Emis (T/yr) 2
108394	m-Cresol	1	X				38.9	3.8	6.3	9.2
108383	m-Xylene	6	X		X		718.1	601.1	583.7	1012
121697	N,N-Dimethyl aniline	1	X				25.6	25.4	45.9	49.5
91203	Naphthalene	2	X	X			1335.9	1853.1	1656.9	1932
0	Nickel Compounds (subsulfide)	16		X	X	X	121.6	132.8	466.5	284.
98953	Nitrobenzene	1	X				26.3	33.1	19.4	19.6
62759	N-Nitrosodimethylamine	1			X					
59892	N-Nitrosomorpholine									
684935	N-Nitroso-N-methylurea									
90040	o-Anisidine	1	X				0.5	0.9	1.1	1.1
95487	o-Cresol	1	X				30.6	19.6	29.8	44.5
95534	o-Toluidine	1	X				5.4	3.7	12.8	23.5
95476	o-Xylene	23	X	X	X	X	864.9	952.3	899.6	979
56382	Parathion						0.3	0.3	0.8	1.6
82688	Pentachloronitrobenzene (Quintobenzene)						0.1	0.1	1.1	0.5
87865	Pentachlorophenol	4	X		X	X	6.2	11.6	5.6	7.1
108952	Phenol	12	X	X	X	X	3165.6	3827.4	5264.1	508?
75445	Phosgene	2	X		X		2.2	2.4	4.1	10.8
7803512	Phosphine									
7723140	Phosphorus	5		X	X	X	11.7	12.1	30.1	9.6
85449	Phthalic anhydride	5	X		X	X	315.9	343.7	325.1	273.
1336363	Polychlorinated biphenyls (PCB's)							0.0		0.1
0	Polycyclic Organic Matter	12		X	X	X				
123386	Propionaldehyde	4	X			X	694.1	494.5	453.8	523.
114261	Propoxur (Baygon)							0.1	0.3	0.1
78875	Propylene dichloride (1,2-Dichloropropane)	1	X				386.7	315.3	616.7	682.
75569	Propylene oxide	16	X	X	X		533.3	680.0	897.1	1482
106445	p-Cresol	1	X				67.8	119.5	127.4	320.
106503	p-Phenylenediamine	1	X				1.8	0.4	2.1	56.9
106423	p-Xylene	3	X		X		2639.2	2969.3	2360.1	315?
91225	Quinoline						22.5	13.8	31.8	24.7
106514	Quinone (1,4-benzoquinone)	1	X				2.1	0.8	0.9	5.7
0	Radionuclides (including radon)									
0	Selenium Compounds	15		X	X	X	18.5	15.3	16.7	14.8
100425	Styrene	15	X	X	X	X	14238.2	15838.3	16650.9	173?
96093	Styrene oxide	1	X				0.8	1.2	0.4	1.2
127184	Tetrachloroethylene (Perchloroethylene)	11	X	X	X	X	8343.7	10822.5	12752.4	157?
7550450	Titanium tetrachloride						16.8	27.2	28.6	39.3
108883	Toluene	39	X	X	X	X	99260.1	116912.8	127718.9	13?
8001352	Toxaphene (chlorinated camphene)									
79016	Trichloroethylene	8	X	X	X	X	17529.2	18949.0	22162.8	2409?
121448	Triethylamine	1	X							
1582098	Trifluralin						5.6	7.8	2.1	1.6
108054	Vinyl acetate	5	X		X		2743.2	2778.4	2699.1	286?
593602	Vinyl bromide (bromoethene)						1.8	5.1	0.4	2.5
75014	Vinyl chloride	4	X		X	X	523.7	567.9	634.4	687.
75354	Vinylidene chloride (1,1-Dichloroethylene)	3	X			X	142.6	151.8	110.3	149.
1330207	Xylenes (isomers and mixture)	23	X	X	X	X	57776.5	69988.4	73743.4	713?
	Total Emis								708443.8	

	IUR per ug/m3	OUR per ug/L	EPA WOE Stat	RfC mg/m3	IARC WOE	Exp. Asses.	Genetic MVV So	G	Toxicity MVT M	C	Data NM S	E	Repr/ Dev Data
	5	6	7	8	9	10	11		12		13		14
108394			C	NV									
108383				NV			-				-		X
121697													
91?03	4.2E-6		C				-				-		
0	4.8E-4		A	UR	1	B	+	-	-+	+	+	-	
98953			D	UR			-				-		X
62759	1.4E-2	1.4E-3	B2		2A			-	+	+	+	+	
59892					2B								
684935			B2										
90040				NV	2B								
95487			C	NV									
95534					2B		-		+-	+-	-	-	
95476				NV			-				-		X
56382			C		3								
82688			UR		3								
87865		3.0E-6	B2	UR	2B				-	+	-		
108952			D	NV		B	-		+		v		
75445				NV									
7803512			D										
7723140			D										
85449													
1336363		2.2E-4	B2		2A		-	-		-	-		X
0			UR			B							
123386				NV									
114261			UR										
78875			UR	6.0E-3	3								X
75569	3.7E-6	6.8E-6	B2	3.0E-2	2A		+	-	+	+	+	+	
106445			C	NV									
106503					3								
106423				NV			-				-		X
91225				NV									
106514				NV									
0			UR			A							
0			D		3								
100425			UR	1.0E+0	2B		+-			+	+		X*
96093					2A		-	-	+	+	+	+	X
127184					2B	B		-	+		-	-	X
7550450													
108883			D	4.0E-1		B	+		-			-	X*
8001352	3.2E-4	3.2E-5	B2		2B			-			+		
79016			UR	UR	3	B	+	-	+-	+-	+	+	X
121448				7.0E-3									
1582098		2.2E-7	C										
108054			UR	2.0E-1	3					+	-		X
593602				3.0E-3	2A						+		
75014	8.4E-5	5.4E-5	A		1	A	+	-	+		+	+	X*
75354	5.0E-5	1.7E-5	C	UR	3	B	-	-	-	-	+	+	X
1330207			D	NV			-			-	-	-	X
		800960.9		863337.1	906614.9	39.0	44.0						

Column Numbers and Footnotes to Table

1 Chemicals emitted from sources that will be regulated within the 2, 4, 7, and 10 year deadlines for maximum achievable control technology standards (# preceding X indicates the # of source categories (#SC) HAP in).

2 Toxic Release Inventory Data (TRI) in tons/year (1988,1989,1990,1991)

5 IUR= Inhalation unit risk estimate per ug/m3; Source is EPA's Integrated Risk Information System (IRIS)

6 OUR= Oral unit risk estimate per ug/l; Source is EPA's IRIS data base

7 WOE= Weight of Evidence classification; Source is EPA's IRIS data base

8 RfC workgroup; verified, on IRIS= conc. given in mg/m3
             verified, not on IRIS= 'v'

9 IARC (International Agency for Research on Cancer) WOE

10 Exposure Assessments:
   A) HAPS with risk assessments done for development of Section 112 standards
   B) HAPS with screening assessments done for listing purposes

11, 12, 13 Information on Genetic Toxicology; Source= Genetic Activity Profiles data base provided by Dr. Michael Waters, EPA's Health Effects Research Lab. (Data as of 1992)
   The + or - represents the overall call for that group which may contain more than one assay. When discrepancies exist within a group, this is indicated by a +- (or-+).
   The first symbol represents the majority call for that group.

   MVV= Mammalian, In Vivo
     So= Somatic cell; G= Germ cell
   MVT= Mammalian, In Vitro
     M= Mutation; C= Chromosome aberration
   NM= Non-mammalian
     S= Salmonella typhi.; E= Eschericia coli

14 Data from Non-cancer Health Effects Database, prepared and provided by Dr. John Vandenberg, EPA's Health Effects Research Lab.
   'X' indicates data available; X* indicates some human data available
   Note: Data includes effects on maternal toxicity: All data from Inhalation exposure

Symbols used: UR= under review
             V= verified
             NA= not available
             NV= not verifiable

IARC vs. EPA: Classification Differences

EPA Modifications to IARC Approach:
1. Considers statistically significant association between an agent and life-threatening benign tumors when evaluation human risk
2. Added "no data available" category
3. Added "no evidence of carcinogenicity" category

By Category

EPA

Group A - Known human carcinogen

Group B - Probable human carcinogen

      B1 - Limited human data

      B2 - Inadequate human data, sufficient animal data

Group C - Possible human carcinogen

IARC

Group 1 - Known human carcinogen

Group 2A - Probable human carcinogen

Group 2B - Possible human carcinogen

Group 3 - Not classifiable as to human carcinogenicity

Group 4 - Probably not carcinogenic to humans

IV. Question 4: What does EPA consider to be the prioritization of the information gathering needs? What criteria would EPA use for determining this prioritization?

IV.A Introduction

Existing data on effects and exposure to the hazardous air pollutants (HAPs) listed under Section 112 have supported a variety of decisions under Title III of the Clean Air Act (CAA). Rules that use these data and additional data collected in a timely fashion will continue to be issued on CAA schedules that extend to the year 2010. Future information gathering on the HAPs will support residual risk decisions, biennial Great Waters reports, urban air toxics reports, and other continuous activities required to administer Section 112 provisions. Interest in the HAPs exists beyond the CAA. Other EPA-administered programs and programs of other agencies address many of the same chemicals and mixtures. Therefore, whatever data are gathered will be gathered with an eye to serving needs beyond Section 112.

The process of prioritizing data collection activities must consider many factors. Decisions for gathering information will have both science and management components. Important considerations include: the types of information needed to make a statutory finding, the current state-of-the-science, priorities given other EPA work, budget constraints, and statutory deadlines. The EPA has not, as yet, made decisions about the extent, mechanism, or timing of data gathering activities. The information presented below generally describes EPA's initial thoughts regarding the gathering of information needed to effectively implement Title III of the CAA.

Under Title IX of the CAA, EPA and other agencies will be looking generally at the research needs for all of the HAPs. This Title provides a forum for planning research to advance the state of the art beyond standard testing. The plans for carrying out this Title are currently being formulated as the Title was added after the FY91 appropriation process was completed.

Overall, the goals by which the priorities and needs can be balanced may be stated as:

- ensure that the data collected meet the requirements of the statutory finding(s) that must be made
- ensure that the data are collected in a timely fashion
- ensure efficient use of resources, given the parallel data gathering efforts of others

- ensure that adequate resources are invested in HAPs that are emitted in significant volumes
- avoid enriching an already rich data base of one HAP at the expense of another HAP of importance

IV.B Criteria for Effects Data-Gathering Plan

The major focus in planning for health and environmental effects data collection activities is to ensure that adequate data are available to conduct the residual risk determinations that will be made under Section 112(f). In order to obtain the data necessary to support these decisions required later in the decade, EPA must begin collection efforts immediately. The Agency anticipates that activities will begin with a ranking of HAPs that takes several factors into account. These factors include:

- promulgation dates of control technology standards
- estimation of the extent to which a particular HAP will contribute to risks resulting from combined HAP emissions from sources in a source category (using effects and exposure data available now)
- importance of a HAP to the Great Waters or Urban Area Source programs
- overlapping priority/interest of other EPA programs or governmental agencies (e.g. timing of ongoing Agency for Toxic Substances and Disease Registry or National Toxicology Program activities)

Decisions on the extent and type of data to be gathered on potential adverse effects associated with exposure to a HAP will also require a balancing of several factors including incorporation of professional judgment on the likelihood that additional data may significantly alter current opinions on the toxicity of a specific HAP. Critical elements will include:

- the richness of the current data base
- the need for data to enable route-to-route extrapolation of existing toxicity data
- the need to expand a data set on an already identified endpoint in order to improve dose-response characterization
- the need to extend the scope of data to cover endpoints other than those previously identified
- the need for research beyond standard test protocols to understand biological fate and transformation or mechanism of action

IV.C Options for Scope of Effects Data-gathering

While the alternatives have not been exhaustively explored, and substantial work remains to be done, there are three general options that are being considered. These options are:

1. Broad Scope. This approach would use staged testing for a large number of HAPs, screening a range of endpoints and proceeding to full endpoint tests as the screening assays indicate.

2. Medium Scope. This option would focus screening tests on those HAPs with the most significant emissions. Testing strategies would be more robust and address critical endpoints (carcinogenicity and developmental toxicity, at a minimum). Other HAPs with significant emissions would be considered under the narrow scope testing identified below.

3. Narrow Scope. Under this alternative, testing would focus on complimenting and making more useable existing data bases. For example, HAPs with significant emissions may be studied to "convert" oral to inhalation data or to elucidate dose-response relationships. This narrow scope testing could include: pharmacokinetics studies, a 90 day subchronic inhalation study, or a repeat of a previous study on an endpoint to better define the dose-response relationship.

IV.D Mechanisms for Obtaining Effects Data

There are a variety of mechanisms that may be accessed for collecting effects data, all of which will likely be employed. Major data gathering efforts are underway that will complement data collected specifically for Section 112 use. For example, the Superfund Amendments and Reauthorization Act of 1986 (SARA) requires the Agency for Toxic Substances and Disease Registry (ATSDR) to prepare toxicity profiles for over 200 pollutants. These profiles identify data gaps and efforts will be put forth to fill these gaps. Of the pollutants studied by ATSDR, 76 are HAPs. A second example is efforts being undertaken by the European communities. They are interested in generating data for a list of chemicals that overlaps the HAPs list. In addition, the National Toxicology Program (NTP) is working with EPA to identify testing and research the NTP can undertake for several HAPs. The EPA's Health and Environmental Research Laboratory (HERL) has ongoing research that addresses several HAPs, as well as urban toxics issues. This laboratory also conducts fundamental research on pharmacokinetics applicable to the HAPs. Additional EPA laboratories are conducting research on environmental fate, ecological effects, etc. Another alternative for collecting data is to access the regulatory test program under the Toxic Substances Control Act (TSCA) to require that industry conduct testing. Finally, the CAA Title IX research program will be pursued for research on HAPs. Making these overlapping efforts work together will be part of any data gathering EPA does on the HAPs.

IV.E Improving Data Bases for Estimating Exposure to HAPs

In addition to developing quantitative relationships between HAP concentrations and health or environmental effects, it is critical that the EPA pursue parallel efforts to support accurate characterization of the levels of exposure associated with sources of the HAPs. In the past, efforts to obtain sufficient information to accurately characterize HAP exposure levels in the vicinity of an industrial source have focused on one pollutant at a time. These efforts have been severely limited by lack of information on the source(s) being evaluated. In lieu of site-specific data for exposure characterization, EPA has settled for "model plant" types of analyses, which rely on only a sampling of data from one type of source and extrapolate exposure estimates to the rest of the source population. These analyses by nature must be very conservative, and therefore tend to overestimate ambient exposure levels due to any one type of industrial source. As a result, these analyses are often criticized by industry as being "overly conservative".

It is clear that the CAA mandate for residual risk analyses (after the implementation of MACT) would require that such analyses be based on site-specific data rather than "model plant" scenarios. These analyses must therefore require more site-specific data than are currently available. In addition, the analyses will differ from past analyses in that they will be directed at assessing the exposure to multiple pollutants being emitted from a source in a particular source category. The EPA must begin now to develop the tools and process for obtaining the necessary data to perform residual risk analyses. While such efforts may build on past efforts, there are several new and challenging aspects that must be addressed, including:

1) Emission levels of each of the HAPs from each source within a source category must be obtained. Since EPA-approved measurement methods are not available for all HAPs, this will entail research and development efforts for both measurement methods and site-specific emission estimation techniques. It is hoped that cooperative efforts can be undertaken with industry to expand the publicly-available expertise in this area.

2) Data are to be obtained on a source category-by-source category basis. Since most currently available data bases are on a pollutant-by-pollutant basis, most of the current data will be inadequate for this purpose.

3) Exact stack, vent, and fugitive emission locations as well as fenceline locations for each facility are crucial for reducing the uncertainty of exposure assessments. Very little data are available in this regard, and it is unclear whether most industries will be willing to provide such data.

4) Development of guidelines is needed to explore the use of monitoring data or other more direct measures of exposure in assessing exposures resulting from emissions of HAPS. Specifically, the use of these data to complement modeling analyses needs to be examined.

5) Development of a user-friendly, easy-access, centralized data base and retrieval system (such as an electronic bulletin board system) may be desirable to provide a convenient vehicle for obtaining the necessary data. Industry input and cooperation in such development would be crucial to its success. Making sure that industry realizes that, without the necessary data, EPA efforts to assess exposure will be "conservative", may provide the needed cooperation of industry. Development of a data base system that is easy to use will substantially reduce the burden on industry as well as reduce the paperwork that would otherwise be necessary for such an information request.

6) Efforts to check and assure the quality of the data obtained for exposure assessments may prove to be a large part of the data gathering process.

7) Efforts to appropriately include population mobility and microenvironment exposures into the overall exposure assessment process have already begun. Sensitivity studies are needed, however, to determine the extent to which such factors can affect the overall exposure and risk assessment results.

8) Inclusion of short-term exposure quantification is important for many HAPs. Some modeling techniques are already available to address this quantification, but data on short-term emissions variability are generally lacking. The extent to which such information becomes available will dictate the extent to which EPA can incorporate such variability in exposure assessments.

9) Concentration measurements to assist in the validation of human exposure modeling results are generally lacking for most HAPs. While validation of air dispersion models in the field has been done, indoor/outdoor partitioning and multiple route exposures have not received the same level of validation efforts. This is an area where more data would be helpful.

EPA welcomes comments and suggestions from the Committee on the plans for improving the accuracy of exposure and risk assessments required to implement the CAA. Of specific interest are the recommendations of the panel for prioritizing the vast amount of work that is required to fill the existing data gaps.

V. Question 5: What does EPA consider to be some of the critical management aspects of risk assessment decision-making that may not be apparent to an outside observer?

The current regulatory process places a number of challenging demands on the risk manager. Depending on the nature of the regulation and the legislative authority, he or she must try to assimilate a variety of analyses—legal, economic, social and scientific—of which risk characterization is only one part. Because of this diversity, the risk manager must relay on the products of experts in a range of disciplines.

In making risk management decisions, there are a number of considerations and factors to be weighed that may not be apparent to outside observers. Some of the factors influencing these decisions are described below.

1. In dealing with scientific issues, the risk manager is typically a generalist with no particular expertise in the area of risk assessment. This places particular requirements on the risk assessment process. Thus, the products of the risk assessment process must be designed to aid these individuals in decision-making. Risk managers are often frustrated by complex discussions of scientific uncertainties (mechanism of action, uncertainty in extrapolation, etc.). Rather they tend to desire bottom-line characterizations of the likelihood and magnitude of potential problems. In many respects, the popularity of the current cancer classification system lies in its ability to characterize the overall weight of evidence by readily-comprehended categories (e.g., known, probable, possible carcinogen) and the presentation of a measure of carcinogenic potency.

The Agency has increased the emphasis placed on the risk characterization component of risk assessment, and is moving toward a more comprehensive examination of the assumptions and uncertainties in risk assessment. The fact remains, however, that communication of the critical elements of a risk assessment to risk managers remains a challenge.

2. Consistency is important. This does not mean that all risk assessments should look the same. But it is important that a consistent terminology be adopted, even if the terminology draws controversy, and that the risk managers understand and can communicate that understanding. Decision-makers build on previous decisions and examples to put current issues in context. If formats or meanings differ from case to case, this process becomes difficult, if not impossible.

3. Risk managers do not expect perfect information. Critics of risk assessment's imperfections must recognize that public policy is often a blunt instrument rather than a surgeon's scalpel. Decisions are often based on broad bands

of uncertainty within which even differences of several fold may not affect the decision.

It is important for both risk managers and critics of risk assessment to avoid pursuing the ideal risk assessment. These individuals must bear in mind the limits of the real world. These limits include time, money, and the state of scientific knowledge.

4. Statutory mandates may place constraints on the development and use of risk characterization data that are not consistent with our understanding of the underlying science. The establishment of risk targets (or bright lines) such as 10^{-6}, for example, have been criticized as not allowing the consideration of weight-of-evidence in decision making. Another example is the requirement that the Agency consider the risk to the "person most exposed" to emissions from an air toxics source. Thus, the statutory framework constrains full consideration of the distribution of risk across the exposed population.

5. Statutes or court action often mandate regulation at a specific time, effectively mandating decision-making based upon available data. This is exacerbated by the fact that the development of robust health and safety data (e.g., well-conducted animal bioassays, epidemiological, or exposure studies) are both resource- and time-intensive.

6. The risk management process is often the focus of considerable outside attention and controversy. This is particularly true where the impacts of decisions are costly, or where they adversely affect well-organized groups. On these circumstances, there is a natural tendency to continue the process of data development and analysis, rather than to make decisions in an atmosphere of uncertainty. While such an environment can cause delay, it can also have the effect of encouraging more rigorous examination of data and careful consideration of options.

7. Persistent requests for information and more studies lead to paralysis by analysis and the waste of limited resources. The risk of inaction is often forgotten. Additional information needs must be balanced against the need to take timely action where it is warranted. This is particularly true in the risk assessment process, where the limitations of the current state of the science often prevent definitive answers, and can encourage continual additional data development. Reviewers of Agency risk assessments must consider the reasonable resource constraints under which the Agency operates.

APPENDIX

B

EPA Memorandum from Henry Habicht

MEMORANDUM

SUBJECT: Guidance on Risk Characterization for Risk Managers
and Risk Assessors

FROM: F. Henry Habicht II
Deputy Administrator

TO: Assistant Administrators
Regional Administrators

INTRODUCTION

This memorandum provides guidance for managers and assessors on describing risk assessment results in EPA reports, presentations, and decision packages. The guidance addresses a problem that affects public perception regarding
the reliability of EPA's scientific assessments and related regulatory decisions.
EPA has talented scientists, and public confidence in the quality of our scientific
output will be enhanced by our visible interaction with peer scientists and through
presentation of risk assessments and underlying scientific data.

Specifically, although a great deal of careful analysis and scientific judgment goes into the development of EPA risk assessments, significant information is often omitted as the results of the assessments are passed along in the
decision-making process. Often, when risk information is presented to the ultimate decision-maker and to the public, the results have been boiled down to a
point estimate of risk. Such "short hand" approaches to risk assessment do not
fully convey the range of information considered and used in developing the
assessment. In short, informative risk characterization clarifies the scientific

basis for EPA decisions, while numbers alone do not give a true picture of the assessment.

This problem is not EPA's alone. Agency contractors, industry, environmental groups, and other participants in the overall regulatory process use similar "short hand" approaches.

We must do everything we can to ensure that critical information from each stage of the risk assessment is communicated from risk assessors to their managers, from middle to upper management, from EPA to the public, and from others to EPA. The Risk Assessment Council considered this problem over many months and reached several conclusions: 1) We need to present a full and complete picture of risk, including a statement of confidence about data and methods used to develop the assessment; 2) we need to provide a basis for greater consistency and comparability in risk assessment across Agency programs; and 3) professional scientific judgment plays an important role in the overall statement of risk. The Council also concluded that Agency-wide guidance would be useful.

BACKGROUND

Principles emphasized during Risk Assessment Council discussions are summarized below and detailed in the attached Appendix.

Full Characterization of Risk

EPA decisions are based in part on risk assessment, a technical analysis of scientific information on existing and projected risks to human health and the environment. As practiced at EPA, the risk assessment process depends on many different kinds of scientific data (e.g., exposure, toxicity, epidemiology), all of which are used to "characterize" the expected risk to human health or the environment. Informed use of reliable scientific data from many different sources is a central feature of the risk assessment process.

Highly reliable data are available for many aspects of an assessment. However, scientific uncertainty is a fact of life for the risk assessment process as a whole. As a result, agency managers make decisions using scientific assessments that are less certain than the ideal. The issues, then, become when is scientific confidence sufficient to use the assessment for decision-making, and should the assessment be used? In order to make these decisions, mangers need to understand the strengths and limitations of the assessment.

On this point, the guidance emphasizes that informed EPA risk assessors and managers need to be completely candid about confidence and uncertainties in describing risks and in explaining regulatory decisions. Specifically, the Agency's risk assessment guidelines call for full and open discussion of uncertainties in the body of each EPA risk assessment, including prominent display of critical uncertainties in the risk characterization. Numerical risk estimates should always be accompanied by descriptive information carefully selected to ensure an objective and balanced characterization of risk in risk assessment reports and regulatory documents.

Scientists call for fully characterizing risk not to question the validity of the assessment, but to fully inform others about critical information in the assessment. The emphasis on "full" and "complete" characterization does not refer to an ideal assessment in which risk is completely defined by fully satisfactory scientific data. Rather, the concept of complete risk characterization means that information that is needed for informed evaluation and use of the assessment is carefully highlighted. Thus, even though risk characterization details limitations in an assessment, a balanced discussion of reliable conclusions and related uncertainties enhances, rather than detracts, from the overall credibility of each assessment.

This guidance is not new. Rather, it re-states, clarifies, and expands upon current risk assessment concepts and practices, and emphasizes aspects of the process that are often incompletely developed. It articulates principles that have long guided experienced risk assessors and well-informed risk managers, who recognize that risk is best described not as a classification or single number, but as a composite of information from many different sources, each with varying degrees of scientific certainty.

Comparability and Consistency

The Council's second finding, on the need for greater comparability, arose for several reasons. One was confusion — for example, many people did not understand that a risk estimate of 10^{-6} for an "average" individual should not be compared to another 10^{-6} risk estimate for the "most exposed individual." Use of such apparently similar estimates without further explanation leads to misunderstandings about the relative significance of risks and the protectiveness of risk reduction actions. Another catalyst for change was the SAB's report, Reducing Risk: Setting Priorities and Strategies for Environmental Protection. In order to implement the SAB's recommendation that we target our efforts to achieve the greatest risk reduction, we need common measurements of risk.

EPA's newly revised Exposure Assessment Guidelines provide standard descriptors of exposure and risk. Use of these terms in all Agency risk assessments will promote consistency and comparability. Use of several descriptors, rather than a single descriptor, will enable us to present a more complete picture of risk that corresponds to the range of different exposure conditions encountered by various populations exposed to most environmental chemicals.

Professional Judgment

The call for more extensive characterization of risk has obvious limits. For example, the risk characterization includes only the most significant data and uncertainties from the assessment (those that define and explain the main risk conclusions) so that decision-makers and the public are not overwhelmed by valid but secondary information.

The degree to which confidence and uncertainty are addressed depends largely on the scope of the assessment and available resources. When special circumstances (e.g., lack of data, extremely complex situations, resource limitations, statutory deadlines) preclude a full assessment, such circumstances should be explained. For example, an emergency telephone inquiry does not require a full written risk assessment, but the caller must be told that EPA comments are based on a "back-of-the-envelope" calculation and, like other preliminary or simple calculations, cannot be regarded as a risk assessment.

GUIDANCE PRINCIPLES

Guidance principles for developing, describing, and using EPA risk assessments are set forth in the Appendix. Some of these principles focus on differences between risk assessment and risk management, with emphasis on differences in the information content of each process. Other principles describe information expected in EPA risk assessments to the extent practicable, emphasizing that discussion of both data and confidence in the data are essential features of a complete risk assessment. Comments on each principle appear in the Appendix; more detailed guidance is available in EPA's risk assessment guidelines (e.g., 51 Federal Register 33992-34054, 24 September 1986).

Like the EPA's risk assessment guidelines, this guidance applies to the development, evaluation, and description of Agency risk assessments for use in regulatory decision-making. This memorandum does not give guidance on the use of completed risk assessments for risk management decisions, nor does it address the use of non-scientific considerations (e.g., economic or societal factors) that are considered along with the risk assessment in risk management and decision-making. While some aspects of this guidance focus on cancer risk assessment, the guidance applies generally to human health effects (e.g., neurotoxicity, developmental toxicity) and, with appropriate modifications, should be used in all health risk assessments. Guidance specifically for ecological risk assessment is under development.

IMPLEMENTATION

Effective immediately, it will be Agency policy for each EPA office to provide several kinds of risk assessment information in connection with new Agency reports, presentation, and decision packages. In general, such information should be presented as carefully selected highlights from the overall assessment. In this regard, common sense regarding information needed to fully inform Agency decision-makers is the best guide for determining the information to be highlighted in decision packages and briefings.

1. Regarding the interface between risk assessment and risk management, risk assessment information must be clearly presented, separate from any non-scientific risk management considerations. Discussion of risk management options should follow, based on consideration of all relevant factors, scientific and non-scientific.

2. Regarding risk characterization, key scientific information on data and methods (e.g., use of animal or human data for extrapolating from high to low doses, use of pharmacokinetics data) must be highlighted. We also expect a statement of confidence in the assessment that identifies all major uncertainties along with comment on their influence on the assessment, consistent with guidance in the attached Appendix.

3. Regarding exposure and risk characterization, it is Agency policy to present information on the range of exposures derived from exposure scenarios and on the use of multiple risk-descriptors (i.e., central tendency, high end of individual risk, population risk, important subgroups, if known) consistent with terminology in the attached Appendix and Agency guidelines.

This guidance applies to all Agency offices. It applies to assessments generated by EPA staff and those generated by contractors for EPA's use. I believe adherence to this Agency-wide guidance will improve understanding of Agency risk assessments, lead to more informed decisions, and heighten the credibility of both assessments and decisions.

From this time forward, presentations, reports, and decision packages from all Agency offices should characterize risk and related uncertainties as described here. Please be prepared to identify and discuss with me any program-specific modifications that may be appropriate. However, we do not expect risk assessment documents that are close to completion to be rewritten. Although this is internal guidance that applies directly to assessments developed under EPA auspices, I also encourage Agency staff to use these principles as guidance in evaluating assessments submitted to EPA from other sources, and in discussing these submissions with me and with the Administrator.

This guidance is intended for both management and technical staff. Please distribute this document to those who develop or review assessments and to your mangers who use them to implement Agency programs. Also, I encourage you to discuss the principles outlined here with your staff, particularly in briefings on particular assessments.

In addition, I expect that the Risk Assessment Council will endorse new guidance on Agency-wide approaches to risk characterization now being developed in the Risk Assessment Forum for EPA's risk assessment guidelines, and that the Agency and the Council will augment that guidance as needed.

The Administrator and I believe that this effort is very important. It furthers our goals of rigor and candor in the preparation, presentation, and use of EPA risk assessments. The tasks outlined above may require extra effort from you, your managers, and your technical staff, but they are critical to full implementation of these principles. We are most grateful for the hard work of your representatives on the RAC and other staff in pulling this document together. I appreciate your cooperation in this important area of science policy, and look forward to our discussions.

Attachment

cc: The Administrator
     Risk Assessment Council

<u>GUIDANCE FOR RISK ASSESSMENT</u>

Section 1. Risk Assessment-Risk Management Interface

Section 2. Risk Characterization

Section 3. Exposure and Risk Descriptors

U.S. Environmental Protection Agency
Risk Assessment Council
November, 1991

SECTION 1. RISK ASSESSMENT —
RISK MANAGEMENT INTERFACE

Recognizing that for many people the term risk assessment has wide meaning, the National Research Council's 1983 report on risk assessment in the federal government (hereafter "NRC report") distinguished between risk assessment and risk management.

> Broader uses of the term [risk assessment] than ours also embrace analysis of perceived risks, comparison of risks associated with different regulatory strategies, and occasionally analysis of the economic and social implications of regulatory decisions—<u>functions that we assign to risk management</u> (emphasis added). (1)

In 1984, EPA endorsed these distinctions between risk assessment and risk management for Agency use (2), and later relied on them in developing risk assessment guidelines (3).

This distinction suggests that EPA participants in the process can be grouped into two main categories, each with somewhat different responsibilities, based on their roles with respect to risk assessment and risk management.

Risk Assessment

One group <u>generates</u> the risk assessment by collecting, analyzing, and synthesizing scientific data to produce the hazard identification, dose-response, and exposure assessment portion of the risk assessment and to characterize risk. This group relies in part on Agency risk assessment guidelines to address science policy issues and scientific uncertainties.

Generally, this group includes scientists and statisticians in the Office of Research and Development, the Office of Pesticides and Toxic Substances and other program offices, the Carcinogen Assessment Verification Endeavor (CRAVE), and the RfD/RfC Workgroups.

Others <u>use</u> analyses produced by the first group to generate site- or media-specific exposure assessments and risk characterizations for use in regulation development. These assessors rely on existing databases (e.g., IRIS, ORD Health Assessment Documents, CRAVE, and RfD/RfC Workgroup documents) to develop regulations and evaluate alternatives.

Generally, this group includes scientists and analysts in program offices, regional offices, and the Office of Research and Development.

Risk Management

A third group <u>integrates</u> the risk characterization with other non-scientific considerations specified in applicable statutes to make and justify regulatory decisions.

Generally, this group includes Agency managers and decision-makers.

Each group has different responsibilities for observing the distinction between risk assessment and risk management. At the same time, the risk assessment process involves regular interaction between each of the groups, with overlapping responsibilities at various stages in the overall process.

The guidance to follow outlines principles specific for those who generate, review, use, and integrate risk assessments for decision-making.

1. Risk assessors and risk managers should be sensitive to distinctions between risk assessment and risk management.

The major participants in the risk assessment process have many shared responsibilities. Where responsibilities differ, it is important that participants confine themselves to tasks in their areas of responsibility and not inadvertently obscure differences between risk assessment and risk management.

Shared responsibilities of assessors and managers include initial decisions regarding the planning and conduct of an assessment, discussions as the assessment develops, decisions regarding new data needed to complete an assessment and to address significant uncertainties. At critical junctures in the assessment, such consultations shape the nature of, and schedule for, the assessment.

For the generators of the assessment, distinguishing between risk assessment and risk management means that scientific information is selected, evaluated, and presented without considering non-scientific factors including how the scientific analysis might influence the regulatory decision. Assessors are charged with (1) generating a credible, objective, realistic, and balanced analysis; (2) presenting information on hazard, dose-response, exposure and risks; and (3) explaining confidence in each assessment by clearly delineating uncertainties and assumptions along with the impacts of these factors (e.g., confidence limits, use of conservative/non-conservative assumptions) on the overall assessment. They do not make decisions on the acceptability of any risk level for protecting public health or selecting procedures for reducing risks.

For users of the assessment and for decision-makers who integrate these assessments into regulatory decisions, the distinction between risk assessment and risk management means refraining from influencing the risk description through consideration of non-scientific factors—e.g., the regulatory outcome — and from attempting to shape the risk assessment to avoid statutory constraints, meet regulatory objectives, or serve political purposes. Such management considerations are often legitimate considerations for the overall regulatory decision (see next principle), but they have no role in estimating or describing risk.

However, decision-makers establish policy directions that determine the overall nature and tone of Agency risk assessments and, as appropriate, provide policy guidance on difficult and controversial risk assessment issues. Matters such as risk assessment priorities, degree of conservatism, and acceptability of particular risk levels are reserved for decision-makers who are charged with making decisions regarding protection of public health.

2. The risk assessment product, that is, the risk characterization, is only one of several kinds of information used for regulatory decision-making.

Risk characterization, the last step in risk assessment, is the starting point for risk management considerations and the foundation for regulatory decision-making, but it is only one of several important components in such decisions. Each of the environmental laws administered by EPA calls for consideration of non-scientific facts at various stages in the regulatory process. As authorized by different statutes, decision-makers evaluate technical feasibility (e.g., treatability, detection limits), economic, social, political, and legal factors as part of the analysis of whether or not to regulate and; if so, to what extent. Thus, regulatory decisions are usually based on a combination of the technical analysis used to develop the risk assessment and information from other fields.

For this reason, risk assessors and managers should understand that the regulatory decision is usually not determined solely by the outcome of the risk assessment. That is, the analysis of the overall regulatory problem may not be the same as the picture presented by risk analysis alone. For example, a pesticide risk assessment may describe moderate risk to some populations but, if the agricultural benefits of its use are important for the nation's food supply, the product may be allowed to remain on the market with ceratin restrictions on use to reduce possible exposure. Similarly, assessment efforts may produce an RfD for a particular chemical, but other considerations may result in a regulatory level that is more or less protective than the RfD itself.

For decision-makers, this means that societal considerations (e.g., costs, benefits) that, along with the risk assessment, shape the regulatory decision should be described as fully as the scientific information set forth in the risk characterization. Information on data sources and analyses, their strengths and limitations, confidence in the assessment, uncertainties, and alternate analyses are as important here as they are for the scientific components of the regulatory decision. Decision-makers should be able to expect, for example, the same level of rigor from the economic analysis as they receive from the risk analysis.

Decision-makers are not "captive of the numbers." On the contrary, the quantitative and qualitative risk characterization is only one of many important factors that must be considered in reaching the final decision — a difficult and distinctly different task from risk assessment per se. Risk management decisions involve numerous assumptions and uncertainties regarding technology, economics and social factors, which need to be explicitly identified for the decision-makers and the public.

SECTION 2. RISK CHARACTERIZATION

EPA risk assessment principles and practices draw on many sources. The environmental laws administered by EPA, the National Research Council's 1983 report on risk assessment (1), the Agency's Risk Assessment Guidelines (3), and various program-specific guidance (e.g., the Risk Assessment Guidance for Superfund) are obvious sources. Twenty years of EPA experience in developing, defending, and enforcing risk assessment-based regulation is another. Together these various sources stress the importance of a clear explanation of Agency processes for evaluating hazard, dose-response, exposure, and other data that provide the scientific foundation for characterizing risk.

This section focuses on two requirements for full characterization of risk. First, the characterization must address qualitative and quantitative features of the assessment. Second, it must identify any important uncertainties in the assessment as part of a discussion on confidence in the assessment.

This emphasis on a full description of all elements of the assessment draws attention to the importance of the qualitative as well as the quantitative dimensions of the assessment. The 1983 NRC report carefully distinguished qualitative risk assessment from quantitative assessments, preferring risk statements that are not strictly numerical.

> The term risk assessment is often given narrower and broader meanings than we have adopted here. For some observers, the term is synonymous with quantitative risk assessment and emphasizes reliance on numerical results. Our broader definition includes quantification, but also includes qualitative expressions of risk. Quantitative estimates of risk are not always feasible, and they may be eschewed by agencies for policy reasons (Emphasis in original) (1)

More recently, an Ad Hoc Study Group (with representatives from EPA, HHS, and the private sector) on Risk Presentation reinforced and expanded upon these principles by specifying several "attributes" for risk characterization.

1. The major components of risk (hazard identification, dose-response, and exposure assessment) are presented in summary statements, along with quantitative estimates of risk, to give a combined and integrated view of the evidence.

2 . The report clearly identifies key assumptions, their rationale, and the extent of scientific consensus; the uncertainties thus accepted; and the effect of reasonable alternative assumptions on conclusions and estimates.

3. The report outlines specific ongoing or potential research projects that would probably clarify significantly the extent of uncertainty in the risk estimation....(4)

Particularly critical to full characterization of risk is a frank and open discussion of the uncertainty in the overall assessment and in each of its components. The uncertainty statement is important for several reasons.

- Information from different sources carries different kinds of uncertainty and knowledge of these differences is important when uncertainties are combined for characterizing risk.
- Decisions must be made on expending resources to acquire additional information to reduce the uncertainties.
- A clear and explicit statement of the implications and limitations of a risk assessment requires a clear and explicit statement of related uncertainties.
- Uncertainty analysis gives the decision-maker a better understanding of the implications and limitations of the assessments.

A discussion of uncertainty requires comment on such issues as the quality and quantity of available data, gaps in the data base for specific chemicals, incomplete understanding of general biological phenomena, and scientific judgments or science policy positions that were employed to bridge information gaps.

In short, broad agreement exists on the importance of a full picture of risk, particularly including a statement of confidence in the assessment and that the uncertainties are within reason. This section discusses information content and uncertainty aspects of risk characterization, while Section 3 discusses various descriptors used in risk characterization.

1. The risk assessment process calls for characterizing risk as a combination of qualitative information, quantitative information, and information regarding uncertainties.

Risk assessment is based on a series of questions that the assessor asks about the data and the implications of the data for human risk. Each question calls for analysis and interpretation of the available studies, selection of the data that are most scientifically reliable and most relevant to the problem at hand, and scientific conclusions regarding the question presented. As suggested below, because the questions and analyses are complex, a complete characterization includes several different kinds of information, carefully selected for reliability and relevance.

a. Hazard identification — What do we know about the capacity of an environmental agent for causing cancer (or other adverse effects) in laboratory animals and in humans?

Hazard identification is a qualitative description based on factors such as the kind and quality of data on humans or laboratory animals, the availability of ancillary information (e.g., structure-activity analysis, genetic toxicity, pharmacokinetics) from other studies, and the weight-of-the evidence from all of these data sources. For example, to develop this description, the issues addressed include:

1. the nature, reliability, and consistency of the particular studies in humans and in laboratory animals;
2. the available information on the mechanistic basis for activity; and
3. experimental animal responses and their relevance to human outcomes.

These issues make clear that the task of hazard identification is characterized by describing the full range of available information and the implications of that information for human health.

     b.   **Dose-Response Assessment** — **What do we know about the biological mechanisms and dose-response relationships underlying any effects observed in the laboratory or epidemiology studies providing data for the assessment?**

The dose-response assessment examines quantitative relationships between exposure (or dose) and effects in the studies used to identify and define effects of concern. This information is later used along with "real world" exposure information (see below) to develop estimates of the likelihood of adverse effects in populations potentially at risk.

Methods for establishing dose-response relationships often depend on various assumptions used in lieu of a complete data base and the method chosen can strongly influence the overall assessment. This relationship means that careful attention to the choice of a high-to-low dose extrapolation procedure is very important. As a result, an assessor who is characterizing a dose-response relationship considers several key issues:

1. relationship between extrapolation models selected and available information on biological mechanisms;
2. how appropriate data sets were selected from those that show the range of possible potencies both in laboratory animals and humans;
3. basis for selecting interspecies dose scaling factors to account for scaling doses from experimental animals to humans; and
4. correspondence between the expected route(s) of exposure and the exposure route(s) utilized in the hazard studies, as well as the interrelationships of potential effects from different exposure routes.

EPA's Integrated Risk Information System (IRIS) is a primary source of this information. IRIS includes data summaries representing Agency consensus on specific chemicals, based on a careful review of the scientific issues listed above.

For specific risk assessments based on data in IRIS and on other sources, risk assessors should carefully review the information presented, emphasizing confidence in the database and uncertainties (see subsection d below). The IRIS statement of confidence should be included as part of the risk characterization for hazard and dose-response information.

C. Exposure Assessment — **What do we know about the paths, patterns, and magnitudes of human exposure and numbers of persons likely to be exposed?**

The exposure assessment examines a wide range of exposure parameters pertaining to the "real world" environmental scenarios of people who may be exposed to the agents under study. The data considered for the exposure assessment range from monitoring studies of chemical concentrations in environmental media, food, and other materials to information on activity patterns of different population subgroups. An assessor who characterizes exposure should address several issues.

1. The basis for values and input parameters used for each exposure scenario. If based on data, information on the quality, purpose, and representativeness of the database is needed. If based on assumptions, the source and general logic used to develop the assumption (e.g., monitoring, modeling, analogy, professional judgment) should be described.

2. The major factor or factors (e.g., concentration, body uptake, duration/frequency of exposure) thought to account for the greatest uncertainty in the exposure estimate, due either to sensitivity or lack of data.

3. The link of the exposure information to the risk descriptors discussed in Section 3 of this Appendix. This issue includes the conservatism or non-conservatism of the scenarios, as indicated by the choice of descriptors.

In summary, confidence in the information used to characterize risk is variable, with the result that risk characterization requires a statement regarding the assessor's confidence in each aspect of the assessment.

d. **Risk Characterization** — **What do other assessors, decision-makers, and the public need to know about the primary conclusions and assumptions, and about the balance between confidence and uncertainty in the assessment?**

In the risk characterization, conclusions about hazard and dose response are integrated with those from the exposure assessment. In addition, confidence about theses conclusions, including information about the uncertainties associated with the final risk summary, is highlighted. As summarized below, the characterization integrates all of the preceding information to communicate the overall meaning of, and confidence in, the hazard, exposure, and risk conclusions.

Generally, risk assessments carry two categories of uncertainty, and each merits consideration. Measurement uncertainty refers to the usual variance that accompanies scientific measurements (such as the range around an exposure estimate) and reflects the accumulated variances around the individual measured values used to develop the estimate. A different kind of uncertainty stems from data gaps — that is, information needed to complete the data base for the assessment. Often, the data gap is broad, such as the absence of information on the effects of exposure to a chemical on humans or on the biological mechanism of action of an agent.

The degree to which confidence and uncertainty in each of these areas is addressed depends largely on the scope of the assessment and the resources available. For example, the Agency does not expect an assessment to evaluate and assess every conceivable exposure scenario for every possible pollutant, to examine all susceptible populations potentially at risk, or to characterize every possible environmental scenario to determine the cause and effect relationships between exposure to pollutants and adverse health effects. Rather, the uncertainty analysis should reflect the type and complexity of the risk assessment, with the level of effort for analysis and discussion of uncertainty corresponding to the level of effort for the assessment. Some sources of confidence and of uncertainty are described below.

Often risk assessors and managers simplify discussion of risk issues by speaking only of the numerical components of an assessment. That is, they refer to the weight-of-evidence, unit risk, the risk-specific dose or the q^{1*} for cancer risk, and the RfD/RfC for health effects other than cancer, to the exclusion of other information bearing on the risk case. However, since every assessment carries uncertainties, a simplified numerical presentation of risk is always incomplete and often misleading. For this reason, the NRC (1) and EPA risk assessment guidelines (2) call for "characterizing" risk to include qualitative information, a related numerical risk estimate and a discussion of uncertainties, limitations, and assumptions.

Qualitative information on methodology, alternative interpretations, and working assumptions is an important component of risk characterization. For example, specifying that animal studies rather than human studies were used in an assessment tells others that the risk estimate is based on assumptions about human response to a particular chemical rather than human data. Information that human exposure estimates are based on the subjects's presence in the vicinity of a chemical accident rather than tissue measurements defines known and unknown aspects of the exposure component of the study.

Qualitative descriptions of this kind provide crucial information that augments understanding of numerical risk estimates. Uncertainties such as these are expected in scientific studies and in any risk assessment based on these studies. Such uncertainties do not reduce the validity of the assessment. Rather, they are highlighted along with other important risk assessment conclusions to inform others fully on the results of the assessment.

2. Well-balanced risk characterization presents information for other risk assessors, EPA decision-makers, and the public regarding the strengths and limitations of the assessment.

The risk assessment process calls for identifying and highlighting significant risk conclusions and related uncertainties partly to assure full communication among risk assessors and partly to assure that decision-makers are fully informed. Issues are identified by acknowledging noteworthy qualitative and quantitative factors that make a difference in the overall assessment of hazard and risk, and hence in the ultimate regulatory decision.

The key word is "noteworthy": information that significantly influences the analysis is retained — that is, noted — in all future presentations of the risk assessment and in the related decision. Uncertainties and assumptions that strongly influence confidence in the risk estimate require special attention.

As discussed earlier, two major sources of uncertainty are variability in the factors upon which estimates are based and the existence of fundamental data gaps. This distinction is relevant for some aspects of the risk characterization. For example, the central tendency and high end individual exposure estimates are intended to capture the variability in exposure, lifestyles, and population. Key considerations underlying these risk estimates should be fully described. In contrast, scientific assumptions are used to bridge knowledge gaps such as the use of scaling or extrapolation factors and the use of a particular upper confidence limit around a dose-response estimate. Such assumptions need to be discussed separately, along with the implications of using alternative assumptions.

For users of the assessment and others who rely on the assessment, numerical estimates should never be separated from the descriptive information that is integral to risk characterization. All documents and presentations should include both; in short reports, this information is abbreviated but never omitted.

For decision-makers, a complete characterization (key descriptive elements along with numerical estimates) should be retained in all discussions and papers relating to an assessment used in decision-making. Fully visible information assures that important features of the assessment are immediately available at each level of decision-making for evaluating whether risks are acceptable or unacceptable. In short, differences in assumptions and uncertainties, coupled with non-scientific considerations called for in various environmental statutes, can clearly lead to different risk management decisions in cases with ostensibly identical quantitative risks; i.e, the "number" alone does not determine the decision.

Consideration of alternative approaches involves examining selected plausible options for addressing a given uncertainty. The key words are "selected" and "plausible;" listing all options, regardless of their merits would be superfluous.

Generators of the assessment should outline the strengths and weaknesses of each alternative approach and as appropriate, estimates of central tendency and variability (e.g., mean, percentiles, range, variance.)

Describing the option chosen involves several statements.

1. A rationale for the choice.
2. Effects of option selected on the assessment.
3. Comparison with other plausible options.
4. Potential impacts of new research (on-going, potential near-term and/or long-term studies).

For users of the assessment, giving attention to uncertainties in all decisions and discussions involving the assessment, and preserving the statement of confidence in all presentations is important. For decision-makers, understanding the effect of the uncertainties on the overall assessment and explaining the influence of the uncertainties on the regulatory decision.

SECTION 3. EXPOSURE ASSESSMENT AND RISK DESCRIPTORS

The results of a risk assessment are usually communicated to the risk manager in the risk characterization portion of the assessment. This communication is often accomplished through risk descriptors which convey information and answer questions about risk, each descriptor providing different information and insights. Exposure assessment plays a key role in developing these risk descriptors, since each descriptor is based in part on the exposure distribution within the population of interest. The Risk Assessment Council (RAC) has been discussing the use of risk descriptors from time to time over the past two years.

The recent RAC efforts have laid the foundation for the discussion to follow. First, as a result of a discussion paper on the comparability of risk assessments across the Agency programs, the RAC discussed how the program presentations of risk led to ambiguity when risk assessments were compared across programs. Because different assessments presented different descriptors of risk without always making clear what was being described, the RAC discussed the advisability of using separate descriptors for population risk, individual risk, and identification of sensitive or highly exposed population segments. The RAC also discussed the need for consistency across programs and the advisability of requiring risk assessments to provide roughly comparable information to risk managers and the public through the use of a consistent set of risk descriptors.

The following guidance outlines the different descriptors in convenient order that should not be construed as a hierarchy of importance. These descriptors should be used to describe risk in a variety of ways for a given assessment's purpose, the data available, and the information the risk manager needs. Use of a range of descriptors instead of a single descriptor enables Agency programs to present a picture of risk that corresponds to the range of different exposure conditions encountered for most environmental chemicals. This analysis, in turn, allows risk managers to identify populations at greater and lesser risk and to shape regulatory solutions accordingly.

EPA risk assessments will be expected to address or provide descriptions of (1) individual risk to include the central tendency and high end portions of the risk distribution, (2) important subgroups of the population such as highly exposed or highly susceptible groups or individuals, if known, and (3) population risk. Assessors may also use additional descriptors of risk as needed when these add to the clarity of the presentation. With the exception of assessments where particular descriptors clearly do not apply, some form of these three types of descriptors should be routinely developed and presented for EPA risk assessments. Furthermore, presenters of risk assessment information should be prepared to routinely answer questions by risk managers concerning these descriptors.

It is essential that presenters not only communicate the results of the assessment by addressing each of the descriptors where appropriate, but they also communicate their confidence that these results portray a reasonable picture of the actual or projected exposures. This task will usually be accomplished by highlighting the key assumptions and parameters that have the greatest impact on the results, the basis or rationale for choosing these assumptions/parameters, and the consequences of choosing other assumptions.

In order for the risk assessor to successfully develop and present the various risk descriptors, the exposure assessment must provide exposure and dose information in a form that can be combined with exposure-response or dose-response relationships to estimate risk. Although there will be differences among individuals within a population as to absorption, intake rates, susceptibility, and other variables such that a high exposure does not necessarily result in a high does or risk, a moderate or highly positive correlation among exposure, dose, and risk is assumed in the following discussion. Since the generation of all descriptors is not appropriate in all risk assessments and the type of descriptor translates fairly directly into the type of analysis that the exposure assessor must perform, the exposure assessor needs to be aware of the ultimate goals of the assessment. The following sections discuss what type of information is necessary.

1. Information about <u>individual</u> exposure and risk is important to communicating the results of a risk assessment.

Individual risk descriptors are intended to address questions dealing with risks borne by individuals within a population. These questions can take the form of:

- Who are the people at the highest risk?
- What risk levels are they subjected to?
- What are they doing, where do they live, etc., that might be putting them at this higher risk?
- What is the average risk for individuals in the population of interest?

The "high end" of the risk distribution is, conceptually, above the 90th percentile of the actual (either measured or estimated) distribution. This conceptual range is not meant to precisely define the limits of this descriptor, but should be used by the assessor as a target range for characterizing "high end risk." Bounding estimates and worst case scenarios[1] should not be termed high end risk estimates.

> The high end risk descriptor is a plausible estimate of the individual risk for those persons at the upper end of the risk distribution. The intent of this descriptor is to convey an estimate of risk in the upper range of the distribution, but to avoid estimates which are beyond the true distribution. Conceptually, high end risk means risks above about the 90th percentile of the population distribution, but not higher than the individual in the population who has the highest risk.

This descriptor is intended to estimate the risks that are expected to occur in small but definable "high end" segments of the subject population. The individuals with these risks may be members of a special population segment or individuals in the general population who are highly exposed because of the inherent stochastic nature of the factors which give rise to exposure. Where no particular differences in sensitivity can be identified within the population, the high end risk will be related to the high end exposure or dose.

In those few cases where the complete data on the population distributions of exposure and doses are available, high end exposure or dose estimates can be represented by reporting exposures or doses at selected percentiles of the distri-

---

[1] High end estimates focus on estimates of the exposure or dose in the actual populations. "Bounding estimates," on the other hand, purposely overestimate the exposure or dose in an actual population for the purpose of developing a statement that the risk is "not greater than...." A "worst case scenario" refers to a combination of events and conditions such that, taken together, produce the highest conceivable risk. Although it is possible that such an exposure, dose, or sensitivity combination might occur in a given population of interest, the probability of an individual receiving this combination of events and conditions is usually small, and often so small that such a combination will not occur in a particular, actual population.

butions, such as 90th, 95th, or 98th percentile. High end exposures or doses, as appropriate, can then be used to calculate high end risk estimates.

In the majority of cases where the complete distributions are not available, several methods help estimate a high end exposure or dose. If sufficient information about the variability in lifestyles and other factors are available to simulate the distribution through the use of appropriate modeling, e.g., Monte Carlo simulation, the estimate from the simulated distribution may be used. As in the method above, the risk manager should be told where in the high end range the estimate is being made by stating the percentile or the number of persons above this estimate. The assessor and risk manager should be aware, however, that unless a great deal is known about exposures and doses at the high end of the distribution, these estimates will involve considerable uncertainty which the exposure assessor will need to describe.

If only limited information on the distribution of the exposure or dose factors is available, the assessor should approach estimating the high end by identifying the most sensitive parameters and using maximum or near maximum values for one or a few of these variables, leaving others at their mean values[2]. In doing this, the exposure assessor needs to avoid combinations of parameter values that are inconsistent, e.g., low body weight used in combination with high intake rates, and must keep in mind the ultimate objective of being within the distribution of actual expected exposures and doses, and not beyond it.

If almost no data are available on the range for the various parameters, it will be difficult to estimate exposures or doses in the high end with much confidence, and to develop the high end risk estimate. One method that has been used in these cases is to start with a bounding estimate and "back off" the limits used until the combination of parameter values is, in the judgment of the assessor, clearly within the distribution of expected exposure, and still lies within the upper 10% of persons exposed. Obviously, this method results in a large uncertainty and requires explanation.

> The risk descriptor addressing central tendency may be either the arithmetic mean risk (Average Estimate) or the median risk (Median Estimate), either of which should be clearly labeled. Where both the arithmetic mean and the median are available but they differ substantially, it is helpful to present both.

The Average Estimate, used to approximate the arithmetic mean, can be derived by using average values for all the exposure factors. It does not neces-

---

[2]Maximizing all variables will in virtually all cases result in an estimate that is above the actual values seen in the population. When the principal parameters of the dose equation (e.g., concentration, intake rate, duration) are broken out into subcomponents, it may be necessary to use maximum values for more than two of these subcomponent parameters, depending on a sensitivity analysis.

sarily represent a particular individual on the distribution. The Average Estimate is not very meaningful when exposure across a population varies by several orders of magnitude or when the population has been truncated, e.g., at some point prescribed distance form a point source.

Because of the skewness of typical exposure profiles, the arithmetic mean is not necessarily a good indicator of the midpoint (median, 50th percentile) of a distribution. A Median Estimate, e.g., geometric mean, is usually a valuable descriptor for this type of distribution, since half the population will be above and half below this value.

2. Information about population exposure leads to another important way to describe risk.

Population risk refers to an assessment of the extent of harm for the population as a whole. In theory, it can be calculated by summing the individual risks for all individuals within the subject population. This task, of course, requires a great deal more information than is normally, if ever, available.

Some questions addressed by descriptors of population risk include:

- How many cases of a particular health effect might be probabilistically estimated in this population for a specific time period?

- For noncarcinogens, what portion of the population are within a specified range of some benchmark level, e.g., exceedance of the RfD (a dose), the RfC (a concentration), or other health concern level?

- For carcinogens, how many persons are above a certain risk level such as 10^{-6} or a series of risk levels such as 10^{-5}, 10^{-4}, etc?

Answering these questions requires some knowledge of the exposure frequency distribution in the population. In particular, addressing the second and third questions may require graphing the risk distribution. These questions can lead to two different descriptors of population risk.

> The first descriptor is the probabilistic number of health effect cases estimated in the population of interest over a specified time period.

This descriptor can be obtained either by (a) summing the individual risks over all the individuals in the population when such information is available, or (b) through the use of a risk model such as carcinogenic models or procedures which assume a linear non-threshold response to exposure. If risk varies linearly with exposure, knowing the mean risk and the population size can lead to an estimate of the extent of harm for the population as a whole, excluding sensitive subgroups for which a different dose-response curve needs to be used.

Obviously, the more information one has, the more certain the estimate of this risk description, but inherent uncertainties in risk assessment methodology place limitations on the accuracy of the estimate. With the current state of the science, explicit steps should be taken to assure that this descriptor is not confused with an actuarial prediction of cases in the population (which is a statistical prediction based on a great deal of empirical data).

Although estimating population risk by calculating a mean individual risk and multiplying by the population size is sometimes appropriate for carcinogen assessments using linear, non-threshold models[3], this is not appropriate for non-carcinogenic effects or for other types of cancer models. For non-linear cancer models, an estimate of population risk must be calculated by summing individual risks. For non-cancer effects, we generally have not developed the risk assessment techniques to the point of knowing how to add risk probability, so a second descriptor, below, is more appropriate.

> Another descriptor of population risk is an estimate of the percentage of the population, or the number of persons, above a specified level of risk or within a specified range of some benchmark level, e.g., exceedance of the RfD or the RfC, LOAEL, or other specific level of interest.

This descriptor must be obtained through measuring or simulating the population distribution.

3. Information about the distribution of exposure and risk for different subgroups of the population are important components of a risk assessment.

A risk manager might also ask questions about the distribution of the risk burden among various segments of the subject population such as the following:

- How do exposure and risk impact various subgroups?
- What is the population risk of a particular subgroup?

Questions about the distribution of exposure and risk among such population segments require additional risk descriptors.

---

[3]Certain important cautions apply. These cautions are more explicitly spelled out in the Agency's Guidelines for Exposure Assessment, tentatively scheduled to be published in late 1991.

> Highly exposed subgroups can be identified, and where possible, characterized and the magnitude of risk quantified. This descriptor is useful when there is (or is expected to be) a subgroup experiencing significantly different exposures or doses from that of the larger population.

These subpopulations may be identified by age, sex, lifestyle, economic factors, or other demographic variables. For example toddlers who play in contaminated soil and certain high fish consumers represent subpopulations that may have greater exposures to certain agents.

> Highly susceptible subgroups can also be identified, and if possible, characterized and the magnitude of risk quantified. This descriptor is useful when the sensitivity or susceptibility to the effects for specific subgroups is (or is expected to be) significantly different from that of the larger population. In order to calculate risk for these subgroups, it will sometimes be necessary to use a different dose-response relationship.

For example, upon exposure to a chemical, pregnant women, elderly people, children, and people with certain illnesses may each be more sensitive than the population as a whole.

Generally, selection of the population segments is a matter of either a priori interest in the subgroup, in which case the risk assessor and risk manager can jointly agree on which subgroups to highlight, or a matter of discovery of a sensitive or highly exposed subgroup during the assessment process. In either case, once identified, the subgroup can be treated as a population in itself, and characterized the same way as the larger population using the descriptors for population and individual risk.

4. Situation-specific information adds perspective on possible future events or regulatory options.

These postulated questions are normally designed to answer "what if" questions, which are either directed at low probability but possibly high consequence events or are intended to examine candidate risk management options. Such questions might take the following form:

- What if a pesticide applicator applies this pesticide without using protective equipment?
- What if this site becomes residential in the future?
- What risk level will occur if we set the standard at100 ppb?

The assumptions made in answering these postulated questions should not be confused with the assumptions made in developing a baseline estimate of exposure or with the adjustments in parameter values made in performing a sensitivity analysis. The answers to these postulated questions do not give information about how likely the combination of values might be in the actual population or about how many (if any) persons might be subjected to the calculated exposure or risk in the real world.

> A calculation of risk based on specific hypothetical or actual combinations of factors postulated within the exposure assessment can also be useful as a risk descriptor. It is often valuable to ask and answer specific questions of the "what if" nature to add perspective to the risk assessment.

The only information the answers to these questions convey is that if conditions A, B, and C are assumed, then the resulting exposure or risk will be X, Y, or Z, respectively. The values for X, Y, and Z are usually fairly straightforward to calculate and can be expresses as point estimates or ranges. Each assessment may have none, one, or several of these types of descriptors. The answers do not directly give information about how likely that combination of values might be in the actual population, so there are some limits to the applicability of these descriptors.

References

1.   National Research Council. Risk Assessment in the Federal Government: Managing the Process. 1983.
2.   U.S. EPA. Risk Assessment and Management: Framework for Decision Making. 1984.
3.   U.S. EPA. Risk Assessment Guidelines. 51 Federal Register 33992-34054. September 24, 1986.
4.   Presentation of Risk Assessment of Carcinogens; Ad Hoc Study Group on Risk Assessment Presentation. American Industrial Health Council. 1989.

APPENDIX

C

Calculation and Modeling of Exposure

This appendix describes some of the mathematical relationships and models used in exposure assessment.

CALCULATION OF EXPOSURE

Assessing exposure to a pollutant requires information on the pollutant concentration at a specific location (microenvironment) and the duration of contact with a person or population. If the concentration of a pollutant to which a person is exposed can be measured or modeled and the time spent in contact with the pollutant is known, exposure is determined from concentration and time. When concentration varies with time, the total exposure from time t_1 to t_2 is given by

$$E = \int_{t_1}^{t_2} C(t)\, dt \,,$$

where E is the exposure of a person to a pollutant at concentration C; C(t) represents the functional relationship of concentration with time t for an interval t_1 through t_2. The average ("time-weighted average") exposure during this interval is $E/(t_2-t_1)$.

It is often assumed that the concentration is constant within a given microenvironment j for some finite interval, Δt_j. Thus, any particular exposure within a given microenvironment e_j is given by

$$e_j = \overline{C}_j \Delta t_j \,,$$

which means that a person stays within the microenvironment with average con-

375

centration $\overline{C}_j$ for the interval Δt_j. A person's total exposure E to an airborne pollutant is the summation over all the microenvironments M in which the person is in contact with the pollutant:

$$E = \sum_{j=1}^{M} e_j = \sum_{j=1}^{M} \overline{C}_j \Delta t_j \,.$$

The latter equation includes the totality of all locations and activities that the person can occupy and engage in.

To obtain the total exposure of a population E_{pop} of N persons, it is necessary to sum the individual exposures E_i of all the persons in the population from $i = 1$ to N:

$$E_{pop} = \sum_{i=1}^{N} E_i = \sum_{i=1}^{N} \sum_{j=1}^{M} \overline{C}_j \Delta t_{ij} \,.$$

Generally, the amount of time spent in each microenvironment is averaged over the exposed population,

$$E_{pop} = N \sum_{j=1}^{M} \overline{C_j \Delta t_{ij}} \,,$$

so that the average population exposure is given by

$$\overline{E_{pop}} = \sum_{j=1}^{M} \overline{C_j \Delta t_j} \,.$$

Thus, it is necessary to estimate the atmospheric concentration of the pollutant to which people are exposed to obtain C_j and their activity patterns to obtain Δt_j.

MODELING OF EXPOSURE

It is often impossible or impractical to measure the exposures of individuals or populations directly, and instead mathematical models are used to estimate exposures. Microenvironmental concentrations are estimated with concentration models, which are based on the physics and chemistry of the environment. The time spend by an individual in a microenvironment with a pollutant is another important input to an exposure model. Population-exposure models combine data representing the time-activity patterns of an entire population with pollutant concentrations.

Gaussian-Plume Models

Gaussian-plume models are used by the Environmental Protection Agency (EPA) to estimate the concentration of a pollutant at locations some distance from an emission source. The models have this name because they represent the plume of emissions from a stack as having a Gaussian, or normal, distribution, with a maximum at the center line, as shown in Figure C-1. The effect of boundaries (such as the ground or an atmospheric inversion cap), multiple emission sources, and deposition can alter the basic Gaussian distribution. Gaussian-plume models have been generalized to consider continuous and intermittent emissions, as well as emissions from points (e.g., concentrated emissions from a stack), areas (e.g., distributed emissions throughout a modeled region, such as home heating), and lines (e.g., roads). Gaussian-plume models have been further extended to complex topographic regions, such as valleys and bodies of water, and to industrial sources. They have also been designed for various temporal averaging periods. A number of Gaussian-plume models, with individual names, correspond to the various mathematical formulations used in the models. A few of the more commonly used Gaussian-plume models are the industrial-source complex long-term (ISCLT) and industrial-source complex short-term (ISCST) models, for long- and short-term averaging times, respectively; LONGZ (basic long-term model); Complex (for complex terrain); and Valley (for valleys). These are parts of the EPA UNAMAP modeling library (see Zannetti, 1990, for a brief description of each one and how to obtain the models).

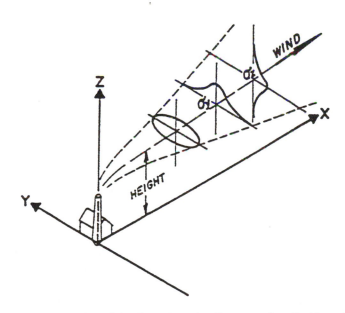

FIGURE C-1 Visualization of the dispersion of pollutants as described by a Gaussian-plume model. Source: Russell, 1988. Reprinted with permission; copyright 1988, Health Effects Institute, Cambridge, Mass.

Gaussian-plume models are among the simplest atmospheric-dispersion models, but they can still involve a number of complexities. For example, many sources emit their effluent at higher than ambient temperatures, so their pollutants tend to rise. The rise is a complex process to describe, requiring the simultaneous consideration of heat and mass transfer, atmospheric turbulence, and source characteristics. Conversely, a pollutant may be emitted without sufficient buoyancy or momentum to be lifted above the wake of turbulent air downwind of a building or a topographic feature. The pollutants can then be caught in the wake and downwashed, increasing the potential exposure. Specific Gaussian-plume models, such as the ISCLT and ISCST models, have been developed for that possibility (EPA, 1987). The ISCLT and ISCST models are often suggested for use in exposure assessment of air pollutants from industrial sources. The Human-Exposure Model (described below), which is used by EPA, also uses the industrial-source complex models. Multiple sources are treated by superimposing the calculated contributions of individual sources. It is possible to include the first-order chemical decay of pollutant species within the Gaussian-plume framework, as well as deposition of both gases and particles.

Although Gaussian-plume models have been used for many years, their results are still subject to considerable error. In many cases, especially far from the source, they are biased to predict high concentrations. Applying Gaussian-plume models in complex terrain (such as hilly areas or areas with tall buildings) leads to even greater uncertainties and can result in significant overprediction and underprediction. Their rather simple formulation makes it difficult to handle complex terrain.

Human-Exposure Model

The HEM, one of the more commonly used models developed for EPA, incorporates a simple Gaussian-plume dispersion model with a fixed-location population model. Although EPA has developed several Gaussian-plume dispersion models for which validation studies have been conducted, the HEM was constructed with a model that incorporates an alternative approach to estimating the horizontal and vertical dispersion rates. The model was then compared with the standard UNIMAP models issued by the EPA Office of Air Quality, Planning, and Standards as part of the National Ambient Air Quality Standards (NAAQS) State Implementation Plan process, and it was found that they generally agreed to within a factor of 3. No comparison with field-measurement data was reported. In the most recent version of the program, the ISCLT model was incorporated as the default dispersion model, so that multiple emission points within the source area could be modeled, rather than aggregating all the emissions at a single point source within the source complex.

It is possible to substitute concentration data from other dispersion models into the HEM. For example, LONGZ was used to model the dispersion of arsenic from the ASARCO smelter in Tacoma, Wash. LONGZ is a complex-terrain model that was optimized to reproduce the sulfur dioxide dispersion from this plant. However, it is not clear that it was adequately modified to take particle deposition into account, and it was found to overpredict the airborne concentration of arsenic by factors of 5-8 for distances of up to 3 km from the plant and factors of 1.6-1.8 for larger distances. Assays of arsenic in urine also suggested that the model substantially overestimated arsenic exposure.

For distributed sources, such as perchloroethylene from dry cleaners, area sources were used with emission rates proportional to area population. The dispersion model was modified to incorporate the additional dispersion that comes from surface roughness and heat-island effects. The correction is included by making some of the parameters depend on the city geographic area.

In the HEM, the population is based on data from the Bureau of the Census (enumeration district/block groups, ED/BGs). An ED/BG is the area containing on average about 800 people and can range from part of a single city block to several hundred square kilometers. The population of each ED/BG is assumed to be at the center of the population's geographic distribution (centroid). The pollutant concentration at that location is interpolated from the results of the dispersion model. The interpolation is logarithmic in the radial direction and linear in the azimuthal direction. The product of the population and the concentration summed over the total area is then the total annual population exposure.

NAAQS Exposure Model

The NAAQS exposure model (NEM) was developed to estimate exposure to the criteria pollutants (e.g., carbon dioxide, CO). In 1979, EPA began to develop this model by assembling a database of human activity patterns that could be used to estimate exposures to outdoor pollutants (Roddin et al., 1979). The data were then combined with measured outdoor concentrations in the NEM to estimate exposures to CO (Biller et al., 1981; Johnson and Paul, 1983). The NEM has recently been modified to include indoor exposures by incorporation of the Indoor Air Quality Model (IAQM) (Hayes and Lundberg, 1985). The IAQM is based on the recursive (stepwise) solution of a one-compartment mass-balance model and incorporates three basic indoor microenvironments: home, office or school, and transportation vehicle. It has been used to estimate distributions of ozone exposures (Hayes and Lundberg, 1985) and to evaluate mitigation strategies for indoor exposures to selected pollutants for five scenarios, such as exposure to CO from a gas boiler in a school (Eisinger and Austin, 1987).

Simulation of Human Air Pollution Exposure (SHAPE) Model

SHAPE (Ott, 1981) is a computerized simulation model that generates synthetic exposure profiles for a hypothetical sample of human subjects; the exposure profiles can be summarized into exposure measures—say, integrated exposures—to estimate the distribution for the exposure measure of interest. The bulk of the model estimates the exposure profile for pollutants attributable to local sources; the contribution of remote sources is assumed to be the same as that of a background site where there is no local source. The total exposure is therefore estimated as the exposure due to local sources plus the ambient concentration at the background site.

For each person in the hypothetical sample, the model generates a profile of activities and pollutant concentrations attributable to local sources over a given period, such as a 24-hour period. The activity profiles are generated by a modified Markov model. A later version of SHAPE can accept given profiles of activities, instead of using simulation to generate the activity profiles. At the beginning of the profile, an initial microenvironment is generated according to a probability distribution with the time spent in it, generated according to a microenvironment-specific probability distribution: each microenvironment has a specific probability distribution for its duration. At the end of the duration, a transition into another microenvironment is generated according to a transition probability distribution with another duration. The procedure is repeated until the end of the given period. For each time unit, such as a minute, spent in a given microenvironment, a pollutant concentration is generated according to a microenvironment-specific probability distribution, and each microenvironment has a specific probability distribution for its pollutant concentration. All random values are generated independently of each other.

Convolution Model

Duan (1981) originally developed the convolution model for the integrated exposure attributable to local sources and later (1987) expanded it for a broader context. In this model, distributions of exposure are calculated from the distributions of concentrations observed in each defined microenvironment and the distribution of time spent in those microenvironments. Thus, distributions of exposure are calculated for a population by assuming that values of concentration and time can be independently drawn from the exposure distributions and combined to yield a series of individual exposures. The exposures can then be summed over time to yield a time-integrated exposure for an individual in the population. Enough cases are drawn to provide a distribution of exposures for the entire population.

Variance-Component Model

The variance-component model assumes that short-term pollutant concentrations comprise two components, a time-varying component and a time-invariant component. If neither the time-varying component nor the time-invariant component is negligible, SHAPE and the convolution method can no longer be used; it is necessary to use the variance-component model, which can incorporate both the time-variant and the time-invariant components. Depending on the needs of the analyst, the two components can be either summed or multiplied to estimate the modeled concentration value. Contaminant concentrations are usually more variable at higher values, so the multiplicative form may often be more realistic.

It is first necessary to determine the distributions of the two components. If random samples of locales belonging to the same microenvironment are available and if there are continuous monitoring data for at least a random sample of locales, it is possible to estimate the distributions of time-varying and time-invariant components of the concentration directly. If integrated personal monitoring data are available, the methods described by Duan (1987) can be applied. Once those distributions are available, exposure distributions are estimated with a computer simulation similar to that in SHAPE. However, instead of generating a contaminant concentration for each time unit independently, as in SHAPE, values of the time-invariant and time-varying components for each time unit are generated and then combined to determine 1-minute average concentrations. The remainder of the simulation is identical with SHAPE.

REFERENCES

Biller, W.F., T.B. Feagans, T.R. Johnson, G.M. Duggan, R.A. Paul, T. McCurdy, and H.C. Thomas. 1981. A general model for estimating exposure associated with alternative NAAQS. Paper No. 81-18.4 in Proceedings of the 74th Annual Meeting of the Air Pollution Control Association, Philadelphia, Pa.

Duan, N. 1981. Micro-Environment Types: A Model for Human Exposure to Air Pollution. SIMS Technical Report No. 47. Department of Statistics, Stanford University, Stanford, Calif.

Duan, N. 1987. Cartesianized Sample Mean: Imposing Known Independence Structures on Observed Data. WD-3602-SIMS/RC. The RAND Corporation, Santa Monica, Calif.

Eisinger, D.S., and B.S. Austin. 1987. Indoor Air Quality: Problem Characterization and Computer Simulation of Indoor Scenarios and Mitigation Strategies. Report No. SYSAPP-87/170. Systems Applications, Inc., San Rafael, Calif.

EPA (U.S. Environmental Protection Agency). 1987. Industrial Source Complex (ISC) User's Guide, 2nd ed., revised, Vols. 1 and 2. EPA 450/4-88-002a,b. U.S. Environmental Protection Agency, Research Triangle Park, N.C.

Hayes, S.R., and G.W. Lundberg. 1985. Further Improvement and Sensitivity Analysis of an Ozone Population Exposure Model. Draft final report to the American Petroleum Institute. Report No. SYSAPP-85/061. Systems Applications, Inc., San Rafael, Calif.

Johnson, T.R., and R.A. Paul. 1983. The NAAQS Exposure Model (NEM) Applied to Carbon Monoxide. EPA-450/5-83-003. Prepared for the U.S. Environmental Agency by PEDCo Environmental Inc., Durham, N.C. under Contract No. 68-02-3390. U.S. Environmental Protection Agency, Research Triangle Park, N.C.

Ott, W. 1981. Computer Simulation of Human Air Pollution Exposures to Carbon Monoxide. Paper 81-57.6. Paper presented at 74th Annual Meeting of the Air Pollution Control Association, Philadelphia, Pa.

Roddin, M.F., H.T. Ellis, and W.M. Siddiqee. 1979. Background Data for Human Activity Patterns, Vols. 1, 2. Draft Final Report prepared for Strategies and Air Standards Division, Office of Air Quality Planning and Standards, U.S. Environmental Protection Agency, Research Triangle Park, N.C.

Russell, A.G. 1988. Mathematical modeling of the effect of emission sources on atmospheric pollutant concentrations. Pp. 161-205 in Air Pollution, the Automobile, and Public Health, A.Y. Watson, R.R. Bates, and D. Kennedy, eds. Washington, D.C.: National Academy Press.

Zannetti, P. 1990. Air Pollution Modeling: Theories, Computational Methods, and Available Software. New York: Van Nostrand Reinhold.

D

Working Paper for Considering Draft Revisions to the U.S. EPA Guidelines for Cancer Risk Assessment

NOTICE

THIS DOCUMENT IS A PRELIMINARY DRAFT. Until formal announcement by the U.S. Environmental Protection Agency is made in the **Federal Register**, the policies set forth in the 1986 **Guidelines for Carcinogen Risk Assessment**, as they are now interpreted, remain in effect. This working paper does not represent the policy of the U.S. Environmental Protection Agency with respect to carcinogen risk assessment.

Office of Health and Environmental Assessment
Office of Research and Development
U.S. Environmental Protection Agency
Washington, D.C.

DISCLAIMER

This document is a draft working paper for review purposes only and does not constitute Agency policy. Mention of trade names or commercial products does not constitute endorsesement or recommendation for use.

CONTENTS

LIST OF FIGURES

AUTHORS AND CONTRIBUTORS

This draft working paper was prepared by an intra-Agency EPA working group chaired by Jeanette Wiltse of the Office of Health and Environmental Assessment.

WORKING PAPER FOR CONSIDERING
DRAFT REVISIONS TO THE
U.S. EPA GUIDELINES FOR CANCER RISK ASSESSMENT

This working paper identifies cancer risk assessment issues that some Agency scientists have been discussing as a basis for possible proposed revisions to EPA's 1986 Guidelines for Carcinogen Risk Assessment. The working paper is being given to other scientists to obtain early comment on the many issues that remain undeveloped or are still under discussion. The working paper is not a proposal. It has not been reviewed or approved by any EPA official, and the proposal that is eventually approved is likely to be very different in many respects from this working paper. When proposed revisions are ready, EPA will publish them in the Federal Register for public comment.

Until formal announcement by the U.S. Environmental Protection Agency is made in the Federal Register, the policies set forth in the 1986 Guidelines for Carcinogen Risk Assessment, as they are now interpreted, remain in effect. This working paper does not represent the policy of the U.S. Environmental Protection Agency with respect to carcinogen risk assessment.

PREAMBLE

The United States Environmental Protection Agency (EPA) 1986 guidelines on carcinogenic risk assessment (51 FR 33992, September 24, 1986)) stated that, "...[a]t present, mechanisms of the carcinogenesis process are largely unknown..." . This is no longer true. The last several years have brought research results at an explosive pace to elucidate the molecular biology of cancer. This new knowledge is only beginning to be applied in generating data about environmental agents. Guideline revisions are intended to be flexible and open to the use of such new kinds of data even though the guidelines cannot fully anticipate the future forms that carcinogenicity testing and research may take. At the same time, the guidelines address assessment of the kinds of data that are the current basis of carcinogenicity assessment as a result of the past two decades of development of the science of risk assessment. Because methods and knowledge are expected to change more rapidly than guidelines can practicably be revised, most of the Agency's development of procedures for cancer risk assessment will henceforth be accomplished through publication of technical work performed under the aegis of the Agency's Risk Assessment Forum. The technical documents of the Forum are developed by a process that engages the general scientific community with EPA scientists. The documents are made available for public examination as well as for scientific peer review through the EPA Science Advisory Board and other groups. The Forum sponsored two workshops in which areas of potential revision to the guidelines were discussed by scientists from public and private groups. (USEPA, 1989a; USEPA, 1991a).

Major Changes from 1986 Guidelines

Revisions in this working paper differ in many respects from the Agency's 1986 guidelines. The reasons for change arise from new research results, particularly about the molecular biology of cancer, and from experience using the 1986 guidelines.

One area of change is increased emphasis on providing characterization discussions for each part of a risk assessment (hazard, dose-response, exposure, and risk assessments). These serve to summarize the assessments with emphasis on explaining the extent and weight of evidence, major points of interpretation and rationale, strengths and weaknesses of the evidence and analysis, and alternative conclusions that deserve serious consideration.

Two other areas of major change are in:

(1) the way the weight of evidence about an agent's[1] hazard potential is expressed; and

---

[1]The term "agent" is used throughout (unless otherwise noted) for a chemical substance, mixture, or physical or biological entity that is being assessed.

(2) approaches to dose-response assessment.

1. To express the weight of evidence for carcinogenic hazard potential, the 1986 guidelines provided tiered summary rankings for human studies and for animal bioassays. These summary rankings of evidence were integrated to place the overall evidence in alphanumerically designated classification groups A through E, Group A being associated with the greatest probability of carcinogenicity. Other experimental evidence played a modulating role for ranking. Considerations such as route of exposure (e.g., oral versus inhalation) and mechanism of action were not explicitly captured in a characterization.

These working revisions take a different approach. The idea of summary ranking of individual kinds of evidence is retained and expanded, but these are integrated differently and expressed in a narrative weight of evidence characterization statement. *{Whether an alphanumerical rating will be a part of this statement is an unresolved issue still under discussion at EPA.}*
The narrative statement is preceded by summary rankings of human observational evidence and of all experimental evidence. The summary ranking for experimental evidence is composed of long-term animal bioassay evidence and all other experimental evidence on biological and chemical attributes relevant to carcinogenicity. This stepwise approach anticipates marshalling evidence and organizing conclusions as analysis proceeds, for convenience of consideration. It also gives explicit weight to certain kinds of experimental evidence that previously were considered in a "modulating" role.
The narrative statement provides a place to describe evidence by route of exposure and to describe the hazard assessment and dose-response implications of mechanism of action data in characterizing the overall weight of evidence about human carcinogenicity.

2. The approach to dose-response assessment is another area of major change. It calls for a stepwise analysis that follows the conclusions reached in the hazard assessment as to potential mechanism of action. Two steps divide the analysis into modeling in the range of observed data and analysis of dose-response below the range of observed data.

{The process for combining all the findings relevant to human carcinogenic potential is a matter of continuing discussion at EPA. This working paper presents one of a number of suggested approaches. The objective is to be integrative and holistic in judging evidence while at the same time giving guidance to junior scientists in various disciplines about how to marshal and present findings.}

{How to use mechanistic information in dose-response assessment is incom-

pletely developed in these working paper. Specific issues are pointed out in later sections.}

Perspectives for Carcinogenicity Assessment

The following paragraphs summarize part of the current picture of the events in the process of carcinogenesis. Most of the research cited was conducted with experimental approaches not commonly used to study environmental agents. Nevertheless, as this picture is elaborated, more experimental approaches will become available for testing specific mechanisms of action of environmental agents. Even before this happens as a general forward step, information currently available for some agents can be interpreted in light of this picture to make informed inferences about the role the agent may play at the molecular level.

Normally, cell growth in tissues is controlled by a complex and incompletely understood process governing the occurrence and frequency of mitosis (cell division) and cellular differentiation. Adult tissues, even those composed of rapidly replicating cells, maintain a constant size and cell number (Nunez et al., 1991). This appears to involve a balance among three cell fates: (1) continued replication or loss of ability to replicate, followed by (2) differentiation to take on a specialized function or (3) programmed cell death (Raff, 1992; Maller, 1991; Naeve et al., 1991; Schneider et al., 1991; Harris, 1990). As a consequence of either the inactivation of processes that lead to differentiation or cell death, replicating cells may have a competitive growth advantage over other cells, and neoplastic growth clonal expansion can result (Sidransky et al., 1992; Nowell, 1976).

The path a cell takes is determined by a timed sequence of biochemical signals. Signal transduction pathways, or "circuits" in the cell, involve chemical signals that bind to receptors, generating further signals in a pathway whose target in many cases is control of transcription of a specific set of genes (Hunter, 1991; Cantley et al., 1991; Collum and Alt, 1990). A cell produces its own constituent receptors, signal transducers, and signals, and is subject to signals produced by other cells, either neighboring ones or distant ones, for instance, in endocrine tissues (Schuller, 1991). In addition to hormones produced by endocrine tissues, numerous soluble polypeptide growth factors have been identified that control normal growth and differentiation (Cross and Dexter, 1991; Wellstein et al., 1990). The cells responsive to a particular growth factor are those that express transmembrane receptors that specifically bind the growth factor.

One can postulate many ways to disrupt this kind of growth control circuit, including increasing or decreasing the number of signals, receptors, or transducers, or increasing or decreasing their individual efficiencies. In fact, human genetic diseases that make individuals cancer-prone involve mutations that appear to have some of these effects (Hsu et al., 1991; Srivastava, 1990; Kakizuka et al., 1991). Tumor cells found in individuals who do not have genetic disease have also been shown to have mutations with these consequences (Salomon et al., 1990; Bottaro et al., 1991; Kaplan et al., 1991; Sidransky et al., 1991). For example, neoplastic cells of individuals with acute promyelocytic leukemia (APL) have a mutation that blocks cell differentiation in myeloblasts that normally give rise to certain white cells in blood. The mutation apparently alters a receptor that normally responds positively to a differentiation signal. Patients with APL involving this mutation have been successfully treated by oral administration of retinoic acid, which functions as a chemical signal that apparently overrides the effect of the mutation, and drives the neoplastic cells to stop replicating and differentiate. This "differentiation therapy" demonstrates the power conveyed by understanding the growth control signals of these cells (Kakizuka et al., 1991; deThe et al., 1991).

Several kinds of gene mutations[2] have been found in human and animal cancers. Among these are mutations in genes termed tumor susceptibility genes. One kind, mutations that amplify positive signals to replicate or avert differentiation, are termed oncogenes (proto-oncogenes in their normal state). Another kind are mutations in genes involved in generating negative growth signals, termed tumor suppressor genes (Sager, 1989). Damage to these two kinds of genes has been found in cells of tumors in many animal and human tissues including the sites of the most frequent human cancers (Bishop, 1991; Malken et al., 1990; Srivastava et al., 1990; Hunter, 1991). The functions and deoxyribonucleic acid (DNA) base sequences of the genes are highly conserved across species in evolution (Auger et al., 1989a, b; Kaplan, 1991; Hollstein et al., 1991; Herschman, 1991; Strausfeld et al., 1991; Forsburg and Nurse, 1991). Some 100 oncogenes and several tumor suppressor genes have thus far been identified; specific functions are known for only a few.

The growth control circuit can also be altered without permanent genetic change by, for example, affecting the responsiveness of signal receptors, the concentration of signals, or the level of gene transcription (Holliday, 1991; Cross and Dexter, 1991; Lewin, 1991). These can come about through mimicry or inhibition of a signal or through physiological changes such as alteration of hormone levels that influence cell growth generally in some tissues.

Current reasoning holds that cell proliferation which results from changes at the level of DNA sequence or DNA transcription, from changes at the level of growth control signal transduction, or from cell replication to compensate for toxic injury to tissue can begin a process of neoplastic change by increasing the number of cells that are susceptible to further events that may lead to uncontrolled growth. Such further events may include, for instance, errors in DNA replication that occur normally at a low background rate or effects of exposure to

---

[2]The term "mutations" includes the following permanent structural changes to DNA: single basepair changes, deletions, insertions, transversions, translocations, amplifications, and duplications.

mutagenic agents. Effects on elements of the growth control circuit, both permanent and transient, probably occur continuously in virtually all animals due to
endogenous causes. Exogenous agents (e.g., radiation, chemicals, viruses) also
are known to influence this process in a variety of ways.

Endogenous events and exogenous causes such as chemical exposure appear
to increase the probability of occurrence of cancer by increasing the probability
of occurrence of effects on one or more parts of the growth control circuit. The
specific effect of one exogenous chemical, aflatoxin B1, on a tumor suppressor
gene has been postulated on the basis of molecular epidemiology. Mutations in
the tumor suppressor gene p53 are commonly found in the more prevalent human cancers, e.g., colon carcinomas, lung cancer, brain and breast tumors (Levine et al., 1991; Malkin et al., 1990). Populations with high exposure to aflatoxin B1 have a high incidence of hepatocellular carcinoma showing a base
change at a specific codon in the p53 gene (Hollstein et al., 1991). However, the
patterns of base changes in this gene that are found in virus-associated hepatocellular carcinomas and at other sites of sporadic tumors showing p53 gene
mutation are different from the pattern found in aflatoxin B1-exposed populations, supporting the postulate that the specific codon change is a marker of the
effect of aflatoxin B1 (Hayward et al., 1991).

Research continues to reveal more and more details about the cell growth
cycle and to shed light on the events in carcinogenesis at the molecular level. As
molecular biology research progresses, it will become possible to better understand the potential mechanisms of action of environmental carcinogens. It has
long been known that many agents that are carcinogenic are also mutagenic.
Recognition of the role of oncogenes and mutations of tumor suppressor genes
has provided specific ideas about the linkage of chemical mutagenesis to the cell
growth cycle. Other agents that are not mutagenic, such as hormones and other
chemicals that are stimulants to cell replication (mitogens), can be postulated to
play their role by acting directly on signal pathways, for example as growth
signals or by disrupting signal transduction (Raff, 1992; McCormick and Campisi, 1991; Schuller, 1991).

While much has been revealed about likely mechanisms of action at the
molecular level, much remains to be understood about tumorigenesis. A cell that
has been transformed, acquiring the potential to establish a line of cells that grow
to a tumor, will probably realize that potential only rarely. The process of
tumorigenesis in animals and humans is a multistep one (Bouk, 1990; Fearon
and Vogelstein, 1990; Hunter, 1991; Kumar et al., 1990; Sukumar, 1989; Sukumar, 1990), and normal physiological processes appear to be heavily arrayed
against uncontrolled growth of a transformed cell (Weinberg, 1989). Powerful
inhibition by signals from contact with neighboring normal cells is one known
barrier (Zhang et al., 1992). Another is the immune system (at least for viral
infection). How a cell with tumorigenic potential acquires additional properties
that are necessary to enable it to overcome these and other inhibitory processes is

unknown. For known human carcinogens studied thus far, there is an often decades-long latency between exposure to carcinogenic agents and development of tumors, which may suggest a process of evolution (Fidler and Radinsky, 1990; Tanaka et al., 1991; Thompson et al., 1989).

The events in experimental tumorigenesis have been described as involving three stages: initiation, promotion, and progression. The initiation stage has been used to describe a point at which a cell has acquired tumorigenic potential. Promotion is a stage of further changes, including cell proliferation, and progression is the final stage of further events in the evolution of malignancy (Pitot and Dragan, 1991). The entire process involves a combination of endogenous and exogenous causes and influences. The individual human's susceptibility is likely to be determined by a combination of genetic factors and medical history (Harris, 1989; Nebreda et al., 1991), lifestyle, diet, and exposure to chemical and physical agents in the environment.

A number of key questions about carcinogenesis have no generic answers—questions such as: How many events are required? Is there a necessary sequence of events? The answers to these questions may vary for different tissues and species even though the nature of the overall process appears to be the same. The fact that the nature of the process appears empirically to be the same across species is the basis for using assumptions that come from general knowledge about the process to fill gaps in empirical data on a particular chemical. Knowledge of the mechanisms that may be operating in a particular case must be inferred from the whole of the data and from principles on which there is some consensus in the scientific community.

Information from studies that support inferences about mechanism of action can have several applications in risk assessment. For human studies, analysis of DNA lesions in tumor cells taken from humans, together with information about the lesions that a putative tumorigenic agent causes in experimental systems, can provide support for or contradict a causal inference about the agent and the human effect (Vahakangas et al., 1992; Hollstein et al., 1991; Hayward et al., 1991).

An agent that is observed to cause mutations experimentally may be inferred to have potential for carcinogenic activity (U.S. EPA, 1991a). If such an agent is shown to be carcinogenic in animals the inference that its mechanism of action is through mutagenicity is strong. A carcinogenic agent that is not mutagenic in experimental systems, but is mitogenic or affects hormonal levels or causes toxic injury followed by compensatory growth may be inferred to have effects on growth signal transduction or to have secondary carcinogenic effects. The strength of these inferences depends in each case on the nature and extent of all the available data.

These differing mechanisms of action at the molecular level have different dose-response implications for the activity of agents. The carcinogenic activity of a direct-acting mutagen should be a function of the probability of its reaching and reacting with DNA. The activity of an agent that interferes at the level of signal pathways with many potential receptor targets should be a function of multiple reactions. The activity of an agent that acts by causing toxicity followed by compensatory growth should be a function of the toxicity.

1. INTRODUCTION

1.1. PURPOSE AND SCOPE OF THE GUIDELINES

The new guidelines will revise and replace EPA Guidelines for Carcinogen Risk Assessment published in 51 FR 33992, September 24, 1986. Through guidelines, EPA provides its staff and decisionmakers with guidance and perspectives necessary to their performing and using risk assessments. Publication of EPA's guidelines also provides basic information about the Agency's approaches to risk assessment for those who participate in Agency proceedings, or in basic research or scientific commentary on the subjects the guidelines cover.

As the National Research Council pointed out in 1983 that there are many questions encountered in the risk assessment process that are unanswerable based on scientific knowledge (NRC, 1983). To bridge the uncertainty that exists in areas where there is no scientific consensus, inferences must be made to ensure that progress continues in the assessment process. While the application of scientific inferences is both necessary and useful, the bases for these inferences must be continually reviewed to assure that they remain consistent with predominating scientific thought.

The guidelines incorporate basic principles and science policies based on evaluation of the currently available information. Certain general assumptions are described that are to be used when data are incomplete. Standard, default assumptions are described in order to maintain consistency and comparability from one assessment to the next. However, these guidelines explain that such assumptions are to be displaced by facts or better reasoning when appropriate data are available. Short of displacement, an analysis of any promising alternatives is expected to be presented alongside default assumptions.

These guidelines serve two policy goals that must be balanced: first, to maintain consistency of procedures that will support regularity in Agency decisionmaking and, second, to be adaptable to advances in science. Each risk assessment must balance these goals. To assist in balancing these and other science policies, the Agency will rely on input from the general scientific community through the Agency's established scientific peer review processes. The Agency will continually adapt its practices to new developments in the science of environmental carcinogenesis, and restate or revise, where appropriate, the principles, procedures, and operating assumptions of the risk assessment process. Changes will be made through either revisions to these guidelines or, more frequently, issuance of documents on scientific perspectives and procedures and science policies that are developed under the aegis of the EPA Risk Assessment Forum.

1.2. TYPES OF DATA USED IN CARCINOGENICITY ASSESSMENT

Under these guidelines all available direct and indirect evidence is considered to assess whether the weight of the combined evidence supports a conclusion about potential human carcinogenicity. Direct evidence for carcinogenicity in humans comes from epidemiological studies of cancer or, in a few instances, from case reports. Other data providing direct evidence can come from long-term animal cancer bioassays. Indirect evidence comes from a variety of information about toxicological and biochemical effects related to carcinogenicity.

The most direct evidence for identifying and characterizing an agent's human cancer hazard potential is from human epidemiologic studies in which cancer is attributed to exposure to a specific agent. These studies are rarely available because the identification and follow up of populations of sufficient size and sufficient exposure to detect underlying risk is rarely feasible. Moreover, exposure to many potential but unidentifiable causative factors is frequent, making statistical attribution of incidence of a cancer to a single agent difficult. Much of the human evidence comes from occupational studies in which workplace exposure to an agent has been high, and the increased incidence of a cancer attributed to the agent has been distinguishable from other potential causes. Studies that are statistically not powerful enough to discern an association between environmental exposure and tumor incidence or to distinguish among potential causative factors are unable to show that an agent is not carcinogenic. Such studies, if well conducted, may nevertheless be used to estimate a "ceiling" on an agent's carcinogenic potency.

Long-term animal cancer bioassays are more frequently available for more agents than are epidemiologic studies. Approximately 400 of these have been conducted by the National Cancer Institute and National Toxicology Program (NTP)(Huff et al., 1988; NTP, 1992) and many additional ones have been conducted by others. The correspondence between positive results in human studies and long-term animal cancer bioassays is high (Tomatis et al., 1989; Rall, 1991) in the limited number of cases in which comparison is possible. In the absence of epidemiologic information, tumor induction in animal assays remains the best single piece of direct evidence on which to evaluate potential human carcinogenic hazard (OSTP, 1985). Results of animal studies have to be carefully analyzed along with other relevant data (such as metabolism and pharmacokinetic data used to compare animals and humans) to evaluate biological significance, causation, and reproducibility of results, and to determine the reasonable inferences about human hazard they support (Allen et al., 1988; Ames and Gold, 1990).

Data on physicochemical characteristics and biological effects of an agent that make it more or less likely to affect processes involved in producing neoplasia provide important evidence supporting influences about carcinogenic potential. These include, for example, the ability to alter genetic information, influences on cell growth, differentiation, and death, and structural and functional analogies to other compounds that are carcinogenic.

1.3. ORGANIZATION OF THE GUIDELINES

These guidelines follow and should be read with two other publications that provide basic information and general principles. These are: Office of Science and Technology Policy (OSTP, 1985) *Chemical Carcinogens: A Review of the Science and its Associated Principles* (50 FR 10371), and National Research Council (NRC, 1983), *Risk Assessment in the Federal Government: Managing the Process* (Washington, DC, National Academy Press). The 1983 NRC document provided the 1986 guidelines with a thematic organization of risk assessment into hazard identification, dose-response assessment, exposure assessment, and risk characterization. This thematic organization has been slightly revised in these guidelines to focus attention on the importance of characterization in each part of the assessment. Nonetheless, the four questions addressed in these four areas remain the same; they are: Can the agent present a carcinogenic hazard to humans? At what levels of exposure? What are the conditions of human exposure? What is the overall character of the risk, and how well do data support conclusions about the nature and extent of the risk?

1.4. APPLICATION OF THE GUIDELINES

The guidelines are to be used within the policy framework already provided by applicable EPA statutes and do not alter such policies. The Guidelines provide general directions for analyzing and organizing available data. They do not imply that one kind of data or another is prerequisite for regulatory action to control, prohibit, or allow the use of a carcinogen.

Regulatory decision making involves two components: risk assessment and risk management. Risk assessment defines the adverse health consequences of exposure to toxic agents. The risk assessments will be carried out independently from considerations of the consequences of regulatory action. Risk management combines the risk assessment with directives of regulatory legislation, together with socioeconomic, technical, political, and other considerations, to reach a decision as to whether or how much to control future exposure to the suspected toxic agents.

2. HAZARD ASSESSMENT

2.1. INTRODUCTION

Hazard assessment covers a wide variety of data relevant to the question, can an agent pose a human carcinogenic hazard? Available data may include: long term animal cancer bioassays and human studies, physical-chemical properties of the agent and its structural relationship to other carcinogens, studies of cellular and molecular interactions and mechanisms of action, and results from toxicological tests and experiments on the bioavailability and transformation of an agent in experimental animals and humans. Hazard assessment results are summarized in a hazard characterization that conveys the nature and impact of available data and appropriate scientific inferences about human carcinogenic hazard.

Experience shows that the nature and extent of information available on each agent is different and can vary from a wealth of epidemiologic data to only physical-chemical properties. Frequently, results from a long-term animal carcinogenesis bioassay are the only direct evidence available for the evaluation. These guidelines follow the assumption that chemicals with evidence to demonstrate carcinogenicity in animal studies are likely to present a carcinogenic hazard to humans under some conditions of exposure (OSTP, 1985). At the same time, there may be mechanistic, physiological, biochemical, or route-of-entry differences which alter the toxicological consequences in humans from those observed in the particular animals tested. When the results of animal testing are extrapolated to humans, effects observed at high continuous exposures are often projected to low or intermittent exposures and results from one route of exposure are often extrapolated to other routes of exposure. The risk analysis must examine each assumption and extrapolation for mechanistic and biological plausibility. The elements of hazard assessment described below are the foundation for these examinations.

The characterization of an agent's carcinogenic human hazard potential depends on the weight of all the relevant evidence. Studies are evaluated according to accepted criteria for study quality, sensitivity, and specificity. These have been described in several publications (Interagency Regulatory Liaison Group, 1979; OSTP, 1985; Peto et al., 1980; Mantel, 1980; Mantel and Haenszel, 1959; Interdisciplinary Panel on Carcinogenicity, 1984; National Center for Toxicological Research, 1981; National Toxicology Program, 1984; U.S. EPA, 1983a, b, c; Haseman, 1984). The hazard characterization describes how likely the agent is to be carcinogenic to humans, including the judgment whether or not the hazard is considered to be contingent on certain conditions of exposure (e.g., oral versus dermal exposure). The characterization summarizes the basis of, and confidence in, inferences drawn from data and the rationale for conclusions about weight-of-evidence; these are accompanied by judgments on issues and uncertainties that cannot be resolved with available information.

The characterization of potential hazard is qualitative. It does not address the magnitude or extent of effects under actual exposure conditions. However, observations and conclusions from the hazard characterization that are relevant to quantitative dose-response analysis are carried forward to the section on quantitative dose-response analysis, and those that are relevant to actual exposure conditions are discussed in the risk characterization.

2.2. INTEGRATING DATA FOR HAZARD ASSESSMENT

The assessment of potential carcinogenic hazard to humans is a process in which many kinds of data are integrated to examine the inferences and conclusions they support. The process is conducted as an interdisciplinary effort.

While the discussion that follows explores data analyses along separate disciplinary lines and provides for making intermediate summaries of human observational data and experimental data, it must be recognized that this is done simply for convenience of organization and marshalling of thought, and the individual analyses are interdependent not separate. Each kind of analysis, from evaluation of human studies to structure-activity relationship analysis, looks to the others for interpretive alliance and perspective. Confidence in conclusions is built upon the overall coherence of inferences from different kinds of data as well as confidence in individual data sets.

For example, in examining the issue of causation as part of human studies analyses, one uses knowledge of the biological activity of the agent in animal systems and of pertinent features of its structure, metabolism and other properties to address issues of biological plausibility of a causal hypothesis. Likewise, where there are no epidemiologic studies and one is examining relevance of animal responses to human hazard potential, one uses human data to address comparative biology of animals and humans with respect to, for instance, metabolism, pharmacokinetics, physiology, and disease history.

2.3. ANALYSIS OF HUMAN DATA

2.3.1. Epidemiologic Studies

Epidemiology is the study of the distribution of a disease in a human population and the determinants that may influence disease occurrences. Epidemiologic studies provide direct information about the response of humans who have been exposed to suspect carcinogens and avoids the need for interspecies extrapolation of animal toxicological data.

2.3.1.1. Exposure Focus

An identification of hazard in a human population depends critically on the exposure assessment, which consists of two components: (a) the qualitative determination of the presence of an agent in the environment and (b) the quantitative assessment. An exposure assessment which includes an attribution of quantified exposure to an individual is considered more precise and will carry more weight in an evaluation of human hazard. In many epidemiologic studies, the populations are selected and studied retrospectively, and the time between exposure and observation of effects is very long because of the latency of cancer. The past exposure is a critical determinant. In an environmental situation, quantitative exposure assessment is usually difficult to achieve due to lack of measures of past exposure. This is one reason why occupational studies where exposure is based on job classification are often used for identifying environmental hazard. Past occupational exposures are usually considered to be at higher levels than those encountered environmentally; therefore, the question whether any identified hazard is pertinent at lower exposure levels needs to be addressed.

Exposure assessment becomes more complicated when the exposure is to a complex mixture of incompletely identified chemicals. In addition, human exposures to agents can occur by more than one route as compared to the controlled exposure regimens used in the animal carcinogenicity studies (c.g., occupational exposure to solvents can occur through inhalation and dermal absorption). The characterization of the patterns of exposure to identify exposure-effect relationships is another consideration. Important exposure measurements in epidemiologic studies include cumulative exposure (sometimes time-weighted), duration of exposure, peak exposure, exposure frequency or intensity, and "dose" rate. Some insight on which measurement of exposure will be the best predictor of a cancer can come from an understanding of the disease process itself.

In epidemiological studies, "biological markers", usually the reaction products of an agent or its metabolite with DNA or a protein or other markers of exposure such as excretion of metabolites in urine have been increasingly considered as reliable measures of exposure. More rarely a marker of *effect* specific to an agent may be found (Vahakangas et al., 1992). Information on the relationship between exposure or effect and markers is often derived from metabolism and kinetic studies in animals. Validation of the relationship with comparative human data is needed to support confidence in use of such markers.

{The generic issue of use biomarkers of exposure and effect is still under consideration.}

2.3.1.2. Types of Epidemiology Studies

Various types of epidemiologic studies or reports can provide useful information for identifying hazards. An important consideration is the validity and representativeness of the studied population with respect to the larger population of interest. Study designs include cohort, case-control, proportionate ratio, clinical trials, and correlational studies. In addition, cluster investigations and case reports, while not constituting studies, may yield useful information under certain situations (e.g., reports associated with exposure to vinyl chloride and diethylstilbestrol). The above designs have well-defined strengths and limitations (Breslow et al., 1980; 1987; Kelsey et al., 1986; Lilienfeld and Lilienfeld, 1979; Mausner and Kramer., 1985; Rothman, 1986).

2.3.2. Elements of Critical Analysis

Aspects of the available human data, which are described in this section, are evaluated to determine whether there is a causal relationship between exposure to the agent and an increase in cancer incidence. Certain elements of analysis are brought to bear on the criteria for causality, which are listed and discussed in Section 2.3.2.5. In general, these elements address the study design and conduct; the ability to sort out the potential role of the agent in question as opposed to other risk factors; assessment of exposure of the study and referent populations to the agent and to other risk factors; and, given all of the above, the statistical power of the study or studies.

2.3.2.1. Exposure

Exposure is the foundation upon which any exposure-effect relationship is evaluated. Often, the exposure is not to a single agent, but to a combination of agents (e.g., exposure to chloromethylmethyl ether and its ever-present contaminant bischloromethyl ether). When exposures occur simultaneously, it is generally assumed that each chemical exposure contributes to the exposure- or exposures-effect relationship.

Exposure can be defined in hierarchical levels. Greater weight will be given to studies where exposures are more precisely defined and can be quantified. The broadest definition of exposure is that inferred for a group of individuals living in a geographic area. At this level, it is not known whether all individuals are exposed to the agent, and if exposed, the patterns and lengths of exposure. The result is a mixture of individuals with higher exposure and those with little or no exposure. This leads to exposure misclassification, which, if random, may result in a study's reduced ability to detect underlying elevations in risk. For the same reasons, exposure as defined by assignment to a broad occupational category in the absence of qualitative or quantitative data yields less useful information on an individual's exposure.

A more recent application in epidemiologic studies is the use of job-exposure matrices to infer semi-quantitative and quantitative levels of exposure to specific agents (Stewart and Herrick, 1991). The job-exposure matrix has been applied to occupational scenarios where at least some current and historical monitoring data exist. In examining exposure levels inferred from a job-exposure matrix, the basis of the monitoring data must be considered—whether data are from routine monitoring or reflect accidental (i.e., higher than average) releases.

Biological markers are indicators of processing within a biological system. Using such a marker as a measure of exposure is potentially the most reliable level of data since the quantity measured is thought to more precisely characterize a biologically available dose, rather than exposure that is the amount of material presented to the individual and is usually inferred from a measurement of atmospheric concentrations (NAS, 1989). Validated markers are the most desirable, i.e., markers which are highly specific to the exposure and those which are highly predictive of disease (Blancato, OHR Biomarker Strategy, cite published paper; Hulka and Margolin, 1992) (e.g., urinary arsenic (Entertine et al., 1987), and alkylated hemoglobin (hemoglobin adducts) from exposure to ethylene oxide (Callemen et al., 1986; van Sittert et al., 1985).

2.3.2.2. *Population Selection Criteria*

The study population and the comparison or referent population are identified and examined to decide whether or not comparisons between populations are appropriate and to determine the extent of any bias resulting from their selection. The ideal referent population would be similar to the study population in all respects except exposure to the agent in question. Potential biases (e.g., healthy worker effect, recall bias, selection bias, and diagnostic bias) and the representativeness of the studied population for a much larger population are addressed.

Generally, the referent population in cohort studies consists of mortality or incidence rates of a larger population (e.g., the U.S. population). The healthy worker bias is specific to occupational cohort studies, and it asserts that an employed population is healthier than the general population (McMichael, 1976). The influence of the healthy worker effect is toward a more favorable mortality in the exposed population; this influence is thought to decrease with increasing age and to have less influence on site-specific cancer rates. The influence of the healthy worker effect is thought to be minimized by the use of an internal comparison group (e.g., incidence or mortality rates of employees who are from the same company, but not among the employees in the study population).

In case-control designs, the potential for differences in recalling past events (recall bias) between the case and control series needs to be evaluated. The characteristics of the control series also need to be discussed. Hospital controls have associated limitations with respect to possible associations with the exposure of interest. Randomly-selected population or community controls are thought to be more like cases in the case series; however, response rates are often lower.

2.3.2.3. Confounding Factors

A confounding variable is a risk factor for the disease under study that is distributed unequally among the exposed and unexposed populations. Adjustment for possibly confounding factors can occur either in the design of the study (e.g., matching on critical factors) or in the statistical analysis of the results. If adjustment within the study data is not possible due to the presentation of the data or because needed information was not collected during the study, indirect comparisons may be made (e.g., in the absence of direct smoking data from the study population, an examination of the possible contribution of cigarette smoking to increased lung cancer risk and to the exposure in question may include information from other sources such as the American Cancer Society's longitudinal studies (Hammond, 1966; Garfinkel and Silverburg, 1991).

In a collection of heterogenous studies possible confounding factors are usually randomly distributed across studies. If consistent increases in cancer risk are observed across the collection of studies, greater weight is given to the agent under investigation as the etiologic factor even though the individual studies may not have completely adjusted for confounding factor.

2.3.2.4. Sensitivity

Epidemiologic studies which consist of a large number of individuals with sufficient exposure to a putative cancer-causing agent and adequate length of time for cancer development or detection are considered to have a greater ability to detect cancer risk. Studies for review, however, do not always fulfill these criteria. In addition, the ability to detect increases in relative risk associated with environmental exposure is very difficult due to heterogeneous exposure regarding both pattern and levels and which potentially bias risk toward the null hypothesis of no effect.

If the underlying risk is actually increased, examination of persons considered at higher risk increases the detection ability of a study. Such examination may include an evaluation of risk among individuals with higher or peak exposure, with greater duration of exposure, or with the longest time since first exposure (to allow for latency of effect), and those of older age, and those with long latencies.

A study in which no increases in risk were observed may be useful for inferring an upper limit on possible human risk. Statistical reanalysis is another approach for examining the sensitivity of results from an individual study (e.g., the dose-response relationship reported in one formaldehyde-exposed cohort (Blair et al., 1986) has been examined by several investigators (Blair et al., 1987; Sterling and Weinkam, 1987; Collins et al., 1988; Marsh, 1992). These further analyses are a reaggregation of exposure groups or an examination of the influence of a subgroup on the disease incidence of the much larger group.

Statistical methods for examining several studies together are frequently applied to the collection of data. These methods, commonly referred to as meta-analysis, are used to contrast and combine results of different studies with the goal of increasing sensitivity. In meta-analysis, study results are evaluated as whether they differ randomly from the null hypothesis of no effect (Mann, 1990); meta-analysis presumes that observed results are not biased. If an underlying effect is not present, the observed results should appear randomly distributed and cancel each other when studies are combined (Mann, 1990). Several important issues are pertinent to meta-analysis. These are controlling for bias and confounding prior to combining studies, criteria for study inclusion, assignment of weights to individual studies, and possible publication and aggregation bias. Greenland, 1987 discusses may of these issues in addition to identifying methodologic approaches.

{Participants at the December 4, 1992, Society for Risk Analysis on cancer risk assessment issues were asked to look at meta-analysis.}

2.3.2.5. *Criteria for Causality*

A causal interpretation is enhanced for studies to the extent that they meet the criteria described below. None of the criteria, with the exception of a temporal relationship, should be considered as either necessary or sufficient in itself to establish causality. These criteria are modelled after those developed by Hill in the examination of cigarette smoking and lung cancer (Rothman, 1986).

a. *Temporal relationship*: This is the single absolute requirement, which itself does not prove causality, but which must be present if causality is to be considered. The disease occurs within a biologically reasonable time frame after the initial exposure. The initial period of exposure to the agent is the accepted starting point in most epidemiologic studies.

b. *Consistency*: Associations are observed in several independent studies of a similar exposure in different populations. This criterion also applies if the association occurs consistently for different subgroups in the same study.

c. *Magnitude of the association*: A causal relationship is more credible when the risk estimate is large and precise (narrow confidence intervals).

d. *Biological gradient*: The risk ratio is correlated positively with increasing exposure or dose. A strong dose-response relationship across several categories of exposure, latency, and duration is supportive although not conclusive for causality given that confounding is unlikely to be correlated with exposure. The absence of a dose-response relationship, however, should not be construed by itself as evidence of a lack of a causal relationship.

e. *Specificity of the association*: The likelihood of a causal interpretation is increased if a single exposure produces a unique effect (one or more cancers also found in other studies) or if a given effect has a unique exposure.

f. *Biological plausibility*: The association makes sense in terms of biological knowledge. Information from animal toxicology, pharmacokinetics, structure-activity relationship analysis and short-term studies of the agent's influence on events in the carcinogenic process are considered.

g. *Coherence*: The cause-and-effect interpretation is in logical agreement with what is known about the natural history and biology of the disease, i.e., the entire body of knowledge about the agent.

2.4. SUMMARY OF HUMAN EVIDENCE

{The process in combining all findings relevant to human carcinogenic potential is an issue for further development. The need for this summarization step for human evidence and the one in Section 2.5 for experimental evidence are open questions at EPA.}

Each epidemiological study is critically evaluated for its relevance with respect to the exposure-effect relationship, exposure assessment such as intensity, duration, time since first exposure, and methodological issues such as study design, selection and characterization of comparison group, sample size, handling of latency, confounders, and bias.

Following critical evaluation, the totality of the weight-of-evidence for human carcinogenicity is assessed and summarized according to one of the following four categories, which are meant to represent a judgment regarding the weight of all of the human evidence even if only one study exists on the subject. Rarely, the judgment can be based on a series of case reports. More likely, the evaluation will involve several studies. Inferences from summary analyses such as meta-analysis can provide support for placement into these categories. In addition, evidence that the agent in question is metabolized to a compound, for which independent human evidence exists, is supportive of the categorization.

The weight a particular study or analysis is given in the evaluation depends on its design, conduct, and avoidance of bias (selection, confounding, and measurement) (OSTP, 1984). Results, both positive and null, are considered in light of the study's rigor. The weight of evidence is based on the plausibility of the association and the conclusiveness of observed findings. Greater plausibility and conclusiveness can be ascribed to an exposure-effect relationship when it can be explained in terms of adherence to the criteria for causality, including coherence with other evidence such as animal toxicology. The plausibility of exposure-effect relationship also can be bolstered or mitigated by evidence of structure-activity relationship analysis with well characterized agents, studies of mechanism of action, understanding of metabolic pathways, and other indirect evidence relevant to human effects. A mixture (e.g., cigarette smoke, coke oven emissions) may be categorized as an agent when causation is ascribed to the mixture, but not to necessarily to its individual components.

2.4.1. Category 1

Plausible evidence exists, and from this evidence a conclusive causal association can be judged. Cause and effect relationships are supported with results from well-designed and conducted studies in which random or nonrandom error can be reasonably excluded.

2.4.2. Category 2

Evidence exists to suggest that causal association is plausible; however, such evidence is not conclusive due to a number of reasons which may include lack of consistency, wide confidence intervals which may or may not include a risk, or absence of an observed dose-response relationship. The effect of random or nonrandom error in individual studies which could influence the risk ratio away from the null is considered minimal. This category covers a broad range of possible weights of evidence. At the top of the category are highly suggestive, but short of convincing data. At the bottom of the category are suggestive but weak data. A statement of the relative position of data in this continuum accompanies the description of the data as Category 2.

2.4.3. Category 3

The body of evidence is inconclusive. The assertion of a causal association is not plausible from the available data in which studies of equal quality have contradictory results in which random or nonrandom error is a more likely explanation for observations of increased risk. This category also applies when no epidemiologic data are available.

2.4.4. Category 4

The available studies are designed with defined ability to detect increases in risk, and resultant risk ratios are precise with tight confidence intervals. Evidence derived from the studies consistently show no positive association between the suspect agent and cancer. The evidence is described as showing no cause and effect relationship at the exposure levels studied. It is not considered to show that the agent is non-carcinogenic under all circumstance unless the evidence is so complete that potential for human carcinogenicity can be eliminated.

2.5. ANALYSIS OF LONG-TERM ANIMAL STUDIES

Long-term animal studies are evaluated to decide whether biologically significant responses have occurred and whether responses are statistically significantly increased in treated versus control animals. The unit of comparison is an experiment of one sex, in one species.

2.5.1. Significance of Response

Evidence for carcinogenicity is based on the observation of biologically and statistically significant tumor responses in specific organs or tissues. Criteria for categorizing the strength of evidence of animal carcinogenicity in bioassays have been established by the National Toxicology Program (NTP, 1987). Animal study results are evaluated for adequacy of design and conduct (40 CFR Part 798). The results are described and biological significance of observed toxicity is evaluated (non-neoplastic endpoints included).

{For EPA's purposes, the criteria for evaluating animal cancer bioassays are still under review, and could be somewhat different from those of NTP. Nevertheless, much of the animal cancer data available to EPA carries the NTP designations of "clear, some, equivocal, or none".}

Interpretation of animal studies is aided by the review of target organ toxicity and other non-neoplastic effects (e.g., changes in the immune and endocrine systems) that may be noted in prechronic or other toxicological studies. Time and dose-related changes in the incidence of preneoplastic and neoplastic lesions may also be helpful in interpreting responses in long-term animal studies.

It is recognized that chemicals that induce benign tumors also frequently induce malignant tumors, and that certain benign tumors may progress to malignant tumors. Benign and malignant tumor incidence are combined for analysis of carcinogenic hazard when scientifically defensible (OSTP, 1985; Principle 8).

The Agency follows the National Toxicology Program framework for combining benign and malignant tumor incidence of a particular site (McConnell, 1986).

Elevated tumor incidences in adequate experiments are analyzed for biological and statistical significance. Generally, a statistical test that shows a positive trend in dose-response at a level of significance of five percent (i.e., the likelihood of false positive results is less than five percent) supports a conclusion that the experiment is positive. If false positive outcomes are a serious concern, the use of a formal multiple comparison adjustment procedure should be considered. No rigid decision rule should be used as substitute for scientific judgment. Other statistical tests may be applied if the trend test is not statistically significant or, for some reason, not applicable for a given experiment. The significance level should be adjusted if multiple comparisons of the same data are made, in order to avoid raising the overall likelihood of false positives (Haseman, 1983, 1990; U.S. FDA, 1987).

Data from all long-term animal studies, positive and negative, are to be considered in the evaluation of carcinogenicity. Different results according to species, sex, or strain, or by route of administration, duration of study or site of effect are not unexpected. The issues are how different results affect the weight of evidence and whether the differences suggest the operation of any particular mechanisms of action or tissue sensitivity that may assist in judging human relevance.

2.5.2. Historical Control Data

{NOTE TO THE READER: The issues of how to consider historical control data and high background tumors are knotty ones. For high background tumors there are varying views, some question relevance, but usually there are insufficient data about the mechanism of action to question its relevance. Others point to the fact that both humans and animals have tissues with high background rates.}

Historical control data often add valuable perspective in the evaluation of carcinogenic responses (Haseman et al., 1984). For the evaluation of rare tumors, even small increases in tumor response over that of the concurrent controls may be significant compared to historical data. Historical data can also identify sites with high spontaneous background in the test strain. Nevertheless, historical control data have limitations as compared to concurrent control data. One limitation is the potential for genetic drift in laboratory strains over time that makes historical data less useful beyond a few years. Other limitations are the differences in pathological examinations at different times and in different laboratories; these are due to changes over time in criteria for evaluating lesions and to variations in preparation techniques and reading of tissue samples between laboratories. Other differences may include biological and health differences in animal strains from different suppliers. Concurrent controls are, for these reasons, more valuable comparison for judging whether observed effects in dosed animals are treatment related.

Comparison of an observed response that appears to be treatment related with historical control data may call the response into question if the observed response is well within the range of historical control data. Whenever historical control data are compared with the current data the reasons should be given for judging the historical control data to be adequately representative of the current expected response background.

2.5.3. High Background Tumor Incidence

Tumor data at sites with high spontaneous background requires special consideration (OSTP, 1985; Principle 9). Questions raised about high background tumors in animals (and humans) are whether they are due to particular genetic predispositions or ongoing proliferative processes that are species-specific prerequisites to a neoplastic response or, on the other hand, represent sensitivities due to biological processes that are alike among species. Answering these questions requires a body of research data beyond the data obtained in standard animal studies. Unless there are research data to establish that such tumor data at a site occur because of a mechanism-of-action that is unique to the species, strain, and sex with the high background, the tumor data are considered, as are other tumor data, in the overall weight of evidence. These data may receive relatively less weight than other tumor data.

2.5.4. Dose Issues

Long-term animal studies at or near the maximum tolerated dose level (MTD) are used to ensure an adequate power for the detection of carcinogenic activity of an agent (NTP, 1984; IARC, 1982). The MTD is a dose which is estimated to produce some minimal toxic effects in a long term study (e.g., a small reduction in body weight), but should not shorten an animal's life span or unduly compromise normal well-being except for chemically induced carcinogenicity (International Life Sciences Institute, 1984; Haseman, 1985). Assays in which the MTD may have been exceeded or may not have been reached require special scrutiny.

Exceedance of the MTD in a study may result in tumorigenesis that is secondary to tissue damage or physiological damage and is more a function of this damage than of the carcinogenic influence of the particular agent tested. Inferences drawn from the study must consider observed non-neoplastic toxicity and the tissues affected, as well as the existence of carcinogenic effects in tissues, or at doses, not affected by the exceedance. Study results at doses that exceed the

MTD can be rejected if toxic damage is so severe as to compromise interpretation.

Null results in long-term animal studies at exposure levels above the MTD may not be acceptable if animal survival is so impaired that the sensitivity of the study is significantly reduced below that of a conventional chronic animal study at the MTD. The import of non-positive studies at exposure levels below the MTD may be compromised by lack of power to detect effects.

2.5.5. Human Relevance

Relevance of tumor responses to human hazard is a judgment that is integral to analysis of bioassay results. The assumption is made under these guidelines that observation of tumors at any animal tissue site supports an inference that humans may respond at some site. This assumption is reexamined as data on the issue become available for specific responses. The Agency will undertake analyses of relevance issues as needed in reports to be published from time to time (e.g., USEPA, 1991b).

If information on the mechanism of tumorigenesis supports the conclusion that a response seen in an animal study is unique to that species or strain, the response is considered to provide no evidence for human hazard potential (U.S. EPA, 1991a). Agency decisions of this kind about particular animal responses are made and published under the aegis of the EPA Risk Assessment Forum. Such mechanistic uniqueness is be differentiated from quantitative differences in dose-response which are not, *per se*, issues of relevance.

2.6. ANALYSIS OF EVIDENCE RELEVANT TO CARCINOGENICITY

Certain structural, chemical, and biological attributes of an agent provide key information about its potential to cause or influence carcinogenic events. These attributes and comparative studies between species provide information to support carcinogenic hazard identification and compare potential activity across species. The following sections provide guidance for inclusion of analyses of these kinds of evidence in hazard identification.

2.6.1. Physical-Chemical Properties

Physical-chemical properties that can affect the agent's absorption, tissue distribution (bioavailability), biotransformation, or chemical degradation in the body are analyzed as part of the overall weight of evidence on hazard potential. These include, but are not limited to: molecular weight, size, and shape; physical state (gas, liquid, solid); water or lipid solubility that can influence retention and tissue distribution; and potential for chemical degradation or stabilization in the body.

Interaction with cellular components and reactivity with macromolecules is a second major area covered. Factors such as molecular size and shape, electrophilicity, and charge distribution are analyzed to decide whether they would facilitate such reactions by the agent.

2.6.2. Structure-Activity Relationships

The role of structure-activity relationship (SAR) analysis in the assessment of the carcinogenic risk of an agent in question is dependent upon the availability and the quality of the toxicological data on the agent. For chemicals with data from reasonably conducted studies, SAR analysis is useful in providing input to determine the probable mechanism of action, which is important for hazard identification and for decisions on the appropriate methodology for quantitative risk assessment. For chemicals with either unsatisfactory or inadequate carcinogenicity data, SAR analysis may be used to generate, bolster, or mitigate the carcinogenic concern for the chemical, depending on the strength of and confidence in the SAR analysis. In addition, SAR analysis can also serve as a guide to evaluate carcinogenic potential of untested chemicals.

Currently, SAR analysis is most useful for chemicals that are believed to produce carcinogenesis, at least initially, through covalent interaction with DNA (i.e., DNA-reactive mutagenic electrophilic or proelectrophilic chemicals) (Ashby and Tennant, 1991; Woo and Arcos, 1989). In analyzing the SAR of DNA-reactive mutagenic chemicals, the following parameters should be considered (Woo and Arcos, 1989):

a. the nature and reactivity of the electrophilic moiety or moieties present;
b. the potential to form electrophilic reactive intermediate(s) through chemical, photochemical; or metabolic activation;
c. the contribution of the carrier molecule to which the electrophilic moiety(ies) is attached;
d. physicochemical properties (e.g., physical state, solubility, octanol-water partition coefficient, half-life in aqueous solution);
e. structural and substructural features (e.g., electronic, stearic, molecular geometric);
f. metabolic pattern (e.g., metabolic pathways and activation and detoxification ratio); and
g. the possible exposure route(s) of the subject chemical.

Following compilation of a carcinogenicity database for structural analogs, the above parameters are used to compare and place the subject chemical as to its carcinogenic potential among its analogs or congeners. In addition, the analysis is supplemented with any available information on the pertinent toxic effects of the compound, its potential metabolites, and its structural analogs. The pertinent toxic effects are those known to contribute to carcinogenesis such as immune suppression or mutagenicity.

Suitable SAR analysis of non-DNA-reactive chemicals and of DNA-reactive chemicals that do not appear to bind covalently to DNA requires knowledge or postulation of the most probable causative mechanism(s) of action (e.g., receptor-mediated, cytotoxicity related) of closely related carcinogenic structural analogs. Examination of the physicochemical and biochemical properties of the subject chemical may then allow one to assess the likelihood that such a mechanism also may be applicable to the chemical in question and to determine the feasibility of conducting SAR analysis based on the mechanism.

2.6.3. Metabolism and Pharmacokinetics

Studies of the absorption, distribution, biotransformation and excretion of agents are used to make comparisons among species to assist in determining the implications of animal responses for human hazard assessment, to support identification of toxicologically active metabolites, to identify changes in distribution and metabolic pathway or pathways over a dose range and between species, and to make comparisons among different routes of exposure.

In the absence of data to compare species, it is necessary to assume that pharmacokinetic and metabolic processes are qualitatively comparable. If data are available (e.g., blood/ tissue partition coefficients and pertinent physiological parameters of the species of interest), physiologically based pharmacokinetic models can be constructed to assist in determination of tissue dosimetry, species-to-species extrapolation of dose, and route-to-route extrapolation (Connolly and Andersen, 1991).

Analyses of adequate metabolism and pharmacokinetic data can be applied toward the following as data permit. Confidence in conclusions is greatest when in vivo data are available.

a. Identifying metabolites and reactive intermediates of metabolism and determining whether one or more of these intermediates are likely to be responsible for the observed effects. This information on the reactive intermediates will support and appropriately focus SAR analysis, analysis of potential mechanisms of action, and, in conjunction with physiologically based pharmacokinetic models, estimation of tissue dose in risk assessment (D'Souza et al., 1987; Krewski el al., 1987).

b. Identifying and comparing the relative activities of relevant metabolic pathways in animals with those in humans. This analysis can give insight on whether extrapolation of results of animal studies to humans will produce useful results.

c. Describing anticipated distribution within the body, and possibly identifying target organs. Use of water solubility, molecular weight, and structure analysis can support inferences about anticipated qualitative distribution and excretion. In addition, describing whether the agent or metabolite of concern will be excreted rapidly or slowly or will be stored in a particular tissue or tissues to be mobilized later can identify issues in comparing species and formulating dose-response assessment approaches.

d. Identifying changes in pharmacokinetics and a metabolic pathway or pathways with increases in dose. These changes may result in the formation and accumulation of toxic products following saturation of detoxification enzymes. These studies have an important role in providing a rationale for dose selection in carcinogenicity studies. In addition, these studies may be important in estimating a dose over a range of high to low exposure for the purpose of dose-response assessment.

e. Determining the bioavailability of different routes of entry by analyzing uptake processes under various exposure conditions. This analysis supports identification of hazard for untested routes of entry. In addition, use of physicochemical data (e.g., octanol-water partition coefficient information) can support an inference about the likelihood of dermal absorption (Flynn, 1990).

In all of the above-listed areas of inquiry, attempts are made to clarify and describe as much as possible the variability to be expected because of differences in species, sex, age, and route of entry. Utilization of pharmacokinetic information takes into account that there may be subpopulations of individuals who are particularly vulnerable to the effects of an agent because of metabolic deficits or pharmacokinetic or metabolic differences (genetically or environmentally determined) from the rest of the population.

2.6.4. Mechanistic Information

{The material in this section is only a start. Substance-specific risk assessments may have little or no data in this category. Even when data are available, there is no standard for what is acceptable or what to expect. If there are no data, we will have to use default assumptions. How much information is enough is difficult to say until testing in this area is more regular.}

"Knowledge of carcinogenic mechanisms is incomplete in all cases. Information on how particular agents are likely to cause cancer may, however, be useful for appreciating more accurately the hazard that such agents pose to humans" (IARC, 1991). Results from short-term toxicological tests and molecular and cellular mechanistic studies are also useful in the interpretation of epidemiological and rodent chronic bioassay data used in hazard identification and characterization. These data may provide guidance for dose-response modelling.

Testing for tumorigenicity is usually done in long-term assays that involve exposure for much of an animal's lifespan.

Data from the long-term animal studies and the toxicity studies preceding them (e.g., evidence of lesion progression, or lack of progression, and hyperplasia at the same site as the neoplasia) may suggest a line of inquiry for further study. Cell necrosis is often an early finding (e.g., 20-90 days) and provides indirect evidence for subsequent tissue regeneration and compensatory growth mechanisms when these events are not directly observed. Other early changes observed during pre-chronic studies range from biochemical changes to altered hormone levels to organ enlargement (hyperplasia) to specific and marked histopathological changes (Hildebrand et al., 1991).

Conventional animal cancer bioassays provide little information on mechanism of action. Short-term animal assays generally have more defined study designs to provide information about potential mechanisms of action. A large number of short-term assays examine biological activities relevant to the carcinogenic process (e.g., mutagenesis, tumor promotion, aberrant intercellular communication, increased cell proliferation, malignant conversion, immunosuppression). In the future, mechanistic-based end points should play an increasing, and perhaps major, role in the assessment of cancer risk.

2.6.4.1. Genetic Toxicity Tests

Information on genetic damaging events induced by an agent is revealing about the possible mechanism of action of a carcinogen. Although the effectiveness of genetic toxicology tests in predicting cancer has been questioned (Brockman and DeMarini, 1988), the ability of these tests to detect mutagenic carcinogens has not been seriously challenged (Brockman and DeMarini, 1988; Prival and Dunkel, 1989; Tennant and Zeiger, 1992; Shelby et al., 1992; Jackson et al., 1992).

Recent studies on oncogenes provide evidence for the linkage between mutation and cancer (Bishop, 1991); activation of protooncogenes to oncogenes can be triggered, for example, by point mutations, DNA insertions, or chromosomal translocation (Bishop, 1991). In addition, the inactivation of tumor suppressor genes (anti-oncogenes) can occur by chromosomal deletion or aneuploidy (chromosome loss), and mitotic recombination (Bishop, 1989; Varmus, 1989; Stanbridge and Vavenee, 1989).

Genetic toxicology tests have been described in various reviews (Brusick, 1990; Hoffman, 1991). The EPA has published various testing requirements and guidelines for detection of mutagenicity (USEPA, 1991a). A useful method to "portray" data graphically, and which provides a reasonable starting point for analysis, is the genetic activity profile (GAP) methodology developed by the USEPA (Garrett et al., 1984; Waters et al., 1988).

Many test systems have been developed to assay agents for their mutagenic potential.[3] These include assays for changes in DNA base pairs of a gene (i.e., gene mutations) and microscopically visible changes in chromosome structure or number. Structural aberrations include deficiencies, duplications, insertions, inversions, and translocation. Other assays that do not measure gene mutations or chromosomal aberrations per se provide some information on an agent's DNA damaging potential (e.g., tests for DNA adducts, strand breaks, repair, or recombination).

Distinguishing a carcinogenic agent as a mutagen or nonmutagen is an important decision point in defining the mechanism of action. To designate a putative carcinogen as a mutagen, there should be confidence that the primary target is DNA. Mutagenic end points that involve stable changes in DNA structure are emphasized because of their relevance to carcinogenesis. These include gene mutations and chromosomal aberrations.

To be of value in cancer risk assessment, genetic toxicology data must meet the demands of scientific scrutiny. A higher level of confidence that a carcinogen is a mutagen is assigned to agents that consistently induce direct structural changes in DNA in a number of test systems. Although important information can be gained from in vitro assays, a higher level of confidence is given to a data set that includes in vivo evidence. In vivo data is emphasized because many agents require metabolic conversion to an active intermediate for biological activity. Metabolic activation systems can be incorporated into in vitro assay; however, they do not always mimic mammalian metabolism perfectly. If available, human genetic toxicity end points relevant to carcinogenesis are important in vivo data.

It is not possible to illustrate all potential combinations of evidence, and considerable judgment must be exercised in reaching conclusion. Certain responses in tests that measure DNA damaging potential (e.g., DNA repair activity, adducts or strand breakage in DNA) other than gene mutations and chromosomal aberrations may provide a basis for raising the level of confidence in designating a carcinogen as mutagenic.

There are many other mechanisms by which agents cause genetic damage secondary to other effects. For example, an agent might interfere with DNA repair or possibly increase DNA damage through an increase in oxidative radical production (Cerutti et al., 1990). Reliance on evidence for induced gene mutations or chromosomal aberrations to define a mutagenic carcinogen is not meant to downplay the importance of these secondary mechanisms or other genetic end points.

Aneuploidy (i.e, a change in chromosome number) may play an important role in the development of some tumors (Kondo et al., 1984; Cavenee et al., 1983; Barrett et al., 1985), but it may result from interactions with cellular com-

---

[3]Ability to induce heritable or stable alterations in DNA structure and content.

ponents (e.g., mitotic apparatus) other than with DNA. For this reason, aneuploidy is not considered evidence for designating a carcinogen as mutagenic. Aneuploidy is important information regarding potential carcinogenicity by other genetic mechanisms and should be factored into the evaluation concerning mechanisms of action.

Because mutagenic carcinogens have been observed to induce tumors across species and at multiple sites, evidence of both mutagenicity and tumor responses in multiple species or sexes significantly increases concern for the human carcinogenic potential of an agent. Absence of mutagenicity in multiple test systems gives insight into alternative mechanisms by which non-mutagenic carcinogens may act. The consideration of alternative non-mutagenic mechanisms does not necessarily provide a basis for discounting positive results in the animal cancer bioassay and thus does not negate the concern for human risk. On the other hand, evidence for non-mutagenicity and the lack of responses in a chronic rodent bioassay increases the confidence that an agent is not a human hazard.

2.6.4.2. Other Short-Term Tests

In addition to genetic toxicity tests, information on increased cell proliferation, cell transformation, aberrant intercellular communication, receptor mediated effects, changes in gene transcription (i.e., events that involve a change in the function of the genome) can provide useful information in the evaluation of mechanism of action and insight into the carcinogenic potential of an agent. It is not possible to describe all the data that might be encountered in a substance-specific assessment. Thus, the most conventional ones or those that are currently emphasized are mentioned as examples.

Cell proliferation plays a key role at each stage in the carcinogenic process and it is well established that increased rates of cell proliferation are associated with increased cancer risk. This increased risk is due to the increased susceptibility of proliferating cells to both spontaneous genetic damage as well as that induced by mutagens. Therefore, mitogenic activity in a mutagenic agent could be expected to further increase the probability of mutagenesis and, therefore, carcinogenesis. Cell proliferation or mutation alone are insufficient to cause neoplasia; further events are required for cells to escape from growth control, to attain the ability to grow independently, and to acquire invasiveness.

Evidence for the increased rate of cell division may be determined by measuring the mitotic index, or by supplying a specific DNA precursor to the cell (e.g., ^{3}H-thymidine or bromodeoxyuridine) and counting the percentage of cells that have incorporated the precursor into the replicating DNA, or by immunodetection of proliferation-specific antigens. These analyses are carried out in vitro, during pre-chronic studies, or as part of the long-term animal cancer bioassay.

Non-mutagenic carcinogens are more likely than mutagenic carcinogens to affect a specific sex or organ. Stable cell populations with a potential for a high rate of cell replication are more often affected than cell populations with a naturally high rate of replication. These properties have been used to develop two stage initiation-promotion studies based on preneoplastic lesions or tumors of the mammary gland, urinary bladder, forestomach, thyroid, kidney, and liver. Such tests provide mechanistic insight as well as supportive evidence for carcinogenicity (Drinkwater, 1990).

Several short-term tests respond to both mutagenic and non-mutagenic carcinogens. Assays for measuring perturbation of gap-junctional intercellular communication may provide and indication of carcinogenicity, especially promotional activity, and provide mechanistic information (Yamasaki, 1990). Cell transformation assays have been widely used for studying mechanistic aspects of chemical carcinogenesis because in vitro cell transformation is considered to be relevant to the in vivo carcinogenic process.

2.6.4.3. Short-Term Assays for Carcinogenesis

In addition to more conventional long-term animal studies, other shorter-term animal models can yield useful information about the carcinogenicity of agents. Some of the more common tests include mouse skin (Ingram and Grasso, 1991), transplacental and neonatal carcinogenesis (Ito, 1989), mammary gland tumor studies and preneoplastic lesions or altered cell foci (e.g., in liver, kidney, pancreas). Currently, increased research emphasis is being put on alternative approaches to the chronic rodent cancer bioassay. As an example, significant progress is being made using fish models (Bailey et al., 1984; Couch and Harshbarger, 1985).

2.6.4.4. Evaluation of Mechanistic Studies

The entire range of data about an agent's physical-chemical properties, structure-activity relationships to carcinogenic agents, and biological activity in vitro and in vivo is reviewed for mechanistic insights. The weight and significance of the observation of carcinogenic activity of the agent in vivo can be greatly influenced by the available data in several areas, all of which should be considered. Discussion should summarize available data on the agent's effects on DNA structure or expression and its effects on the cell cycle. Types of information to be considered include: whether the agent is a mutagenic or a non-mutagenic carcinogen, specific effects on proto-oncogenes or tumor suppressor genes and DNA transcription, and structural or functional analogies to agents with the above effects.

Information demonstrating effects on the cell cycle would include: mitogenesis, effects on differentiation, effects on cell death (apoptosis), tissue damage resulting in compensatory cell proliferation, receptor-mediated effects on growth-signal transduction, and structural or functional analogies to agents with the above effects.

Information demonstrating effects on cell interaction might include: effects on contact inhibition of growth, intracellular communication, or immune reactions, and structural or functional analogies to agents with these effects.

These are not intended to be exclusive of other pertinent data not specifically listed. In addition, available data on the comparative pharmacokinetics and metabolism of the agent in animals and humans is assessed to consider whether similar mechanisms of action may be operating in humans and animals. (A similar summarization of evidence has been reported by IARC, 1991).

In evaluating carcinogenic potential and mechanism of action, analyses and conclusions based on short-term tests are accompanied by a discussion of the level of confidence that can be applied to all the data. The level of confidence is based on the following (not necessarily exclusive) factors: (a) the spectrum of endpoints relevant to carcinogenesis and the number of studies used for detecting each end point and consistency of the results obtained in different test systems and different species, (b) in vivo as well as in vitro observations, (c) the consistency and concordance of test results, (d) reproducibility of the results within a test system, (e) existence of a dose-response relationship, and (f) whether the tests are conducted in accordance with appropriate protocols agreed upon by experts in the field. For, example, a high level of confidence in describing the potential influence of an agent on carcinogenic events is based on results covering a number of events relevant to stages of carcinogenesis, a number of studies including in vivo tests showing consistent trends and good concordance. A low confidence data set is one that was sparse or has incongruous results and no clear data trends.

The strength of an hypothesis about mechanism of action generated by analysis of data in the above areas should be described by the following criteria:

a. The operation of the mechanism in carcinogenesis must have been explained by a body of research data and have been generally accepted in the scientific community as a mechanism of carcinogenesis;

b. There must be a body of experimental data that show how the agent in question participates in the mechanism of action. In the absence of data about the mechanism of action of an agent, decisions are made using default assumptions:

c. That animal effects are relevant to human effects; and

e. That the agent affects carcinogenesis with dose and response relating linearly at low exposure.

Both of these science policy assumptions are supported by current knowledge of carcinogenic processes, in the absence of better data. Each assumption must be examined in substance-specific risk assessments and replaced or joined by alternative analysis when adequate scientific data exist.

2.7. SUMMARY OF EXPERIMENTAL EVIDENCE

{Criteria and examples for categorization of experimental evidence are major issues, particularly the weight of evidence contribution of research data of new kinds of genes and signal transduction pathways of growth control.}

A summary is made of all the experimental evidence that is relevant to human carcinogenic potential.

The confidence of an agent is potentially carcinogenic for humans increases as the number of animal species, strains, or number of experiments and doses showing a carcinogenic response increases. It also increases as the number of tissue sites affected by the agent increases and as the time to tumor occurrence or time to death with tumor decreases in dose-related fashion. Confidence also increases as the proportion of tumors that are malignant increases with dose and if the observed tumor types are historically rare in the species.

{The appropriate use of molecular biological data in the overall weight of evidence is a question. The strength of inferences to be drawn from data such as tumor susceptibility or gene effects is an unsettled issue.}

The weight of other experimental evidence increases or decreases the weight of findings relevant to human hazard in the following ways listed below. Findings in vivo add to the weight of evidence more rapidly than in vitro findings.

- physical-chemical properties and structural or functional analogies can support inferences of potential carcinogenicity;
- results in a number of short-term studies that are consistent can support inferences about potential human effects;
- evidence of mutagenic effects on proto-oncogenes or tumor suppressor genes;
- evidence of effects on cell growth signal transduction affecting cell division, differentiation; or cell death; and
- induction of neoplastic behavioral characteristics in cells in culture or in vivo.

The summarization of experimental evidence refers only to the weight of evidence that an agent may or may not be carcinogenic in humans, not the dose-response relationship, which is the subject of a separate analysis.

The following four categories are used to summarize all of the experimental data relevant to inferences about human carcinogenic potential of an agent. Tumor responses that the Agency has found to be not relevant for inferring human hazard are not given weight. Other responses whose relevance is unresolved are noted in the categorization of evidence. Categorization is a matter of scientific judgment, and the descriptions below are to be used as guidance in making that judgment, not as absolute criteria.

2.7.1. Category 1

The following examples illustrate persuasive evidence of carcinogenic potential. Other combinations of data also may be persuasive. In prospect, continued research on the role of agents in mutations of proto-oncogenes and tumor suppressor genes and related research on receptor-mediated effects on growth control genes also may provide persuasive data.

Examples:

1. Long-term animal experiments showing increased malignant and benign tumors
   a. when the increased incidence of tumors is in more than one species or in more than one experiment (i.e., results are complicated with different routes of administration, or affect a range of dose levels)
      - at multiple sites, or
      - at a limited number of sites with a supporting weight of evidence from structure-activity analysis, or available short-term tests;
   b. when there is a response to an unusual degree in a single experiment with regard to high incidence of a low-incidence background tumor, unusual site or type of tumor, or early age at onset
      - with a dose-related increase in a highly malignant tumor or in early death with cancer, or
      - with a supporting weight of evidence from structure activity analysis or from available short-term studies; or
   c. in more than one experiment, at a single site
      - with a highly supportive weight of evidence from SAR analysis and numerous consistent findings of effects on carcinogenic processes in short-term studies, or
      - with a dose-related increase in tumor malignancy.

2. Evidence that an agent is readily converted to a metabolite for which independent human or animal evidence is categorized as Group 1 and data are supportive of like pharmacokinetic disposition, or short-term studies of the agent are comparable in result with those of the metabolite.

3. Short-term experiments that demonstrate an agent's influence on carcinogenic processes in vivo consistent with in vitro studies, SAR, and physical-chemical properties that are highly supportive of carcinogen activity. These are supported

by studies showing comparable metabolism and pharmacokinetics between study species and humans.

2.7.2. Category 2

Examples for this category include:

1. A long-term animal experiment or experiments showing increased incidence of malignant tumors or combined malignant and benign tumors that falls short of the weight for categorization as Category 1.

2. Evidence that an agent is readily converted to a metabolite for which independent human or animal evidence is Category 2 and data are supportive of like pharmacokinetic disposition, or short-term studies of the agent are comparable in result with those of the metabolite.

3. Short-term studies and other evidence as described in 2.6.4.4. together with data supporting the likelihood of comparability in metabolism and pharmacokinetics between species.

2.7.3. Category 3

The experimental evidence does not support a conclusion either way about potential carcinogenicity because:
*   too few data are available;
*   evidence is limited to tumorigenicity and is found *solely* in studies in which the manner of administration (e.g., injection) or other aspects of study protocol present difficulties of interpretation; or
*   evidence of carcinogenicity is found at a single animal site in one species and sex in one or more experiments; the response is weak and without characteristics that give weight to a conclusion about potential human carcinogenicity.

For example, data are inconclusive if experimental data apart from the animal response do not support any positive inference about the agent's carcinogenic potential and if the animal response has a consistent pattern of most of the following characteristics:
*   At least two species have been tested, and the tumor response is seen only at the highest dose, in one sex, and one species.
*   The tumor incidence is predominantly benign and is seen only in one target organ.
*   The tumor is recognized as a common tumor type in that species, strain, and sex. In addition, the observed tumor rate, although statistically

significant in the experiment, is at or near the upper range of the historical control incidence.

- The tumors do not cause death in the affected animals during the duration of the study and do not appear sooner in the treated animals than in the controls.

Such evidence may add some weight to results of the human studies.

2.7.4. Category 4

This summarization would apply when no increased incidence of neoplasms has been observed in at least two well-designed and well-conducted animal studies in different species including both sexes. The exposures are specified and the implication is that either the agent is not carcinogenic or the studies had insufficient power to detect an effect.

2.8. HUMAN HAZARD CHARACTERIZATION

Evidence from all of the elements of hazard assessment are drawn together for an overall characterization of potential human hazard as indicated in Figure 1.

2.8.1. Purpose and Content of Characterization

The major lines of observational human evidence and experimental evidence and reasoning are clearly described. Major judgments made in the face of conflicting data are particularly highlighted and explained, as are the assumptions or inferences made to address gaps in information. The strengths and weaknesses of the available data are described and related to resulting confidence in the characterization. The hazard characterization addresses not only the question of carcinogenic properties, but also, as data permit, the question of the conditions (dose, duration, route) under which these properties may be expressed.

To provide a basis for combining hazard and environmental exposure data in the final risk characterization, the hazard characterization points to differences expected according to route of exposure, if such differences can be determined. The assumption is made that the hazard is not route-specific, if this is reasonable and not contradicted by existing data. Information about the plausible mechanism or mechanisms of action is characterized and its implications for dose-response assessment are explained, including conditions of dose and duration.

2.8.2. Weight of Evidence for Human Carcinogenicity

{NOTE TO THE READER: The question as to whether to abandon our al-

*phanumerical system entirely or merge it with a narrative statement has not
been decided. We may retain labels of A, B, C, etc., labels for weight of
evidence groups.}*

A brief narrative statement is used to summarize the weight of evidence. It
incorporates judgment about data from all elements of hazard assessment. A
summary statement cannot resolve data interpretation issues; it can only focus
judgments and help convey them. The purpose is to give the risk manager a
sense of the evidence and of the risk assessor's confidence in the data and their
interpretation for the assessment of human carcinogenicity potential and to allow
comparison of weight of evidence judgments from case to case. A weight of
evidence conclusion incorporates judgments both about overall confidence in a
set of data as a basis for drawing conclusions and about the consistency and
congruence of inferences supported by the set of data.

A weight of evidence conclusion is based both observational data from human studies and experimental data. All of the elements of analysis included in
hazard assessment form the basis of judgment. The summarizations of experimental evidence and human evidence are ingredients for a weight of evidence
statement. Note that animal tumor responses that the Agency considers not
relevant for inferring human hazard are not weighed. However, unresolved questions about relevance are all noted and considered in the statement.

As the first step, a decision is made on whether the evidence is adequate or
not adequate for characterization. "Not adequate" means that the existing data
are inadequate overall to support a conclusion because either there are too few
data or the data are flawed due to experimental design or conduct, or because
findings are not substantial enough to support inferences either way about potential human carcinogenicity. Typically, human or experimental data that are in
Category 3 would be considered as not adequate for characterization.

If the evidence is adequate for a weight-of-evidence determination, it is
described within a narrative statement. The narrative statement explains the
weight of evidence by summarizing the content and contribution of individual
lines of evidence and explaining how they combine to form the overall weight of
evidence. The statement highlights the quality and extent of data and the congruence, or lack of congruence, of inferences they support. The statement also
highlights default assumptions used to address gaps in knowledge.

The statement gives the weight of evidence by route of exposure, pointing
out the basis of anticipated differences and whether the default assumption supporting extrapolation of hazard potential between routes has been used and is
appropriate. Anticipated potency differences by route are pointed out, based on
comparatively poor to ready absorption by different routes (see § 2.6.3. Metabolism and Pharmacokinetics).

The statement discusses the data implications for mechanism of action. It
recommends a general approach or approaches for dose-response assessment in

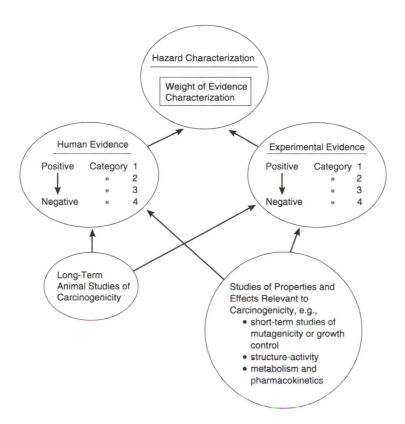

FIGURE 1

accordance with what the hazard data imply about the nature of dose-response below the range of observation of available studies. A weight of evidence for hazard by any mechanism is characterized. Thus, for example, an agent that is estrogenic and not likely to cause permanent genetic changes is characterized as a carcinogenic hazard, with any limitations of dose being explained in the narrative statement. The quantitative dose-response estimation or shape of the dose-response curve does not affect the weight of evidence for hazard.

The statement notes whether its source is an individual EPA office or an EPA consensus. The overall conclusion is noted by use of one of the following descriptors: **"known," "highly likely,"** or **"likely"** to be a human carcinogen; **"some evidence"** or **"not likely to be a human carcinogen at exposure levels studied or alternately under conditions of environmental exposure."** These descriptors fall along a continuum of likelihood that an agent has human carcinogenic potential. More than one descriptor may apply to a single agent if the weight of evidence differs by route of administration. Also, two descriptors may be applied if the evidence for a route is judged to fall between two descriptors. These standard descriptors are provided for the purpose of maintaining consistency of expression of conclusions from case to case. The text of the narrative statement as a whole is the primary means of conveying information on the weight of evidence.

2.8.2.1. Descriptors

{The number of descriptor categories for total weight of evidence is a continuing issue. The evidence is along a continuum. How many descriptors are needed to represent the continuum? What are the criteria for establishing them?}

Explanations of the general levels of evidence associated with descriptors in terms of the summarizations of evidence made in the course of a hazard assessment are as follows:

"Known" to be carcinogenic in humans is a statement that evidence is convincing (Category 1) that the agent has observed carcinogenic effects in humans by a specified route or routes of exposure.

"Highly likely" is a statement that:

1. there is persuasive experimental evidence of carcinogenicity (Category 1) and suggestive human evidence (Category 2), or
2. there is persuasive experimental evidence (Category 1) showing a very strong animal response (multiple tumor sites in more than one species), or
3. an agent is known to be a carcinogen in humans by one route of exposure (known) is also absorbed by another route, making carcinogenic effects " highly likely" by the second route.

"Likely" is a statement that:

1. there is persuasive experimental evidence (Category 1), or
2. there is suggestive evidence from human data (Category 2) with experimental evidence (Category 2) that supports the likelihood that the human effects seen were due to the agent in question.

"Some evidence" is a statement that:

1. there is experimental evidence (Category 2), or
2. suggestive human evidence (Category 2).

However, the totality of the evidence is weak because findings are inconsistent, or there are many gaps in the data.

"Not likely to be a human carcinogenic at exposure levels studied or alternately, under conditions of environmental exposure" is a statement that:

1. human evidence has been summarized as no evidence at exposure levels studied (Category 4), and there are no positive animal findings, or
2. experimental evidence has been summarized as no evidence at exposure levels studied (Category 4), and there are no positive human findings, or
3. the occurrence of carcinogenic effects is not expected for a particular route of human environmental exposure (oral, dermal, inhalation) because the agent is not absorbed by that route, or
4. the mechanism of carcinogenicity of an agent operates only at doses above the range of plausible environmental exposure, e.g., carcinogenesis as a secondary effect of another effect that occurs only at high doses, or
5. the occurrence of carcinogenic effects depends on administration of the agent in a manner that has no parallel with plausible environmental exposure, e.g., injection of polymers.

This descriptor is explained in the narrative statement as being applicable only to the specific exposure levels studied or environmental exposure conditions which are given in the statement.

2.8.2.2. *Examples of Narrative Statements*

Compound X

Following review of all available data relevant to the potential human carcinogenic hazard of X (CAS # 000001), the .... Office of EPA concludes that X is not likely to be carcinogenic to humans by any route of exposure at environmental levels. This determination is based on experimental evidence. No human studies on X are available for evaluation. The evidence supporting this finding is the animal response.

With dietary administration, X caused a statistically significant increase in the incidence of urinary bladder hyperplasia and tumors (urinary bladder transitional cell papillomas and carcinomas) in male but not in female Charles River CD rats at high dose levels (>30,000 ppm). The tumors were seen only at dose levels producing calculi in the kidneys, ureters and the urinary bladder. The presence of the urinary bladder calculi was associated with a decrease in the urinary pH. The urinary bladder calculi were almost always associated with urinary bladder hyperplasia (>90%). A major metabolite of X did not cause any increase in tumor incidence in another bioassay in rats. X was not carcinogenic in mice in well-conducted experiments.

The in vivo (mouse micronucleus test) and in vitro (in bacteria and yeast) short term- studies on X indicate with medium confidence that X is not genotoxic. Structure-activity-relationship analysis reveals no chemicals which are related to X and also induce tumors. It is concluded that the tumor response in male rats was secondary to stone formation at high doses, and may be a phenomenon unique to the male rats. No dose-response analysis is recommended unless a high-dose environmental exposure to humans is discovered.

Compound Y

Following review of all available data relevant to the potential human carcinogenic hazard of Y (CAS # 000002), EPA concludes that Y is likely to be carcinogenic to humans by all routes of exposure. This determination is based on experimental evidence. No human studies are available for evaluation. The strongest lines of evidence supporting findings on Y are animal experiments and structure-activity relationships.

Rodent studies showed statistically significant increases in the incidence of liver tumors (hepatocellular adenomas and carcinomas combined) in two strains of mice, in two independent and adequately conducted studies. The increases of liver tumors occurred at high and low doses. Y also produced a statistically significant increase in stomach tumors (papillomas) in both male and female mice at a dose also producing significant mortality and reduced body weight (-18% to -23% throughout the study) and the presence of white foci and ulcers in the stomach of occasional animals.
Y, administered orally, did not induce tumors in F344 rats in an adequately conducted study. Data from acute inhalation toxicity and dermal absorption studies show that Y is absorbed by both dermal and inhalation exposure.

Y caused gene mutations and chromosome aberrations in *D. melanogaster* and DNA damage in yeast, but it did not induce mutagenic effects in either in vitro or in vivo mammalian systems. The mutagenicity data set is of low confidence, and it neither supports nor contradicts inferences about carcinogenicity. In addition, it does not suggest a mechanism of action.

Structure-activity relationship analysis shows that Y is very closely related in structure to eight other chemicals, all of which produce liver tumors in mice, rats, or both.

Based upon the above analysis, it is suggested that the dose-response analysis employ a default assumption of linearity at low dose and consider the liver tumor in mice as an appropriate endpoint.

3. DOSE-RESPONSE ASSESSMENT

3.1. PURPOSE AND SCOPE OF DOSE-RESPONSE ASSESSMENT

Dose-response assessment tests the hypothesis that an agent has produced an effect and portrays the relationship between the agent and the response elicited. In risk assessments, dose and response observations from experimental or epidemiological studies are often projected to much lower exposure levels encountered in the environment.[4] In addition, the mathematical models used for extrapolation are based on general assumptions about the nature of the carcinogenic process. These assumptions may be untested for the particular agent being evaluated (Kodell, in press). If the dose-response relationship is developed from an experimental animal study, it also must be extrapolated from animals to humans. Because of these inherent uncertainties, projections well outside the range of the observed data are treated as bounding estimates, not as true values.

Information that shows a comparable pharmacokinetic and metabolic response to an agent in humans and animals greatly increases confidence in the dose-response analysis. Data suggesting that an agent works through a common mechanism of action in humans and animals also greatly increases confidence in the low dose extrapolation. In the absence of such data, default approaches provide upper-bound estimates of response at low doses, with a lower limit as small as zero at very low doses.

In the absence of dose-response data on members of a class of agents, it may be possible to construct a set of toxicity equivalence factors (TEF) to be used to

---

[4]For this discussion, "exposure" means contact of an agent with the outer boundary of an organism. "Applied dose" means the amount of an agent presented to an absorption barrier and available for absorption; "internal dose" means the amount crossing an absorption barrier (e.g., the exchange boundaries of skin, lung, and digestive tract) through uptake processes; and the amount available for interaction with an organ or cell is the "delivered dose" for that organ or cell. For more detailed discussion see Exposure Assessment Guidelines __ FR ____ (1992).

quantify dose-response by reference to an already-characterized member of the class.

3.2. Elements of Dose-Response Assessment

The elements of dose response analysis include selection of response data and dose data, followed by a stepwise dose-response analysis. The first step in the dose-response analysis is fitting of the data in the range of study observation; the second step, if needed, is extrapolation of the dose-response relationship to the range of the human exposure of interest.

A dose-response assessment should take advantage of available data to support a more confident analysis. When data gaps exist, assumptions based on current knowledge about the biological events in carcinogenesis and pharmacokinetic processes are used.

3.2.1. Response Data

Appropriate response data, as well as mechanistic information from the hazard characterization, are applied in the dose-response assessment. The quality of the data and their relevance to human exposure are important selection considerations.

If adequate positive human epidemiologic data are available, they are usually the preferred basis for analysis. Positive data are analyzed to estimate response to environmental exposure in the observed range. (USEPA, 1992a). Extrapolation to lower environmental exposure ranges is carried out, as needed. If adequate exposure data exist in a well-designed and well-conducted epidemiologic study that detects no effects, it may be possible to obtain an upper-bound estimate of the potential risk. Animal-based estimates, if available, are also presented, and the animal results are compared with the upper-bound estimate from human data for consistency.

When animal studies are used, response data from a species that responds most like humans should be used, if information to this effect exists. When an agent was tested in several experiments involving different animal species, strains, and sexes at several doses and different routes of exposure, the following approach to selecting the data sets is generally used:

a. The tumor incidence data are separated into data sets according to organ site and tumor type.
b. All biologically and statistically acceptable data sets are examined.
c. Data sets are analyzed with regard to route of exposure.
d. A judgment is reached based on biological criteria as to which set or sets best represents the body of data for the purpose of estimating human response. This judgment is augmented with judgment as to the statistical suitability of the data for modeling in the experimental data

range. The hazard characterization is the point of reference for the initial judgment. The following characteristics of a data set favor its selection.
- high quality of study protocol and execution;
- malignant neoplasms;
- earlier onset of neoplasm;
- greater number of data points to define the relationship of dose and response;
- background incidence in test animal is not unusually high;
- most sensitive-responding species are used; or
- data on a related effect (e.g., DNA adduct formation) or mechanistic data to augment the tumor.

Appropriate options for presenting results include use of a single data set, combining data from different experiments (Stiltler et al., 1992), showing a range of results from more than one data set, representing total response in a single experiment by combining animals with tumors or a combination of these options. The rationale for selecting an approach is presented, including the biological and statistical considerations involved. The objective is to provide a best judgment of how to represent the observed data.

Benign tumors are usually combined with malignant tumors for risk estimation if the benign tumors are considered to have the potential to progress to associated malignancies of the same histogenic origin. (McConnell, 1986). When tumors are thus combined, the contribution to the total risk of benign tumors is indicated. The issue of how to consider the contribution of the benign tumors should be discussed in the dose-response characterization and risk characterization.

Data on certain endpoints related to tumor induction may be used to extend dose-response analysis below the relatively high dose range in which tumors are observable. These data permit extension of the curve-fitting analysis (Swenberg et al., 1987) and may provide parameters for applying a mechanism-based model (US EPA Dioxin Assessment, 1992c). Data might include information on receptor binding, DNA adduct formation, physiological effects such as disruption of hormone activity, or agent-specific alterations in cell division rates. In considering whether such endpoints can be applied, key issues are confidence that the data reflect carcinogenic effects of the agent and that these have been well measured with a dose-effect trend.

3.2.2. Dose Data

Regardless of the source, animal experiments or epidemiologic studies, several questions need to be addressed in arriving at an appropriate measure of dose. One question is whether data are sufficient to estimate internal dose or delivered dose. Part of this question is whether the parent compound, a metabolite, or both agents are closer in a metabolic pathway to a carcinogenic form.

The delivered dose to target is the preferred measure of dose. In practice, there may be little or no information on the concentration or identity of the active agent at a site of action; thus, being able to compare the applied and delivered doses between routes and species is an ideal that is rarely attained. Even so, incorporating data to the extent possible is desirable.

Even if pharmacokinetic and metabolic data are sufficient to derive a measure of delivered dose to the target, the dose-response relationship is also affected by kinetics of reactions at the target (pharmacodynamics) and by other steps in the development of neoplasia. With few exceptions, these processes are currently undefined.

The following discussion assumes that the analyst will have data of varying detail in different cases about pharmacokinetics and metabolism. Approaches to limited data are outlined as well as approaches and judgments for more sophisticated analysis based on additional data.

3.2.2.1. Base Case — Few Data

Where there are insufficient data available to define the equivalent delivered dose between species, it is assumed that delivered doses at target tissues are directly proportional to applied doses. This assumption rests on the similarities of mammalian anatomy, physiology, and biochemistry generally observed across species. This assumption is more appropriate at low applied dose concentrations where sources of nonlinearity, such as saturation or induction of enzyme activity, are less likely to occur.

The default procedure is to scale daily applied doses experienced for a lifetime in proportion to body weight raised to the 3/4 power ($W^{3/4}$). Equating exposure concentrations in parts per million units for air, food, or water is an alternative version of the same default procedure because daily intakes of these are in proportion to $W^{3/4}$. The rationale for this factor rests on the empirical observation that rates of physiological processes consistently tend to maintain proportionality with $W^{3/4}$. A more extensive discussion of the rationale and data supporting the Agency's adoption of this scaling factor can be found in (USEPA, 1992b).

The differences in biological processes among routes of exposure (oral, inhalation, dermal) can be great, due to, for example, first pass effects and differing results from different exposure patterns. There is no generally applicable method for accounting for these differences in uptake processes in quantitative route-to-route extrapolation of dose-response data in the absence of good data on the agent of interest. Therefore, route-to-route extrapolation of dose data will be based on a case-by-case analysis of available data. When good data on the agent itself are limited, an extrapolation analysis can be based on expectations from physical chemical properties of the agent, properties and route-specific data on structurally analogous compounds, or in vitro or in vivo uptake data on the agent. Route-to-route uptake models may be applied if model parameters are suitable for the compound of interest. Such models are currently considered interim methods; further model development and validation is awaiting the development of more extensive data (see generally, Gerrity and Henny, 1990).

3.2.2.2. *Pharmacokinetic Analyses*

Physiologically based mathematical models are potentially the most comprehensive way to account for pharmacokinetic processes affecting dose. Models build on physiological compartmental modeling and attempt to incorporate the dynamics of tissue perfusion and the kinetics of enzymes involved in metabolism of an administered compound.

A comprehensive model requires the availability of empirical data on the carcinogenic activity contributed by parent compound and metabolite or metabolites and data by which to compare kinetics of metabolism and elimination between species. A discussion of issues of confidence accompanies presentation of model results (Monro, 1991). This includes considerations of model validation and sensitivity analysis that stress the predictive performance of the model. Another assumption made when a *delivered* dose measure is used in animal-to-human extrapolation of dose-response data is that the pharmacodynamics of the target tissue(s) will be the same in both species. This assumption should be discussed, and confidence in accepting it should be considered in presenting results.

Pharmacokinetic data can improve dose-response assessment by accounting for sources of change in proportionality of applied-to- internal dose or to delivered dose at various levels of applied dose. Many of the sources of potential nonlinearity involve saturation or induction of enzymatic processes at high doses. An analysis that accounts for nonlinearity (for instance, due to enzyme saturation kinetics) can assist in avoiding over estimation or under estimation of low dose if extrapolation is from a sublinear or supralinear part of the experimental dose-response curve. (Gillette, 1983). Pharmacokinetic processes tend to become linear at low doses, an expectation that is more robust than low-dose linearity of response (Hattis, 1990). Thus, accounting for nonlinearities allows better description of the shape of the curve at higher levels of dose, but cannot determine linearity or nonlinearity of *response* at low dose levels (Lutz, 1990; Swenberg et al., 1987).

3.2.2.3. *Additional Considerations for Dose in Human Studies*

The applied dose in a human study has uncertainties because of the exposure fluctuations that humans experience compared with the controlled exposures received by animals on test. In a prospective cohort study, there is opportunity to monitor exposure and human activity patterns for a period of time that supports estimation of applied dose (USEPA, 1992a). In a retrospective cohort study, exposure is based on human activity patterns and levels reconstructed from historical data, contemporary data, or a combination of the two. Such reconstruction is accompanied by analysis of uncertainties considered with sensitivity analysis in the estimation of dose (Wyzga, 1988; USEPA, 1986). These uncertainties can also be assessed for any confounding factor, for which a quantitative adjustment of dose-response data is made (USEPA, 1984).

Exposure levels of groups of people in the study population often are represented by an average when they are actually in a range. The full range of data are analyzed and portrayed in the dose-response analysis when possible (USEPA, 1986).

The cumulative dose of an agent is commonly used when modeling human data. This can be done, as in animal studies, with a default assumption in the absence of data that support a different dose surrogate. Given data of sufficient quality, dose rate or peak exposure can be used as an alternative surrogate to cumulative dose.

3.3. SELECTION OF QUANTITATIVE APPROACH

Because risks at relatively low exposure levels generally cannot be measured directly either by animal experiments or by epidemiologic studies of reasonable sample size, a number of mathematical models have been developed to extrapolate from high to low dose. Different extrapolation models may fit the observed data reasonably well but may lead to large differences in the projected risk at lower doses. As was pointed out by OSTP (1985 see Principle 26), no single mathematical procedure is recognized as the most appropriate for low-dose extrapolation in carcinogenesis. Low-dose extrapolation procedures use either mechanistic or empirical models. When sufficient biological information exists to identify and describe a mechanism of action, low-dose extrapolation may be based on a mathematical representation of the mechanism. When the mechanism is unknown or information is limited, low-dose is derived from an empirical fit of a curve compatible with the available information.

If a carcinogenic agent acts by accelerating the same carcinogenic process that leads to the background occurrence of cancer, the added effect on the population at low doses marginally above background level is expected to be linear. Above background level, the population response may continue to be linear in the case of an agent acting directly on DNA, or the population response may be influenced by individual variability in sensitivity to phenomena such as disruption of hormone homeostasis or receptor-mediated activity. If the agent acts by a mechanism with no endogenous counterpart, a population response threshold may exist (Crump et al., 1976; Peto, 1978; Hoel, 1980; Lutz, 1990). The

Agency reviews each assessment as to the evidence on carcinogenesis mechanisms and other biological or statistical evidence that indicates the suitability of a particular extrapolation model. When longitudinal data on tumor development are available, time-to-tumor or survival models may be used and are preferred. In all cases, a rationale is included to justify the use of the chosen model.

The goal in choosing an approach is to achieve the closest possible correspondence between the approach and the view of the agent's mechanism of action developed in the hazard assessment. If the hazard assessment describes more than one mechanism as plausible and persuasive given the data available, corresponding alternative approaches for dose-response analysis are considered.

3.3.1. Analysis in the Range of Observation

In portraying dose response in the range of observed data, analyses incorporate as much reliable information as possible. Pharmacokinetic data or interspecies scaling is used to derive human-equivalent measures of the animal-administered dose. The empirical response data analyzed include tumor incidence data augmented, if possible, by incidence data on effects leading to the tumor response, e.g., DNA adduct or other effect-marker data (Swenberg, 1987).

Dose-response models span a hierarchy that reflects an ability to incorporate different kinds of information. If data to support it are available, a mechanism-based procedure is the preferred approach for modeling. A mechanism-based procedure is explicitly devised to reflect biological processes. Theoretical values for parameters, e.g., theoretical cell proliferation rates, are not used to enable application of a mechanism-based model (Portier, 1987). If such data are absent, a mechanism-based model is not used. An example of a mechanism-based model is the receptor mediated toxicity model for dioxin, under development at EPA (U.S. EPA, 1992c).

Dose-response models based on general concepts of a mechanism of action are next in amount of information required. For a specific agent, model parameters are obtained from laboratory studies. Examples are the two-stage models of initiation, clonal expansion, and progression developed by Moolgavkar et al. (1981) and Chen et al. (1991). Such models require extensive data to build the form of the model as well as to estimate how well it conforms with the observed carcinogenicity data.

Empirical models, which do not incorporate information about mechanism of action, form the rest of the hierarchy. Among these, time-to-tumor models incorporate longitudinal information on tumor development. Simple quantal models use only the final incidence at each dose level. The linearized multistage procedure is an example of an empirical model.

If a mechanism-based model is judged to be not suitable, the analysis uses an empirical model whose underlying parameters correspond to the putative mechanism of action identified in the hazard characterization. A multistage model (Zeise et al., 1987) structured with time to response as the random variable is appropriate when time is the dominant factor for probability of response. This is the approach when available information described in the hazard characterization is consistent with an assumption that there is no threshold of response for individuals. When the probability of effect is due to the distribution of thresholds for individuals in the population, a model considering dose as the random variable may be used. This may be considered an appropriate approach when the mechanism has been identified as one such as disruption of hormone homeostasis.

{The issue of appropriate dose-response models is still under discussion at EPA.}

Ordinarily, models are expected to provide an adequate fit to the observed dose-response information. The outcome of most tests of goodness of fit to the observations is not an effective means of discriminating among models that all provide an adequate fit. Although a model may adequately fit the observed dose-response information, all models have limitations in their ability to describe the underlying processes and make projections outside the observed information. A prime consideration is the potential for model error, that is the possibility that a model might appear to fit the observed data but be based on an inadequate mathematical description of the true underlying mechanism. This is especially crucial when making inferences outside the range of observation, as alternative models may provide an adequate fit to the observed information but have substantially different implications outside the range of observation.

Sometimes an inadequate fit might be improved by incorporating more information. For example, data in which there is high mortality may be poorly fit unless competing risks of death by toxicity are taken into consideration with time-to-tumor information and survival adjustments. If an adequate fit cannot be obtained, it may be necessary to give less weight to the observations most removed from low-dose risk., e.g,, from the highest dose level in a study with several dose levels.

Statistical considerations can affect the precision of model estimates. These include the number and spacing of dose levels, sample sizes, and the precision and accuracy of dose measurements. Sensitivity analysis can be performed to describe the sensitivity of the model to slight variations in the observed data. A large divergence between upper and lower confidence bounds indicates that the model cannot make precise projections in that range. All of these considerations are important in determining the range in which a model is supported by data.

With the recent expansion of readily available computing capacity, computer-intensive methods are being adapted to create simulated biological data that are comparable with the observed information. These simulations can be used for sensitivity analysis, for example, to analyze how small, plausible variations in the observed data could affect the risk estimates. These simulations can also provide information about experimental uncertainty in risk estimates, including a distribution of risk estimates that are compatible with the observed data. Because these simulations are based on the observed data, they cannot, however, assist in evaluating the extent to which the observed data as a whole are idiosyncratic rather than typical of the true state of risks.

The lowest reliable area of a curve is identified as a result of the data modeling. This point is generally at the level of not less than a 1.0 percent response if only animal tumor response data are available. (This 1.0 percent response level is about an order of magnitude below the potential power of a standard rodent study to detect effects.) The lowest reliable area may be extended below a 1.0 percent response if based on a more powerful study, on combined studies, or on joining the analysis of tumor response data with data on other markers of effect. This lowest reliable area provides an estimate that can be used for comparision with similar analyses of the observed range of noncancer effects of an agent (USEPA, 1991f).

3.3.2. Extrapolation

Using the lowest reliable point from the first step of analysis as a point of departure, the preferred approach for this second step of analysis still is a mechanism-based model, if data support it. If a mechanism-based model has been used to portray the observed data, the question in this step is whether confidence in the model extends to using it for extrapolation. If data are insufficient to support a mechanism-based model, extrapolation is done by a default procedure whose parameters reflect the general mechanism or mechanisms of action considered to be supported by the available biological information.

If the mechanism of action being considered leads to an expected linear dose-response relationship, the linearized multistage model or a model-free approach may be appropriate (Gaylor and Kodell, 1980; Krewski, 1984; Flamm and Winbush, 1984).

The mechanism of action being considered may project that the dose-response relationship in the population is most influenced by the differences in sensitivities. In this case, a model including tolerance distribution parameters may be used to provide estimates of the proportion of the population at risk for specific doses of interest, e.g., 1/1000, 1/10000 lifetime risk levels. This approach requires data for a mathematical portrayal of the distribution.

{NOTE: The appropriate empirical modeling approaches for extrapolation are an undecided issue when a putative mechanism of action has been recognized but data are not supportive of a mechanism-based model. Further technical analysis and discussion are necessary before this section can be completed.}

Alternatively, the mechanism may be one that involves a population threshold. In these cases, extrapolation is not made. Instead, a "margin of exposure" presentation is made in the risk characterization. The margin of exposure in this context is the lowest reliable dose-response area from observed data divided by the environmental dose level of interest.

3.3.3. Issues for Analysis of Human Studies

Issues and uncertainties arising in dose-response assessment based on epidemiological studies are analyzed in each case.   Several sources of uncertainty need to be addressed in the dose-response analysis. Consideration needs to be given to the data on the exposure and mortality experience of the study population and of the population that will represent the background incidence of the neoplasm(s) involved. In this area, there are potentials for mistakes or uncertainty in the data or adjustments to the data concerning the occurrence or level of exposure of the population members, mortality experience of a population, incomplete follow-up of individuals, exposure (or not) of individuals to confounding causes, or consideration of latency of response. These are assessed by analyzing the sensitivity of dose-response study results to errors where data permit. Other kinds of uncertainty can occur because of small sample size which can magnify the effects of misclassification or change assumptions about statistical distribution that underlie tests of statistical significance (Wyzga, 1988). These uncertainties are discussed. Where possible, analyses of the sensitivity of results to the potential variability in the data in these areas are performed.

The suitability of various available mathematical procedures for quantifying risk attributed to exposure to the study agent is discussed. These methods (e.g., absolute risk, relative risk, excess additive risk) account differently for duration of exposure and background risk, and one or more can be used in the analysis as data permit. The use of several of these methods is encouraged when they can be used appropriately in order to gain perspectives on study results.

3.3.4. Use of Toxicity Equivalence Factors

A toxicity equivalence factor (TEF) procedure is one used to derive quantitative dose-response estimates for agents that are members of a category or class of agents. TEFs are based on shared characteristics that can be used to order the class members by carcinogenic potency when cancer bioassay data are inadequate for this purpose (USEPA, 1991c). The ordering is by reference to the characteristics and potency of a well-studied member or members of the class. Other class members are indexed to the reference agent(s) by one or more shared characteristic to generate their TEFs. The TEFs are usually indexed at increments of a factor of 10. Very good data may permit a smaller increment to be used. Shared characteristics that may be used are, for example, receptor-binding characteristics, results of assays of biological activity related to carcinogenicity, or structure-activity relationships.

TEFs are generated and used for the limited purpose of assessment of agents or mixtures of agents in environmental media when better data are not available. When better data become available for an agent, its TEF should be replaced or revised.

Guiding criteria for the successful application of TEFs are (USEPA, 1991c):

1. A demonstrated need. A TEF procedure should not be used unless there is a clear need to do so.
2. A well-defined group of chemicals.
3. A broad base of toxicological data.
4. Consistency in relative toxicity across toxicological endpoints.
5. Demonstrated additivity between toxicities of group members for assessment of mixtures.
6. A mechanistic rationale.
7. Consensus among scientists.

3.4. DOSE-RESPONSE CHARACTERIZATION

The conclusions of dose-response analysis are presented in a characterization section. Because alternative approaches may be plausible and persuasive in selecting dose data, response data, or extrapolation procedures, the characterization presents the judgments made in such selections. The results for the approach or approaches chosen are presented with a rationale for the one(s) that is considered to best represent the available data and best correspond to the view of the mechanism of action developed in the hazard assessment.

The exploration of significant uncertainties in data for dose and response and in extrapolation procedures is part of the characterization. They are described quantitatively if possible through sensitivity analysis and statistical uncertainty analysis. If quantitative analysis is not possible, significant uncertainties are described qualitatively. Dose-response estimates are appropriately presented in ranges or as alternatives when equally persuasive approaches have been found.

Numerical dose-response estimates are presented to one significant figure and qualified as to whether they represent central tendency or plausible upper-bounds on risk or, in general, as to whether the direction of error is to overestimate or under estimate risk. For example, the straight line extrapolation used as a default is typically considered to place a plausible upper- bound on risk at low doses. On the other hand, a tolerance distribution model used as a default to portray risk-specific response distribution of the population may greatly underestimate risks if the mechanism is in fact a linear, nonthreshold one. (Krewski, 1984).

In cases, where a mechanism has been identified that has special implications for early-life exposure, differential effects by sex, or other concerns for sensitive subpopulations, these are explained. Similarly, any expectations that high dose-rate exposures may alter the risk picture for some portion of the population are described. These and other perspectives are recorded to guide exposure assessment and risk characterization.

4. EXPOSURE ASSESSMENT

Guidelines for exposure assessment of carcinogenic and other agents are published in USEPA, 1992a. The exposure characterization is a key part of the exposure assessment; it is the summary explanation of the exposure assessment. The exposure characterization

a. provides a statement of purpose, scope, level of detail, and approach used in the assessment;
b. presents the estimates of exposure and dose by pathway and route for individuals, population segments, and populations in a manner appropriate for the intended risk characterization;
c. provides an evaluation of the overall quality of the assessment and the degree of confidence the authors have in the estimates of exposure and dose and the conclusions drawn; and
d. communicates the results of exposure assessment to the risk assessor, who can then use the exposure characterization, along with the characterization of the other risk assessment elements, to develop a risk characterization.

In general, the magnitude, duration, and frequency of exposure provide fundamental information for estimating the concentration of the carcinogen to which the organism is exposed. These data are generated from monitoring information, modeling results, and or reasoned estimates. An appropriate treatment of exposure should consider the potential for exposure via ingestion, inhalation, and dermal penetration from relevant sources of exposures, including multiple avenues of intake from the same source.

Special problems arise when the human exposure situation of concern suggests exposure regimens, e.g., route and dosing schedule that are substantially different from those used in the relevant animal studies. The cumulative dose received over a lifetime, expressed as average daily exposure prorated over a lifetime, is an appropriate measure of exposure to a carcinogen particularly for an agent that acts by damaging DNA. The assumption is made that a high dose of a carcinogen received over a short period of time is equivalent to a corresponding low dose spread over a lifetime. This approach becomes more problematic as the exposures in question become more intense but less frequent, especially when there is evidence that the agent acts by a mechanism involving dose-rate effects.

5. CHARACTERIZATION OF HUMAN RISK

5.1. PURPOSE

The risk characterization is prepared for the purpose of communicating results of the risk assessment to the risk manager. Its objective is to be an appraisal of the science that the risk manager can use, along with other decisionmaking resources, to make public health decisions. A complete characterization presents the risk assessment as an integrated picture of the analysis of the hazard, dose response, and exposure. It is the risk analyst's obligation to communicate not only summaries of the evidence and results, but also perspectives on the quality of available data and the degree of confidence to be placed in the risk estimates. These perspectives include explaining the constraints of available data and the state of knowledge about the phenomena studied.

5.2. APPLICATION

A risk characterization is a necessary part of any Agency report on risk, whether the report is a preliminary one prepared to support allocation of resources toward further study or a comprehensive one prepared to support regulatory decisions. Even if only parts of a risk assessment (hazard and dose-response analyses for instance) are covered in a document, the risk characterization will carry the characterization to the limits of the document's coverage.

5.3. CONTENT

Each of the following subjects should be covered in the risk characterization.

5.3.1. Presentation and Descriptors

The presentation of the results of the assessment should fulfill the aims as outlined in the purpose section above. The summary draws from the key points of the individual characterizations of hazard, dose response, and exposure analysis performed separately under these guidelines. The summary integrates these characterizations into an overall risk characterization (AIHC, 1989).

The presentation of results clearly explains the descriptors of risk selected to portray the numerical estimates. For example, when estimates of individual risk are used or population risk (incidence) is estimated, there are several features of such estimates that risk managers need to understand. They include, for instance, whether the numbers represent average exposure circumstances or maximum potential exposure. The size of the population considered to be at risk and the distribution of individuals' risks within the population should be given. When risks to a sensitive subpopulation have been identified and characterized, the explanation covers the special characterization of this population.

5.3.2. Strengths and Weaknesses

The risk characterization summarizes the kinds of data brought together in the analysis and the reasoning upon which the assessment rests. The description conveys the major strengths and weaknesses of the assessment that arise from availability of data and the current limits of understanding of the process of cancer causation. Health risk is a function of the three elements of hazard, dose response, and exposure. Confidence in the results of a risk assessment is, thus, a function of confidence in the results of the analyses of each element. The important issues and interpretations of data are explained, and the risk manager is given a clear picture of consensus or lack of consensus that exists about significant aspects of the assessment. Whenever more than one view of the weight of evidence or dose-response characterization is supported by the data and the policies of these guidelines, and when choosing between them is difficult, the views are presented together. If one has been selected over another, the rationale is given; if not, both are presented as plausible alternative results. If a quantitative uncertainty analysis of data is appropriate, it is presented in the risk characterization; in any case, qualitative discussion of important uncertainties is appropriate.

6. REFERENCES

Allen, B. C.; Crump, K. S.; Shipp, A. M. (1988) Correlation between carcinogenic potency of chemicals in animals and humans. Risk Anal. 8: 531-544.

American Industrial Health Council, (1989) Presentation of risk assessment of carcinogens. Report of an Washington, D.C. ad hoc study group on risk assessment presentation.

Ames, B. N.; Gold, L. S. (1990) Too many rodent carcinogens: mitogenesis increases mutagenesis. Science 249: 970-971.

Ashby, J.; DeSerres, F. J.; Shelby, M. D.; Margolin, B. H.; Isihidate, M. et al., eds. (1988) Evaluation of short-term tests for carcinogens: report of the International Programme on Chemical Safety's collaborative study on in vivo assays. Cambridge, United Kingdom: Cambridge University. v. 1,2; 431 pp., 372 pp.

Ashby, J.; Tennant, R. W. (1991) Definitive relationship among chemical structure, carcinogenicity and mutagenicity for 301 chemicals tested by the U.S. NTP. Mutat. Res. 257: 229-306.

Auger, K. R.; Carpenter, C. L.; Cantley, L. C.; Varticovski, L. (1989a) Phosphatidylinositol 3-kinase and its novel product, phosphatidylinositol 3-phosphate, are present in *Saccharomyces cerevisiae*. J. Biol. Chem. 264: 20181-20184.

Auger, K. R.; Sarunian, L. A.; Soltoff, S. P.; Libby, P.; Cantley, L. C. (1989b) PDGF-dependent tyrosine phosphorylation stimulates production of novel polyphosphoinositides in intact cells. Cell 57: 167-175.

Aust, A. E. (1991) Mutations and Cancer. In: Li, A. P. and Heflich, R. H., eds., Genetic Toxicology. Boca Raton, Ann Arbor, and Boston: CRC Press. p. 93-117.

Bailey et al. (1984)

Barrett, J. C.; Oshimura, C. M.; Tanka, N.; and Tsutsui, T. (1985) Role of aneuploidy early and late stages of neoplastic progression of syrian hamster embryo cell in culture. In: Dellarco, V. L.; Voytek, P. E. and Hollaender, eds., Aneuploidy: etiology and mechanisms, New York: Plenum, in press.

Birner et al., (1990) Biomonitoring of aromatic amines. III: Hemoglobin binding and benzidine and some benzidine congeners. Arch. Toxicol. 64(2):97-102

Bishop, J. M. (1991) Molecular themes in oncogenesis. Cell 64:235-248.

Bishop, J. M. (1989) Oncogens and clinical cancer. In: Weinberg, R.A. (ed). Oncogens and the molecular origins of cancer. Cold Spring Laboratory Press. pp.327-358.

Blair, A.; Stewart, P.; O'Berg, M.; Gaffey, W.; Walrath, J.; Ward, J.; Bales, R.; Baplan, S.; Cubit, D. (1986) Mortality among industrial workers exposed to formaldehyde. J. Natl.Cancer Inst. 76: 1071-1084.

Blair A.; Stewart, P. A.; Hoover, R. N.; Fraumeni, J. F.; Walrath, J.; O'Berg, M.; Gaffey, W. (1987) Cancers of the nasopharynx and oropharynx and formaldehyde exposure. (Letter to the Editor.) J. Natl.Cancer Inst. 78: 191-192.

Blair, A.; Stewart, P. A. (1990) Correlation between different measures and occupational exposure to formaldehyde. Am. J. Epidemiol. 131:570-516.

Blancato, J. ( )

Bottaro, D. P.; Rubin, J. S.; Faletto, D. L.; Chan, A. M. L.; Kmieck, T. E.; Vande Woude, G. F.; Aaronson, S. A. (1991) Identification of the hepatocyte growth factor receptor as the c-*met* proto-oncogene product.

Bouk, N. (1990) Tumor angiogenesis: the role of oncogenes and tumor suppressor genes. Cancer Cells 2: 179-183.

Breslow, N. E.; and Day, N.E. (1980) Statistical methods in cancer research. Vol.I - The analysis of case-control studies. IARC Scientific Publication No. 32. Lyon, France: International Agency for Research on Cancer.

Bressac, B.; Kew, M.; Wands, J., Oztuk, M. (1991) Selective G to T mutations of p53 in hepatocellular carcinoma from southern Africa. Nature 429-430.

Brockman, H. E.; DeMarini, D. M. (1988) Utility of short-term test for genetic toxicity in the aftermath of the NTP's analysis of 73 chemicals. Environ. Molecular Mutagen. 11: 121-435.

Brusick, __. (1990)

Callemen, C.J.; Ehrenberg, L.; Jansson, B.; Osterman-Golkar, S.; Segerback, D.; Svensson, K; Wachtmeister, C.A. (1978) Monitoring and risk assessment by means of alkyl groups in hemoglobin in persons occupationally exposed to ethylene oxide. J. Environ. Pathol. Toxicol. 2:427-442.

Callemen et al. (1986)

Cantley, L. C.; Auger, K. R.; Carpenter, C.; Duckworth, B.; Graziani, A.; Kapeller, R.; Soltoff, S. (1991) Oncogenes and signal transduction. Cell 64: 281-302.

Castagna, M.; Takai, Y.; Kaibuchi, K.; Sano, K.; Kikkawa, U.; Nishizuka, Y. (1982) Direct activation of calcium-activated, phospholipid-dependent protein kinase by tumor-promoting phorbol esters. J. Biol. Chem. 257: 7847-7851.

Cavenee, W. K.; Dryja, T. P.; Phillips, R. A.; Benedict, W. F.; Godbout, R.; Gallie, B. L.; Murphree, A. L.; Strong, L.C.; White, R. L. (1983) Expression of recessive alleles by chromosome mechanisms in reteiroblastoma. Nature 303: 779-784.

Cerutti, P.; Larsson, R.; and Krupitza, G. (1990) Mechanisms of carcinogens and tumor progression. Harris, C.C.; Liiotta, L. A., eds. New York: Wiley-Liss. pp. 69-82.

Chen, C.; Farland, W. (1991) Incorporating cell proliferation in quantitative cancer risk assessment: approaches, issues, and uncertainties. In: Butterworth, B.; Slaga, T.; Farland, W.; McClain, M. (eds.) Chemical induced cell proliferation: Implication for risk assessment. Wiley-Liss.

Collum, R. G.; Alt, F. W. (1990) Are *myc* proteins transcription factors? Cancer Cells 2: 69-73.

Connolly, R. B.; Andersen, M. E. (1991) Biological based pharmacodynamic models: tools for toxicological research and risk assessment. Ann. Rev. Pharmacol. Toxicol. 31: 503-523.

Coouch and Harsbarger (1985)

Cross, M.; Dexter, T. (1991) Growth factors in development, transformation, and tumorigenesis. Cell 64: 271-280.

Crump, K. S.; Hoel, D. G.; Langley, C. H.; Peto, R. (1976) Fundamental carcinogenic processes and their implications for low dose risk assessment. Cancer Res. 36: 2973-2979.

deThe, H.; Lavau, C.; Marchio, A.; Chomienne, C.; Degos, L.; Dejean, A. (1991) The PML-RAR_ fusion mRNA generated by the t (15;17) translocation in acute promyelocytic leukemia encodes a functionally altered RAR. Cell 66: 675-684.

Drinkwater, N. R. (1990) Experimental models and biological mechanisms for tumor promotion. Cancer Cells,, 2(1):8.

D'Souza, R. W.; Francis, W. R.; Bruce, R. D.; Andersen, M. E. (1987) Physiologically based pharmacokinetic model for ethylene chloride and its application in risk assessment. In: *Pharmacokinetics in risk assessment*. Drinking Water and Health. Vol 8. Washington, DC: National Academy Press.

Fearon, E.; Vogelstein, B. (1990) A genetic model for colorectal tumorigenesis. Cell 61:959-767.

Fidler, I. J.; Radinsky, R. (1990). Genetic control of cancer metastasis. J. Natl. Cancer Inst. 82: 166-168.

Flamm, W. G.; Winbush, J. S. (1984) Role of mathematical models in assessment of risk and in attempts to define management strategy. Fundam. Appl. Toxicol. 4: S395-S401.

Flynn, G. L. (1990) Physicochemical determinants of skin absorption. In: Gerrit, T. R.; Henry, C. J. (eds.) Principles of route to route extrapolation for risk assessment. New York, NY: Elsevier Science Publishing Co. pp. 93-127.

Forsburg, S. L.; Nurse, P. (1991) Identification of a G1-type cyclin pug1+ in the fission yeast *Schizosaccharomyces pombe*. Nature 351: 245-248.

Garfinkel, L; Silverberg, E. (1991) Lung cancer and smoking trends in the United States over the past 25 years. Cancer 41: 137-145.

Garrett, N. E.; Stack, H. F.; Gross, M. R.; Waters, M. D. (1984) An analysis of the spectra of genetic activity produced by known or suspected human carcinogens. Mutat. Res. 134: 89-111.

Gaylor, D. W.; Kodell, R. L. (1980) Linear interpolation algorithm for low-dose risk assessment of toxic substances. J. Environ. Pathol. Toxicol. 4: 305-312.

Gaylor, D. W. (1988) Quantitative risk estimation. Adv. Mod. Environ. Toxicol. 15: 23-43.

Gerrity, T. R.; Henry, C., eds. (1990) Principles of route to route extrapolation for risk assessment. New York, NY: Elsevier Science Publishing Co.

Gillette, J. R. (1983) The use of pharmacokinetics in safety testing. Safety evaluation and regulation of chemicals 2. 2nd Int. Conf., Cambridge, MA: pp. 125-133.

Greenland, S. (1987) Quantitative methods in the review of epidemiologic literature. Epidemiol. Rev. 9:1-29.

Hammand, E. C. (1966) Smoking in relation to the death rates of one million men and women. In: Haenxzel, W., ed. Epidemiological approaches to the study of cancer and other chronic diseases. National Cancer Institute Monograph No. 19, Washington, DC.

Harris, C. C. (1989) Interindividual variation among humans in carcinogen metabolism, DNA adduct formation and DNA repair. Carcinogenesis 10: 1563-1566.

Harris, H. (1990) The role of differentiation in the suppression of malignancy. J. Cell Sci. 97: 5-10.

Haseman, J. K. (1990) Use of statistical decision rules for evaluating laboratory animal carcinogenicity studies. Fundam. Appl. Toxicol. 14: 637-648.

Haseman, J. K. (1985) Issues in carcinogenicity testing: Dose selection. Fundam. Appl. Toxicol. 5: 66-78.

Haseman, J. K. (1984) Statistical issues in the design, analysis and interpretation of animal carcinogenicity studies. Environ. Health Perspect. 58: 385-392.

Haseman, J. K.; Huff, J.; Boorman, G.A. (1984) Use of historical control data in carcinogenicity studies in rodents. Toxicol. Pathol. 12: 126-135.

Haseman, J. K. (1983) Issues: A reexamination of false-positive rates for carcinogenesis studies. Fundam. Appl. Toxicol. 3: 334-339.

Hattis, D. (1990) Pharmacokinetic principles for dose-rate extrapolation of carcinogenic risk from genetically active agents. Risk Anal. 10: 303-316.

Hayward, N. K.; Walker, G. J.; Graham, W.; Cooksley, E. (1991) Hepatocellular carcinoma mutation. Nature 352: 764.

Herschman, H. R. (1991) Primary response genes induced by growth factors or promoters. Ann. Rev. Biochem. 60: 281-319.

Hildebrand, B.; Grasso, P.; Ashby, J.; Chamberlain, M.; Jung, R.; van Kolfschoten, A.; Loeser, E.; Smith, E.; Bontinck, W. J. (1991) Validity of considering that early changes may act as indicators for non-genotoxic carcinogenesis. Mutat. Res. 248: 217-220.

Hoel, D. G. (1980) Incorporation of background in dose-response models. Fed. Proc., Fed. Am. Soc. Exp. Biol. 39: 73-75.

Hoffmann, G. R. (1991) Genetic toxicology. Casarett and Doull's Toxicology: The Basic Science of Poison. Pergamon Press. Fourth Edition, pp. 201-225

Holliday, R. (1991) Mutations and epimutations in mammalian cells. Mutat. Res. 250: 351-363.

Hollstein, M.; Sidransky, D.; Vogelstein, B.; Harris, C. C. (1991) p53 mutations in human cancers. Science 253: 49-53.

Hsu, I. C.; Metcaff, R. A.; Sun, T.; Welsh, J. A.; Wang, N. J.; Harris, C. C. (1991) Mutational hotspot in human hepatocellular carcinomas. Nature 350: 427-428.

Huff, J. E.; McConnell, E. E.; Haseman, J. K.; Boorman, G. A.; Eustis, S. L. et al. (1988) Carcinogenesis studies: results from 398 experiments on 104 chemicals from the U.S. National Toxicology Program. Ann. N.Y. Acad. Sci. 534: 1-30.

Hulka, B. S.; Margolin, B. H. (1992) Methodological issues in epidemiologic studies using biological markers. Am. J.Epidemiol. 135: 122-129.

Hunter, T. (1991) Cooperation between oncogenes. Cell 64: 249-270.

Ingram, A. J.; Grasso, P. (1991) Evidence for and possible mechanism of non-genotoxic carcinogenesis in mouse skin. Mutat. Res. 248: 333-340.

Interagency Regulatory Liaison Group (IRLG). (1979) Scientific basis for identification of potential carcinogens and estimation of risks. J. Natl. Cancer Inst. 63: 245-267.

International Agency for Research on Cancer. (1982) IARC Monographs on the evaluation of the carcinogenic risk of chemicals to humans, Suppl. 4. Lyon, France: IARC.

International Agency for Research on Cancer (1991) Mechanisms of carcinogenesis in risk identification: A consensus report of an IARC Monographs working group. IARC Internal Technical Report No. 91/002. Lyon, France: IARC.

International Life Sciences Institute. (1984) The selection of doses in chronic toxicity/carcinogenic studies. In: Grice, H. C., ed. Current issues in toxicology. New York: Springer-Verlag, pp. 6-49.

Ito, N.; Imaida, K.; Hasegawa, R.; Tsuda, H. (1989) Rapid bioassay methods for carcinogens and modifiers of hepatocarcinogenesis; CRC Critical Review in Toxicology. 19(4): 385-415.

Jackson, M. A.; Stack, H. F.; Waters, M. D.(1992) The genetoxic toxicology of putative nongenotoxic carcinogens; Mutat. Res. (in press).

Kaplan, D. R.; Hempstead, B. L.; Martin-Zanca, D.; Chao, M. V.; Parada, L. F. (1991) The *trk* proto-oncogene product:a signal transducing receptor for nerve growth factor. Science 252: 554-558.

Kakizuka, A.; Miller, W. H.; Umesono, K.; Warrell, R. P.; Frankel, S. R.; Marty, V.V.V.S.; Dimitro-vsky, E.; Evans, R.M. (1991). Chromosomal translocation t(15;17) in human acute promyelo-cyte leukemia fuses RAR∝ with a novel putative transcription factor, pml. Cell 66: 663-674.

Kelsey, J. L.; Thompson, W. D.; Evans, A. S. (1986) Methods in observational epidemiology. New York: Oxford University Press.

Kodell, R. L.; Park, C. N. Linear extrapolation in cancer risk assessment. ILSI Risk Science Institute (in press).

Kondo, K. R.; Chilcote, R.; Maurer, H. S.; Rowley, J. D. (1984) Chromosome abnormalities in tumor cells from patients with sporadic Wilms' tumor. Cancer Res. 44: 5376-5381

Krewski, D.; Murdoch, D.J.; Withey, J. R. (1987). The application of pharmacokinetic data in carci-nogenic risk assessment. In: Pharmacokinetics in risk assessment. Drinking Water and Health. Volume 8. Washington, DC: National Academy Press. pp. 441-468

Krewski, D.; Brown, C.; Murdoch, D. (1984) Determining "safe" levels of exposure: Safety factors of mathematical models. Fundam. Appl. Toxicol. 4: S383-S394.

Kumar, R.; Sukumar, S.; Barbacid, M. (1990) Activation of *ras* oncogenes preceding the onset of neoplasia. Science 248: 1101-1104.

Levine, A. J.; Momand, J.; Finlay, C. A. (1991) The p53 tumor suppressor gene. Nature 351: 453-456.

Lewin, B. (1991) Oncogenic conversion by regulatory changes in transcription factors. Cell 64: 303-312.

Lilienfeld, A. M.; Lilienfeld, D. (1979) Foundations of epidemiology. 2nd ed. New York: Oxford University Press.

Lutz, W. K. (1990) Dose-response relationship and low doseextrapolation in chemical carcinogene-sis. Carcinogenesis 11: 1243-1247.

Malkin, D.; Li, F. P.; Strong, L. C.; Fraumeni, J. F., Jr.; Nelson, C. E.; Kim, D. H.; Kassel, J.; Gryka, M. A.; Bischoff, F. Z.; Tainsky, M. A.; Friend, S. H. (1990) Germ line p53 mutations in a familial syndrome of breast cancer, sarcomas, and other neoplasms. Science 250: 1233-1238.

Maller, J. L. (1991) Mitotic control. Curr. Opin. Cell Biol. 3:269-275.

Mann, C. (1990) Meta-analysis in a breech. Science 249: 476-480.

Marsh, G. M.; Stone, R. A.; Henderson, V. L. (1992) A reanalysis of the National Cancer Institute study on lung cancer mortality among industrial workers exposed to formaldehyde. J. Occup. Med. 34: 42-44.

Mantel, N.; Haenszel, W. (1959) Statistical aspects of the analysis of data from retrospective studies of disease. J. Natl. Cancer Inst. 22: 719-748.

Mantel, N. (1980) Assessing laboratory evidence for neoplastic activity. Biometrics 36: 381-399.

Mausner, J. S.; Kramer, S. (1985) Epidemiology, 2nd ed. Philadelphia: W. B. Saunders Company.

McConnell, E. E.; Solleveld, H. A.; Swenberg, J. A.; Boorman, G. A. (1986) Guidelines for combin-ing neoplasms for evaluation of rodent carcinogenesis studies. J. Natl. Cancer Inst. 76: 283-289.

McCormick, A.; Campisi, J. (1991) Cellular aging and senescence. Curr. Opin. Cell Biol. 3: 230-234.

McMichael, A. J. (1976) Standardized mortality ratios and the "healthy worker effect": Scratching beneath the surface. J. Occup. Med. 18: 165-168.

Moolgavkar, S. H.; Knudson, A. G. (1981) Mutation and cancer: a model for human carcinogenesis. J. Natl. Cancer Inst. 66: 1037-1052.

Monro, A. (1992) What is an appropriate measure of exposure when testing drugs for carcinogenic-ity in rodents? Toxicol. Appl. Pharmacol. 112:171-181.

Naeve, G. S.; Sharma, A.; Lee, A. S. (1991) Temporal events regulating the early phases of the mammalian cell cycle. Curr. Opin. Cell Biol. 3: 261-268.

National Academy of Sciences (NAS). (1989) Biological markers in pulmonary toxicology. Washington, DC: National Academy Press.

National Research Council. (1983) Risk assessment in the Federal government: managing the process. Committee on the Institutional Means for Assessment of Risks to Public Health, Commission on Life Sciences, NRC. Washington, DC: National Academy Press.

National Toxicology Program. (1984) Report of the Ad Hoc Panel on Chemical Carcinogenesis Testing and Evaluation of the National Toxicology Program, Board of Scientific Counselors. Washington, DC: U.S. Government Printing Office. 1984-421-132: 4726.

Nebert, D. W. (1991) Polymorphism of human CYP2D genes involved in drug metabolism; possible relationship to individual cancer risk. Cancer Cells 3: 93-96.

Nebreda, A. R.; Martin-Zanca, D.; Kaplan, D. R.; Parada, L. F.; Santos, E. (1991) Induction by NGF of meiotic maturation of *xenopus* oocytes expressing the *trk* proto-oncogene product. Science 252: 558-561.

Nowell, P. (1976) The clonal evolution of tumor cell populations. Science 194: 23-28.

Nunez, G.; Hockenberry, D.; McDonnell, J.; Sorenson, C. M.; Korsmeyer, S. J. (1991). Bcl-2 maintains B cell memory. Nature 353: 71-72.

Office of Science and Technology Policy (OSTP). (1985) Chemical carcinogens: review of the science and its associated principles. Federal Register 50: 10372-10442.

Peto, R. (1978) Carcinogenic effects of chronic exposure to very low levels of toxic substances. Environ. Health Perspect. 22: 155-161.

Peto, R.; Pike, M.; Day, N.; Gray, R.; Lee, P.; Parish, S.; Peto, J.; Richard, S.; Wahrendorf, J. (1980) Guidelines for simple, sensitive, significant tests for carcinogenic effects in long-term animal experiments. In: Monographs on the long-term and short-term screening assays for carcinogens: a critical appraisal. IARC Monographs, Suppl. 2. Lyon, France: International Agency for Research on Cancer. pp. 311-426.

Pitot, H.; Dragan, Y. P. (1991) Facts and theories concerning the mechanisms of carcinogenesis. FASEB J. 5: 2280-2286.

Portier, C. (1987) Statistical properties of a two-stage model of carcinogenesis. Environ. Health Perspect. 76: 125-131.

Prival, M. J.; Dunkel, V.C. (1989) Reevaluation of the mutagenicity and carcinogenicity of chemicals previously identified as "false positives' in the *Salmonella typhimurium* mutagenicity assay. Environ. Molec. Mutag. 13(1):1-24.

Raff, M. C. (1992) Social controls on cell survival and cell death. Nature 356: 397-400.

Rall, D. P. (1991) Carcinogens and human health: part 2. Science 251: 10-11.

Rothman, K. T. (1986) Modern Epidemiology. Boston: Little, Brown and Company.

Salomon, D. S.; Kim, N.; Saeki, T.; Ciardiello, F. (1990) Transforming growth factor _ - an oncodevelopmental growth factor. Cancer Cell 2: 389-397.

Sager, R. (1989) Tumor suppressor genes: the puzzle and the promise. Science 246: 1406-1412.

Schneider, C.; Gustincich, S.; DelSal, G. (1991) The complexity of cell proliferation control in mammalian cells. Curr. Opin. Cell Biol. 3: 276-281.

Schuller, H. M. (1991) Receptor-mediated mitogenic signals and lung cancer. Cancer Cells 3: 496-503.

Shelby et al. (1989)

Sidransky, D.; Mikkelsen, T.; Schwechheimer, K.; Rosenblum, M. L.; Cavanee, W.; Vogelstein, B. (1992) Clonal expansion of p53 mutant cells is associated with brain tumor progression. Nature 355: 846-847.

Sidransky, D.; Von Eschenbach, A.; Tsai, Y.C.; Jones, P.; Summerhayes, I.; Marshall, F.; Paul, M.; Green, P.; Hamilton, P.F.; Vogelstein, B. (1991) Identification of p53 gene mutations in bladder cancers and urine samples. Science 252: 706-710.

Srivastava, S.; Zou, Z.; Pirollo, K.; Blattner, W.; Chang, E. (1990) Germ-line transmission of a mutated p53 gene in a cancer-prone family with Li-Fraumeni syndrome. Nature 348(6303): 747-749.

Stiltler et al. (1992)

Starr, T. B. (1990) Quantitative cancer risk estimation for formaldehyde. Risk Anal. 10: 85-91.

Stanbridge, E. J.; Cavenee, W. K. (1989) Heritable cancer and tumor suppressor genes: A tentative connection. In: Weinberg, R. A. (ed.) Oncogenes and the molecular origins of cancer. p. 281.

Sterling, T.D.; Weinkam, J.J. (1987) Reanalysis of lung cancer mortality in a National Cancer Institute study of "mortality among industrial workers exposed to formaldehyde": additional discussion. J. Occup. Med. 31: 881-883.

Stewart, P. A.; Herrick, R. F. (1991) Issues in performing retrospective exposure assessment. Appl. Occup. Environ. Hygiene.

Strausfeld, U.; Labbe, J. C.; Fesquet, D.; Cavadore, J. C. Dicard, A.; Sadhu, K.; Russell, P.; Dor'ee, M. (1991) Identification of a G1-type cyclin puc1+ in the fission yeast *Schizosaccharomyces pombe*. Nature 351: 242-245.

Sukumar, S. (1989) *ras* oncogenes in chemical carcinogenesis. Curr. Top. Microbiol. Immunol. 148: 93-114.

Sukumar, S. (1990) An experimental analysis of cancer: role of *ras* oncogenes in multistep carcinogenesis. Cancer Cells 2: 199-204.

Swenberg, J. A.; Richardson, F. C.; Boucheron, J. A.; Deal, F. H.; Belinsky, S. A.; Charbonneau, M.; Short, B. G. (1987) High to low dose extrapolation: critical determinants involved in the dose-response of carcinogenic substances. Environ. Health Perspect. 76: 57-63.

Tanaka, K.; Oshimura, M.; Kikiuchi, R.; Seki, M.; Hayashi, T; Miyaki, M. (1991) Suppression of tumorigenicity in human colon carcinoma cells by introduction of normal chromosome 5 or 18. Nature 349: 340-342.

Tennant, R. W.; Zeiger, E. (1992) Genetic toxicology: the current status of methods of carcinogen identification. Environ. Health Perspect. (in press)

Thompson, T. C.; Southgate, J.; Kitchener, G.; Land, H. (1989) Multistage carcinogenesis induced by *ras* and *myc* oncogenes in a reconstituted organ. Cell 56: 917-3183.

Tomatis, L.; Aitio, A.; Wilbourn, J.; Shuker, L. (1989) Jpn. J. Cancer Res. 80: 795-807.

U.S. Environmental Protection Agency. (1983a) Good laboratory practices standards—toxicology testing. Federal Register 48: 53922.

U.S. Environmental Protection Agency. (1983b) Hazard evaluations: humans and domestic animals. Subdivision F. Springfield, VA: NTIS. PB 83-153916.

U.S. Environmental Protection Agency. (1983c) Health effects test guidelines. Springfield, VA: NTIS. PB 83-232984.

U.S. Environmental Protection Agency. (1984) Estimation of the public health risk from exposure to gasoline vapor via the gasoline marketing system. Washington, DC: Office of Health and Environmental Assessment.

U.S. Environmental Protection Agency. (1986) Health assessment document for beryllium. Washington, DC: Office of Health and Environmental Assessment. EPA/600/

U.S. Environmental Protection Agency. (1986a) The risk assessment guidelines of 1986. Washington, DC: Office of Health and Environmental Assessment. EPA/600/8-87/045.

U.S. Environmental Protection Agency. (1989a) Interim procedures for estimating risks associated with exposures to mixtures of chlorinated dibenzo-*p*-dioxins and -dibenzofurans (CDDs and CDFs) and 1989 update. Washington, DC: Risk Assessment Forum. EPA/625/3-89/016.

U.S. Environmental Protection Agency. (1989b) Workshop on EPA guidelines for carcinogen risk assessment. Washington, DC: Risk Assessment Forum. EPA/625/3-89/015.

U.S. Environmental Protection Agency. (1989c) Workshop on EPA guidelines for carcinogen risk assessment: use of human evidence. Washington, DC: Risk Assessment Forum. EPA/625/3-90/017.

U.S. Environmental Protection Agency. (1991a) Pesticide Assessment Guidelines: Subdivision F, hazard evaluation: human and domestic animals. Series 84, Mutagenicity. PB91-158394, 540/09-91-122. Office of Pesticide Programs.

U.S. Environmental Protection Agency. (1991b) Alpha-2u-globulin: association with chemically induced renal toxicity and neoplasia in the male rat. Washington, DC: Risk Assessment Forum. EPA/625/3-91/019F.

U.S. Environmental Protection Agency. (1991c) Workshop report on toxicity equivalency factors for polychlorinated biphenyl congeners. Washington, DC: Risk Assessment Forum. EPA/625/3-91/020.

U.S. Environmental Protection Agency. (1991f) Guidelines for developmental toxicity risk assessment. Federal Register 56(234): 63798-63826.

U.S. Environmental Protection Agency. (1992a) Guidelines for exposure assessment. Washington, DC: Federal Register 57(104): 22888-22938

U.S. Environmental Protection Agency. (1992b) Draft Report: A cross-species scaling factor for carcinogen risk assessment based on equivalence of $mg/kg^{3/4}/day$. Washington, DC: Federal Register 57(109): 24152-24173.

U.S. Environmental Protection Agency. (1992c, August) Health assessment for 2,3,7,8-tetrachlorodibenzo-*p*-dioxin (TCDD) and related compounds (Chapters 1 through 8). Workshop Review Drafts. EPA/600/AP-92/001a through 001h.

U.S. Food and Drug Administration. (December 31, 1987) Federal Register 52: 49577 et. seq.

Vahakangas et al. (1992)

Van Sittert. N.J.; De Jong, G.; Clare, M.G.; Davies, R.; Dean, B.J. Wren, L.R. Wright, A.S. (1985) Cytogenetic, immunological, and hematological effects in workers in an ethylene oxide manufacturing plant. Br. J. Indust. Med. 42:19-26.

Varmus, H. (1989) An historical overview of oncogenes. Cold Spring Harbor Laboratory Press, pp. 3-44.

Waters et al. (1988)

Weinberg, R. A. (1989) Oncogenes, antioncogenes, and the molecular bases of multistep carcinogenesis. Cancer Res. 49: 3713-3721.

Wellstein, A.; Lupu, R.; Zugmaier, G.; Flamm, S. L.; Cheville, A.L.; Bovi, P. D.; Basicico, C.; Lippman, M. E.; Kern, F.G. (1990) Autocrine growth stimulation by secreted Kaposi fibroblast growth factor but not by endogenous basic fibroblast growth factor. Cell Growth Differ. 1: 63-71.

Woo, Y. T. and Arcos, J. C. (1989) Role of structure-activity relationship analysis in evaluation of pesticides for potential carcinogenicity. ACS Symposium Series No. 414. Carcinogenicity and pesticides: principles, issues, and relationship. Ragsdale, N. N.; Menzer, R. E. eds. pp. 175-200.

Wyzga, R. E. (1988) The role of epidemiology in risk assessments of carcinogens. Adv. Mod. Environ. Toxicol. 15: 189-208.

Yamasaki, H. (1990) Gap junctional intercellular communication and carcinogenesis. Carcinogenesis 11:1051-1058.

Zeise, L.; Wilson, R.; Crouch, E. A. C. (1987) Dose-response relationships for carcinogens: a review. Environ. Health Perspect. 73: 259-308.

Zhang, K.; Papageorge, A.G.; Lowry, D.R. (1992) Mechanistic aspects of signalling through ras in NIH 3T3 cells. Science 257: 671-674.

E

Use of Pharmacokinetics to Extrapolate From Animal Data to Humans

INTRODUCTION

In classical toxicology, the issue of extrapolation of (usually) animal data to human applications is phrased as:

- Dose to dose (usually high dose in animals to low dose for applications).
- Route to route (e.g., ingestion vs. inhalation).
- Species to species (animal or cell culture to humans).

Pharmacokinetics (PK) can aid in understanding information and in predicting outcomes with respect to the absorption, disposition, metabolism, and excretion of chemicals. Traditionally, analysis has been done empirically, with direct use of the data at hand, and possibly with the aid of simple mathematical models that use overall mass balances. More recently, compartmental models based on chemical transfer in and out of body organs, or even portions of organs, have been developed to describe and predict relationships between administered dose and biologically effective concentrations of parent compounds or metabolites in critical target tissues. These models, which are based on the anatomy and physiology of mammals and use the vast amount of published comparative physiologic data, are known as physiologically based pharmacokinetic (PBPK) models. Details are given in a review by Bischoff (1987).

Each of the three main kinds of extrapolation is briefly described below.

Dose to Dose

PBPK permits reasonable extrapolation from one dose to another, if adequate information on physicochemical properties, physiology, pharmacology, and biochemistry is available. That is not often the case, with less being known as one moves along the list from physiochemical properties to biochemistry; however, PBPK models clearly reveal what data they require and thus what experiments will be needed to make them useful. If the dynamic processes modeled by the PBPK approach are all directly proportional to administered concentrations, then the extrapolation can be relatively straightforward. However, this is not often the case, especially at higher doses, where saturation of metabolic or clearance processes can occur. Despite those difficulties, there are many examples in the literature where useful PBPK analyses have been undertaken. Although PBPK analyses do not always directly address the question of pharmacodynamics (how the biologically effective dose to a critical target tissue is related to toxic response in that tissue), such analyses might provide insight pertinent to this question.

Route to Route

Two broad categories of route-specific toxicity need to be considered: "noncorrosive" and "corrosive." In the former, a chemical enters the body by some route and exerts its effect in the interior of the body; it must enter the blood circulation before it has its effect. In the latter, a very active chemical can have a direct effect at the point of entry, such as high levels of formaldehyde in the case of the rat, nitric acid on skin, or ethylene dibromide at the tip of a gavage tube. Some compounds, such as ethylene dibromide, can be both corrosive and noncorrosive.

Most toxicants are noncorrosive, and knowledge of relevant physiology and pharmacology can permit extrapolation between routes of exposure, because the important information is the concentration in the blood and the transport to and uptake at the site of action. There could still be route-to-route differences, e.g., if the peak concentration after exposure determines toxicity. For example, absorption might be faster (and thus the peak higher) for intravenous than for oral exposure. PBPK models are useful, because they permit estimation of peak concentrations.

Species to Species

Species-to-species extrapolation is one of the most useful aspects of PBPK, because all mammals have the same macrocirculatory anatomy and much is known about the comparative dimensions of their physiologic characteristics—organ volumes, blood flow rates, some clearances, etc. The basic data are usually presented as a function of body weight raised to some fractional power, W^b, with $b = 0.7$-1.0 (so-called "allometric scaling"). This aspect is relatively straightforward. However, other aspects can be more complicated, particularly those involving metabolism. For instance, there might be qualitative differences between species, such as the presence or absence of a given enzyme, that would result in a (potentially dose-dependent) difference in metabolic capacity and make their metabolism different.

F
Uncertainty Analysis of
Health Risk Estimates

INTRODUCTION

Large uncertainties are typically associated with the estimation of public health risks associated with air toxic emissions. Such uncertainties arise in: (1) the formulation of the models used to simulate the fate and transport of chemicals in the environment and the foodchain, public exposure, dose, and health risks; and (2) the estimation of the parameter values used as input values to these models.

The uncertainty due to model formulation can be reduced to some extent by using models that provide a more comprehensive treatment of the relevant physico-chemical processes. (It should be noted, however, that as a model becomes more comprehensive, the input data requirements may increase substantially; and while the uncertainty associated with the model formulation decreases, the uncertainties associated with the input parameters may increase). Seigneur et al. (1990) provide some guidance on the selection of mathematical models for health risk assessment with various levels of accuracy in their formulation. We focus here on the uncertainties due to the input parameters for a given health risk assessment model.

Prepared by Christian Seigneur and Elpida Constantinou, ENSR Consulting and Engineering, 1320 Harbor Bay Parkway, Alameda, California 94501, and Thomas Permutt, Department of Epidemiology and Preventive Medicine, University of Maryland, 655 West Baltimore Street, Baltimore, Maryland 21202 for Leonard Levin, Electric Power Research Institute, 3412 Hillview Avenue, Palo Alto, California 94303. November 1992. Document Number 24600-009-510.

Uncertainties in parameter values arise for three reasons. First, the value may have been measured, in which case some imprecision is associated with the process of measurement. In the context of this report, however, errors of measurements are likely to be insignificant compared to other kinds of uncertainty. Second, the value may have been measured, but under circumstances other than those for which it must be applied. In this case, additional uncertainty arises from the variation of the parameter in time and space. Third, the value may not have been measured at all, but estimated from relationships with other quantities that are known or measured. In this case, uncertainty in the parameter of interest arises form both uncertainty in the quantities that are measured and from uncertainty about the estimating relationship.

By characterizing the uncertainties in the input parameters of a model and studying the effects of variation in these parameters on the model predictions, we can estimate the part of the uncertainty in the predictions that is due to uncertainty in the inputs.

Uncertainty can be characterized by a probability distribution. That is, the value of a parameter is not known exactly, but, for example, it might be thought to lie between 90 and 100 with probability 0.5, between 85 and 120 with probability 0.75, and so on. Sometimes such probability distributions can be usefully summarized by a few parameters, such as the mean and standard deviation. Uncertainties in the input parameters propagate through the model to produce probability distributions on the output parameters. Figure 1 presents a schematic description of the uncertainty propagation through a model.

We present here a structured methodology for the parameter uncertainty analysis of health risk estimates. The methodology involves: (1) a sensitivity analysis of the model used to perform the health risk calculations, (2) the determination of probability distributions for a number of selected input parameters (i.e., the ones identified as the most influential to the output variable); and (3) the propagation of the uncertainties through the model.

This methodology is applied here to the uncertainty analysis of the carcinogenic health risks estimated as due to the emissions of a coal-fired plant.

UNCERTAINTY ANALYSIS METHODOLOGY

Overview

A health risk assessment model combines a number of models to simulate the transport and fate of chemicals in air, surface water, surface soil, groundwater and the foodchain. Concentrations calculated by the fate and transport models are used by exposure-dose models to calculate the doses to exposed individuals, which are then used to calculate health risks.

For the description of the uncertainty analysis methodology, we consider each individual model component as a function Y (dependent variable) of a number of parameters $X_1, X_2,...X_n$ (independent variables). In summary, we view this methodology as consisting of the following 5 steps:

Step 1: Sensitivity analysis of the health risk assessment model: This analysis allows one to determine the influential parameters of the model, i.e., those that need to be included in the uncertainty analysis.

Step 2: Parameterization of the health risk assessment by construction of response surface models: This parameterization allows one to simplify the uncertainty propagation and therefore allows the incorporation of a large number of parameters in the analysis.

Step 3: Selection of probability distribution for the input parameters.

Step 4: Propagation of the parameter uncertainties: This task is performed with the parameterized version of the model and provides the uncertainties in the model outputs.

Step 5: Analysis of the probability distribution of the risk estimates.

A more detailed description of each individual step of the methodology is presented in the following paragraphs.

Sensitivity Analysis

Mathematical models describing physical phenomena are often composed of relatively complex sets of equations involving a large number of input parameters. However, some of these parameters do not have any significant influence on the health risk calculated by the model; i.e., the model output is not sensitive to the values of these input parameters. Therefore, such parameters that do not affect the health risk values significantly, do not need to be known with great accuracy and the uncertainty analysis should focus on those parameters to which the calculated health risks are most sensitive.

The sensitivity analysis allows us to determine the parameters to which the model is most sensitive. These parameters will be called the *influential parameters*.

When dealing with a complex model, such as a multimedia health risk assessment model, sensitivity analysis should be performed for each individual model component as well as for the overall model.

The sensitivity of the model output (i.e., the dependent variable) to a model input parameter can be measured by the ratio of the change in the model output to the perturbation in the input parameter. We define this ratio as the sensitivity index, *SI*. For parameter i:

$$SI_i = \frac{\Delta Y}{\Delta X_i}$$

where ΔX_i is the perturbation in the input parameter, and ΔY is the corresponding change in the model output. In order to compare the sensitivity index for various input parameters, it is appropriate to use a dimensionless representation of the sensitivity index:

$$SI_i^* = \frac{\Delta Y / \bar{Y}}{\Delta X_i / \bar{X}_i} = \frac{\Delta Y^*}{\Delta X_i^*}$$

where $\bar{X}_i$ and $\bar{Y}$ are the mean or some other reference values of the variables, X_i and Y, respectively; and ΔY^* and ΔX_i^* refer to normalized perturbations.

Two characteristics of the sensitivity index must be noted:

- The value of the sensitivity index is a function of the value of and the perturbation in the input parameter except for cases where the relationship between the model output variable and the input parameter is linear.
- The value of the sensitivity index may be a function of the value of the other model input parameters except for cases where the relationship between the model output and the input parameters is linear.

Even though the sensitivity index, as defined above, sufficiently describes the effect on the model result for a given change in the input parameter, it does not provide a measure of the range of variation in the model output, given the expected range of variation of the input parameter. In other words, a parameter that has a high sensitivity index, may have little effect on the model output if that parameter can only have a very small variation. The height of a power plant stack is an example of a parameter that has significant effect on atmospheric ground-level concentrations but has a small uncertainty. For the case of the power plant studied here, a 100% change in stack height caused a 73% change in the resulting concentrations. Since the uncertainty in stack height, however, can only be due to measurement error, it is not expected to be more than ±2%. Consequently, the actual influence of this parameter on the model result is very small.

We define the uncertainty index as a measure of uncertainty associated with a parameter X_i. Although several definitions of this uncertainty index are possible, we select one that is objective in a statistical sense by using the standard deviation and the mean of the parameter. The uncertainty index is defined as follows:

$$UI_i = \frac{\sigma_i}{\overline{X}_i}$$

where: σ_i is the standard deviation of the parameter distribution

$\overline{X}$ is the mean value of parameter X_i

The uncertainty index, consequently provides a measure of the expected variation of parameter X_i over its range of probable values.

The combination of the model sensitivity to a parameter, and the uncertainty in that parameter provides the information required to assess which parameters need to be included in the uncertainty analysis. We define a sensitivity/uncertainty index as follows:

$$I_j = (SI_i)(UI_i) = \frac{\Delta Y^*}{\Delta X_i^*} \frac{\sigma_i}{\overline{X}_i}$$

The sensitivity/uncertainty index, therefore constitutes a representative measure of the influence that a parameter has on the model results and can, thus, be used to select the model influential parameters to be included in the uncertainty analysis. Figure 2 presents a schematic description of the steps followed in the sensitivity analysis procedure.

Even though the concept of the standard deviation of a parameter was used in the definitions of the uncertainty and sensitivity/uncertainty indexes, it is rather unlikely that actual standard deviations will be available for all the parameters examined in the sensitivity analysis. Since the sensitivity analysis is a screening procedure whose goal is to minimize the number of parameters included in the final uncertainty analysis, it is generally appropriate to use other measures that are more readily available to characterize the variability of a parameter. For example, the expected range of variation can be used instead of an actual standard deviation.

Parameterization of the Mode—Response Surface Construction

A multimedia health risk assessment model typically involves a large number of input parameters and comprises several individual models for simulating fate and transport, exposure, dose, and health effects. Such a model can be computationally very demanding and performing an uncertainty analysis for a large number of parameters may, therefore, not be feasible. It is, therefore, necessary to parameterize the various model components in order to reduce the magnitude of the computations. This model parameterization can be achieved by constructing response surfaces.

A response surface is a simplified version of the actual model which can be used efficiently in the uncertainty analysis as a replacement of the real model. In the case of simple analytical models in the form of a single equation (e.g., dose models) only minor simplifications need to be made for the construction of the response surface. Such simplifications can be accomplished by factoring out of each of the terms of the equation the selected influential parameters, and representing the remaining part of the term by a lumped parameter calculated from previous results of the model. So, the response surface can be of the following form:

$$rY = \sum_{i=1}^{m} \left(X_{i1}^{P_{i1}} \dots X_{1K_i}^{P_1K_i} \right) A_i$$

where: m = Number of terms in dependent variable expression
K_i = Number of independent variables included in term i
A_i = Calculated lumped parameter of term i

In the case of complex models (e.g., environmental transport models) the response surface can be developed using the following procedure: For all influential parameters of a model component select a number K of parameter sets $X_i = (X_{i1}....X_{In})$ $i = 1,K$ and perform experimental runs of the actual complex model. Then use the pairs of parameter sets $X_1...X_k$ and corresponding model results $Y_1....Y_k$ to construct the response surface.

A simple example of a response surface can be that of the atmospheric transport model, in which the air concentration, C_a can be expressed in terms of four independent influential parameters and six constant parameters as follows:

$$C_a = Q_e F_1 F_2$$
where: $F_1 = A_1 + A_2 V_s = A_3 V_3^2$
$F_2 = A_4 + A_5(T_s - T_a) + A_6(T_s - T_a)^2$

where: C_a = Chemical concentration in air
Q_e = Chemical emission rate
V_s = Stack exit velocity
T_s = Stack exit temperature
T_a = Ambient air temperature
A_i = Calculated constant parameters (functions of meteorological data, source characteristics and environmental setting)

A response surface is a parameterization of the model that allows one to calculate the model results with considerably less computations. However, a response surface is typically specific to a given model application. That is, some of the case study characteristics (e.g., meteorology, hydrology) are implicitly included in the constant parameters of the response surfaces.

Selection of the Probability Distributions for the Input Parameters

Once the influential parameters have been identified, and the response surfaces for each model component constructed, probability distributions must be selected to represent each one of the parameters.

As was mentioned previously, a parameter value can be either directly measured or indirectly estimated through an estimation procedure which usually involves fitting of a curve through a set of experimental points.

In the case of a directly measured parameter, the uncertainty results from uncertainty in the measurement process, and this can sometimes be estimated from repeated measurements. Often, however, the amount of available data is not enough to produce meaningful histograms or probability plots. What is usually available is a range of values within which the true value of a parameter is expected to lie, and possibly a most likely, or range of most likely values for the parameter. In this case, it is left to our judgment and experience to decide what probability distribution is appropriate.

In the case of parameters estimated indirectly through curve fitting (e.g., bioconcentration factors, and cancer potency factors) uncertainty results from both statistical errors in fitting the curve, which can be estimated by statistical procedures, and uncertainty about the form of the curve, which is a matter of judgment.

The development of parameter probability distributions through a combination of *a priori* expert judgment along with current information in the form of available direct or indirect measurements is known in statistical theory as the Bayesian method. If the measurements are direct, precise, and numerous enough to sufficiently describe the variation pattern of the parameters, then the *a priori* judgment may have little or no influence on the resulting probability distributions. Conversely, if the measurements are indirect and imprecise, then the *a priori* judgment may be of great importance.

Propagation of the Model Uncertainties

At this step, the response surfaces developed for the different model components can be combined in a single spreadsheet that performs the function of the overall risk assessment model in a simplified fashion for the case study considered. Several techniques exist to develop a probability distribution in the model output given probability distributions in the model input parameters. Monte-Carlo and Latin hypercube simulations are standard examples of such techniques. In the special cases where the probability distributions are similar and simple (e.g., normal), the probability distribution in the model output can be calculated analytically. For the general case where no simple analytical approach can be used, however, the spreadsheet model can then be coupled to one of several commercial software packages (e.g., @Risk; Palisade Corp., 1991), which uses the specified probability distributions of the parameters together with the spreadsheet calculations to generate a set of synthetic model results.

Analysis of the Probability Distribution of the Model Health Risk Estimates

If the number of replications for the probabilistic synthetic simulations is large enough, the synthetic results can be statistically analyzed to yield a reasonably reliable probability distribution of the dependent variable (i.e., health risk). If the uncertainty analysis procedure was performed correctly, this probability distribution should represent a more complete and realistic characterization of the anticipated health risks, as it provides a range of possible values accompanied by their corresponding likelihoods instead of a single, deterministic point estimate.

DESCRIPTION OF THE MULTIMEDIA HEALTH RISK ASSESSMENT MODEL

In this section, we present a description of the multimedia health risk assessment model which was used in the application. This model was developed by combining a number of individual models that handle the fate and transport of chemicals in air, surface water, surface soil, groundwater, and the foodchain, into an integrated multimedia model. A brief description of its individual components, and its overall structure are provided in the following paragraphs. A detailed description is provided by Constantinou and Seigneur (1992).

The model consists of the following nine distinct components:

- Atmospheric fate and transport model
- Deposition model
- Overland model
- Surface water fate and transport model
- Vadose zone fate and transport model
- Groundwater fate and transport model
- Foodchain fate and transport model
- Exposure and dose model
- Health risk model

The multimedia health risk assessment model combines all the model components into a single computer program that takes as input emission stream characteristics and environmental physical parameters, and calculates the resulting carcinogenic and noncarcinogenic health effects. Figure 3 represents the general calculation steps that lead to the final model results.

DETERMINISTIC HEALTH RISK ASSESSMENT

The boiler of the power plant studied in this application is a 680 MW unit which burns high-sulfur bituminous coal. Four chemicals with listed carcinogenic effect were sampled from the 200m high stack of the facility. Stack air emissions were the only emissions considered in the analysis. Liquid and solid waste discharges were ignored.

The study area examined in the present application was defined to be the area within a 50 km radius of the power plant. This area was divided into 40 subregions by a concentric grid. The major surface water bodies included in the area include a river and a large lake. For the health effect calculations, all public water supply was considered to come from the river and all fish supply was considered to come from the lake.

Carcinogenic and noncarcinogenic health effects were calculated in each of the subregions considered in the study area. The results subject to the uncertainty analysis presented in this report correspond to the carcinogenic health effects in the subregion of maximum risk. Noncarcinogenic risks are not addressed here.

The carcinogenic chemicals detected in the stack air emissions were chromium, arsenic, cadmium, and benzene. The corresponding chemical emission rates were estimated to be 1.08×10^{-2}, 4.4×10^{-4}, 5.39×10^{-4}, and 1.4×10^{-2} g/s, respectively. Since chemical speciation for chromium was not available, the corresponding health effect calculations were performed based on the assumption that total chromium emissions consisted of 5% Cr(VI) and 95% Cr(III). The cumulative carcinogenic lifetime risk from all chemicals and pathways in the subregion of maximum risk was calculated to be 2.2×10^{-8}.

Chromium (VI) and arsenic were calculated to be the two major contributors to carcinogenic risk with contributions of 59 and 32%, respectively. Cadmium contributed 8%, and the contribution of benzene was 1%. Among the three exposure pathways considered in the analysis, inhalation was calculated to be the major contributor, with a contribution of 85%. Ingestion ranked second with 15%, and dermal absorption had an insignificant contribution of 0.4%.

Produce was calculated to be the foodchain component which contributed the most to the ingestion risk, with a contribution of 92%. Fish and soil ingestion had small contributions of 5 and 3%, respectively, and drinking water had an insignificant contribution of 0.3%.

It should be noted that among the four carcinogenic chemicals included in the analysis, only arsenic and benzene are considered to be carcinogenic through noninhalation pathways. Since benzene's contribution was very small, benzene was not included in the uncertainty analysis. Even though arsenic is considered carcinogenic through the ingestion pathway, no cancer potency value is currently tabulated (October, 1992) for this pathway in the Integrated Risk Information System (IRIS) database. The value used for this parameter in the deterministic health risk assessment was the most recent value listed in IRIS.

UNCERTAINTY ANALYSIS

Sensitivity Analysis

Sensitivity analysis of the individual model components as well as the overall multimedia health risk assessment model was performed to help identify the influential parameters. A total of 49 parameters were examined, and 22 were selected to be included in the final uncertainty analysis, based on their calculated sensitivity/uncertainty indexes. Table 1 provides a list of all parameters examined, together with their corresponding symbols and units.

Sensitivity/uncertainty indexes of the input parameters were derived for each of the individual model components as well as for the overall risk assessment model. The resulting indexes for the three chemicals included in this analysis are summarized in Table 2.

Model Simplification

Using the influential parameters selected, response surfaces were constructed for each of the multimedia health risk assessment model components. In the case of simple models such as the foodchain, exposure-dose, and risk models, the response surfaces were constructed manually by factoring out the influential parameters, and representing the remaining parts of the equations by constants calculated based on the model results. In the case of the more complex environmental transport models, additional sensitivity runs were performed by varying the influential parameters within their assumed range of variation.

In the case of the atmospheric transport model, ISC-LT, four influential parameters were identified: the chemical emission rate (Q_e), the stack exit velocity (V_s), the stack exit temperature (T_s), and the ambient air temperature (T_a). The influences of T_s and T_a were found to be correlated, as what affected the results was the difference between the stack and the ambient temperature, $(T_s - T_a)$ and not their absolute values. Consequently, the two parameters were treated together as one, the temperature difference $(T_s - T_a)$.

Multiple runs were performed by varying V_s and $(T_s - T_a)$ within their assumed range of variation to identify their individual and combined effect on the resulting maximum ground level chemical concentration. Both parameters were found to affect the model results in an exponentially decaying way (i.e., resulting concentrations decreased exponentially for higher values of V_s and $(T_s - T_a)$) and their variation patterns were fit at the reference point (i.e., parameter values used in the deterministic calculations) by second degree polynomials.

The combined response surface for both parameters was derived by combining their individual curves. It should be noted that this approach is only valid for the range of perturbations considered in this analysis. For larger ranges of parameter variation the model response surface for two or more influential parameters should be derived through multiple regression, where the variation of the model results is examined simultaneously for all parameters. Figure 4 provides a graphical presentation of the derived ISC-LT response surface.

The complete set of equations of the simplified multimedia health risk assessment model is presented below:

- Atmospheric Transport Model (Component 1):

$$C_a = \alpha Q_e F_1 F_2 A_{11}$$

where:
$$F_1 = A_{12} + A_{13}V_s + A_{14}V_s^2$$
$$F_2 = A_{15} + A_{16}(T_s - T_a) + A_{17}(T_s - T_a)^2$$

where: C_a= Ground-level air concentration; Q_e= Chemical emission rate; α = Chemical speciation fraction (applies only to chromium case); V_s= Stack exit velocity; T_s= Stack exit temperature; T_a= Ambient temperature; A_{1j}= Constant j for model component 1

- Deposition Model (Model Component 2):

$$DR = C_a(V_d + V_w A_{21})A_{22}$$

where: DR = Chemical deposition rate; V_d= Dry deposition velocity; V_w = Wet deposition velocity; A_{2j}= Constant j for model component 2

- Overland Model (Model Component 3):

$$L_{ss} = DR(1-OR_f)A_{31}$$

where: L_{ss}= Surface soil chemical load; OR_f= Fraction of deposited chemical attributed to overland runoff; A_{3j} = Constant j for model component 3

- Soil Transport Model (Model Component 4)

$$C_s = L_{ss}d_s^{-1}(EST + ED/2)p_b^{-1}A_{41}$$

where: C_s = Surface soil concentration; d_s = Surface soil depth; EST = Exposure starting time; ED = Exposure duration; p_b = Soil bulk density; A_{4j} = Constant j for model component 4

- Foodchain Model—Plants (Model Component 5):

$$C_p = C_s BCF_p$$

where: BCF_p = Soil-to-plant bioconcentration factor

- Dose Model (Model Component 6):

(1) Inhalation

$$D_1 = C_s IR\ ED\ BW^{-1}A_{61}$$

where: D_1 = Inhalation dose; IR = Inhalation rate; BW = Body Weight; A_{6j} = Constant j for model component 6

(2) Ingestion

$$D_2 = (C_p INR_p + A_{62})ED\ BW^{-1}A_{63}$$

where: D_2 = Ingestion dose; INR_p = Plant ingestion rate

- Risk Model (Model Component 7)

$$R = D_1 CPF_1 + D_2 CPF_2 + A_{71}$$

where: R = Total Carcinogenic risk; CPF_1 = Inhalation cancer potency factor; CPF_2 = Ingestion cancer potency factor; A_{7j} = Constant j for model component 7

It should be noted that the full set of equations presented above applies only to the arsenic case. Cadmium and chromium are not considered carcinogenic through noninhalation pathways. Consequently, only the atmospheric transport, inhalation dose, and inhalation risk equations apply to their case.

Probability Distribution Selection

Evaluation of the probability distributions of the 22 influential parameters of the model was performed on the basis of available statistical data, literature value ranges, and personal judgment. The selected probability distributions, and the information which the distribution types and parameters where based on are summarized in Table 3.

In a health risk assessment, the uncertainty associated with the health effect parameters (i.e., cancer potency factors in the case of the present application) is of major importance. The EPA recommended values for these parameters are usually derived based on limited animal or epidemiological studies the conditions of which may differ significantly from the conditions for which these values will be applied in a risk assessment.

In the case of epidemiological studies, uncertainty is associated with high-to-low dose extrapolation and factors related to secondary exposures, diet, and hygiene of the population under study. The data are fitted by an assumed model, and the maximum lifetime estimate (MLE) is usually recommended for use by EPA.

In the case of animal studies, uncertainty is associated with interspecies as well as high-to-low dose extrapolations. Due to the additional uncertainty of interspecies extrapolation in this case an upper bound value (95th percentile) is usually recommended for use by EPA.

The cancer potency factors for the three chemicals included in this application were all derived based on epidemiological studies.

In the case of arsenic inhalation, the CPF derivation was based on two separate U.S. smelter worker populations (EPA, 1984a). The data collected were analyzed by five different investigators who derived five different CPF values, using a linear nonthreshold model. The EPA recommended value was derived by obtaining the geometric mean of the individually derived CPFs. In this application we chose to represent the uncertainty of the arsenic inhalation CPF by a uniform distribution extending over the range of values provided by the above mentioned five investigators.

In the case of arsenic ingestion, the CPF derivation was based on an epidemiological study of a Taiwanese population exposed to high arsenic concentrations in drinking water (EPA, 1984a). Analysis of these data resulted in a CPF which was later deemed overconservative by EPA, after comparison with the results of limited U.S. studies. Its high value was attributed to underestimation of the exposure of the Taiwanese population, and the lack of consideration of the populations's poor diet and hygiene in the analysis. The value was then scaled by EPA to a lower value that would be more representative for a U.S. population. Due to the uncertainties associated with its derivation, this CPF was recently removed from IRIS until a more reliable value could be derived. In this application we chose to represent the uncertainty of the arsenic ingestion CPF by a triangular distribution with lower bound equal to zero, most likely value equal to the scaled value most recently listed in IRIS (September, 1991), and upper bound equal to the value originally derived from the Taiwanese population data.

In the case of cadmium inhalation, the CPF derivation was based on epidemiological data of a U.S. smelter worker population (EPA, 1985). The EPA-recommended value was determined by fitting a linear nonthreshold model to these data. A 90% confidence interval based only on statistical consideration was constructed around the MLE. In this application we chose to represent uncertainty associated with the cadmium inhalation CPF by a normal probability distribution with mean equal to the maximum likelihood estimate and standard deviation calculated from the estimated confidence interval.

In the case of chromium (VI) inhalation, the CPF was based on a U.S. population of chromate plant workers (EPA, 1984b). The EPA-recommended value was derived by fitting a two-stage model to the data. A lower bound and an upper bound were constructed around the MLE to account for the possibility of underestimation or overestimation of the exposure due to lack of consideration of poor hygiene, smoking habits, and chemical speciation in the analysis. In this application, we chose to represent the uncertainty associated with the chromium (VI) CPF by a triangular distribution defined by the best estimate, upper, and lower bounds.

Monte Carlo Analysis—Health Risk Probability Distribution

The derived response surfaces were combined in a simplified spreadsheet model which was coupled to the software package @RISK (Palisade Corporation, 1991) which performed the propagation of the input parameter uncertainties through the model. A Monte Carlo analysis with 5000 iterations of the simplified model was performed to produce a synthetic set of carcinogenic health risks associated with the studied coal-fired power plant. Statistical analysis of the synthetic results yielded a probability distribution for the risk. The risk value calculated in the deterministic risk assessment (2.2×10^{-8}) was estimated to be at the 83rd percentile of the derived probability distribution. The statistical parameters of this distribution are summarized below:

- Mean (expected value), $\mu = 1.5 \times 10^{-8}$ (i.e., 68% of the deterministic value)

- Mode (most probable value), $M_o = 2.5 \times 10^{-9}$
- Standard Deviation, $\sigma = 3.4 \times 10^{-8}$
- Skewness, $\gamma = 13.6$ (i.e., positively skewed-right tail)
- Percentiles: 5%, $F_{0.05} = 1.2 \times 10^{-9}$
                       25%, $F_{0.25} = 3.2 \times 10^{-9}$
     Median   50%, $F_{0.5} = 6.9 \times 10^{-9}$
                       75%, $F_{0.75} = 1.6 \times 10^{-9}$
                       95%, $F_{0.95} = 5.1 \times 10^{-9}$

The derived probability density plot is presented in Figure 5.

CONCLUSION

A general methodology for the performance of sensitivity/uncertainty analysis was presented. The methodology was applied to a multimedia health risk assessment model. A case study of a coal-fired power plant was used as the basis of this application. The uncertainty of the carcinogenic risk associated with the power plant emissions was examined. The results indicated that the deterministic risk value calculated in the original risk assessment study was a conservative estimate, corresponding to a higher risk percentile on the estimated risk probability distribution.

ACKNOWLEDGMENTS

This work was performed under contract No 3081-1 with the Electric Power Research Institute, Palo Alto, California. The authors gratefully acknowledge the continuous support of the EPRI project manager, Dr. Leonard Levin.

REFERENCES

Constantinou, E. and C. Seigneur. A Mathematical Model for Multimedia Health Risk Assessment. Submitted for publication in Environmental Software, 1992.

Palisade Corporation. Risk Analysis and Simulation Add-In for Microsoft Excel, Windows or Apple Macintosh Version, Release 1.02 User's Guide, March, 1991.

Seigneur, C., C. Whipple, et al. Formulation of Modular Mathematical Model for Multimedia Health Risk Assessment, American Institute of Chemical Engineers Summer Meeting, San Diego, California, August, 1990.

U.S. Environmental Protection Agency. Updated Mutagenicity and Carcinogenicity Assessment of Cadmium. Addendum to the Health Assessment Document for Cadmium (May 1981). EPA/600/8-81/023, June, 1985.

U.S. Environmental Protection Agency. Health Assessment Document for Inorganic Arsenic. Environmental Criteria and Assessment Office, Research Triangle Park, NC. EPA/600/8-83-021F, 1984a.

U.S Environmental Protection Agency. Health Assessment for Chromium - Final Report. EPA/600/8-83-014F, August, 1984b.

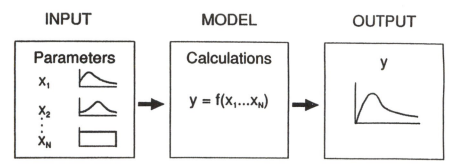

FIGURE 1 Uncertainty Propagation

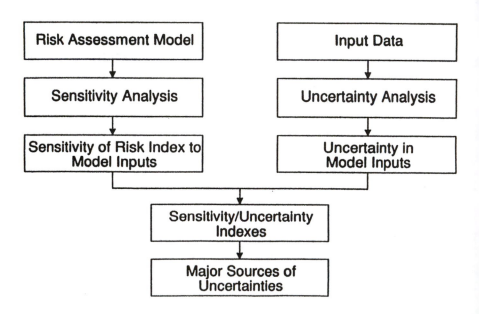

FIGURE 2 Sensitivity Analysis Summary

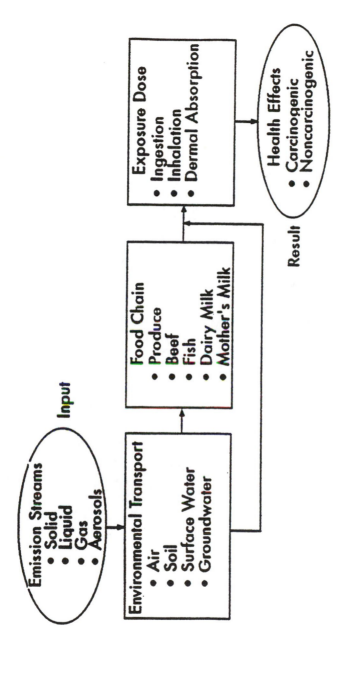

FIGURE 3 Multimedia Health Risk Assessment Model

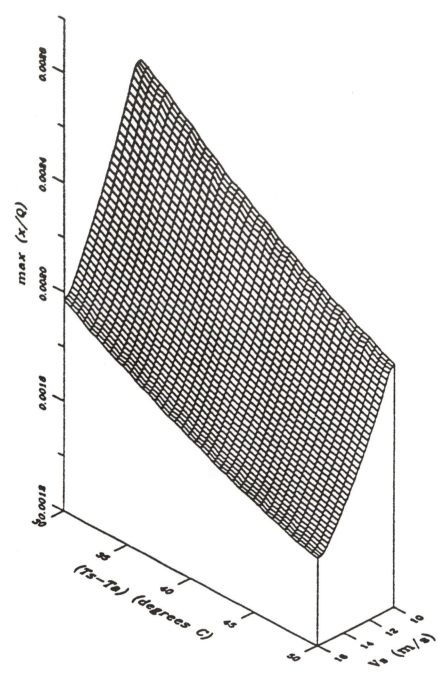

FIGURE 4 ISC-LT Response Surface

471

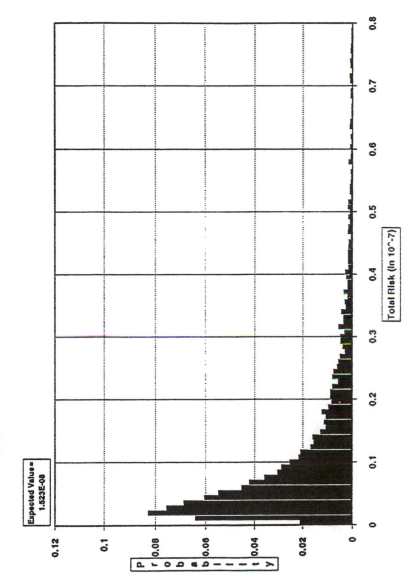

FIGURE 5 Total Carcinogenic Risk Probability Distribution

TABLE 1 Parameter Reference List

Parameter Index	Parameter Description	Parameter Symbol (Units)
1	Chemical Emission Rate - Arsenic	Q_{e1} (g/s)
2	Chemical Emission Rate - Cadmium	Q_{e2} (g/s)
3	Chemical Emission Rate - Chromium	Q_{e3} (g/s)
4	Chemical Speciation Fraction	α (-)
5	Stack Height	H_s (m)
6	Stack Exit Temperature	T_s (K)
7	Stack Exit Velocity	V_s (m/s)
8	Stack Diameter	D_s (m)
9	Ambient Temperature	T_a (K)
10	Mixing Height	h_m (m)
11	Arsenic Dry Deposition Velocity	V_d (m/s)
12	Wet Deposition Velocity	V_w (m/s)
13	Fraction of Time with Rain	R_f (-)
14	Fraction of Chemical in Overland Runoff	OR_f (-)
15	River Discharge	Q_r (m³/s)
16	Arsenic Chemical Decay Coefficient in River	K_r (d⁻¹)
17	Lake Water Exchange Rate	Q_L (m³/s)
18	Arsenic Chemical Decay Coefficient in Lake	K_L (d⁻¹)
19	Surface Soil Depth	d_s (m)
20	Exposure Duration	ED (years)
21	Exposure Starting Time	EST (years)
22	Cation Exchange Capacity	CEC (meq/100cc)
23	Arsenic Chemical Decay Coefficient in Soil	KDES (d⁻¹)
24	Soil Permeability	k_p (cm²)
25	Soil Porosity	θ
26	Soil Bulk Density	ρ_b (kg/m³)
27	Chemical Plant Interception Fraction	IF (-)

TABLE 1 Continued

Parameter Reference List

Parameter Index	Parameter Description	Parameter Symbol (Units)
28	Weathering Elimination Rate	K_{el} (d^{-1})
29	Crop Density	CD (kg/m^2)
30	Arsenic Soil-to-Plant Bioconcentration Factor	BCF_p (-)
31	Arsenic Water-to-Fish Bioconcentration Factor	BCF_f (-)
32	Inhalation Rate	IR (m^3/d)
33	Plant Ingestion Rate	INR_p (kg/d)
34	Soil Ingestion Rate	INR_s (kg/d)
35	Water Ingestion Rate	INR_w (l/d)
36	Fish Ingestion Rate	INR_f (kg/d)
37	Skin Surface Area Exposed to Soil	SA_s (cm^2)
38	Soil Absorption Factor	ABS (-)
39	Soil Adhesion Factor	AF (mg/cm^2)
40	Soil Exposure Frequency	EF_s (d/yr)
41	Skin Surface Area Exposed to Water	SA_w (cm^2)
42	Arsenic Permeability Constant	PC (cm/hr)
43	Water Exposure Frequency	EF_w (d/yr)
44	Body Weight	BW (kg)
45	Inhalation Cancer Potency Factor - Arsenic	CPF_{11} (kg-d/mg)
46	Ingestion Cancer Potency Factor - Arsenic	CPF_{21} (kg-d/mg)
47	Dermal Absorption Cancer Potency Factor - Arsenic	CPF_{31} (kg-d/mg)
48	Inhalation Cancer Potency Factor - Cadmium	CPF_{12} (kg-d/mg)
49	Inhalation Cancer Potentcy Factor - Chromium (VI)	CPF_{13} (kg-d/mg)

TABLE 2 Sensitivity/Uncertainty Indexes

Model Component	Output Variable Description	Output Variable Symbol	Parameter Symbol	Model Component Sensitivity/Uncertainty Indexes			Overall Model Sensitivity/Uncertainty Indexes		
				Chromium (VI)	Arsenic	Cadmium	Chromium (VI)	Arsenic	Cadmium
ISCLT	Air Concentration	C_a	$Q_{e\,(1,\,2,\,3)}$	0.91	0.38	1.19	0.91	0.38	1.19
			α	0.60	NA	NA	0.60	NA	NA
			H_s	-0.03	-0.03	-0.03	-0.03	-0.03	-0.03
			T_s	-0.07	-0.07	-0.07	-0.07	-0.07	-0.07
			V_s	-0.14	-0.14	-0.14	-0.14	-0.14	-0.14
			D_s	-0.03	-0.03	-0.03	-0.03	-0.03	-0.03
			T_a	0.07	0.07	0.07	0.07	0.07	0.07
			h_{in}	0.02	0.02	0.02	0.02	0.02	0.02
Deposition	Deposition Rate	DR	V_d	NA	0.76	NA	NA	0.35	NA
			V_w	NA	0.24	NA	NA	0.11	NA
			R_f	NA	0.05	NA	NA	0.02	NA
Overland	Surface Soil Load	L_{ss}	OR_f	NA	-0.18	NA	NA	-0.07	NA
WTRISK	Surface Water Concentration	C_{sw}	Q_r	NA	-0.09	NA	NA	-0.002	NA

TABLE 2 Continued

Model Component	Output Variable Description	Output Variable Symbol	Parameter Symbol	Model Component Sensitivity/Uncertainty Indexes			Overall Model Sensitivity/Uncertainty Indexes		
				Chromium (VI)	Arsenic	Cadmium	Chromium (VI)	Arsenic	Cadmium
WTRISK (Cont'd)	Surface Water Concentration (Cont'd)		K_r	NA	0.0	NA	NA	0.0	NA
			Q_L	NA	-0.09	NA	NA	-0.002	NA
			K_L	NA	0.0	NA	NA	0.0	NA
SESOIL	Surface Soil Concentration	C_{ss}	d_s	NA	-0.67	NA	NA	-0.29	NA
			ED	NA	0.20	NA	NA	0.09	NA
			EST	NA	1.43	NA	NA	0.64	NA
			CEC	NA	0.0	NA	NA	0.00	NA
			KDES	NA	0.0	NA	NA	0.00	NA
			k_p	NA	0.0	NA	NA	0.00	NA
			Θ	NA	0.0	NA	NA	0.00	NA
			ρ_b	NA	-0.18	NA	NA	-0.08	NA
AT123D	NA	NA	NA	NA	NA	NA	NA	NA	NA
Food Chain	Plant Concentration	C_v	IF	NA	0.00	NA	NA	0.00	NA

475

TABLE 2 Continued

Model Component	Output Variable Description	Output Variable Symbol	Parameter Symbol	Model Component Sensitivity/Uncertainty Indexes			Overall Model Sensitivity/Uncertainty Indexes		
				Chromium (VI)	Arsenic	Cadmium	Chromium (VI)	Arsenic	Cadmium
Food Chain (Cont'd)	Plant Concentration (Cont'd)		K_{ol}	NA	0.00	NA	NA	0.00	NA
			CD	NA	0.00	NA	NA	0.00	NA
			BCF_p	NA	0.5	NA	NA	0.21	NA
	Fish Concentration	C_f	BCF_f	NA	0.5	NA	NA	0.01	NA
Dose	Inhalation Dose	D_1	IR	0.40	0.40	0.40	0.21	0.21	0.21
	Ingestion Dose	D_2	INR_p	NA	1.32	NA	NA	0.60	NA
			INR_s	NA	0.05	NA	NA	0.02	NA
			INR_w	NA	0.00	NA	NA	0.00	NA
			INR_f	NA	0.02	NA	NA	0.01	NA
	Dermal Absorption Dose	D_3	SA_s	NA	0.40	NA	NA	0.005	NA
			ABS	NA	0.50	NA	NA	0.007	NA
			AF	NA	0.50	NA	NA	0.007	NA

TABLE 2 Continued

Model Component	Output Variable Description	Output Variable Symbol	Parameter Symbol	Model Component Sensitivity/Uncertainty Indexes			Overall Model Sensitivity/Uncertainty Indexes		
				Chromium (VI)	Arsenic	Cadmium	Chromium (VI)	Arsenic	Cadmium
Dose (Cont'd)	Dermal Absorption Dose (Cont'd)		EF_s	NA	0.50	NA	NA	0.007	NA
			SA_w	NA	0.00	NA	NA	0.00	NA
			PC	NA	0.00	NA	NA	0.00	NA
			EF_w	NA	0.00	NA	NA	0.00	NA
	All Doses	D_1, D_2, D_3	BW	-0.28	-0.28	-0.28	-0.28	-0.28	-0.28
Risk	Total Carcinogenic Risk	R	$CPF_{1(1, 2, 3)}$	3.39	0.39	0.93	3.39	0.39	0.93
			$CPF_{2(1, 2, 3)}$	NA	1.91	NA	NA	1.91	NA
			$CPF_{3(1, 2, 3)}$	NA	0.01	NA	NA	0.01	NA

Note: The above listed Overall Model Senssitivy/Uncertainty Indexes correspond to the carcinogenic risk of each individual chemical.

TABLE 3 Probablity Distribution Selection

Influential Parameter Index	Parameter Symbol	Probability Distribution	
		Type	Parameters
1	Q_{e1} - Arsenic	Lognormal	$\mu' = -7.76, \sigma' = 0.39$
2	Q_{e2} - Cadmium	Lognormal	$\mu' = -7.73, \sigma' = 1.06$
3	Q_{e3} - Chromium	Lognormal	$\mu' = -5.25, \sigma' = 1.30$
4	α-Chrom (VI)	Normal	$\mu = 0.06, \sigma = 0.015$
5	T_s	Uniform	a = 318, b = 328
6	V_s	Uniform	a = 10.4, b = 15.6
7	T_a	Uniform	a = 278, b = 288
8	V_d	Normal	$\mu = 0.005, \sigma = 0.0025$
9	V_w	Normal	$\mu = 0.1, \sigma = 0.05$
10	OR_f	Normal	$\mu = 0.47, \sigma = 0.047$
11	d_s	Lognormal	$\mu' = -3.10, \sigma' = 0.75$
12	ED	Lognormal	$\mu' = 2.24, \sigma' = 0.58$
13	EST	Uniform	a = 0, b = 100
14	ρ_b	Normal	$\mu = 1550, \sigma = 175$
15	BCF_p	Uniform	a = 0.01, b = 0.05
16	IR	Triangular	a = 14, m = 22, b = 30
17	INR_p	Lognormal	$\mu' = -2.58, \sigma' = 0.61$
18	BW	Normal	$\mu = 71.5, \sigma = 17.0$
19	CPF_{11} - Arsenic	Uniform	a = 4.4, b = 26.7
20	CPF_{21} - Arsenic	Triangular	a = 0, m = 1.75, b = 15.0
21	CPF_{12} - Cadmium	Normal	$\mu = 6.3, \sigma = 2.93$
22	CPF_{13} - Chromium	Triangular	a = 10.5, m = 42.1, b = 295.2

Explanations: The above-listed probability distribution types are defined as follows:

- Uniform [a, b]
  where: a = minimum value
          b = maximum value

- Triagular [a, m, b]
  where: a = minimum value
          m = most likely value

          b = maximum value

- Normal (μ, σ)
  where: μ = distribution mean
          σ = distribution standard deviation

- Lognormal (μ', σ')
  where: μ' = mean of underlying normal distribution
          σ' = standard deviation of underlying normal distribution

G

Improvement in Human Health Risk Assessment Utilizing Site- and Chemical-Specific Information: A Case Study

Del Pup, J.,[1] Kmiecik, J.,[2] Smith, S.,[3] Reitman, F.[1]

1.0 INTRODUCTION

The U.S. Environmental Protection Agency (EPA) has classified 1,3-butadiene (butadiene) as a B2 ("probable") human carcinogen.[4] Conservative screening level cancer risk estimates reported by EPA to rank sources and prioritize regulatory action associated emissions of butadiene from the Texaco Chemical Company, Port Neches, Texas facility with a maximum individual risk of 1 in 10. Although the agency emphasized that these screening level estimates should be viewed only as rough estimates of the relative risks posed by the facility under evaluation, and should not be interpreted to represent an absolute risk of developing cancer, the risk estimate generated a high level of concern. In this paper we provide a discussion of results of an effort to use site-specific data, species differences in the metabolism of butadiene, the Monte Carlo procedure, and other factors to estimate risk to the community. The effect of some of these factors is profound. For example, using this information, the range of risks at the closest residence is estimated to be 1 in 10,000,000 to 3 in 10,000. This range of

---

[1]Texaco, Inc.

[2]Texaco Chemical Company

[3]Radian Corporation

[4]EPA classifies chemicals for which there is sufficient evidence for carcinogenicity in experimental animals and inadequate or no evidence for carcinogenicity in humans as Group B2, "probable human carcinogens."

uncertainty is driven largely by species differences in butadiene uptake and metabolism used in the slope factor.

The purpose of this study is twofold:

1) to address the concern posed by the EPA screening level risk assessment by increasing the precision of estimates of the risks potentially posed by butadiene from the facility
2) to demonstrate a process whereby site specific data is utilized in place of regulatory default assumptions to provide a more scientifically credible estimate.

It is neither the intent of this paper to evaluate any cause and effect relationship between 1,3 butadiene exposure and cancer in humans, nor to provide the most scientifically defensible cancer potency estimates for 1,3-butadiene. Risks referred to in this paper are hypothetical estimates useful for regulatory purposes. These estimates assume as a matter of regulatory policy that a low-dose linear carcinogenic response to butadiene occurs in humans. Actual risks would be zero if butadiene is not carcinogenic to humans at these exposure levels.

Texaco initiated this evaluation in 1990 (Radian Corporation, 1990). That assessment focused on increasing the precision of the EPA screening level risk estimates based on more realistic representation of emissions, dispersion and exposure after completion of the Butadiene Modernization Project. This project centered around changing the extraction solvent used in the distillation process and in changing the "once-through" cooling water system to a recirculating cooling tower system in order to reduce butadiene emissions. Although based on site-specific information wherever possible, the risk assessment noted several sources of uncertainty that impacted interpretation of the risk estimates. Primary sources of uncertainty were identified as estimated emissions rates, assumptions and algorithms associated with dispersion modeling analysis, assumptions used to calculate inhalation exposure, and the theoretical estimate of the carcinogenic potency of butadiene, if any, in humans.

The Butadiene Modernization Project, now largely completed, has resulted in a process that is cleaner from both a product purity and environmental perspective. Butadiene emissions have been reduced more than 90 percent. Repeating the prior EPA screening level analysis predicts a maximum individual cancer risk after completion of this project in the range of 5-10 in 1000 based on a 70 year exposure to the maximum predicted annual-average ground level concentration 200 meters from the center of the plant. The current study was initiated to reexamine some of the sources of uncertainty in the risk estimates and to update the risk estimates, using the most site-specific and chemical-specific information available (Radian 1992a) The resulting risk estimates range from 3 in

10,000 to 1 in 10,000,000 at the nearest residence, which are much lower than EPA's original 1 in 10 risk estimate. In addition, we also provide estimates of risk to the nearest residence and school using the Monte Carlo analyses. These provide central tendencies which result in even lower estimates of risk.

The health risk analysis undertaken by the author improves upon the EPA-generated health risk assessment by reevaluating assumptions pertinent to determining the maximum exposed individual risk and risks at various locations in the community. Risks were characterized for the conventional "worst case" 70-year exposure, the 30-year upper bound exposure, the 9-year average residential exposure, and the 95th percentile fraction of life exposed (FLE) based on national human activity pattern distributions. Assumptions used in the development of the EPA-sanctioned unit risk factor for butadiene and impact on the magnitude of risk using alternate unit risk factor assumptions were also evaluated. The assessment also evaluated differences between ground-level concentrations predictions by the Industrial Source complex Long-Term (ISCLT) and the Industrial Source Complex Short-Term (ISCST) atmospheric dispersion models. In addition, results using two meteorological data sets for the area and various decay coefficients for butadiene were evaluated.

This study addresses many of the issues, assumptions, and uncertainties inherent to inhalation pathway risk assessments. However, it should be noted that the analyses, conducted for the current study are site-specific and, therefore, the results may not be applicable to other source configurations, meteorological data sets, or other receptor populations. The study is intended to illustrate a process by which human risk assessments can be improved by using available site-and chemical-specific information.

2.0 EMISSION STATEMENTS

The facility produces butadiene by solvent extraction from a crude C4 stream. The process involves distilling the extracted butadiene to remove heavy ends and final polishing to obtain a butadiene product with purity of 99.7%. Potential sources of butadiene emissions included equipment components in the process units, tank farms, and on the product loading racks; cooling towers; process flares; the dock flare; steam boilers; wastewater treatment plant; the cracking unit; and the butadiene sphere. The butadiene emission estimates were based primarily on actual process data and source-specific information, and on Air Control Board and/or EPA approved emission factors.

It is recognized there are other butadiene sources in the Port Neches area (e.g. butadiene emissions from other area facilities). These other sources of butadiene emissions were not included in the analysis.

3.0 ENVIRONMENTAL FATE AND TRANSPORT MODELING

Atmospheric fate and transport is usually assessed using a mathematical atmospheric dispersion model. Industrial Source Complex (ISC) Models are classified as "preferred" models in the EPA's "Guidelines on Air QUality Models (Revised), 1987 (EPA, 1987). Two versions of the ISC model are available. Both the Industrial Source Complex Short Term (ISCST) and the Industrial Source Complex Long-Term (ISCLT) model are steady state Gaussian plume models preferred for use with industrial complexes with flat terrain such as that found in the area of the facility.

3.1 Industrial Source Complex Model Comparisons

The ISCST model is designed for use in predicting concentrations using averaging periods form one hour to one year. This model utilizes discrete hourly meteorological data. The ISCLT model is designed for use in predicting annual-average concentrations. This model utilized meteorological data in the format of a STAR summary. The STAR summary is a joint-frequency distribution of wind speed, wind direction, and stability classification, processed from discrete hourly observations. The use of this meteorological data summary enables the ISCLT dispersion model program to calculate ambient concentrations much faster than ISCST because dispersion calculations are performed for a small number of meteorological categories rather than for every hour of the year. The ISCLT and ISCST use identical equations for calculating ambient concentrations, with the exception of several changes necessary for the incorporation of the STAR summary.

A model comparison using site-specific inputs revealed fairly good agreement between long-term and short-term results (Radian, 1992b). A 12.5% higher maximum off-property concentration was predicted using the long-term model, but the average concentration of all receptor locations predicted by both models were identical. Given the good agreement between the models, the requirement of evaluating butadiene for long-term or chronic effects, and the faster model execution time, the ISCLT was chosen for this analysis.

3.2 Effects of the use of Atmospheric Decay Coefficients in ISCLT

The ISCLT model provides a mechanism to account for pollutant removal by physical or chemical processes. There are three main chemical reactions which were considered important to evaluating atmospheric concentrations of butadiene, including: 1) reaction with hydroxyl radical ($\cdot OH$);2) reaction with ozone (O_3); and reaction with nitrogen trioxide radical ($\cdot NO^3$) (EPA, 1983). The reaction with $\cdot OH$ is dominant during the day while reaction with $\cdot NO_3$ is dominant at night. Ozone reactivity occurs during the day and night. All reactions are temperature dependent, with butadiene residence times being greater during the winter months and dependent on the chemical species available for reaction in the particular airshed of interest.

Annualized pollutant decay factors were developed by Radian Corporation for use with the ISCLT model based on site-specific temperatures and airshed data estimates. The decays were annualized to address the long-term or chronic exposure aspects of the study. Due to the low solubility of butadiene, physical removal processes such as pollutant incorporation into clouds and rain were not considered to be important pollutant degradation processes and were not considered in this analysis.

Figures 3-1, 3-2, 3-3, and 3-4 illustrate concentration isopleths for no decay, low decay, median decay, and high decay of butadiene, respectively.These results indicate that the inclusion of pollution decay in the transport and fate analysis of butadiene has only minimal effects on predicted ground-level concentrations near the facility. However, as distance from the facility increases, inclusion of butadiene decay in the fate and transport analysis significantly decreases predicted ground-level concentrations.

3.3 Alternative Meteorological Data Set Comparison for the ISCLT Model

Two quality-assured sets of meteorological data were evaluated for use in this analysis: 1) a 14-year composite annual joint frequency distribution of wind speed, wind direction, and stability class (STAR) data processed from the National Weather Service (NWS) hourly surface observations at the County Airport, located approximately four miles from the plant boundary; and 2) a two-year composite STAR data set processed from 1990 and 1991 Regional Planning Commission (RPC) continuous observations at another County Airport location, approximately three miles form the plant boundary. The RPC data were selected for use in the majority of the analyses due to the continuous nature of the observations and the use of measured mixing heights. However, to examine the sensitivity of the risk estimates to changes in the meteorological data set, the ISCLT dispersion model was run with identical inputs, varying only the meteorological data. At nearby locations, predicted concentrations using RPC data were 25 to 100% higher than predicted concentrators using the NWS data. Using the RPC data, concentrator isopleths would extend farther to the east and are more rounded. using the NWS data, the isopleths would show more of a north-south bias (Radian, 1992a).

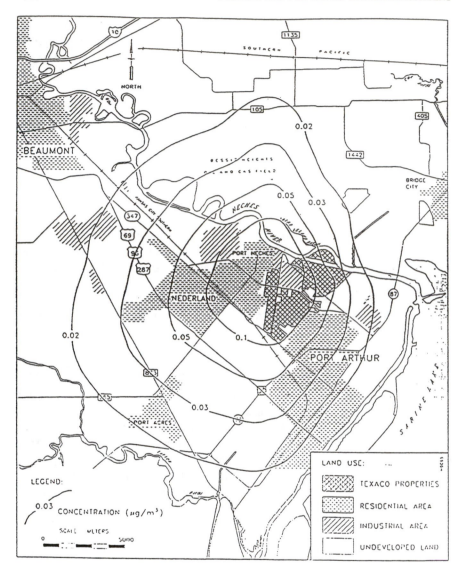

FIGURE 3-1 Concentration Isopleths ($\mu g/m^3$). No Butadiene Decay

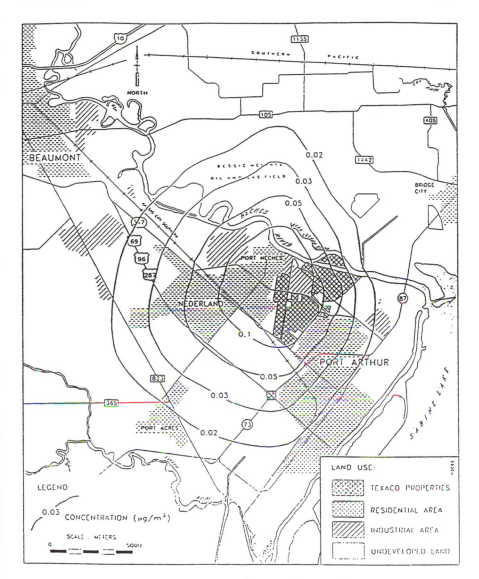

FIGURE 3-2 Concentration Isopleths ($\mu g/m^3$). Low Butadiene Decay

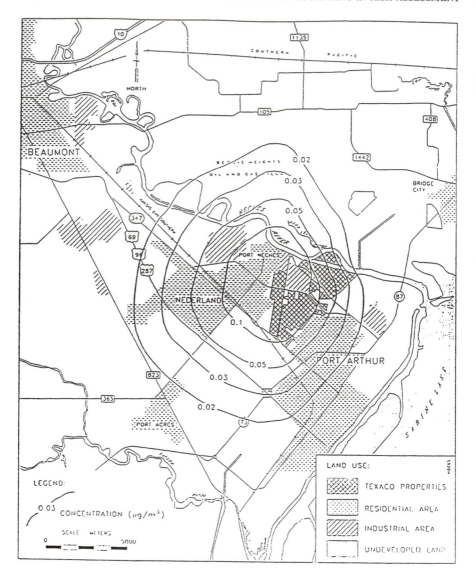

FIGURE 3-3 Concentration Isopleths ($\mu g/m^3$). Median Butadiene Decay

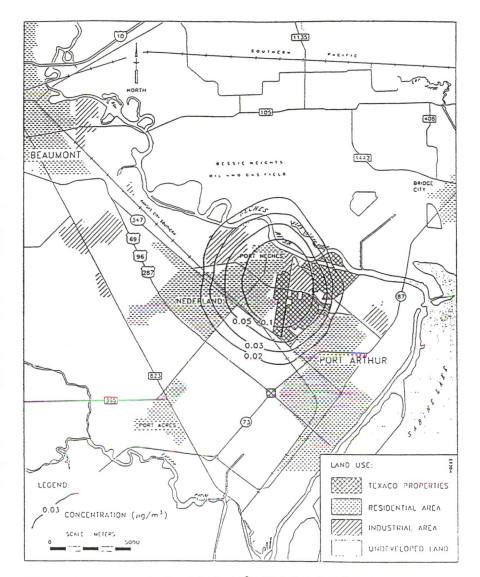

FIGURE 3-4 Concentration Isopleths (μg/m³). High Butadiene Decay

4.0 HUMAN HEALTH ASSESSMENT

Risk characterization involves integrating exposure and toxicity information into quantitative and qualitative expressions of potential health risk. For potential carcinogens such as butadiene, risk can be characterized by estimating the potential for carcinogenic effects or by estimated ambient air concentration with health-based ambient guidelines or standards.

To characterize potential carcinogenic effects, estimated risks that an individual will develop cancer over a lifetime of exposure to butadiene were calculated from projected intakes and the cancer slope factor. The cancer slope factor converts estimated daily intakes directly to an estimate of incremental risk as follows:

Dose (mg/kg-day) × Cancer Slope Factor $(mg/kg-day)^{-1}$
= Lifetime Excess Cancer Risk

The slope is often an upper 95th percentile confidence limit of the probability of response based upon experimental animal data and an assumption of linearity in the low-dose portion of dose-response curve. Therefore, the carcinogenic risk estimate will generally be an upper-bound estimate, indicating that the "true-risk", if any, will probably not exceed the risk estimates based on the slope factor and is likely to be less than that predicted.

Individuals may be exposed to chemical in air by inhalation of chemicals in the vapor phase or adsorbed to particulate. Dermal absorption of vapor phase chemicals such as butadiene is considered to be lower than inhalation intakes and, therefore, was not quantified in this risk assessment (EPA, 1989). Inhalation of airborne vapor-phase chemicals can be quantified using the following formula:

$$\text{Intake (mg/kg-day)} = \frac{CA \times IR \times ET \times EF \times ED}{BW \times AT}$$

where: CA = Contamination Concentration in Air (mg/m^3);
      IR = Inhalation Rate $(m^3/hour)$;
      ET = Exposure Time (hours/day);
      EF = Exposure Frequency (days/year);
      ED = Exposure Duration (years);
      BW= Body Weight (kg); and
      AT = Average Time (period over which exposure is averaged—days)

Lifetime exposure must be evaluated to determine cancer risk. To provide a conservative analysis of lifetime community exposure, the exposed population (represented by an average 70 kg adult) has been assumed to inhale (at an average rate of 20 m^3/day) predicted ground-level concentrations continuously, 24 hours/day, 365 days/year, for a 70 year exposure duration. More recently, EPA has employed "reasonable maximum" assumptions of 24 hours/day, 350 days/year for 30 years.

4.1 Characterization of Risk

To characterize the risks, both the health variables and the exposure variables were combined under three scenarios, the Base Case, Worst Case and Best Case (Table 4.1). For example, the Worst Case includes inputs that reflect a highly conservative approach whereas the Base Case and Best Case make use of different levels of sophistication in the utilization of site-specific data, exposure assumptions, and recent biological data on the uptake and metabolism of butadiene.

The ISCLT model calculates an ambient concentration at each point (or receptor) provided in the model input. Receptor placement was designed to identify the location of the maximum off-property concentration. Additional receptors were also placed at the nearest residences and the nearest school complexes in several directions. Therefore, concentrations at several locations of special interest were determined. Table 4-2 summarizes the Base Case maximum individual risk calculations for each of the nearby receptor locations. Risk estimates at the closest residences were 1 in 10,000. Risk estimates at the location of maximum off-property concentration were about 5 times higher. Estimated risks at the school locations were lower, ranging from 7 in 100,000 to 4 in 1,000,000. This can be compared with the approximate 1 in 4 background risk of developing fatal cancer in the U.S. population (Harvard School of Public Health, 1992). Refinements to this assessment were made by evaluating additional variables impacting on the risk estimates. Some of these, particularly the slope factor, have a high level of uncertainty.

4.1.1   Effect of Exposure Assumptions

Realistically, very few people remain in the same location for a lifetime. To account for exposure durations less than a lifetime, the following formula can be used to quantify the Lifetime Average Daily Exposure (LADE) (Price et al 1991):

$$\text{LADE (mg/kg-day)} = \frac{CA \times IR \times FLE}{BW}$$

where: CA = Contaminant Concentration in Air (mg/m^3);
         IR = Inhalation rate (m^3/day);
         FLE = Fraction of Life Exposed (unitless); and
         BW = Body Weight (kg)

TABLE 4-1
KEY VARIABLES THAT DESCRIBE THREE CASES

Variables	Base Case	Worst Case	Best Case
Meteorological Data			
1 SETPC	✓	✓	
2 NWS			✓
Butadiene Decay			
1. No Decay		✓	
2. Low Decay			
3. Medium Decay	✓		
4. High Decay			✓
Exposure Assumptions			
1. Traditional Worst Case (24 hrs/day, 365 days/yr for 70 yrs)		✓	
2. Reasonable Maximum (24hrs/day, 350 days/yr for 30 yrs)[1]	✓		
3. 95th Percentile Fraction of Life Exposed (based on national human activity pattern distributions)			
4. Average Fraction of Life Exposure (based on national human activity pattern distributions)			✓
Butadiene Slope Factor			
1. EPA Slope Factor	✓	✓	
2. EPA Slope Factor Adjusted by a Factor of 30			
3. EPA slope Factor Adjusted by a Factor of 590			✓

[1]As defined by U.S. EPA, 1989

National statistics are available on the upper-bound (30 years) and average (9 years) number of years spent by individuals at one residence (EPA 1989, 1991). The "upper-bound" value was used as the exposure duration when calculating the reasonable maximum residential exposures. An exposure frequency of 350 days/year was used, assuming 15 days/year are spent away from home. Assuming a 70 year lifetime, the FLE is an average of 0.12 and a reasonable maximum of 0.41.

A Point Source Exposure Model (PSEM) was used to characterize the distri-

TABLE 4-2
ESTIMATED MAXIMUM INDIVIDUAL CANCER RISK AT
IDENTIFIED RECEPTOR LOCATIONS FOR THE BASE CASE.

RECEPTOR LOCATION[a]	MAXIMUM INDIVIDUAL CANCER RISK FOR THE BASE CASE[b,c,d]
Off Property Max (West Property Line)	5E-04 (5 in 10,000)
Residence 1	1E-04 (1 in 10,000)
Residence 2	1E-04 (1 in 10,000)
School 1	5E-05 (5 in 100,000)
School 2	7E-05 (7 in 100,000)
School 3	5E-06 (5 in 1,000,000)
School 4	4E-06 (4 in 1,000,000)

[a]The location of these receptors in relation to the facility is identified in Figure 3-1

[b]Based on EPA "reasonable maximum" inhalation exposures (EPA, 1989) and the EPA potency slope of 1.8 $(mg/kg-day)^{-1}$. Reasonable maximum exposure assumptions based on residential exposure patterns are assumed at all locations.

[c]Range of risks from zero to stated value

[d]Backround fatal cancer risk in the U.S. is approximately 1 in 4.

bution of exposures received over a long-term period based on information on mobility, mortality, and daily activity patterns (Price et al 1991). PSEM models the time of residence in the zone of impact and the amount of time spent at home as variables that yield a probability density function for the FLE. The model predicted that, on average, individuals live in their current house for 16.5 years and spend 18 hours per day at home. The average value of the FLE calculated by PSEM using national statistics was 0.16. The median value was 0.12, and the 95th percentile of the distribution was 0.42. Inhalation exposures for all receptor locations were calculated based on residential exposure assumptions, using these FLE values. It is assumed for this report that 1,3-butadiene concentrations are the same inside and outside the home. No attempt has been made here to validate this assumption.

Several exposure scenarios were examined in this assessment including: 1) "worst case " (24 hours/day, 365 days/year for 70 years); 2) "reasonable maximum" (24 hours/day, 350 days/year for 30 years); 3) 95th percentile FLE based on national human activity pattern distributions, and 4) average FLE based on national human activity pattern distributions. Figure 4-1, 4-2, and 4-3 illustrate the areal extent encompassed by several risk levels using traditional "worst case", reasonably maximum, and average exposure assumptions. As indicated in the figures, the areal extent encompassed by specific risk levels is very sensitive to changes in the time, frequency, and duration of exposure.

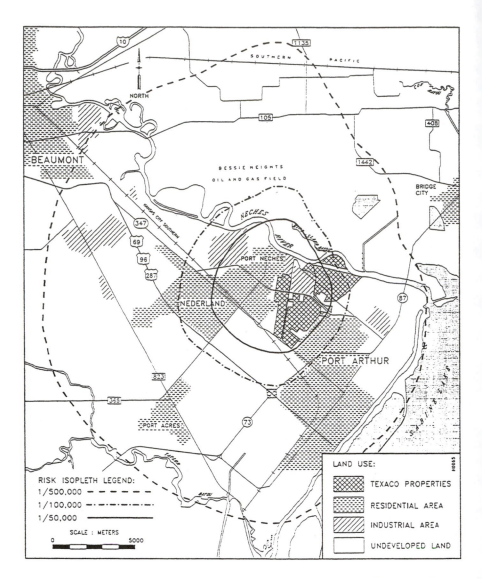

FIGURE 4-1 Area Encompassed by Specific Risk Levels–Traditional Worst Case Exposure Assumptions

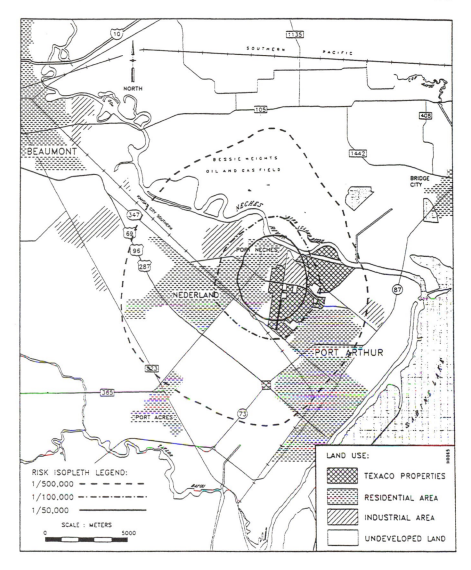

FIGURE 4-2 Area Encompassed by Specific Risk Levels–Reasonable Maximum Exposure Assumptions (Base Case)

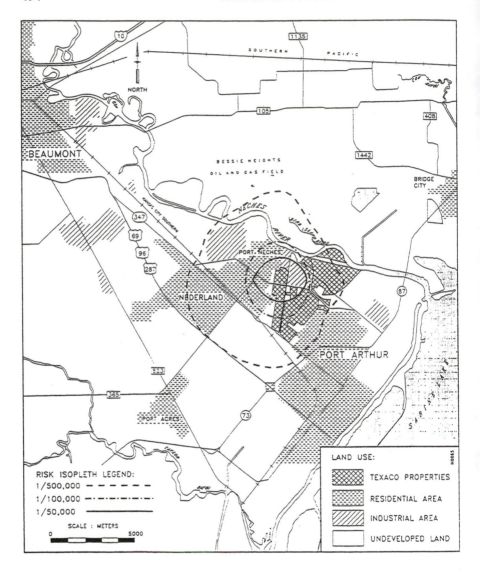

FIGURE 4-3. Area Encompassed by Specific Risk Levels-Average Exposure Assumptions

4.1.2 Effect of Cancer Slope Factor Assumptions

The EPA-sanctioned slope factor for butadiene of 1.8 $(mg/kg\text{-}day)^{-1}$ was used in all previous analyses (IRIS 1992). In the current analysis, however, risk estimates were also generated using alternative slope factors based on research that the EPA slope factor may overpredict risks to the human population.[5] Cancer slope factors can be converted to unit risk estimates to determine the risk per unit air or water concentration. The inhalation unit risk can be calculated by dividing the slope factor by 70 kg (average body weight for an adult) and multiplying by 20 m^3/day (adult average inhalation rate) assuming a 70 year exposure period (EPA 1989,1991).

EPA calculated an inhalation risk estimate for butadiene of 2.8-04 $(\mu g.m^3)^{-1}$, based on an absorption factor of 54%, which was derived from preliminary results of an absorption study conducted in mice and sponsored by the National Toxicology Program (NTP). The procedure for determining animal-to-equivalent human dose was adjusted to account for the fact that at high concentrations, the internal dose (mg/kg) is not directly proportional to external concentrations. A final report of the NTP study has been published and differs significantly from the preliminary results (Bond et al. 1986). Results from the final report suggested that butadiene retention by mice in the initial study may have been overestimated by a factor of five. Based on these data, risk estimates derived using EPA-sanctioned values for butadiene should be adjusted downward by approximately a factor of five. Based on the discussion published in EPA's Integrated risk Information System (IRIS, 1992), EPA used an absorption factor of 54% in calculating a slope factor. IRIS states that differences between the retention of butadiene reported in the initial and final study have been accounted for in EPA's calculations. Assuming this is correct, there is no need to make adjustments in risk estimates based on the EPA value. However, the animal upper-limit slope factors are identical to those published by the EPA in 1985, suggesting that this correction has not been made (EPA, 1985). If the correction was not made, the downward adjustment by a factor of five is appropriate.

The respiratory systems of humans differ from experimental animals in many ways. These differences result in variations in air flow, deposition of inhaled agents, as well as the retention of that agent. The dose of partially soluble vapors, such as butadiene, is proportional to oxygen consumption. Oxygen consumption is, in turn, proportional to (body weight) and is also proportional to the

---

[5]A cohort epidemiologic study of workers employed at this facility between 1943 and 1979 showed a statistically significant deficit for all causes of death and all cancers. There was, however, a statistically significant excess of deaths from lymphosarcoma. This was concentrated in workers employed less than 10 years and first employed prior to 1946 (Divine, 1990).

solubility of the gas in body fluids, which is expressed as an absorption coefficient for the gas. In the absence of experimental evidence to the contrary, the absorption coefficient is assumed to be the same for all species. Therefore, butadiene exposure concentrations (in ppm) used in animal studies were assumed to be equivalent to the same concentration in humans. However, smaller animals have higher minute respiratory volumes per unit of body weight to supply their relatively larger requirements for oxygen. Since the dose of butadiene (by inhalation) is proportional to oxygen consumption, species with higher minute respiratory volumes would be expected to have larger body burden of the chemical.

Studies have been conducted which indicate that nonhuman primates absorb considerably less butadiene than mice (Dahl et al. 1990). At 10 ppm, mice retain approximately 6.6-fold more butadiene than monkeys. The human species is much more closely related to the monkey than the mouse, both physically and anatomically. Therefore, primate retention data should be used as a basis for estimating retention by humans. On this basis, risk estimates derived from EPA sanctioned toxicity values should be adjusted downward by a factor of six.

In quantitative risk modeling, internal concentrations of butadiene were used as a measure of dose. However, in doing so, species differences in metabolism of butadiene were ignored. In studies sponsored by NTP (Dahl et al. 1990), mice were shown to attain approximately 590-fold higher blood levels of the monoepoxide (a DNA-reactive and mutagenic metabolite of butadiene, assumed to be a toxic metabolite) than did primates.[6] Based on the assumption that humans metabolize butadiene in a manner that is more closely related to nonhuman primates, humans should be approximately 590-fold less sensitive to butadiene's carcinogenic effects than mice. Therefore, estimates of risk should be adjusted by a factor of 590 to account for species differences in metabolism of butadiene. Use of the internal concentration of the monoepoxide would obviate the need to adjust for difference in retention of inhaled butadiene.

The available comparative studies suggest that the equivalent potency of butadiene in humans could be substantially less than that used as the basis for EPA's calculated cancer slope factor. Based on the available data, the slope factor could be adjusted downward (i.e., to indicate lower potency for humans) by a factor of 30 (5 x 6 based on current retention data for the mouse and mouse/primate differences in retention) to 590 (based on mouse/primate differences in blood levels of the monoepoxide). Since risks change proportionally to changes in the butadiene slope factor, the risks using the alternative slope factors are lowered by a factor of 30 to 590. Figures 4-4. 4-5 and 4-6 illustrate the way in

---

[6]Metabolites were tentatively identified, based on co-distillation with standards.

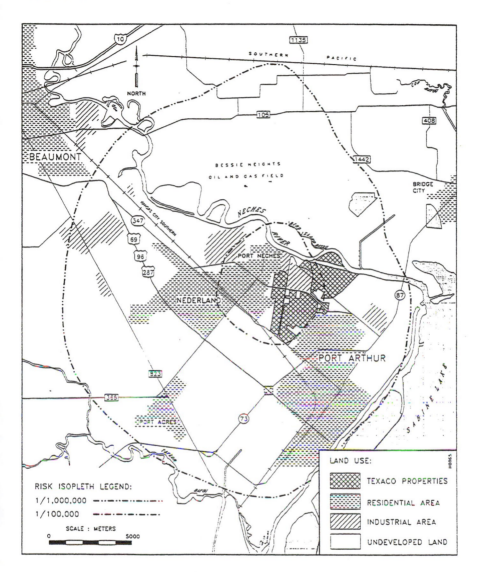

FIGURE 4-4. Area Encompassed by Specific Risk Levels-EPA Slope Factor (Base Case)

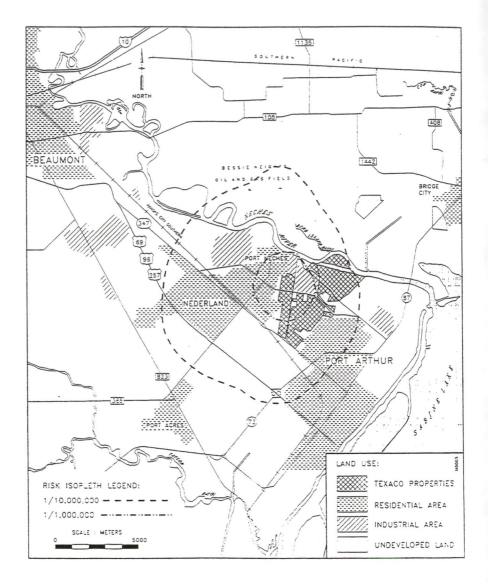

FIGURE 4-5. Area Encompassed by Specific Risk Levels-EPA Slope Factor Adjusted by a Factor of 30

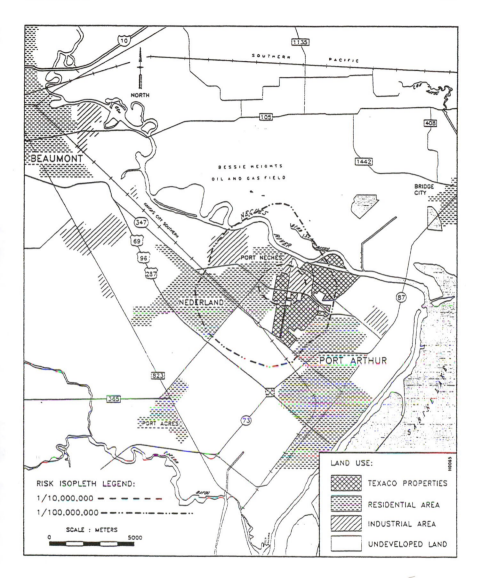

FIGURE 4-6. Area Encompassed by Specific Risk Levels-EPA Slope Factor Adjusted by a Factor of 590

which changes in the butadiene slope factor affect the area encompassed by specific risk levels.

5.0 Probabilistic Monte Carlo Simulation

Risk estimates resulting from a series of "worst case" assumptions can be expected to overestimate actual risk. However, there is no way for the regulator, industry representatives, the potentially exposed population, or other interested parties to interpret the degree of conservatism. EPA risk assessments are expected to address the range of risk including the central tendency and high end portion of the risk distribution (EPA, 1992). In addition, they are expected to include a statement of confidence in the risk assessment itself. Stochastic analysis of risk provides a distribution of estimated risks based on the use of probability density functions for input parameters instead of single point estimates.

Monte Carlo simulation calculates risk through numerous iterations using randomly generated values from the defines probability functions. The resulting distribution of risk estimates makes greater use of the scientific evidence and data related to exposure and theoretical risk without sacrificing conservatism. Monte Carlo avoids compounding of "worst case"assumptions and uncertainty, and provides quantitative information on the uncertainty in the risk values.

The shape of the distribution and the range between low and high end estimates portray the uncertainties incorporated in the assessment and can be used to interpret the level of confidence in the assessment. A narrow range between 5th and 95th percentile of the distribution implies a low level of overall uncertainty and, consequently, a high level of confidence in the assessment. A broad range implies a high level of uncertainty.

In this assessment, the range in the risk estimates from the 5th and 95th percentile at the closest residence was 4 in 100,000,000 to 2 in 10,000 (Radian, 1992b). This range spans almost four orders of magnitude, indicating a very high level of uncertainty. The range in estimated risk from the 5th to the 95th percentile at the closest schools was 5 in 10,000,000,000 to 6 in 1,000,000. This range spans more than four orders of magnitude. The slightly greater span in the risk range at this location results form the greater potential influence of butadiene decay in the atmosphere as the distance from the facility increases. Therefore, the level of confidence in the estimates of risk associated with butadiene at the facility can only be described as low.

6.0 CONCLUSIONS

A number of variables examined in this risk assessment significantly impacted the final theoretical risk estimates. These variables included: 1) the meteorological data used in transport and fate modeling; 2) butadiene decay factors; 3) exposure time, frequency, and duration; and 4) the slope factor for butadiene.

Base Case estimates were developed including inputs for key variables that are relatively conservative. The sensitivity of Base Case estimates to varying inputs for these key variables was evaluated. The Base Case predicted risks in the range of 1 in 10,000 at the nearest residences, and 4 in 1,000,000 to 7 in 100,000 at the nearest schools. Worst Case estimates were only two to three times higher than Base Case estimates.

Best Case estimates, which provide an additional measure of the level of uncertainty associated with the estimates, ranged from more than three to four orders of magnitude lower than Worst Case estimates. The butadiene slope factor contributes almost three orders of magnitude to the theoretical risk estimates separating the Worst Case and Best Case scenarios. While the butadiene decay factor did not significantly affect the risk estimates at nearby locations, this effect was location dependent. The Base Case risk estimates (1 in 10,000 at the nearest residences) represents an upper-bound to the risk associated with the butadiene emissions from the facility. The "true risk" is unlikely to be higher, and is most likely lower. An examination of some of the key variables that influence estimates of theoretical risk indicates that the maximum individual risk at the nearest residences may be as low as 1 in 10,000,000. Risk estimates in this report should be considered in comparison to the approximate 1 in 4 background fatal cancer risk in the U.S. population. In all cases the risk would be zero if butadiene is not carcinogenic in humans at prevailing exposure levels.

REFERENCES

1.   Bond, J.A., Dahl, A.R., Henderson, R.F., Dutcher, J.S., Mauderly, J.L., Birnbaum, L.S., 1986. Species differences in the disposition of inhaled butadiene. *Toxicol. Appl. Pharmacol.* 84: 617-627, 1986.

2.   Dahl, A.R., Bechtold, W.E., Bond. J.A., Henderson, R.F., Muggenburg, B.A., Sun, J.D., and Birnbaum, L.S., 1990. Species Differences in the Metabolism and Disposition of Inhaled 1,3-Butadiene and Isoprene. *Environ. Health Perspect.* 86: 65-69, 1990.

3.   Divine, B.J. 1990 An Update on Mortality Among Workers at a 1,3-Butadiene Facility - Preliminary Results. Environmental Health Perspectives 86: 119-128.

4.   Harvard School of Public Health, The Center for Risk Analysis. Annual Report. 1992.

5.   Integrated Risk Information System (IRIS), 1992. U.S. EPA Information Network On-Line Database. Data retrieved January, 1992.

6.   Price, P.S., Sample, J., Streiter, R., 1991. PSEM. A model of Long-Term Exposures to Emissions from Point Sources. Presented at the 84th annual meeting of the Air and Waste Management Association, Vancouver, British Columbia, June 16-21, 1991. 91-172.3.

7.   Radian Corporation, 1990. Site-Specific Evaluation of Potential Cancer Risk Associated with 1,3-Butadiene Emissions from the Texaco, Port Neches Facility. Prepared for Texaco Chemical Company, May 14, 1990.

8.   Radian Corporation, 1992a. Site-Specific Evaluation of Potential Cancer Risk Associated with 1,3-Butadiene Emissions from the Texaco, Port Neches Facility. Prepared for Texaco Chemical Company, April 16, 1992.

9.   Radian Corporation, 1992b. Technical Memorandum from Randy Parmley, Radian Corporation to Jim Kmiecik, Texaco Chemical Company on "Summary of ISCST and ISCLT Model Comparison," dated 31 January 1992.

10. Radian Corporation 1992c. Site-Specific Evaluation of Potential Cancer Risk Associated with 1,3-Butadiene Emissions from the Texaco, Port Neches Facility. Addendum I-Probabilistic Monte Carlo Simulation. August 10, 1992.

11. U.S. Environmental Protection Agency (EPA), 1983. Health and Environmental Effects Profile for 1,3-Butadiene. EPA/60/0-84/120. May, 1983.

12. U.S. Environmental Protection Agency (EPA), 1985. Mutagenicity and Carcinogenic Assessment of 1,3-Butadiene. EPA/600/8/85-004f.

13. U.S. Environmental Protection Agency (EPA), 1987. Industrial Source Complex (ISC) Dispersion Model. Addendum to User's Guide. 1987.

14. U.S. Environmental Protection Agency (EPA), 1989. Risk Assessment Guidance for Superfund. Volume 1 Human Health Education Manual (Part A). Interim Final. EPA 540/1-89-002. December, 1989.

15. U.S. Environmental Protection Agency (EPA), 1991. Risk Assessment Guidance for Superfund. Volume 1 Human Health Evaluation Manual. Supplemental Guidance. Standard Default Exposure Factors. March 25, 1991.

H-1
Some Definitional Concerns About Variability

Each of the three major types of variability (temporal, spatial, and interindividual) can be characterized in three ways, as follows (these examples are all related to human variability in susceptibility, although other examples are possible):

• Variability is (or can be modeled sufficiently precisely as though it were) is) either *discrete* or *continuous*. For example, albinos are many times more sensitive to sunlight than other members of the population, so a (dichotomous) discrete assumption might well be appropriate here. In contrast, because body weights vary continuously, the cancer risk per unit dose of a substance cannot be modeled dichotomously without the loss of much of information.

• Variability is *identifiable* or *unidentifiable*. Albinism is a good example of identifiable variability, whereas the extent of a person's ability to detoxify a particular active metabolic intermediate might not be discernible without invasive testing, and hence is unidentifiable for most of the population.

• Identifiable variability is *dependent* on or *independent of* additional variable characteristics that society deems salient. For example, some factors that cause genetic predisposition to the carcinogenic effect of chemicals are correlated with race, sex, or age. If society deems that those who are predisposed already deserve special attention because of the other factors, the importance of the variability is heightened. But some kinds of identifiable variability, such as body weight and phenylketonuria, are more "value-neutral" or are uncorrelated with any relevant characteristic.

H-2

Individual Susceptibility Factors

One way to categorize the many different factors that affect susceptibility to cancer is to divide them into qualitatively different classes along two strata—according to the prevalence of each factor in the human population and according to the degree to which the factor can alter susceptibility. Finkel (1987) noted that most of the factors that are very common (Table H-1) tend to confer only marginal increases in relative risk on those affected (less than a doubling of susceptibility). Many of the other predisposing factors, long recognized as conferring extremely high relative risks, also tend to be quite uncommon (see Table H-2).

However, several important determinants of cancer susceptibility might well be neither rare nor of minor importance to people, and some speculate that this might be quite important for societal risk assessment. This section discusses five factors that might be among the most significant.

CARCINOGEN METABOLISM

Most chemical carcinogens require metabolic activation to exert their oncogenic effects, and the amount of carcinogen produced depends on the action of competing activation and detoxification pathways,. Interindividual variation in carcinogen metabolism is therefore an important determinant of cancer susceptibility.

Chemical carcinogens are metabolized by a wide variety of soluble and membrane-bound enzymes. Multiple forms of human cytochrome P450 (CYP) are involved in the oxidative metabolism of chemical carcinogens, such as poly-

TABLE H-1   Examples of Common Predisposing Factors

Predisposing Factor	Mechanism Influencing Susceptibility to Cancer
A. Temporal Factors[a]	
• Circadian rhythms	
• Changing ingestion and inhalation characteristics during life	
• Depression and stress	
B. Nutritional Factors[b]	
• Vitamin A and iron deficiencies	May increase susceptibility to carcinogenic hydrocarbons
• Dietary-fiber intake	Insufficient intake may increase residence time of carcinogens in contact with epithelium of digestive tract
• Alcohol intake	May affect susceptibility through effect on liver
C. Concurrent Diseases[c]	
• Respiratory tract infections and bronchitis	May predispose lungs to cancer by disturbing pulmonary clearance or promoting scarring
• Viral diseases, e.g., Hepatitis B	May activate proto-oncogenes and cause liver necrosis and regeneration
• Hypertension	May increase the potential for DNA damage in peripheral lymphocytes

[a]Data from Fraumeni, 1975; Borysenko, 1987.
[b]Data from Calabrese, 1978.
[c]Data from Warren and Weinstock, 1987.

cyclic aromatic hydrocarbons (PAHs). Interindividual variation by a factor of several thousand has been observed in placental aryl hydrocarbon hydroxylase (AHH) activity, which is catalyzed by CYP1A1; some of this variability is under direct genetic control, but variations also result from an enzyme induction process due to maternal exposure to environmental carcinogens, such as tobacco smoke. A genetic polymorphism in CYP1A1 in which an amino acid substitution in the heme-binding region of the protein increases catalytic activity of PAHs has been linked to enhanced susceptibility to squamous cell carcinoma of the lung in cigarette-smokers (Nakachi et al., 1991). Japanese with the suscepti-

TABLE H-2   Examples of Rare Predisposing Factors[a]

Predisposing Factor	Mechanism Influencing Susceptibility to Cancer
• Ataxia-telangiectasia	Chromosomal fragility, causing sensitivity to agents that increase genetic recombination
• Bloom's syndrome	Hypermutability
• Chediak-Higashi syndrome	Depletion of "natural killer" cells that combat incipient malignancies
• Down's syndrome trisomy 21	Tenfold excess leukemia risk
• Duncan's disease	Lymphoma in those infected by Epstein-Barr virus
• Epidermodysplasia verruciformis	Skin carcinoma associated with chronic infection with human papilloma virus
• Familial polyposis coli	Mutation in APC tumor suppressor gene leads to benign colonic growths that are predisposed to malignant transformation
• Fanconi's anemia	Possible deficiency of enzymes that scavenge active oxidizing species
• Glutathione reductase deficiency	Very high excess risk of leukemia
• Hereditary retinoblastoma	Predisposition to retinal cancer due to mutation of one allele of a tumor suppressor gene
• Li-Fraumeni syndrome	Germline mutation in the p53 tumor suppressor gene predisposes to multiple carcinomas and sarcomas
• X-linked agammaglobulinemia	Immune deficiency, predisposing to leukemia
• Xeroderma pigmentosum	Inability to repair some kinds of DNA damage, predisposing to skin cancer caused by ultraviolet radiation

[a]Data from Swift et al., 1991; Orth, 1986; Kinzler et al., 1991; Nishisho et al., 1991; Groden et al., 1991; Cleaver, 1968; Friend et al., 1986; Harris, 1989.

ble genotype had an odds ratio of 7.3 (95% confidence interval, 2.1-25.1) at a low level of cigarette-smoking; the difference in susceptibility between genotypes was diminished at high levels of smoking, and that suggests that interindividual variation may be especially important for risk-assessment purposes when "low" exposures are involved. The frequencies of this and other genetic polymorphisms of enzymes involved in carcinogen metabolism may vary among ethnic groups.

CYP2D6 activity is polymorphic and has been linked to lung-cancer risk (Ayesh et al., 1984; Caporaso et al., 1990). CYP2D6 hydroxylates xenobiotics, such as debrisoquine (an antihypertensive drug) and a tobacco-specific N-nitrosamine. A person's polymorphic phenotype is inherited in an autosomal recessive manner. The rate of 4-hydroxylation of debrisoquine varies by a factor of several thousand, and lung-, liver-, or advanced bladder-cancer patients are more likely to have the extensive-hydroxylator phenotype than noncancer controls. In a case-control study of lung cancer in the United States (Caporaso et al., 1990), the extensive-hydroxylator phenotype had a greater cancer risk (odds ratio, 6.1; 95% confidence interval, 2.2-17.1) than poor-hydroxylator phenotype. The increase in risk was primarily for histologic types other than adenocarcinoma. British workers who have the extensive-hydroxylator phenotype and who are exposed to high amounts of asbestos or PAHs have an increased risk of lung cancer (odds ratio, 18.4; 95% confidence interval, 4.6-74 and 35.3; 95% confidence interval, 3.9-317, respectively) (Table H-3) (Caporaso et al., 1989). CYP2D6 might activate chemical carcinogens in tobacco smoke, such as some N-nitrosamines, or perhaps inactivate nicotine, the addictive component of tobacco smoke, so as to decrease its steady-state concentration and lead to an increase in smoking. A person with the extensive-hydroxylator phenotype might thus be at greater cancer risk. Another hypothesis is that an allele of the CYP2D6 gene is in linkage disequilibrium with another gene that influences cancer susceptibility.

The N-acetylation polymorphism is controlled by two autosomal alleles at a single locus in which rapid acetylation is the dominant trait and slow acetylation the recessive trait. Both slow acetylation and rapid acetylation of carcinogenic aromatic amines have been proposed as cancer risk factors. The slow-acetylator phenotype has been linked to occupationally induced bladder cancer in dye workers exposed to large amounts of N-substituted aryl compounds (Cartwright et al., 1982). The rapid-acetylator phenotype was more common in two of three studies of colon-cancer cases (Lang et al., 1986; Ladero et al., 1991; Ilett et al., 1987).

Wide interindividual differences in enzymes that detoxify carcinogens are also found. For example, competing detoxifying enzymes are found at each step in the metabolic pathway of benzo[a]pyrene activation to electrophilic diol-epoxides. A recent study of several of the enzymes involved in benzo[a]pyrene metabolism confirmed previous observations by showing a more than 10-fold person-to-person variation in enzyme activities and presented indirect evidence that tobacco smoke induced many of these enzymes (Petruzzelli et al., 1988). Genetic control of the presumed detoxification of benzo[a]pyrene by conversion to water-soluble metabolites has also been reported (Nowak et al., 1988).

Glutathione S-transferases (GST) are multifunctional proteins that catalyze

TABLE H-3 Examples of Interactions Between Inherited Cancer Predisposition and Environmental Carcinogens

Candidate Gene	Condition	Examples of Cancer Site	Environmental Carcinogens	Odds Ratio (95% confidence interval)	Reference
XPAC	Xeroderma pigmentosum	Skin	Sunlight	>1,500	Cleaver, 1968
Unknown	Epidermodysplasia verruciformis	Skin	Sunlight and human papillomavirus	(#30% of affected people)	Orth, 1986
CYP2D6	Extensive-hydroxylator phenotype	Lung	Tobacco smoke	6.1(2.2-17.1)	Caporaso et al., 1990
			Asbestos	18.4 (4.6-74)	Caporaso et al., 1989
			PAH[a]	35.3 (3.9-317)	Caporaso et al., 1989
YP1A1	Extensive-metabolic phenotype[b]	Lung	Tobacco smoke	7.3 (2.1-25.1)	Nokachi et al., 1991
Ha-ras	Restriction-fragment length polymorphisms (rare alleles)	Lung[c]	Tobacco smoke	4.2(1.1-16)	Sugimura et al., 1990
NAT2	Slow-acetylator phenotype (recessive inheritance)	Bladder	Aromatic amine dyes	16.7 (2.2-129)	Cartwright et al., 198
NAT2	Rapid-acetylator phenotype (dominant inheritance)	Colon	Unknown	1.4 (0.6-3.6) 4.1 (1.7-10.3)	Lang et al., 1986 Ilett et al., 1987
CYP1A1 GSTI	Metabolic balance between activation and detoxification	Lung[d]	Aromatic hydrocarbons	9.1(3.4-24.4)	Hayashi et al., 1992 Seidegard et al., 1986 Hayashi et al., 1992
GST1	Metabolic balance between activation and detoxification	Lung[e]	Aromatic hydrocarbons	3.5(1.1-10.8)	Seidegard et al., 1986

[a]Polycyclic aromatic hydrocarbon.
[b]Increased prevalence in Japanese.
[c]Nonadenocarcinoma lung cancer in African-Americans.
[d]Squamous cell carcinoma in Japanese.
[e]Adenocarcinoma.

the conjugation of glutathione to electrophiles, including the ultimate carcinogenic metabolite of benzo[a]pyrene, and are considered to be one means of detoxifying carcinogenic PAHs. The three isoenzymes of GST (α-ϵ, μ, and π) vary in their substrate specificity, tissue distribution, and activities among individuals. Expression of GST-μ is inherited as an autosomal dominant trait, and people with low GST-μ activity might be at greater risk of lung cancer caused by cigarette-smoking (Seidegard et al., 1986, 1990). In addition, an interaction between GST-μ and CYP1A1 genotypes has been observed (Hayashi et al., 1992). People with a homozygous deficient GST-μ genotype and a CYP1A1 genetic polymorphism in the heme-binding region of this cytochrome P450 enzyme have an increased risk of squamous cell carcinoma of the lung (odds ratio, 9.07; 95% confidence interval, 3.38-24.4) and adenocarcinoma of the lung (odds ratio, 3.45; 95% confidence interval, 1.10-10.8).

DNA-ADDUCT FORMATION

DNA adducts are one form of genetic damage caused by chemical carcinogens and might lead to mutations that activate proto-oncogenes and inactivate tumor-suppressor genes in replicating cells. The steady-state concentrations of the adducts depend on both the amount of ultimate carcinogen available to bind and the rate of removal from DNA by enzymatic repair processes. The genomic distributions of adduct formation and repair are nonrandom and are influenced by both DNA sequence and chromatin structure, including protein-DNA interactions that prevent electrophilic attack of the DNA by the active form of the carcinogen.

Although the major DNA adducts are qualitatively similar for the chemical carcinogens so far studied in the in vitro models, quantitative differences have been found among people and among various tissue types. The differences due to interindividual variation and intertissue variation within an individual in formation of DNA adducts have a range of a factor of about 10-150 among humans. The interindividual distribution is generally unimodal (i.e., a curve with a single peak), and the variation is similar in magnitude to that found in pharmacogenetic studies of drug metabolism (Harris, 1989).

DNA-REPAIR RATES

DNA-repair enzymes modify DNA damage caused by carcinogens in reactions that generally result in the removal of DNA adducts. Studies of cells from donors with xeroderma pigmentosum have been particularly important in expanding understanding of DNA excision repair and its possible relationship to risk of cancer. The rate, but not the fidelity, of DNA repair can be determined by measuring unscheduled DNA synthesis and removal of DNA adducts; substantial interindividual variation in DNA repair rates has been observed (Setlow, 1983). The fidelity of DNA repair could also vary among people, and recent advances in the identification of mammalian DNA-repair genes and their molecular mechanisms should soon provide an opportunity to investigate the fidelity of repair by excision. In addition to severe decreases in excision-repair rates in the cells of individuals with the recessive genetic conditions xeroderma pigmentosum cells, an approximately 5-fold variation among people in unscheduled DNA synthesis induced by UV exposure of lymphocytes in vitro has been found in the general population (Setlow, 1983). DNA repair might involve tens of enzymes and cofactors, and genetic polymorphisms of the genes encoding these repair enzymes could be responsible for the variation among both persons and groups.

Interindividual variation has been noted in the activity of O^6-alkyldeoxyguanine-DNA alkyltransferase; this enzyme repairs alkylation damage to O^6-deoxyguanine. Wide variations (a factor of about 40) in this DNA-repair activity have been observed between persons in different types of tissues (Grafstrom et al., 1984; D'Ambrosio et al., 1984, 1987), and fetal tissues exhibit only about 20-50% as much activity as the corresponding adult tissues (Myrnes et al., 1983).

A unimodal distribution of repair rates of benzo[a]pyrene diolepoxide-DNA adducts has been observed in human lymphocytes in vitro (Oesch et al., 1987). The interindividual variation was substantially greater than the intraindividual variation, and this suggests a role of inherited factors. The influence of those variations in DNA-repair rates in determining tissue site and risk of cancer in the general population remains to be determined.

SYNERGISTIC EFFECTS OF CARCINOGENS

People who have been exposed to one type of carcinogen might be at increased risk of cancer when exposed, simultaneously or in sequence, to another type (Table H-4). Cigarette smokers, already at greater risk of lung cancer than nonsmokers, are at even greater risk if they are occupationally exposed to asbestos (Selikoff and Hammond, 1975; Saracci, 1977) or radon (Archer, 1985). Recently, a synergistic effect between hepatitis B virus and aflatoxin B_1 in the risk of hepatocellular carcinoma has been described (Ross et al., 1992).

AGE

Children exposed to carcinogens might be at higher risk of cancer than adults (NRC, 1993; ILSI, 1992). Studies of atomic-bomb survivors and persons irradiated for the treatment of cancer have found the risk of future cancers of breast, lung, stomach, thyroid, and connective tissues to be greater when exposure is at lower ages (Fry, 1989). On the other hand, the elderly may be at increased susceptibility to other carcinogenic stimuli, cue to diminished immune surveillance, exposure to multiple drugs, or simply to a larger accumulation of DNA damage that places some cells at high risk of initiation from one more "hit" to the genetic material.

TABLE H-4   Examples of Synergistic Effects Among Chemical, Physical, and Viral Carcinogens.

Cancer Type	Carcinogens	Odds Ratio (95% Confidence Interval)	Reference
Liver	Hepatitis B virus + aflatoxin B_1 exposure	4.8 (1.2-19.7) 60 (6.4-561.8)	Ross et al., 1992
Esophagus	Tobacco smoke + alcoholic beverages	5.1 (–) 44.4 (–)	Tuyns et al., 1977
Mouth	Tobacco smoke + alcoholic beverages	2.4 (–) 15.5 (–)	Rothman and Keller, 1972
Lung	Tobacco smoke + occupational asbestos exposure	8.1 (5.2-12.0) 92.3 (59.2-137.4)	Selikoff and Hammond, 1975 Saracci, 1977

REFERENCES

Archer, V.E. 1985. Enhancement of lung cancer by cigarette smoking in uranium and other miners. Pp. 23-37 in Cancer of the Respiratory Tract: Predisposing Factors. Carcinogenesis—A Comprehensive Survey, Vol. 8, M.J. Mass, D.G. Kaufman, J.M. Siegfried, V.E. Steele, and S. Nesnow, eds. New York: Raven Press.

Ayesh, R., J.R. Idle, J.C. Ritchie, M.J. Crothers, and M.R. Hetzel. 1984. Metabolic oxidation phenotypes as markers for susceptibility to lung cancer. Nature 312:169-170.

Borysenko, J. 1987. Psychological variables. Pp. 295-313 in Variations in Susceptibility to Inhaled Pollutants: Identification, Mechanisms, and Policy Implications, J.D. Brain, B.D. Beck, A.J. Warren, and R.A. Shaikh, eds. Baltimore, Md.: The Johns Hopkins University Press.
Calabrese, E.J. 1978. Methodological Approaches to Deriving Environmental and Occupational Health Standards. New York: Wiley Interscience.

Caporaso, N.E., R.B. Hayes, M. Dosemeci, R. Hoover, R. Ayesh, M. Hetzel, and J. Idle. 1989. Lung cancer risk, occupational exposure, and the debrisoquine metabolic phenotype. Cancer Res. 49:3675-3679.

Caporaso, N.E., M.A. Tucker, R.N. Hoover, R.B. Hayes, L.W. Pickle, H.J. Issaq, G.M. Muschik, L. Green-Gallo, D. Buivys, S. Aisner, J.H. Resau, B.F. Trump, D. Tollerud, A. Weston, and C.C. Harris. 1990. Lung cancer and the debrisoquine metabolic phenotype. J. Natl. Cancer Inst. 82:1264-1272.

Cartwright, R.A., R.W. Glashan, H.J. Rogers, R.A. Ahmad, D. Barham-Hall, E. Higgins, and M.A. Kahn. 1982. The role of N-acetyltransferase phenotypes in bladder carcinogenesis: A pharmacogenetic epidemiological approach to bladder cancer. Lancet 2:842-846.

Cleaver, J.E. 1968. Defective repair replication of DNA in xeroderma pigmentosum. Nature 218:652-656.

D'Ambrosio, S.M., G. Wani, M. Samuel, and R.E. Gibson-D'Ambrosio. 1984. Repair of O^6-methylguanine in human fetal brain and skin cells in culture. Carcinogenesis 5:1657-1661.

D'Ambrosio, S.M., M.J. Samuel, T.A. Dutta-Choudhury, and A.A. Wani. 1987. O^6-methylguanine-DNA methyltransferase in human fetal tissues: Fetal and maternal factors. Cancer Res. 47:51-55.

Finkel, A. 1987. Uncertainty, Variability, and the Value of Information in Cancer Risk Assessment. D. Sc. Dissertation. Harvard School of Public Health, Harvard University, Cambridge, Mass.

Fraumeni, J.F., Jr., ed. 1975. Persons at High Risk of Cancer: An Approach to Cancer Etiology and Control. Proceedings of a conference, Key Biscayne, Florida, December 10-12, 1974. New York: Academic Press.

Friend, S.H., R. Bernards, S. Rogelj, R.A. Weinberg, J.M. Rapaport, D.M. Albert, and T.P. Dryja. 1986. A human DNA segment with properties of the gene that predisposes to retinoblastoma and osteosarcoma,. Nature 323:643-646.

Fry, R.J.M. 1989. Principles of carcinogenesis: Physical. Pp. 136-148 in Cancer: Principles and Practice of Oncology, Vol. 1, V.T. DeVita, S. Hellman, and S.A. Rosenberg, eds. Philadelphia: Lippincott.

Grafstrom, R.C., A.E. Pegg, B.F. Trump, and C.C. Harris. 1984. O^6-alkylguanine-DNA alkyltransferase activity in normal human tissues and cells. Cancer Res. 44:2855-2857.

Groden, J., A. Thliveris, W. Samowitz, M. Carlson, L. Gelbert, H. Albertsen, G. Joslyn, J. Stevens, L. Spirio, and M. Robertson 1991. Identification and characterization of the familial adenomatous polyposis coli gene. Cell 66:589-600.

Harris, C.C. 1989. Interindividual variation among humans in carcinogen metabolism, DNA adduct formation, and DNA repair. Carcinogenesis 10:1563-1566.

Hayashi, S.I., J. Watanabe, and K. Kawajiri. 1992. High susceptibility to lung cancer analyzed in terms of combined genotypes of P450IA1 and Mu-class glutathione S-transferase genes. Jpn. J. Cancer Res. 83:866-870.

Ilett, K.F., B.M. David, P. Detchon, W.M. Castleden, and R. Kwa. 1987. Acetylation phenotype in colorectal carcinoma. Cancer Res. 47:1466-1469.

ILSI (International Life Sciences Institute). 1992. Similarities and Differences Between Children and Adults: Implications for Risk Assessment, P.S. Guzelian, C.J. Henry, and S.S. Olin, eds. Washington, D.C.: ILSI Press.

Kinzler, K.W., M.C. Nilbert, K.L. Su, B. Vogelstein, T.M. Bryan, D.B. Levy, K.J. Smith, A.C. Preisinger, P. Hedge, D. McKechnie, R. Finniear, A. Markham, J. Groffen, M.S. Boguski, S.F. Altschul, A. Horii, H. Ando, Y. Miyoshi, Y. Miki, I. Nishisho, and Y. Nakamura. 1991. Identification of FAP locus genes from chromosome 5q21. Science 253:661-665.

Ladero, J.M., J.F. Gonz_lez, J. Ben_tez, E. Vargas, M.J. Fern_ndez, W. Baki, and M. Diaz-Rubio. 1991. Acetylator polymorphism in human colorectal carcinoma. Cancer Res. 51:2098-2100.

Lang, N.P., D.Z.J. Chu, C.F. Hunter, D.C. Kendall, T.J. Flammang, and F.F. Kadlubar. 1986. Role of aromatic amine acetyltransferase in human colorectal cancer. Arch. Surg. 121:1259-1261.

Myrnes, B., K.E. Giercksky, and H. Krokan. 1983. Interindividual variation in the activity of O^6-methylguanine-DNA methyltransferase and uracil-DNA glycosylase in human organs. Carcinogenesis 4:1565-1568.

Nakachi, K., K. Imai, S. Hayashi, J. Watanabe, and K. Kawajiri. 1991. Genetic susceptibility to squamous cell carcinoma of the lung in relation to cigarette smoking dose. Cancer Res. 51:5177-5180.

Nishisho, I., Y. Nakamura, Y. Miyoshi, Y. Miki, H. Ando, A. Horii, K. Koyama, J. Utsunomiya, S. Baba, P. Hedge, A. Markham, A.J. Krush, G. Petersen, S.R. Hamilton, M.C. Nilbert, D.B. Levy, T.M. Bryan, A.C. Preisinger, K.J. Smith, L. Su, K.W. Kinzler, and B. Vogelstein. 1991. Mutations of chromosome 5q21 genes in FAP and colorectal cancer patients. Science 253:665-669.

Nowak, D., U. Schmidt-Preuss, R. Jorres, F. Liebke, and H.W. Rudiger. 1988. Formation of DNA adducts and water-soluble metabolites of benzo[a]pyrene in human monocytes is genetically controlled. Int. J. Cancer 41:169-173.

NRC (National Research Council). 1993. Issues in Risk Assessment: I. Use of the Maximum Tolerated Dose in Animal Bioassays for Carcinogenicity. Washington, D.C.: National Academy Press.

Oesch, F., W. Aulmann, K.L. Platt, and G. Doerjer. 1987. Individual differences in DNA repair capacities in man. Arch. Toxicol. 10(Suppl.):172-179.

Orth, G. 1986. Epidermodysplasia verruciformis: A model for understanding the oncogenicity of human papillomaviruses. Ciba Found. Symp. 120:157-174.

Petruzzelli, S., A.M. Camus, L. Carrozzi, L. Ghelarducci, M. Rindi, G. Menconi, C.A. Angeletti, M. Ahotupa, E. Hietanen, A. Aitio, R. Saracci, H. Bartsch, and C. Giuntini. 1988. Long-lasting effects of tobacco smoking on pulmonary drug-metabolizing enzymes: A case-control study on lung cancer patients. Cancer Res. 48:4695-4700.

Ross, R.K., J.M. Yuan, M.C. Yu, G.N. Wogan, G.S. Qian, J.T. Tu, J.D. Groopman, Y.T. Gao, and B.E. Henderson. 1992. Urinary aflatoxin biomarkers and risk of hepatocellular carcinoma. Lancet 339:943-946.

Rothman, K., and A. Keller. 1972. The effect of joint exposure to alcohol and tobacco on risk of cancer of the mouth and pharynx. J. Chronic Dis. 25:711-716.

Saracci, R. 1977. Asbestos and lung cancer: An analysis of the epidemiological evidence on the asbestos-smoking interaction. Int. J. Cancer 20:323-331.

Seidegard, J., R.W. Pero, D.G. Miller, and E.J. Beattie. 1986. A glutathione transferase in human leukocytes as a marker for the susceptibility to lung cancer. Carcinogenesis 7:751-753.

Seidegard, J., R.W. Pero, M.M. Markowitz, G. Roush, D.G. Miller, and E.J. Beattie. 1990. Isoenzyme(s) of glutathione transferase (class Mu) as a marker for the susceptibility to lung cancer: A follow up study. Carcinogenesis 11:33-36.

Selikoff, I.J., and E.C. Hammond. 1975. Multiple risk factors in environmental cancer. Pp. 467-484 in Persons at High Risk of Cancer: An Approach to Cancer Etiology and Control, J.F. Fraumeni, ed. New York: Academic Press.

Setlow, R.B. 1983. Variations in DNA repair among humans. Pp. 231-254 in Human Carcinogenesis, C.C. Harris and H.N. Autrup, eds. New York: Academic Press.

Sugimura, H., N.E. Caporaso, G.L. Shaw, R.V. Modali, F.J. Gonzalez, R.N. Hoover, J.H. Resau, B.F. Trump, A. Weston, and C.C. Harris. 1990. Human debrisoquine hydroxylase gene polymorphisms in cancer patients and controls. Carcinogenesis 11:1527-1530.

Swift, M., D. Morrell, R.B. Massey, and C.L. Chase. 1991. Incidence of cancer in 161 families affected by ataxia-telangiectasia. N. Engl. J. Med. 325:1831-1836.

Tuyns, A.J., G. P_quignot, and O.M. Jensen. 1977. Esophageal cancer in Ille-et-Vilaine in relation to levels of alcohol and tobacco consumption. Risks are multiplying. Bull. Cancer (Paris) 64:45-60.

Warren, A.J., and S. Weinstock. 1988. Age and preexisting disease. Pp. 253-268 in Variations in Susceptibility to Inhaled Pollutants: Identification, Mechanisms, and Policy Implications, J.D. Brain, B.D. Beck, A.J. Warren, and R.A. Shaikh, eds. Baltimore, Md.: The Johns Hopkins University Press.

I

This appendix is split into three parts. The first discusses aggregate risk of occurrence of one or more nonthreshold, quantal, toxic end points caused by exposure to multiple agents (assuming independent actions). The second is a summary assessment of independence in interanimal tumor-type occurrence in the NTP rodent-bioassay database. The third discusses methods for aggregating uncertainty and interindividual variability in predicted risk.

APPENDIX
I-1

Aggregate Risk of Nonthreshold, Quantal, Toxic End Points Caused by Exposure to Multiple Agents (Assuming Independent Actions)

The aggregate increased probability P of occurrence of any of n (presumed) nonthreshold end points caused by exposure to an environmental mixture of m toxic agents may be conveniently expressed under a few general assumptions. First, assume that the m agents are present in an environmental mixture at corresponding concentrations C_i, where $i=1,2,...,m$, each of which produce, in exposed people, corresponding lifetime, time-weighted average biologically effective dose rates D_{ij}, each causing one or more of n quantal (all or none) toxic end points T_j, where $j=1,2,...,n$ (see Figure I-1). Let O_{ij} denote the occurrence of a particular jth end point T_j induced by effective dose rate D_{ij}, and assume that T_j has a background occurrence probability of $p_j = \text{Prob}(O_{ij} | D=0)$ for total effective dose D due to all relevant agents and that O_{ij} may arise only by events independent of those giving rise to either the background incidence rate of T_j or to events O_{gh} for any g and h such that $g \neq i$, $1 \leq g \leq m$, $h \neq j$, and $1 \leq h \leq n$. Finally, for very small values of D_{ij}, assume that the corresponding increased probability of occurrence of the T_j is defined by an independent "one-hit" (nonthreshold, low-dose linear) function of D_{ij}. In the following, $\cup$, $\cap$, and the overbar denote the logical union, intersection, and negation operations, respectively.

It follows from the stated assumptions and definitions that a D_{ij}-induced increased probability P_{ij} of T_j occurrence, conditional on its independent background rate p_j, is:

$$P_{ij} = \text{Prob}(O_{ij}) = \frac{\text{Prob}(T_j | D = D_{ij}) - p_j}{1 - p_j}, \tag{1}$$

m Agents in Mixture   n Toxic Endpoints

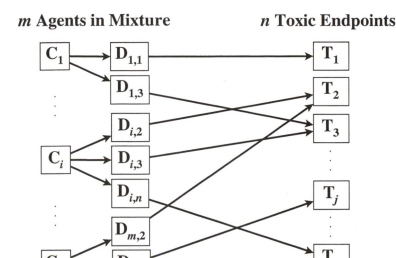

Key: C = Concentration of agent, D = Effective dose in target tissue, T = Toxic endpoint.

FIGURE I-1   Multiple agents associated with multiple toxic end points.

in which q_{ij} (the linear coefficient in dose) is the parameter characterizing the "potency" (or low-dose increased occurrence probability per unit dose) of compound i for inducing end point T_j. Under the stated assumptions, $P_{ij}=q_{ij}=0$ for any jth end point T_j that is unaffected by D_{ij} alone, regardless of concurrent doses from any other agents. The quantity of interest—aggregate increased probability P of occurrence of any of the n end points caused by any of the m toxic agents—may therefore be expressed as

$$P = \text{Prob}\left(O_{1,1} \cup O_{1,2} \cup \ldots \cup O_{1,n} \cup O_{2,1} \cup \ldots \cup O_{m,1} \cup \ldots \cup O_{m,n}\right), \quad (2)$$

which, by de Morgan's rule, may be rewritten as

$$P = \text{Prob}\left(\overline{O}_{1,1} \cap \overline{O}_{1,2} \cap \ldots \cap \overline{O}_{1,n} \cap \overline{O}_{2,1} \cap \ldots \cap \overline{O}_{m,1} \cap \ldots \cap \overline{O}_{m,n}\right), \quad (3)$$

from which, by Equation 1 and the independence assumption, it follows that

$$P = 1 - \prod_{i=1}^{m} \prod_{j=1}^{n} \left(1 - P_{ij}\right), \quad (4)$$

$$P = 1 - \exp\left(-\sum_{i=1}^{m} \sum_{j=1}^{n} q_{ij} D_{ij}\right). \quad (5)$$

For very small values of P ($\ll 1$) relevant to environmental regulatory concern, P is well approximated by

$$P \approx \sum_{i=1}^{m} \sum_{j=1}^{n} q_{ij} D_{ij} \; . \tag{6}$$

If no information is available concerning target-tissue-specific pharmacokinetics, D_{ij} is sometimes taken to be either the absorbed dose rate (e.g., milligrams of agent i absorbed per kilogram of body weight per day) or a whole-body surrogate for effective dose (e.g., estimated milligrams of agent i metabolized per kilogram of body weight per day), that is, a measure of dose identical for all of the particular toxic end point(s) considered. In this case, $D_{ij} = D_i$ is independent of j for any given ith agent, such that Equation 6 may be rewritten as

$$P \approx \sum_{i=1}^{m} Q_i D_i \; , \tag{7}$$

in which Q_i is the sum of q_{ij} for j ranging from 1 to n and represents the aggregate potency of agent i for inducing at least one of the n end points considered.

When applying relations like those represented by Equations 5-7, q, Q, D, and hence P may represent quantities subject to uncertainty or interindividual variability characterized by different probability distributions. If distributed variates are involved, a meaningful confidence bound on P cannot generally be obtained by performing the indicated summations with the same bound on all values of q, Q, and D. In the special case that, say, Q_i and D_i in Equation 7 are all independent and m is sufficiently large, the estimate of P will tend to be normally distributed; however, asymptotic normality is not likely to be useful in situations involving relatively small m and n. If a statistical upper confidence bound is desired is desired for P, Monte Carlo procedures will therefore generally be needed.

I-2

Independence in Inter-Animal Tumor-Type Occurrence in the NTP Rodent-Bioassay Database

Animal cancer bioassay data have been used as the basis for estimating carcinogenic potency (i.e., increased risk per unit dose at very low doses) of a chemical to which a human of average cancer susceptibility might be exposed over a lifetime (Anderson et al., 1983; EPA, 1986, 1992). The bioassay data available may indicate that multiple tumor types are induced in exposed bioassay animals. In this case, it is generally desired to estimate the aggregate cancer potency exhibited by the compound in the bioassay animals, that is, the effectiveness in the experimental animals of the compound in eliciting any one or more of the elevated tumor types. The estimated aggregate cancer potency in bioassay animals may then be used to extrapolate a corresponding potency of that compound in a human of average susceptibility (EPA, 1986, 1992). Neither this interspecies extrapolation nor the issue of human interindividual variability in cancer susceptibility (discussed in Chapter 10) are the subject of this appendix (I-2). Rather, this appendix focuses on the extent of tumor-type correlations in bioassay animals, which in turn bears on the question of how properly to estimate the aggregate cancer potency of a compound exhibited in bioassay animals for a compound that induces multiple tumor types.

One approach to estimating aggregate cancer potency in bioassay animals has been to apply a dose-response model to tumor-incidence rates with the numerators defined as the number of animals with one or more of the histologically distinct and significantly elevated tumor types (EPA, 1986). By this procedure, either a control or a dosed animal with multiple tumor types counts the same as an animal with only a single tumor type. If the tumor types occur in a statistically independent fashion among the bioassay animals tested, it follows that this procedure may under- or over-estimate true aggregate potency because it has the effect of randomly excluding tumor-response information concerning both control and dosed animals (Bogen, 1990).

If potency is estimated using a multistage model (which is in effect a one-hit model at very low doses), and if tumor types assort independently among the animals tested, the statistical problem raised by EPA's tumor-pooling approach is avoided completely if aggregate potency is instead estimated as the sum of tumor-type-specific (that is, independent-end-point-specific) potencies (see Appendix I-1). This alternative to EPA's procedure, however, depends on the validity of the independence assumption regarding tumor-type occurrence within bioassay animals, which is the subject of this appendix (I-2).

In some of the few studies that have focused on tumor-type associations within individual animals, a few significant associations have been noted, mostly negative associations involving one or two specific tumor types among associated pairs. Significant ($p<0.05$) age- and treatment-adjusted associations for five of 21 sex-specific pairs from six tumor types investigated were reported by Breslow et al. (1974) for experiments involving over 4000 CF-1 mice exposed to DDT, urethane or nothing: negative associations between lymphomas and each of hepatomas (males), lung adenomas (males and females), and mammary and ovarian tumors (females), and a positive association between lymphomas and bone tumors (males). (Upon adjustment for multiple significance tests (Wright, 1992), the association between lymphomas and mammary tumors observed in that study may not be significant at the 0.05 level.) Breslow et al. (1974) suggested that the negative lymphoma-related associations, except perhaps those involving liver tumors, were all likely to be spurious, "due to the relative rapidity with which lymphomas tend to kill their bearers." A significant negative association between lymphomas and liver tumors (but not lung tumors) in 1478 similarly exposed CF-1 mice was later confirmed, even after accounting for the relatively rapid lymphoma lethality by use of serial sacrifice information (Wahrendorf, 1983). A significant negative correlation between malignant lymphoma and proliferative hepatocellular lesions at death/sacrifice was also found among 1858 male ICI mice (Young and Gries, 1984). Haseman (1983) also noted this significant negative correlation in raw tumor-incidence data for F344 rats from 25 National Toxicology Program (NTP) bioassays (not analyzed at the level of individual animals).

The most comprehensive study of this type, involving an examination of age- and treatment-adjusted associations between (66 possible) pairs of 12 tumor types at death/sacrifice in 3813 gamma-irradiated female BALB/c mice, reported 21 significant ($p<0.05$) positive or negative associations, 10 of which were negative and involved reticular tumors considered to be rapidly lethal and generally also involved other tumors considered to be lethal in the animals studied; most of these 10 associations were considered to be spurious due to the effect of lethality (Storer, 1982). The remaining associations considered significant generally were positive and involved endocrine-related tumors (Harderian, mammary, adrenal, and pituitary tumors), and none of these involved liver tumors. Aside from associations involving reticular sarcomas, and after appropriate statistical adjustment for multiple tests of significance (Wright, 1992), only three of 55 remaining possible associations reported by Storer (1982) appear to be significant at a 0.05 level, all involving Harderian-gland tumors, which, along with ovary, adrenal and pituitary tumors, were all considered to be nonlethal in the animals studied. A recent study of liver-tumor and reticulum-cell-sarcoma incidence in 1004 gamma-irradiated female C3H mice supported a significant negative correlation of these tumor types, even after adjustment for the relative lethality of the reticular tumors using cause-of-death information available in that study (Mitchell and Turnbull, 1990).

In other smaller studies, an assumption of independence in tumor-types at death/sacrifice was shown to be consistent with ED01 data on four different tumor types in 366 control female BALB/c mice and six tumor types elevated in 193 such mice exposed to 2-acetylaminofluorine (Finkelstein and Schoenfeld, 1989), as well as with Hazelton Laboratory data on three different tumor types elevated in a total of 142 male albino rats exposed to dibromochloropropane (Bogen, 1990).

No comprehensive study of animal-specific tumor-type occurrences at death/sacrifice has been conducted using the extensive set of available NTP rodent-bioassay data, on which most cancer-potency assessment for environmental chemicals is currently based. This report presents the results of such an analysis (Bogen and Seilkop, 1993) conducted on behalf of the National Research Council's Committee on Risk Assessment for Hazardous Air Pollutants.

DATA DESCRIPTION

Tumor-type associations among individual animals were examined for both control and treated animals using pathology data from 62 B6C3F1 mouse studies and 61 F/344N rat studies obtained from a readily available subset of the NTP carcinogenesis bioassay database. Most studies were 2-year studies, although a few were shorter (e.g., 15 months). Separate analyses were conducted for the four sex/species combinations (male and female mice, male and female rats) corresponding to the compounds and species indicated in Table I-1. Analysis was confined to the following common tumor types (occurring at a rate >5%):

Rats:	Adrenal gland:	medulla pheochromocytomas (benign or malignant)
	Thyroid gland:	C-cell adenomas or carcinomas
	Pituitary gland:	carcinomas or adenomas
	Mammary gland:	fibromas, fibroadenomas, carcinomas, or adenomas
	Leukemia:	lymphocytic, monocytic, mononuclear, or undifferentiated.

Mice: Lung: alveolar/bronchiolar adenomas or
                           carcinomas
            Liver: hepatocellular adenomas, hepatocellular
                           carcinomas, and hepatoblastomas
            Lymphoma: histiocytic, lymphocytic, mixed, NOS, or
                           undifferentiated.

TABLE I-1   NTP Studies from Which Data Were Used

CHEMICAL	MICE[a]	RATS[a]
O-Chlorobenzalmalononitrile (CS-2)		X
1,2,3-Trichloropropane	X[b]	X[c]
1,3-Butadiene (butadiene)	X	
2,4-Diaminophenol Dihydrochloride	X	X
2,4-Dichlorophenol	X	X
3,3'-Diemthoxybenzidine Dihydrochloride		X[c,d]
3,3'-Dimethylbenzidine Dihydrochloride		X[c,d]
4,4-Diamino-d,d-Stilbenedisulfonic Acid	X	X
4-Hydroxyacetanilide	X	X
4-Vinyl-1-cyclohexene Diepoxide	X	X
Allyl glycidyl ether	X	
Benzaldehyde	X	X
Benzyl Acetate	X	X
Bromoform	X	X
γ-Butyrolactone	X	X
C.I. Acid Red 114		X[c,d]
C.I. Direct Blue 15		X[c,d]
Carvone	X	
Chloramine	X	X
Chloroacetophenone	X	X
p-Chloroaniline	X	X
p-Chlorobenzalmalononitrile (CS-2)	X	
C.I. Pigment Red 23	X	X
C.I. Pigment Red 3	X	X
Coumarin	X	X
DL-Amphetamine sulfate	X	X
Dichlorvos	X	X
Dihydrocoumarin	X	X
Dimethoxane	X	X
Diphenylhydantoin	X	X
Ephinephrine HCl	X	X
Ethyl chloride	X	X
Ethylene glycol	X	
Ethylenethiourea	X	X
Firemaster FF-1 Polybrominated Biphenyl	X	X
Furan	X[e]	X
Furfural	X	X
HC Yellow 4	X	X

TABLE I-1   Continued

CHEMICAL	MICE[a]	RATS[a]
Hexachloroethane		X
Hydroquinone	X	X
Managanese sulfate	X	X
Mercuric chloride	X	X
Methyl bromide	X	
Monochloroacetic acid	X	X
N-Methylolacrylamide	X[b]	X
Naphthalene	X	
p-Nitroaniline	X	
Nitrofurantoin	X	X
o-Chlorobenzalmalononitrile (CS-2)	X	
o-Nitroanisole	X	X
Ochratoxin A		X
p-Nitroaniline	X	
Pentachloroanisole	X	X
Pentachlorophenol, Dowicide CD-7	X[e]	
Pentachlorophenol, Technical grade	X	
Pentaerythritol Tetranitrate	X	X
Phenylbutazone	X	X
Polysorbate 80	X	X
Probenecid	X	X
Quercetin		X
Resorcinol	X	X
Rhodamine 6G	X	X
Roxarsone	X	X
Sodium Azide		X
Sodium Fluoride	X	X
Succinic Anhydride		X
Talc	X	
Tetranitromethane	X	X
Titanocene dichloride		X
Toluene	X	X
Triamterene	X	X
Tris(2-chloroethyl) phosphate	X	X
Vinyl toluene	X	X

[a]NTP studies from which treated-animal data, in addition to control-animal data, were taken for use in this analysis are indicated by a superscript.

[b]Liver and Harderian-gland effects in treated animals.

[c]Zymbal-gland and clitoral or preputial-gland effects in treated animals.

[d]Liver and skin effects in treated animals.

[e]Liver and adrenal-gland effects in treated animals.

Source: Bogen and Seilkop, 1993.

Analyses of correlations between tumor occurrence in treated animals were based on subsets of the control-animal data, comprising studies for which the NTP declared "clear evidence" of an effect at multiple sites and for which pairs of such effects were exhibited in more than one study (resulting in the use of five rat studies and four mouse studies). The treated animal studies involved tumor types that differed from the control-animal studies, namely, adenomas or carcinomas of: liver, Zymbal's gland, clitoral /preputial gland, and skin in rats; and liver, adrenal and Harderian gland in mice. In both control- and treated-animal analyses, evidence of associations from individual studies were pooled as described below.

STATISTICAL METHODS

Associations among statistically significantly elevated tumor-types within individual animals may pertain either to tumor onset probabilities or to prevalence at death/sacrifice or to both. It is well known that associations present at death/sacrifice may differ, sometimes substantially, from those relating to tumor onset, and that the former may be heavily influenced by the latter as a result of the time-dependent action of competing risks (Hoel and Walburg, 1972; Breslow et al., 1974; Wahrendorf, 1983; Lagakos and Ryan, 1985). For example, if the onset probabilities of two different tumor types are statistically independent, but in addition both are rapidly lethal, then there is little probability of their joint occurrence within an individual animal and thus their prevalence at death/sacrifice will be negatively correlated. This fact was the basis for concluding probable "spurious" negative correlations involving rapidly lethal tumor types in previous assessments of tumor-type associations in rodents (Breslow et al., 1974; Storer, 1972).

Unambiguous detection of associations in onsets of different tumor types requires either serial-sacrifice information or animal- and tumor-specific lethality information (Hoel and Walburg, 1972; Wahrendorf, 1983; Lagakos and Ryan, 1985; Mitchell and Turnbull, 1990), neither of which is available for the NTP data analyzed here. Thus, the present analysis was primarily restricted to an assessment of age-adjusted correlations in tumor-types present at death/sacrifice. This approach provides definitive information on onset (as well as terminal prevalence) correlations only if all tumor types are incidental to fatality. However, as described below, a crude assessment of onset-probability correlations was also conducted using information on tumor lethality obtained from the data studied.

Evaluation of the correlations between occurrences between pairs of tumor types in individual animals observed at death/sacrifice was based on age-adjustment of information from 24 previous similar studies (Breslow et al., 1974; Storer, 1982; Young and Gries, 1984; Finkelstein and Schoenfeld, 1989). Five survival-age strata within each study were used: (1) first 365 days, (2) 366-546 days (1.5 years), (3) 547-644 days (~1.75 years), (4) 644-terminal sacrifice (~2 years), (5) terminal sacrifice. Further stratification addressed the inclusion of the highest two dose groups. Thus, the potential number of analytical strata (i.e., 24 times 2 (the number of dose levels). The method of Mantel and Haenszel (1959) was used to combine results from stratum-specific contingency tables and to assess two-tailed significance of overall associations between tumor occurrences. Overall correlations are represented as the weighted averages of corresponding stratum-specific measures, using the numbers of animals in the strata as weights. Adjusted p-values accounting for multiple tests of a zero-correlation null hypothesis were obtained for all control and all treated rats and mice using Hommel's modified Bonferroni procedure (Wright, 1992).

In the absence of serial sacrifice or lethality information, associations between onsets of pairs of tumor types in individual NTP-bioassay animals were evaluated using two crude techniques. First, a separate correlation analysis was undertaken as above, but using only terminal sacrifice data. This approach provides definitive information on onset (as well as terminal prevalence) correlations only if no animals die prior to terminal sacrifice, but may nevertheless provide meaningful information if a sufficiently large fraction of animals survive until sacrifice. The second approach used was the three-by-three contingency-table method for detection of disease-onset associations devised by Mitchell and Turnbull (1990), which requires lethality determinations for each tumor occurrence in each animal. When in doubt regarding such lethality, Mitchell and Turnbull (1990) recommend that it would be prudent to classify a particular occurrence as lethal, because while doing so falsely may reduce the power of the test, the null distribution will not be affected. Thus, the Mitchell-Turnbull test was applied under the assumption that all occurrences of a given tumor type were lethal for all plausibly lethal tumor types. Tumor-type lethality was investigated using Mann-Whitney U statistics comparing survival times of tumor-bearing and tumor-free animals, where all study-specific results for a given control or treated species and sex were combined to form an overall test by summing these U statistics and dividing this sum by the square root of the sum of the corresponding variances.

RESULTS AND DISCUSSION

The results of our analysis of correlations in incidence at death/sacrifice of tumor types in control rats and mice are summarized in Table I-2. These results indicate four significant ($p^* < 0.05$) but small correlations among 20 sex/tumor-type-pairs investigated in rats (pituitary vs. leukemia in both sexes, and mammary vs. leukemia or pituitary in females—where all those involving leukemia were negative), and no similarly significant correlations among 12 sex/tumor-type-pairs investigated in mice. Corresponding results for treated rats and mice are summarized in Table I-3. Significant ($p^* < 0.05$) but again generally quite small correlations appear present for two of 12 sex/tumor-type-pairs investigated

TABLE I-2   Correlations Between Tumor Prevalence at Death/Sacrifice in Control Groups

SPECIES Tumor Types	Sex	Corr.	n	p-value	Adjusted p*-value
RATS					
Adrenal x Leukemia	Females	0.060	2794	0.017	0.272
	Males	0.025	2786	0.257	1
Adrenal x Thyroid	Females	0.041	2692	0.138	1
	Males	−0.024	2593	0.342	1
Thyroid x Leukemia	Fernales	−0.032	2942	0.120	1
	Males	−0.045	2827	0.076	1
Pituitary x Leukemia	Females	−0.158	3057	C0.001	<0.020
	Males	−0.080	2990	<0.001	<0.020
Mammary x Leukemia	Females	−0.074	3088	<0.001	<0.020
	Males	−0.025	3045	< 0.001	1
Mammary x Pituitary	Females	0.076	3057	<0.001	<0.020
	Males	0.027	2990	0.301	1
Pituitary x Thyroid	Females	−0.002	2916	0.982	1
	Males	0.026	2784	0.254	1
Pituitary x Adrenal	Females	−0.029	2770	0.268	1
	Males	−0.010	2739	0.659	1
Mammary x Adrenal	Females	−0.015	2794	0.597	1
	Males	0.008	2786	0.835	1
Mammary x Thyroid	Females	−0.011	2942	0.642	1
	Males	0.008	2827	0.846	1
MICE					
Liver x Lung	Females	−0.003	3058	0.978	0.204
	Males	−0.022	3011	0.322	1
Liver x Lymphoma	Females	−0.029	3059	0.185	1
	Males	−0.053	3014	0.017	0.204
Lung x Lymphoma	Females	−0.054	3071	0.018	0.204
	Males	−0.008	3016	0.791	1
Pituitary x Lung	Females	0.014	2898	0.592	1
	Males	0.025	2725	0.879	1
Pituitary x Liver	Females	0.020	2891	0.393	1
	Males	−0.074	2724	0.307	1
Pituitary x Lymphoma	Females	−0.041	2899	0.058	0.580
	Males	0.011	2727	0.806	1

Source: Bogen and Seilkop, 1993.

TABLE I-3   Correlations Between Tumor Prevalence at Death/Sacrifice in
Chemically Affected Groups

SPECIES Tumor Types	Sex	Corr.	n	p-value	Adjusted p~value
RATS					
Liver x Zymbal's Gland	Females	0.005	498	0.927	0.961
	Males	0.004	499	0.961	0.961
Zymbal's Gland x	Females	−0.117	590	0.012	0.12
Clitoral/Preputial Gland	Males	−0.152	577	0.003	0.033
Skin x Zymbal's Gland	Females	−0.065	500	0.272	0.961
	Males	−0.071	500	0.213	0.961
Liver x Skin	Females	0.041	498	0.630	1
	Males	0.172	498	0.002	0.24
Liver x Clitoral/	Females	0.034	488	0.658	1
Preputial Gland	Males	−0.078	487	0.187	1
Skin x Clitoral/	Females	0.023	489	0.762	1
Preputial Gland	Males	0.027	487	0.735	1
MICE					
Liver x Adrenal Gland	Females	0.153	194	0.245	0.49
	Males	0.257	196	0.076	0.228
Liver x Harderian Gland	Females	0.236	190	0.004	0.016
	Males	0.024	191	0.889	0.889

Source: Bogen and Seilkop, 1993.

in treated rats (Zymbal's vs.preputial gland and liver vs. skin tumors in males)
and for one of four sex/tumor-type-pairs investigated in treated mice (liver vs.
Harderian gland in females), where the liver-related correlations were both positive.

Terminal-sacrifice animals represented 66 to 68% of all the control mice
and 53 to 63% of all control rats referred to in Table I-2. Analysis of tumor-
type-prevalence correlations in only these animals revealed only a single significant ($p^* < 0.05$) correlation, that between mammary and pituitary tumors in
female rats ($r=0.080$, $p^*=0.013$). Thus, the latter positive (albeit quite small)
correlation may pertain to onset as well as prevalence-at-death/sacrifice correla-

tions, whereas the negative leukemia-related correlations noted above for all control rats did not persist in terminal-sacrifice animals. This finding could be explained by relative lethality associated with rodent leukemia/lymphoma, which has been noted in previous studies (Breslow et al., 1974; Wahrendorf, 1983; Young and Gries, 1984; Portier et al., 1986). Terminal-sacrifice animals represented only 14 to 16% of all the treated rats and 20 to 55% of all treated mice referred to in Table I-3. Correlation analyses for these treated animals yielded no significant (p* < 0.05) correlations, which sheds less light on tumor-onset associations given the greater non-representativeness of these animals.

Our examination of differences in survival time in animals with particular tumors vs. tumor-free animals revealed a few significant differences in control and treated rats. Leukemia in both sexes of control F344 rats studied was associated with a significant reduction in mean survival time (p<0.001). However, this reduction was rather modest: 75% of leukemia -bearing animals lived until the 23rd month of the studies and 50% lived until terminal sacrifice. In contrast, 75% of the leukemia-free animals survived until terminal sacrifice. Thus, any effect of leukemia lethality in inducing negative correlations with other cancers is likely to be small.

There was also evidence that Zymbal's gland tumors in treated rats resulted in reduced survival times (males, p<0.001; females, p=0.003), where the median survival times were reduced by about four months in males (546 vs. 427 days— reduction for more striking than that for leukemia in control males) and by about one month in females. When leukemia and Zymbal's gland tumors in animals dying before terminal sacrifice were assumed to be lethal and all other tumor types incidental, the Mitchell-Turnbull test yielded similar results to those obtained using the unmodified age-stratified analysis. In particular, it provided strong evidence that the small, negative associations between leukemia and pituitary-gland tumors in control rats were not due to chance or to differential lethality (males, p<10-9; females, p=0.000057), and it indicated the same regarding the small, negative associations between Zymbal's-gland tumors and preputial-/clitoral-gland tumors in treated rats ((males, p=0.009; females, p=0.002).

In summary, no evidence was found for any large correlation in either the onset probability or the prevalence-at-death/sacrifice of any tumor-type pair investigated in control and treated rats and mice, although a few of the small correlations present were statistically significant. This finding must be qualified to the extent that tumor-type onset correlations were measured indirectly given the limited nature of the data analyzed. Taken together, these findings indicate that tumor-type occurrences in B6C3F1 mice and F344 rats used in the NTP bioassays analyzed were in most cases nearly independent, and that departures from independence, where they did occur, were small.

REFERENCES

Anderson, E.L., and the Carcinogen Assessment Group of the U.S. Environmental Protection Agency. 1983. Quantitative approaches in use to assess cancer risk. Risk Anal. 3:277-295.

Bogen, K.T. 1990. Uncertainty in Environmental Health Risk Assessment. New York: Garland Publishing.

Bogen, K.T., and S. Seilkop. 1993. Investigation of Independence in Inter-animal Tumor type Occurrence within the NTP Rodent-Bioassay Database. Report prepared for the National Research Council Committee on Risk Assessment for Hazardous Air Pollutants, Washington, D.C.

Breslow, N.E., N.E. Day, L. Tomatis, and V.S. Turusov. 1974. Associations between tumor types in a large-scale carcinogensis study. J. Natl. Cancer Inst. 52:233-239.

EPA (U.S. Environmental Protection Agency). 1986. Guidelines for carcinogen risk assessment. Fed. Regist. 51(Sept. 24):33992-34003.

EPA (U.S. Environmental Protection Agency). 1987. Risk Assessment Guidelines of 1986. EPA/600/8-87-045. U.S. Environmental Protection Agency, Washington, D.C.

EPA (U.S. Environmental Protection Agency). 1992. Guidelines for exposure assessment. Fed. Regist. 57(May 29):22888-22938.

Finkelstein, D.M. and D.A. Schoenfeld. 1989. Analysis of multiple tumor data for a rodent experiment. Biometrics 45:219-230.

Haseman, J.K. 1983. Patterns of tumor incidence in two-year cancer bioassay feeding studies in Fischer rates. Fundam. Appl. Toxicol. 3:1-9.

Hoel, D.G., and H.E. Walburg, Jr. 1972. Statistical analysis of survival experiments. J. Natl. Cancer Inst. 49:361-372.

Lagakos, S.W., and L.M. Ryan. 1985. Statistical analysis of disease onset and lifetime data from tumorigencity experiments. Environ. Health Perspect. 63:211-216.

Mantel, N., and W. Haenszel. 1959. Statistical aspects of the analysis of data from retrospective studies of disease. J. Natl. Cancer Inst. 22:719-748.

Mitchell, T.J., and B.W. Turnbull. 1990. Detection of associations between diseases in animal carcinogenicity experiments. Biometrics 46:359-374.

Portier, C.J., J.C. Hedges and D.G. Hoel. 1986. Age-specific models of mortality and tumor onset for historical control animals in the National Toxicology Program's carcinogenicity experiments. Cancer Res. 46:4372-4378.

Storer, J.B. 1982. Associations between tumor types in irradiated BALB/c female mice. Radiation Res. 92:396-404.

Wahrendorf, J. 1983. Simultaneous analysis of different tumor types in a long-term carcinogenicity study with scheduled sacrifices. J. Natl. Cancer Inst. 70:915-921.

Wright, S.P. 1992. Adjusted p-values for simultaneous inference. Biometrics 48:1005-1013.

Young, S.S., and C.L. Gries. 1984. Exploration of the negative correlation between proliferative hepatocellular lesions and lymphoma in rats and mice-establishment and implications. Fundam. Appl. Toxicol. 4:632-640.

I-3

Aggregation of Uncertainty and Variability

This appendix illustrates why a distinction between uncertainty and interindividual variability within input variates must be maintained, if a quantitative characterization of uncertainty in population risk or in individual risk is sought. Two types of mathematical model used to predict risk are considered here for an exposed population of size n. The first model is a simple one in which a predicted low level of exposure-related increased risk R is well approximated by the product of U (a purely uncertain variate) and V (a purely heterogeneous variate that models interindividual variability).

$$
\begin{aligned}
R &= 1 - \exp\{-(U_1\, U_2\, U_3)\, (V_1\, V_2\, V_3)\} \\
&\approx (U_1\, U_2\, U_3)\, (V_1\, V_2\, V_3) \text{ for } R \ll 1, \\
&\approx UV \text{ for } R \ll 1,
\end{aligned}
\tag{1}
$$

where U_i and V_i represent uncertain and heterogeneous variates, respectively, for $i = 1,2,3$. That is, for a given value of i, V_i models the set of n particular (known or assumed) quantities pertaining to n individuals in the population at risk, whereas U_i models (in this case, using a single, uncertain multiplicative factor) the uncertainty associated with each one of those n quantities; this type of distinction is explained further by Bogen and Spear (1987) and Bogen (1990). In the present simple model, for example, U_1 and V_1 might refer to lifetime time-weighted average exposure, U_2 and V_2 to biologically effective dose per unit exposure, and U_3 and V_3 to cancer "potency" (increased cancer risk per unit biologically effective dose as dose approaches zero). In this case, V_3 would model interindividual variability in susceptibility to dose-induced cancer.

A more complicated risk model assumes that risk R equals some more general function $H(\mathbf{U},\mathbf{V})$ of the vectors $\mathbf{U}$ and $\mathbf{V}$ of purely uncertain and purely heterogeneous variates, respectively. In the following discussion, an overbar denotes the expectation operation with respect to all heterogeneous variates ($\mathbf{V}$) associated with the overbarred quantity and angle-brackets, $\langle \, \rangle$, shall denote the expectation operation with respect to all uncertain variates ($\mathbf{U}$) associated with the bracketed quantity (that is, $\overline{R} = \mathrm{E}_\mathbf{V}(R)$ and $\langle R \rangle = \mathrm{E}_\mathbf{U}(R)$, where E is the expectation operator). Also, $F_X(x)$ shall denote the cumulative probability that $X \leq x$, for some particular value x of any given variate X.

POPULATION RISK

Population risk, N, is the number of additional cases associated with predicted risk R. By definition, N is an uncertain variate, not a heterogeneous one. Uncertainty in N, however, is often ignored under the assumption that it is necessarily small in relation to the expected value of N for large n. For example, in its recent radionuclide-NESHAPS uncertainty analysis, EPA (1989, p. 7-6) stated that

> Because population risks represent the sum of individual risks, uncertainties in the individual risks tend to cancel each other out during the summing process. As a result, the uncertainty in estimates of population risk is smaller than the uncertainty in the estimates of the risks associated with the individual members of the population. Because of this, [our] uncertainty analysis is limited to the uncertainty in risks to an individual.

This assumption is clearly false, as is demonstrated by a comparison of the case (a) of n identical but extremely uncertain individual risks with the case (b) of n identical individual risks all equal to the known constant (i.e., completely certain value) r, for large n. Uncertainty in population risk in case (a) must remain extremely large independent of n, whereas in case (b) the cumulative probability distribution function (cdf) for the ratio N/n is simply a normalized binomial distribution that has smaller and smaller variances around the true value r as $n \rightarrow \infty$. The key point is that in the relationship between n uncertain individual risks and the corresponding uncertain population risk, many of the *uncertain* characteristics of each of the individual risks are not independent, but rather reflect quantities such as potency-parameter estimation error or model-specification error that pertain *identically* or in much the same way to *all* individuals at risk, and thus do not in any sense "cancel out" upon summation.

The uncertain magnitude of population risk N (i.e., the predicted number of cases) is well approximated for large n by the uncertainty quantity $n\,\overline{R}$ where for the simple risk model $\overline{R} = \mathbf{U}\overline{\mathbf{V}}$ and for the more complicated risk model $\overline{R} \approx H(\mathbf{U},\overline{\mathbf{V}})$ as a first-order approximation (Bogen and Spear, 1987). For large n and $0 \leq j \leq n$, $F_N(j)$ is generally well approximated by the expected Poisson probability for the compound-Poisson variate with uncertain parameter $n\,\overline{R}$; for example, $F_{\underline{N}}(0) = \int_0^1 e^{-nr} dF_{\bar{R}}(r)$ (Bogen and Spear, 1987). The expected value, $\langle N \rangle = n \langle \bar{R} \rangle$, of population risk has traditionally has been used in defining risk-acceptability criteria addressing N; however, criteria intended to be conservative with respect to uncertainty associated with N ought logically to refer to some upper confidence bound on N, rather than to its expected value.

INDIVIDUAL RISK

Predicted risk R, as defined above, is a variate that clearly may reflect both uncertainty and interindividual variability. It is tempting to assume that predicted risk to a given individual—say, the person with the jth highest risk among n at risk (for some j with $1 \leq j \leq n$) at some specified level of confidence with respect to uncertainty—might be calculated directly from predicted risk R without distinguishing between uncertain and heterogeneous variates. Indeed, "uncertainty analyses" are often conducted (e.g., see Appendices F and G) in which Monte Carlo techniques are used to approximate F in a way that treats all variates in the same manner, without distinguishing those that are uncertain from those that represent interindividual variability. Except for the trivial case in which $n=1$, $F_R(r)$ calculated in this manner can *only* be interpreted as the cdf pertaining to risk to an individual *sampled at random* from the entire population by which $F_R(r)$ was developed.

More typically, regulators might be interested in the (uncertain) risk $\bar{R}$ to an individual who is at average risk relative to others (which is directly related to population risk as described above); more conservatively, interest might lie in the cdf, $F_{R_{(j)}}(r)$, pertaining to (uncertain) risk $R_{(j)} = R_{(qn)}$ to a jth highest or qth quantile (i.e., $100q$th percentile with respect to variability, not uncertainty) person at risk, where $q = j/n$ and q might, for example, be some upper-bound value such as 0.99. In the most conservative risk assessment, interest is focused on uncertain risk $R_{(n)}$ to the person at greatest risk ($q = 1$). Clearly, $R_{(j)} = R$ only if *all* people incur *identical* (although perhaps uncertain) risks.

When both heterogeneous and uncertain variates are involved in the model used to predict R, the cdf for $R_{(j)}$ might be difficult to calculate. Some possible approaches are discussed below. If all heterogeneous variates are modeled with distributions truncated at the right-hand tail, $R_{(n)}$ may be approximated simply by using the maximal values of those variates. Thus, in the simple case, $R_{(n)} = U\text{Max}(V)$, and in the complicated case, $R_{(n)} \approx H(\mathbf{U}, \text{Max}(\mathbf{V}))$ as a first-order approximation. If truncated distributions are not used for all heterogeneous variates, in which careful and detailed analysis will be needed. Whether or not truncated distributions are used for all input variables, the approximations will be overconservative, perhaps highly so.

$R_{(j)}$ may be described as a *compound* order-statistic, in the sense that the cdf for $R_{(j)}$ has two sources of uncertainty: uncertainty associated with the combined impact of all the uncertain variates used to model R, and the more conventional order-statistic uncertainty associated with sampling the jth highest individual value of R from among a total of n different (but also uncertain) values where these differences arise from all the heterogeneous (as opposed to the uncertain) variates used to model R. For the simple risk model, assuming V and U are statistically independent, it follows that $R_{(j)} = U V_{(j)}$ where $V_{(j)}$ is itself an order-statistic and hence is an *uncertain* quantity that has the following cdf (Kendall and Stuart, 1977):

$$F_{V_{(j)}}(V) = \sum_{k=j}^{n} \binom{n}{k} F_V(v)^j 1 - F_V(v)^{n-j} = I_{F_V(v)}(j, n-j+1).$$ (9)

where $F_V(v)$ is the cdf modeling heterogeneity in V and where I is the incomplete Beta function. In the case of $j=n$, $F_{V_{(n)}}(v)=\{F_V(v)\}^n$.

The median value of $V_{(n)}$ is thus the $2^{-1/n}$th quantile (i.e., the $100(2^{-1/n})$th percentile) of V, which is approximately the $\{1-[Ln(2)/n]\}$th quantile of V for $n>9$. The "characteristic" value of $V_{(n)}$ is defined as the $[1-(1/n)]$th quantile of V, which is: the value of V with an "exceedance probability" of $1/n$, the value of V expected to be less than or equal to $V_{(n)}$, the 0.368th (i.e., the e^{-1}st) quantile of $V_{(n)}$, and (generally) also the modal or most likely value of $V_{(n)}$ (Ang and Tang, 1984).

For the more complicated risk model, the orderd risk $R_{(j)}$ for some or all j may not exist in an unambiguous sense because the cdfs characterizing uncertainty, e.g., in risk R_h and R_k for some particular individuals h and k may intersect one another at one or more probability levels other than 0 and 1 (Bogen, 1990). Although it is always possible to estimate the jth highest "upper-bound" risk among n such R-values (corresponding to n samples from V) all evaluated at some prespecified uncertainty-quantile (Bogen, 1990), this approach is generally difficult or impractical to implement by Monte Carol methods for complicated risk models involving both uncertainty and variability. In contrast, it is relatively simple to estimate the jth highest value of expected risk, $\langle R \rangle_{(j)}$; for example for $j=n$ this value is, as noted above, generally most likely to be $F_{\langle R \rangle}^{-1}(1-n^{-1})$, where $\langle R \rangle \approx H(\langle U \rangle, V)$ may be used as a first-order approximation (see Bogen and Spear, 1987). The ratio $\rho_n = [F_{\langle R \rangle}^{-1}(1-n^{-1})]/\langle \overline{R} \rangle$ may thus serve to characterize the magnitude of interindividual variability (or "inequity") in expected individual risks for a population of size n.

Note that unidentifiable person-to-person variability (that is, known values of a quantity that is known to differ among individuals but which values cannot each be assigned to specific individuals) is, for practical purposes, equivalent to pure uncertainty pertaining to those values insofar as the characterization of individual risk is concerned. However, the real distinction between unidentifiable person-to-person variability and true uncertainty is revealed by their different impacts on estimated population risk. In particular, if all other contributions to risk are equal, any positive amount of person-to-person variability in some determinant of risk such as susceptibility—regardless of its identifiability—will always result in a smaller variance (and thus greater certainty) in corresponding estimated population risk than that resulting from an identically distributed risk determinant whose distribution instead reflects pure uncertainty. For example, if two persons face certain but different risks equal to 0 and 1, respectively (regardless of whether it is known who faces which risk), then the expectation and variance of predicted cases are 1 and 0, respectively; here one case will arise with absolute certainty. However, if both persons face a single uncertain risk equal to 0 or 1 with probability 0.5 and 0.5, respectively, then the expected value of predicted cases is again 1, but its variance is in this case 1; here 0, 1, or 2 cases will arise with probability 0.25, 0.5, and 0.25, respectively.

In general, if n persons face n known risks $p_j = 1, 2, \ldots n$, having mean $E(p) > 0$ and variance $Var(p) > 0$, then it is well known that, regardless of who faces which particular risk, the expectation and variance of the number N of anticipated cases are $E(N) = nE(p)$ and $Var(N) = n\{E(p)[1-E(p)] - Var(p)\}$, respectively. Now consider the analogous case in which interindividual variability is replaced by pure uncertainty. In this case, all persons face a common but uncertain risk p that is distributed identically to p_j (i.e., $Prob(p=p_j) = 1/n$, $j = 1, 2, \ldots n$) and hence has the same mean $E(p)$ and variance $Var(p)$ (where in this case these moments are with respect to uncertainty, not interindividual variability). For this case, it is straightforward to show that again $E(N) = nE(p)$, but that here $Var(N) = n\{E(p)[1-E(p)] + (n-1)Var(p)\}$, which exceeds the previous expression for the variance of N by the quantity $n^2 Var(p)$.

SUMMARY

In summary, $F_{\bar{R}}(r)$ (characterizing uncertainty in risk to the average person and, approximately, in population risk) and $F_{(R)}(r)$ (characterizing interindividual variability in expected risk) are both easily estimated, even in cases involving complex risk models with uncertain and interindividually variable parameters. These estimates may generally be sufficient for regulatory decision-making purposes seeking to address both uncertainty in population risk and differences in individual risk. For example, suppose risk-acceptability criteria were desired to ensure that imposed individual lifetime risks are both de minimis and not grossly inequitable and that 70-year population risk is most likely zero cases. An example of corresponding quantitative criteria might be that the relations $F_{\bar{R}}(10^{-6}) > 0.99$, $\rho_n < 10^3$, and $F_N(0) < 0.50$ should all apply.

REFERENCES

Ang, A. H-S., and W.H. Tang. 1984. Probability Concepts in Engineering Planning and Design, Vol. II: Decision, Risk, and Reliability. New York: John Wiley & Sons.

Bogen, K.T. 1990. Uncertainty in Environmental Health Risk Assessment. New York: Garland Publishing.

Bogen, K.T., and R.C. Spear. 1987. Integrating uncertainty and interindividual variability in environmental risk assessment. Risk Anal. 7:427-436.

EPA (U.S. Environmental Protection Agency). 1989. Risk Assessments Methodology. Environmental Impact Statements: NESHAPS for Radionuclides. Background Information Document, Volumes 1 and 2. EPA/520/1-89-005 and EPA/520/1-89-006-1. Office of Radiation Programs, U.S. Environmental Protection Agency, Washington, D.C.

Kendall, M., and A. Stuart. 1977. The Advanced Theory of Statistics, 4th ed., Vol. 1. New York: Macmillan.

APPENDIX

J

A Tiered Modeling Approach for Assessing the Risks Due to Sources of Hazardous Air Pollutants

David E. Guinnup

U.S. ENVIRONMENTAL PROTECTION AGENCY
Office of Air Quality Planning and Standards
Technical Support Division
Research Triangle Park, NC 27711

February 1992

537

TABLE OF CONTENTS

FIGURES

Number

TABLES

Number

540   SCIENCE AND JUDGMENT IN RISK ASSESSMENT

1.0 INTRODUCTION

1.1 Background and Purpose

Title III of the Clean Air Act Amendment of 1990 (CAAA) sets forth a framework for regulating major sources of hazardous (or toxic) air pollutants which is based on the implementation of MACT, the maximum achievable control technology, for those sources. Under this framework, prescribed pollution control technologies are to be installed without the *a priori* estimation of the health or environmental risk associated with each individual source. The regulatory process is to proceed on a source category-by-source category basis, with a list of source categories to be published by the end of 1991, and a schedule for their regulation to be published a year later. After the implementation of MACT, it will be incumbent on the United States Environmental Protection Agency (EPA) to assess the residual health risks to the population near each source within a regulated source category. The results of this residual risk assessment will then be used to decide if further reduction in toxic emissions is necessary for each source category (refer to §112(f) of the CAAA). These decisions will hinge primarily on a determination of the lifetime cancer risk for the "maximum exposed individual" for each source as well as the determination of whether the exposed population near each source is protected from noncancer health effects with an "ample margin of safety". The determination of lifetime cancer risk involves the estimation of long-term ambient concentrations of toxic pollutants whereas the determination of noncancer health effects can involve the estimation of long-term and short-term ambient concentrations.

Since the measurement of long-term and short-term ambient concentrations for each toxic air pollutant (189 pollutants as listed in §112(b)) in the vicinity of each source is a prohibitively expensive task, it is envisioned that the process of residual risk determination would involve performing analytical simulations of toxic air pollutant dispersion for all sources (or a subset of sources) within each source category. Such simulations will subsequently be coupled with health effects information and compared to available data to quantify human exposure, cancer risk, noncancer health risks, and ecological risks.

In addition to mandating the residual risk assessment process, the CAAA provide for the exemption of source categories and pollutants from the MACT-based regulatory process if it can be demonstrated that the risks associated with that source category or pollutant are below specified levels of concern. EPA-approved risk assessments would need to be performed to justify such an exemption, and the CAAA provide for petition processes to approve or deny claims that a source category or a specific pollutant should not be subject to regulation.

The purpose of this document is to provide guidance on the use of EPA-approved procedures which may be used to assess risks due to the atmospheric dispersion of emissions of hazardous air pollutants. It is likely that the techniques described herein will be useful with respect to several decision-making processes associated with the implementation of CAAA Title III (e.g., petition to add or delete a pollutant from the list of hazardous air pollutants, petition to delete a source category from the list of source categories, demonstration of source modification offsets, etc.). In addition, the procedures may serve as the basis for the residual risk determination processes described above. The guidance addresses the estimation of long-term and short-term ambient concentrations resulting from the atmospheric dispersion of known emissions of hazardous air pollutants, and subsequently addresses the techniques currently used to quantify the cancer risks and noncancer risks associated with the predicted ambient concentrations. It describes a tiered approach which progresses from simple conservative screening estimates (provided in the form of lookup tables) to more complex modeling methologies using computer models and site-specific data. In addition to providing guidance to assist in the CAAA Title III implementation process, it is being provided to the general public to assist State and local air pollution control agencies as well as sources of hazardous pollutants in their own assessment of the impacts of these sources.

While the methods described herein comprise the most up-to-date means for assessing the impacts of sources of toxic air pollution, they are subject to future revision as new scientific information becomes available, possibly as a result of the risk assessment methodology study being conducted by the National Academy of Sciences (NAS) under mandate of section 112(o) of the CAAA (report due to Congress from NAS in May, 1993)

1.2 Risk Assessment in Title III

As mentioned above, several provisions of CAAA Title III describe the need to consider ambient concentration impacts and their associated health risks in establishing the regulatory processes for sources of toxic air pollutants. Specifically, these are:

1. A pollutant may be deleted via a petition process from the list of hazardous or toxic pollutants subject to regulation if the petition demonstrates (among other things) that "ambient concentrations.....of the substance may not reasonably be anticipated to cause any adverse effects to the human health." (§112(b)(3)(C))

2. A pollutant may be added to the list if a petition demonstrates that "ambient concentrations...of the substance are known or may reasonably be

anticipated to cause   to cause adverse effects to human health."
(§112(b)(3)(B))

3. An entire source category may be deleted form the list of source categories subject to regulation if a petition demonstrates, for the case of carcinogenic pollutants, that "no source in the category...emits (carcinogenic) air
pollutants in quantities which may cause a lifetime risk of cancer greater
than one in one million to the individual in the population who is most
exposed to emissions of such pollutants from the source," (§112(c)(9)(B)(i))
and, for the case of noncarcinogenic yet toxic pollutants, that "emissions
from no source in the category...exceed a level which is adequate to protect
public health with an ample margin of safety and no adverse environmental
effect will result from emissions from any source." (§112(c)(9)(B)(ii))

4. Within eight years after a source category has been subject to a MACT
regulation, EPA must determine whether additional regulation of that source
category is necessary based on an assessment of the residual risks associated
with the sources in that category. Based on such an assessment, additional
regulation of the source category is deemed necessary if "promulgation of
such standards is required in order to provide an ample margin of safety to
protect the public health" with respect to noncancer health effects, or if the
MACT standards "do not reduce lifetime excess cancer risks to the individual most exposed to emissions from a source in the category or subcategory
to less than one in one million" with respect to carcinogens, or if a determination is made "that a more stringent standard is necessary to prevent....an
adverse environmental effect." (§112(f)(2)(A))

In the context of these provisions, decisions are to be made based on whether or not the predicted impact of a source exceeds some level of concern. For
comparison to specified levels of concern, source impacts are quantified in four
ways:

1. lifetime cancer risk;
2. Chronic noncancer hazard index;
3. acute noncancer hazard index, and;
4. frequency of acute hazard index exceedances.

These impact measures are discussed in more detail in the next few paragraphs. It is worth noting at this point that insofar as knowledge is available
regarding the effects of specific hazardous pollutants on the environment, it may
be possible to use ecological hazard index values to quantify such impacts. Such
calculations would proceed on a track which is parallel to the calculation of
health hazard index values.

For carcinogenic pollutants, the level of concern is the risk of an individual contracting cancer by being exposed to the ambient concentrations of that pollutant over the course of a lifetime, or lifetime cancer risk. For the purposes of §112(c), the criterion specified in the CAAA is 1 in 1,000,000 lifetime cancer risk for the most exposed individual, or the individual exposed to the highest predicted concentrations of a pollutant. (For other purposes, the lifetime cancer risk specifying the level of concern may be higher or lower.) Lifetime cancer risks are calculated by multiplying the predicted annual ambient concentrations (in $\mu g/m^{3)}$ of a specific pollutant by the unit risk factor or unit risk estimate (URE)[1] for that pollutant, where the unit risk factor is equal to the upper bound lifetime cancer risk associated with inhaling a unit concentration (1 $\mu g/m^3$) of that pollutant. Since predicted annual pollutant concentrations around a source vary as a function of position, so do lifetime cancer risk estimates. Thus, decisions involving whether the impact of a source or group of sources is above some level of concern typically focus on the highest predicted concentration (and hence the highest predicted lifetime cancer risk) outside the facility fenceline. The EPA has developed unit risk factors for a number of possible, probable, or known human carcinogens, and will be developing additional cancer unit risk factors as more information becomes available. For the purposes of this document, cancer risks resulting from exposure to mixtures of multiple carcinogenic pollutants will be assessed by summing the cancer risks due to each individual pollutant, regardless of the type of cancer which may be associated with any particular carcinogen.[2]

For pollutants causing noncancer health effects from chronic or acute exposure, the levels of concern are chronic and acute concentration thresholds, respectively, which would be derived from health effects data, taking into account scientific uncertainties. For purposes of estimating potential long-term impacts of hazardous air pollutants, EPA has derived for some pollutants (and will derive for others) chronic inhalation reference concentration (RfC)[1] values, which are defined as estimates of the lowest concentrations of a single pollutant to which the human population can be exposed over a lifetime without appreciable risk of deleterious effects. For purposes of specific chronic noncancer risk assessment, EPA may designate the RfC value, or some fraction or multiple thereof, as the appropriate long-term noncancer level of concern. For purposes of specific acute noncancer risk assessment, the EPA may designate acute reference thresholds as the appropriate short-term noncancer level of concern. For the purposes of this document, long-term noncancer levels of concern will be referred to as chronic concentration thresholds, and short-term noncancer levels of concern will be referred to as acute concentration thresholds. For ease of implementation, acute concentration thresholds will be designated for 1-hour averaging times. This does not necessarily mean that exposure data indicate deleterious health effects from exposure times of 1 hour, but rather that the 1-hour acute concentration threshold has been derived such that it is protective of the exposure duration of concern.

The risk with respect to long- or short-term deleterious noncancer health effects associated with exposure to a pollutant or group of pollutants is quantified by the <u>hazard index</u>. The chronic noncancer hazard index is calculated by dividing the modeled annual concentration of a pollutant by its chronic concentration threshold value. The acute noncancer hazard index is calculated by dividing the modeled 1-hour concentration of a pollutant by its acute concentration threshold value. If multiple pollutants are being evaluated, the (chronic or acute) hazard index at any location is calculated by dividing each predicted (annual or 1-hour) concentration at that location by its (chronic or acute) concentration threshold value and summing the results.[2] If the hazard index is greater than 1.0, this represents an <u>exceedance</u> of the level of concern at that location. For pollutants which can cause deleterious health effects from acute exposures, exceedances of a level of concern may occur at any location and at any time throughout the modeling period. Thus, the frequency with which any location experiences an exceedance also becomes a measure of the risk associated with a modeled source. Frequency of acute hazard index exceedances is only addressed by the most refined analysis methods referred to in this document.

Information on UREs and RfCs is accessible through the Integrated Risk Information System (IRIS), EPA Environmental Criteria and Assessment Office (ECAO) in Cincinnati, Ohio, (513) 569-7254.

1.3 Overview of Document

This document is divided into three major sections, each section addressing a different level of sophistication in terms of modeling, referred to as "tiers". The first tier is a simplified screening procedure in which the user can estimate maximum off-site ground-level concentrations without extensive knowledge regarding the source and without the need of a computer. The second tier is a more sophisticated screening technique which requires a bit more detailed knowledge concerning the source being modeled and, in addition, requires the execution of a computer program. The third tier involves site-specific computer simulations with the aid of computer programs and detailed source parameters. Since the effects of toxic air pollutants may be of concern from both a long-term and a short-term perspective, each tier is divided into two parts. The first part addresses dispersion modeling to assess long-term ambient concentrations (important from a cancer-causing or chronic noncancer effects standpoint) and the second addresses dispersion modeling for the estimation of short-term concentrations (important from an acute toxicity perspective).

It should be noted that this document is intended to be used in conjunction with the User's Guides for the models described: SCREEN,[3] TOXST,[4] and TOXLT.[5] It is not intended to replace or reproduce the contents of these documents. In addition, the reader may wish to consult the "Guideline on Air Quality Models (Revised)"[6] for more detailed information on the consistent application of air quality models. Modelers may also wish to use the EPA's TSCREEN[7] modeling system to assist in the Tier 2 computer simulation of certain toxic release scenarios. It should be noted, however, that toxic pollutant releases which TSCREEN treats as heavier-than-air are <u>not</u> to be modeled using techniques described herein. Atmospheric dispersion of such pollutants requires a more refined analysis, such as those described in Reference 8. Model codes, user's guides, and associated documentation referred to in this document can be obtained through the Technology Transfer Network (TTN) of the EPA's Office of Air Quality Planning and Standards (OAQPS), and access information is provided in Appendix A.

The modeling tiers are designed such that the concentration estimates from each tier should be less conservative than the previous one. This means that, for a given situation, a Tier 1 modeled impact should be greater than, or more conservative than, the Tier 2 modeled impact, and the Tier 2 modeled impact should be more conservative than the Tier 3 modeled impact. Progression from one tier of modeling to the next thus involves the use of levels of concern, as defined above. For example, if the results of a Tier 1 analysis indicates an exceedance of a level of concern with respect to either (1) the maximum predicted cancer risk, (2) the maximum predicted chronic noncancer hazard index, or (3) the maximum predicted acute hazard index, the analyst may wish to perform a Tier 2 analysis. If all three of these impact measures are below their specified levels of concern, there should be no need to perform a more refined simulation, and thus, there should be no need to progress to the next tier of modeling. Since the establishment of levels of concern for each specific hazardous air pollutant is not a part of this effort, this document will refer to generic levels of concern, and users will need to consult subsequent EPA documents to determine the specific levels of concern for their particular pollutant or pollutant mixture and for the particular purpose of their modeling efforts.

1.4 <u>General Modeling Requirements, Definitions, and Limitations</u>

This document describes modeling methologies for point, area, and volume sources of atmospheric pollution. A <u>point source</u> is an emission which emanates from a specific point, such as a smokestack or vent. An <u>area source</u> is an emission which emanates from a specific, well-defined surface, such as a lagoon, landfarm, or open-top tank. Sources referred to as having "fugitive" emissions (e.g., multiple leaks within a specific processing area) are typically modeled as area sources. The methods used in this document are generally considered to be applicable for assessing impacts of a source from the facility fenceline out to a 50 km radius of the source or sources to be modeled. There is no particular upper or lower limit on emission rate value for which these techniques apply.

For the purposes of this document, "source" means the same thing as "release", and "air toxic" means the same as "hazardous air pollutant". It should be noted that "area source" as defined in the previous paragraph is not the same as the "area source" defined by the CAAA. Modeling techniques described in this document are specifically intended for use in the simulation of a finite number of well-defined sources, not for simulation of a large number of ill-defined small sources distributed over a large region, as might well be the case for some "area sources" specified in the CAAA. Simulation of the acute and chronic impacts of such area sources may utilize the RAM model[9] and the CDM 2.0 model,[10] respectively. Consult the "Guideline on Air Quality Models (Revised)"[6] for additional information. The reader should note that relatively small, well-defined groups of sources, however, may be modeled using the techniques described herein.

This document does not address the simulation of facilities located in complex terrain. Those interested in modeling facilities with possible complex terrain effects are directed to consult the "Guideline on Air Quality Models (Revised)"[6] or their EPA Regional Office modeling contact for assistance in this area (see listing Appendix B).

In order to conduct an impact assessment, it is necessary to have estimates of emission rates of each pollutant from each source or release point being included in the assessment. Emission rates may be best estimated from experimental measurements or sampling, where such test methods are available. Alternatively, mass balance calculations or use of emission factors developed for specific types of processes may be used to quantify emission rates. The procedures discussed in this document do not address the emission estimation process. Guidance for source-specific emission rate estimation and emission test methods is available in other EPA documentation (e.g., see References 11 through 15). Additional information concerning specific emission measurement techniques is available through the OAQPS TTN (see Appendix A).

Since many sources of hazardous air pollutants are intermittent in nature (e.g., batch process emissions), the techniques in this document have been developed to allow the treatment of intermittent sources as well as continuous types of sources. It is important to understand the different treatment of emission rates for both types of sources when carrying out either the analysis of a long-term impact or a short-term impact. In a long-term impact analysis, the emission rate used for modeling is based on the amount of pollutant emitted over a 1-year period, regardless of whether the emission process is a continuous or intermittent one. In addition, to assess the worst-case impact of a source or group of sources, long-term emission rates used in model simulations should reflect the emission rates for a plant or process which is operating at full design capacity. In a short-term impact analysis, the emission rate used for modeling is based on the maximum amount of pollutant emitted over a 1-hour period, during which the source is emitting. The Tier 1 and Tier 2 procedures evaluate the combined worst-case impacts of intermittent sources as if they are all emitting at the same time, whereas the Tier 3 procedures incorporate a more realistic treatment of intermittent sources by turning them on and off throughout the simulation period according to user-specified frequency of occurrence of each release. This frequency of occurrence should reflect the normal operating schedule of the source when operating at maximum design capacity.

In addition to emission rate estimates, it is necessary to have quantitative information about the sources to conduct a detailed impact assessment. Tier 1 analyses require information about the height of the release above ground level and the shortest distance from the release point to the facility fenceline. Higher tiers of analysis require additional information including, but not limited to:

Stack height
Inside stack diameter
Exhaust gas exit velocity
Exhaust gas exit temperature
Dimensions of structures near each source
Dimensions of ground-level area sources
Exact release and fenceline location
Exact location of receptors for determining worst-case impacts
Land use near the modeled facility
Terrain features near the facility
Duration of short-term release
Frequency of short-term release

Where appropriate, this document will address the best means of obtaining these input data. In some more complex cases, the modeling contact at the nearest EPA Regional Office may need to be consulted for specific modeling guidance (see listing in Appendix B).

Depending on the specific purpose of the impact assessment, it may be difficult for the modeler to decide which sources (or release points) and which pollutants should be included in a particular analysis or simulation. Since these questions pertain to the particular purposes for which the impact assessment is being performed, they are not addressed by this document. Instead, this document refers to and provides guidance for modeling various scenarios including single-source, multiple-source, single-pollutant, and multiple-pollutant scenarios. Subsequent EPA documents will address the questions of which sources and which pollutants should be included in an impact analysis for a specific regulatory purpose.

2.0 TIER I ANALYSES

2.1 Introduction

Tier 1 analysis of a stationary source (or group of sources) of toxic pollutant(s) is performed to address the question of whether or not the source has the potential to cause a significant impact. This "screening" analysis is performed by using tables of lookup values to obtain the "worst-case" impact of the source being modeled. The analysis is performed to assess both the potential long- and short-term impacts of the source. If the predicted screening impacts are less than the appropriate levels of concern, no further modeling is indicated. If the predicted screening impacts are above any levels of concern, further analysis of those impacts at a higher Tier may be desireable to obtain more accurate results.

The Tier 1 "lookup tables" have been created as tools which may be easily used to estimate conservative impacts of sources of toxic pollutants with a minimal amount of information concerning those sources. The normalized annual and 1-hour concentration tables were created based on conservative simulations of toxic pollutant sources with the EPA's SCREEN model.[3] In this context, "conservative" simulations use conservative assumptions regarding meteorology, building downwash, plume rise, etc. Conservative annual concentrations were derived from SCREEN 1-hour estimates using the conservative multiplication factor of 0.10.

2.2 Long-term Modeling

Long-term modeling of toxic or hazardous air pollutants is aimed at the estimation of annual average pollutant concentrations to which the public might be exposed as the result of emissions from a specific source or group of sources. From the EPA regulatory viewpoint, this "public" does not include employees of the facility responsible for the emissions (this is the jurisdiction of the Occupational Safety and Health Agency, OSHA). Thus, the impact assessment focuses on estimating concentrations "off-site", or outside the facility boundary. For carcinogens, the calculation of cancer risk proceeds by multiplying annual concentrations by pollutant-specific cancer potency factors derived from health effects data. The impacts of pollutants with chronic noncancer effects are generally assessed by comparing predicted annual concentrations with chronic threshold concentrations which again derived from experimental health data. For the purposes of protecting the general public against "worst-case" pollutant concentrations, the analysis is focused on predicting the worst-case, or maximum annual average concentrations.

2.2.1   Maximum Annual Concentration Estimation

A long-term tier 1 analysis requires the following information:

1. annual average emission rate of each pollutant from each source included in the simulation (T/yr). These emissions do not have to be continuously emitted, but rather should represent the total amount of pollutant which is generated by this source in a year. Note that the tons used in this regard are English tons (1 T. = 2000 lb.) Also note that, for Tier 1 analyses, the emission rate from an area source represents the total emissions from the area, not the emissions per square unit area.

2. height of the release point above ground (m), for each point source.

3. source types (point or area). Point sources typically include exhaust vents (pipes or stacks), or any other type of release that causes toxic materials to enter the atmosphere from a well-defined location, at a well-defined rate. Area sources may also be well-defined, but differ from point sources in that the extent over which the release occurs is substantial.

4. maximum horizontal distance across each area source (m).

5. nearest distance to property line (m). Concentration estimates are needed at locations that are accessible to the general public. This is typically taken to be any point at or beyond the property-line of a facility. Estimate the distance from the point of each release to the <u>nearest</u> point on the fenceline. (This need not be the same fenceline point for each release). If the source is characterized as an area source, this distance should be measured from the nearest <u>edge</u> of the area source, not from the center.

Once these five items are determined for each release (or source), screening estimates of normalized maximum annual concentrations resulting from each release are obtained from Table 1 using the following procedure.

1. For an area source, select the "side length" in the table (10m, 20m, 30m) which is less than or equal to the maximum horizontal distance across the source.

2. For a point source, select the largest "emission height" in the table (0m, 2m, 5m, 10m, 35m, or 50m) that is less than or equal to the estimated height of release.

3. Select the largest distance in the table (10m, 30m, 50m, 100m, or 200m) that is less than or equal to the nearest distance to the property-line.

4. Take the appropriate normalized maximum annual concentration for this release height and distance from the table, and multiply by the emission rate

TABLE 1. NORMALIZED MAXIMUM ANNUAL CONCENTRATIONS ($\mu g/m^3$)/(T/yr)

Source type[a]	Emission height, m	Side length,[b] m	Normalized maximum concentrations at or beyond[c]					
			10m	30m	50m	100 m	200 m	500 m
A	0	10	9.56E+2	3.02E+2	1.64E+2	6.48E+1	2.32E+1	5.53E+0
A	0	20	5.15E+2	1.83E+2	1.07E+2	4.78E+1	1.91E+1	5.04E+0
A	0	30	3.51E+2	1.31E+2	7.92E+1	3.74E+1	1.61E+1	4.58E+0
P	0	—	5.41E+3	7.92E+2	3.25E+2	9.67E+1	2.91E+1	6.08E+0
P	2	—	1.87E+2	1.42E+2	1.35E+2	7.28E+1	2.64E+1	5.96E+0
P	5	—	9.62E+1	7.46E+1	5.18E+1	2.72E+1	1.48E+1	5.18E+0
P	10	—	2.77E+1	2.44E+1	2.11E+1	1.36E+1	7.17E+1	2.88E+0
P	20	—	6.91E+0	4.52E+0	4.52E+0	3.80E+0	2.44E+0	1.06E+0
P	35	—	2.26E+0	2.26E+0	1.13E+0	1.11E+0	8.98E-1	4.41E-1
P	50	—	1.11E+0	1.10E+0	1.11E+0	4.69E-1	4.23E-1	2.53E-1

[a]Source type P=Point Source, type A=Area source
[b]Side length of square area source
[c]Distance downwind of an area source indicates distance from downwind edge of the area source.

of each toxic substance (t/yr) in the release to obtain the concentration estimate (μg/m$^{3)}$). DO NOT INTERPOLATE TABLE VALUES.

For example, consider the situation in which a toxic pollutant A is released at a
rate of 11.6 T/yr from a vent-pipe that is 40m tall, and which is attached to a
building that is 4m tall, 10m long, and 5m wide. The nearest boundary of the
facility is located 65m from the pipe. A value of 35m should be selected for the
emission height, because all larger entries in the table exceed the actual height of
release of 40m. Concentrations should be estimated for a distance of 50m,
because once again, all greater entries in the table exceed the actual distance of
45m. The appropriate normalized maximum annual concentration is 1.13 (μg/
m^3)(T/yr). Multiplying by the emission rate of 14.6 T/yr results in a maximum
annual concentration estimate for screening purposes equal to 16.5 μg/m^3.

2.2.2   Cancer risk assessment

Once the maximum annual concentration has been estimated for each release being modeled, upper bound lifetime individual cancer risk may be estimated by mutiplying the maximum annual concentration estimates of each carcinogenic pollutant by the unit cancer risk factor for that pollutant and then
summing results. This approach assumes that all cancer risks are additive,
regardless of the organ system which may be affected. It should be noted that
this approach assumes that all worst-case impacts occur at the same location.
While this assumption may not be very realistic, it does help to insure that Tier 1
results are conservative, and, therefore protective of the public.

As an example of this approach, suppose one is simulating a plant which
emits 2 pollutants A and B, through 4 different stacks such that pollutant A is
released from stacks 1 and 2, and pollutant B is released from stacks 2, 3, and 4.
In this example, stack 1 is the same as that described in the example above.
After going through the above procedure to estimate the maxium annual concentrations of each pollutant from each stack, the results are:

Source	Compound	Max impact
Stack 1	Pollutant A	16.5 μg/m^3
Stack 2	Pollutant A	5.49 μg/m^3
Stack 2	Pollutant B	2.35 μg/m^3
Stack 3	Pollutant B	4.13 μg/m^3
Stack 4	Pollutant B	24.9 μg/m^3

Suppose that the unit cancer risk factors for pollutants A and B are know to be
1.0×10^{-7} and 2.0×10^{-7} (μg/m^3)$^{-1}$, respectively. The Tier 1 maximum cancer

risk is calculated for the individual releases and pollutants and summed as follows:

Source	Compound	Max impact	Max risk
Stack 1	Pollutant A	16.5 µg/m^3	1.65×10^{-4}
Stack 2	Pollutant A	5.49 µg/m^3	5.49×10^{-7}
Stack 2	Pollutant B	2.35 µg/m^3	4.70×10^{-7}
Stack 3	Pollutant B	4.13 µg/m^3	8.26×10^{-7}
Stack 4	Pollutant B	24.9 µg/m^3	4.98×10^{-4}
		Total risk	8.48×10^{-4}

If we are assessing the impact of this group of sources in relation to the CAAA specificed level of concern of 1×10^{-4} lifetime cancer risk, and since the maximum Tier 1 risk is greater than the CAAA specified concern level of 1×10^{-4}, this source warrants further modeling on the basis of cancer risk (note that this does not rule out the need to investigate acute or chronic nonccancer risk).

2.2.3 Chronic Noncancer Risk Assessment

For all pollutants which pose a chronic noncancer threat to health, an assessment of the magnitude of this threat is made using the hazard index approach. The chronic noncancer hazard index is calculated by summing the maximum annual concentrations for each pollutant divided by the chronic threshold concentration value for that pollutant. if the calculated hazard index is greater than 1.0, the release or releases being simulated may pose a threat to the public, and further modeling may be indicated. It should again be noted that, for the sake of erring conservatively, this approach assumes that the worst-case impacts of all releases occur at the same location.

As an example of the above procedure, suppose that pollutants A and B in the example above pose a chronic noncancer health risk, and their respective chronic concentration threshold values are 20.0 and 5.0 µg/m^3, respectively. The chronic noncancer hazard index would be formulated as follows:

Source	Compound	Max. Impact	Hazard Index
Stack 1	Pollutant A	16.5 µg/m^3	0.825
Stack 2	Pollutant A	5.49 µg/m^3	0.275
Stack 2	Pollutant B	2.35 µg/m^3	0.470
Stack 3	Pollutant B	4.13 µg/m^3	0.826
Stack 4	Pollutant B	24.9 µg/m^3	4.980
		Total Hazard index	7.376

SCIENCE AND JUDGMENT IN RISK ASSESSMENT

In this case, one of the individual hazard index values exceeds 1.0, the total hazard index for this modeled facility exceeds 1.0, and further modeling at a higher Tier may be desired.

2.3 Short-term Modeling

Since short-term modeling of toxic or hazardous air pollutants is aimed at the estimation of 1-hour average pollutant concentrations to which the public might be exposed as the result of emissions from a specific source or group of sources. Again, from the EPA regulatory viewpoint, this "public" does not include employees of the facility responsible for the emissions (this is the jurisdiction of OSHA). Thus, the impact assessment focuses on estimating concentrations "off-site", or outside the facility boundary. From the short-term perspective, the health effects of most concern vary, but they are those which create detrimental health effects as the result of short-term exposure to toxic pollutants. The risks associated with such exposures are generally assessed by comparing 1-hour predicted concentrations with acute threshold concentrations which are derived from experimental health data. For the purposes of protecting the general public against "worst-case" pollutant concentrations, the analysis is focused on predicting the worst-case, or maximum 1-hour average concentrations.

2.3.1 Maximum Hourly Concentration Estimation

A short-term Tier 1 analysis requires the following information:

1. maximum 1-hour average emission rate of each pollutant from each source included in the simulation (g/s). If the release is a continuous, constant-rate emission, then this value is equivalent to the release rate for long-term modeling, except that it is expressed in g/s instead of T/yr. (To convert from T/yr to g/s, divide by 34.73; to convert from g/s to T/yr, multiply by 34.73) If the release is intermittent, such as a batch process, this value is equivalent to the maximum number of grams emitted during any hour when the release is occurring divided by 3600. Again note that, for Tier 1 analyses, the emissions from an area source represent the total emissions from that source, not just the emissions per unit area surface.

2. height of each release above ground (m), for point sources.

3. source types (point or area). Point sources typically include exhaust vents (pipes or stacks), or any other type of release that causes toxic materials to enter the atmosphere from a well-defined location, at a well-defined rate. Area sources may also be well-defined, but differ from point sources in that the extent over which the release occurs is substantial.

4. maximum horizontal distance across each area source (m).

5. nearest distance to property-line (m). Concentration estimates are needed at locations that are accessible to the general public. This is typically taken to be any point at or beyond the property-line of a facility. Estimate the distance from the point of each release to the <u>nearest</u> point on the fence-line. (This need not be the same fenceline point for each release). If the source is characterized as an area source, this distance should be measured from the nearest <u>edge</u> of the area source, rather than from the center of the area source.

Once these five items are determined for each release, screening estimates of maximum 1-hour average concentrations resulting from each release are obtained from Table 2 using the following procedure.

1. For an area source, select the "side length" in the table (10m, 20m, 30m) which is less than or equal to the maximum horizontal distance across the source.

2. For point sources, select the largest "emission height" in the table (0m, 2m, 5m, 10m, 35m, or 50m) that is less than or equal to the estimated height of release.

3. For each source, select the largest distance in the table (10m, 20m, 50m, 100m, or 200m) that is less than or equal to the nearest distance to the property-line.

4. Take the normalized maximum 1-hour concentration for this release and fenceline distance, and multiply by the emission rate of each toxic pollutant (g/s) in the release to obtain the maximum off-site 1-hour average concentration estimates ($\mu g/m^3$). <u>DO NOT INTERPOLATE TABLE VALUES.</u>

For example, again consider the situation in which toxic material A is released from a vent-pipe that is 40m tall, and which is attached to a building that is 4m tall, 10m long, and 5m wide. The nearest boundary of the facility is located 65m from the pipe. For the short-term assessment, it has been determined that the maximum emissions of A that can occur during any hour of the year is 1800g, therefore the emission rate for short-term assessment is 1800g/3600s = 0.50g/s. A value of 35m is again selected for the emission height, because all larger entries in the table exceed the actual height of release. Concentrations are estimated for a distance of 50m, because once again, all greater entries in the table exceed the actual distance of 65m. The appropriate normalized maximum 1-hour average concentration is 3.94E = 2 ($\mu g/m^3$)/(g/s). Multiplying by the emission rate of 0.50g/s results in a maximum hourly concentration estimate for screening purposes equal to 197 $\mu g/m^3$.

556

TABLE 2. NORMALIZED MAXIMUM 1-HOUR AVERAGE CONCENTRATIONS $(\mu g/m^3)/(g/s)$

Source type [a]	Emission height, m	Side length,[b] m	Normalized maximum concentrations at or beyond[c]					
			10 m	30m	50m	100 m	20 m	500 m
A	0	0	3.32E+5	1.05E+5	5.70E+4	2.25E+4	8.07E+3	1.92E+3
A	0	20	1.79E+5	6.36E+4	3.72E+4	1.66E+4	6.62E+3	1.75E+3
A	0	30	1.22E+5	4.54E+4	2.75E+4	1.30E+4	5.59E+3	1.59E+3
P	0	—	1.88E+6	2.75E+5	1.13E+5	3.36E+3	1.01E+4	2.11E+3
P	2	—	6.51e+4	4.92E+4	4.69E+4	2.53E+4	9.18E+3	2.07E+3
P	5	—	3.34E+4	2.59E+4	1.80E+4	9.44E+3	5.13E+3	1.80E+3
P	10	—	9.61E+3	8.49E+3	7.36E+3	4.71E+3	2.49E+3	1.00E+3
P	20	—	2.45E+3	1.57E+3	1.57E+3	1.32E+3	8.46E+2	3.67E+2
P	35	—	7.84E+2	7.84E+2	3.94E+2	3.85E+2	3.12E+2	1.53E+2
P	50	—	3.84E+2	3.84E+2	3.84E+2	1.63E+2	1.47E+2	8.77E+2

[a]Source type P=Point Source, type A=Area source
[b]Side length of square area source
[c]Distance downwind of an area source indicates distance from downwind edge of the area source.

2.3.2   Acute Hazard Index Assessment

For all pollutants which pose a threat to health based on acute exposure, an assessment of the magnitude of this threat is made using the acute hazard index approach, similar to that used in chronic noncancer risk assessment. In this case, however, the acute hazard index is calculated by summing the maximum 1-hour concentrations for each pollutant divided by the acute concentration threshold value for that pollutant. It should again be noted that, for the sake of erring conservatively, this approach assumes that the worst case impacts of all releases can occur simultaneously at the same location. Similar to the chronic risk assessment, if the calculated hazard index is greater than 1.0, the release or releases being simulated may pose a threat to the public, and further modeling at a higher Tier may be indicated.

As an example of the acute hazard index approach, consider the same plant being simulated in Section 2.2.2, but this time the maximum 1-hour concentrations are determined using the procedure in Section 2.3.2 to be the following:

Source	Compound	Max. 1-hr impact
Stack 1	Pollutant A	197 μg/m^3
Stack 2	Pollutant A	257 μg/m^3
Stack 2	Pollutant B	110 μg/m^3
Stack 3	Pollutant B	301 μg/m^3
Stack 4	Pollutant B	367 μg/m^3

Further suppose that pollutants A and B pose health problems from acute exposures with acute threshold concentration values of 200 and 100 μg/m^3, respectively. The acute hazard index is calculated as follows:

Source	Compound	Max. 1-hr impact	Hazard Index
Stack 1	Pollutant A	197 μg/m^3	0.985
Stack 2	Pollutant A	257 μg/m^3	1.285
Stack 2	Pollutant B	110 μg/m^3	1.100
Stack 3	Pollutant B	301 μg/m^3	3.010
Stack 4	Pollutant B	367 μg/m^3	3.670
		Total Hazard Index	10.050

In this case, 4 of the individual hazard index values exceeds 1.0, the total hazard index for the modeled plant exceeds 1.0, and further modeling at a higher Tier may be desired.

3.0 TIER 2 ANALYSES

3.1 Introduction

Tier 2 analysis of a stationary source (or group of sources) of toxic pollutant(s) may be desired if the results of a Tier 1 analysis indicate an exceedance of a level of concern with respect to one or more of the following: (1) the maximum predicted cancer risk; (2) the maximum predicted chronic noncancer hazard index, or; (3) the maximum predicted acute hazard index. Note that in situations where only one or two of the Tier 1 criteria are exceeded, only those analyses which exceed the Tier 1 criteria may need to be performed at the higher Tier. For example, if the Tier 1 analysis showed cancer risk and chronic noncancer risks to be of concern while the acute risk analysis showed no cause for concern, only long-term modeling for cancer risk and noncancer risk may need to be performed at Tier 2. Tier 2 analyses are slightly more sophisticated than Tier 1 analyses, and therefore require additional input information as well as a computer for their execution. Tier 2 analyses are structured around the EPA's SCREEN model and its corresponding documentation entitled "Screening Procedures for Estimating the Air Quality Impact of Stationary Sources."[3] The SCREEN model source code and documentation is available through the OAQPA TTN (see Appendix A).

Again, similar to the Tier 1 analysis, if any of the predicted model impacts from Tier 2 are above the appropriate levels of concern, further modeling is indicated at a higher Tier.

3.2 Long-term Modeling

Long-term Tier 2 modeling utilizes the SCREEN[3] model to estimate 1-hour maximum concentrations, and then utilizes a conservative conversion factor to derive maximum annual concentration values from the SCREEN predictions.[16,17] These maximum annual concentration estimates are used to assess cancer risk and chronic noncancer risk exactly as in Section 2.2.2 and 2.2.3 of this document.

3.2.1   Maximum Annual Concentration Estimation

In addition to the information required to perform a Tier 1 analysis, a Tier 2 analysis requires the following information:

1. the inside diameter of the stack at the exit point (m).
2. the stack gas exit velocity (m/s)
3. the stack gas exit temperature (K)

4. a determination of whether the area surrounding the modeled facility is urban or rural. This is usually assessed on the basis of land use in the vicinity of the facility.

Refer to the "Guideline on Air Quality Models (Revised)"[6] for additional guidance on this determination.

5. downwash potential. Downwash effects must be included in dispersion estimates for point (stack) sources wherever the point of release is located on the roof of a building or structure, or within the lee of a nearby structure. The potential for downwash is determined in the following way. First, estimate the heights and maximum horizontal dimensions[*] of the structures nearest the point of release. For each structure, determine which of these two dimensions is <u>less</u>, and call this length L. If the structure is less than 5L away from the source, then this structure may cause downwash. For every structure satisfying this criterion, calculate a height by multiplying L by 1.5, and adding this to the actual height of the structure. If any calculated height exceeds the height of the release, then downwash calculations must be made for that release.

Once these items are determined for each release being modeled, estimates of maximum concentrations from each release are obtained through individual SCREEN runs for each release. Recommendations for each SCREEN run are as follows:

1. The emission rates used for Tier 1 long-term modeling should be converted from T/yr to g/s (divide T/yr by 34.73). Area source emission rates should be converted to $g/s/m^2$ by dividing the total area of the source.

2. Choose the default atmospheric temperature of 293K

3. For each release, exercise the automated distance array choosing as the minimum receptor distance the appropriate nearest fenceline distance for that release, and choosing 50 km as the maximum receptor distance. The maximum concentration for that release will then be chosen as the maximum at or beyond the nearest fenceline distance.

4. The option for flagpole receptors should not be used.

---

[*]Note: The maximum horizontal dimension is defined as the largest possible alongwind distance the structure could occupy.

5. For each release, the maximum 1-hour concentration should be noted.

6. Maximum annual concentrations should be calculated for each release by multiplying predicted maximum 1-hour concentrations by 0.08.

As an example of the Tier 2 long-term analysis, consider Stack 1 from the Tier 1 example. To consider downwash possibilities, the maximum horizontal dimension is first estimated as $\{(10m)^2 + (5m)^2\}^{1/2} = 11.2m$. The dimension L is then 4m, and the maximum stack height for which downwash is possible would be $4m + 1.5 \times 4m = 10m$. Since the actual stack height is 40m, downwash need not be considered in the SCREEN simulation. The emission rate specified in the example of 14.6 T/yr is converted to g/s to be used in the SCREEN simulation, resulting in an emission rate of $14.6/34.73 = 0.42$ g/s. In addition to the actual stack height (40m) and minimum fenceline distance (65m), input parameters for the SCREEN simulation are:

Inside stack diameter	0.5m
Stack gas exit velocity	5.6m/s
Stack gas exit temperature	303 K
Plant Location	urban

The results from the SCREEN simulation indicates that the maximum 1-hour concentration at or beyond 65m is 32.5 $\mu g/m^3$, occurring 165m downwind. Using the recommended conversion factor of 0.09, the maximum annual concentration is estimated at 2.6 $\mu g/m^3$ (this value can be contrasted with the Tier 1 estimation of 16.5 $\mu g/m^3$).

3.2.2 Cancer Risk Assessment

Maximum annual concentrations for all releases of carcinogens should be multiplied by the appropriate unit cancer risk factor and summed to estimate the maximum cancer risk. It should be noted that this approach, as in Tier 1, presumes that all worst-case impacts occur at the same location. While this assumption may not be very realistic, it does help insure that the results of a Tier 2 analysis are conservative and therefore protective of the public. More receptor-specific risk calculations are addressed in the Tier 3 analyses.

Borrowing again from the Tier 1 example, maximum annual impacts for each source and pollutant combination are estimated using the SCREEN model. Risk estimates are then made by summing the risk due to each release, regardless of downwind distance to maximum impact. The results are:

Source	Compound	Max. Impact	Max risk
Stack 1	Pollutant A	2.60 µg/m³	2.60×10^{-7}
Stack 2	Pollutant A	1.34 µg/m³	1.34×10^{-7}
Stack 2	Pollutant B	0.58 µg/m³	1.16×10^{-7}
Stack 3	Pollutant B	0.62 µg/m³	1.24×10^{-7}
Stack 4	Pollutant B	3.70 µg/m³	7.40×10^{-7}
		Total Risk	1.38×10^{-6}

For this example, the maximum lifetime cancer risk estimated using the Tier 2 methods is a factor of 6 lower than that estimated in the Tier 1 analysis. However, the cancer risk level still exceeds 1×10^{-6}, indicating that modeling at a higher Tier may be desireable.

3.2.3 Chronic Noncancer Risk Assessment

As in Tier 1, maximum annual concentrations are divided by their chronic concentration threshold values and summed to calculate the hazard index values. Again, this approach conservatively assumes that all worst-case impacts occur at the same location.

Continuing with the example, the chronic noncancer hazard index is recalculated using the Tier 2 estimated long-term impacts. Threshold concentration values for chronic noncancer effects again are taken as 20.0 and 5.0 µg/m³ for pollutants A and B, respectively. The following results:

Source	Compound	Max. Impact	Hazard Index
Stack 1	Pollutant A	2.60 µg/m³	0.130
Stack 2	Pollutant A	1.34 µg/m³	0.067
Stack 2	Pollutant B	0.58 µg/m³	0.116
Stack 3	Pollutant B	0.62 µg/m³	0.124
Stack 4	Pollutant B	3.70 µg/m³	0.740
	Total Hazard Index		1.177

The chronic noncancer hazard index estimated in Tier 2 is a good deal less than that estimated for the same sources in Tier 1. Even though none of the individual source/pollutant combinations exceeds a chronic threshold concentration value, the total hazard index exceeds 1.0, and further analysis at Tier 3 is indicated for chronic noncancer effects.

3.3 Short-term Modeling

Short-term Tier 2 modeling utilizes the SCREEN[3] model to estimate 1-hour maximum concentrations directly. These maximum 1-hour concentration estimates are used to assess acute hazard index values exactly as in Section 2.3.2 of this document.

3.3.1   Maximum Hourly Concentration Estimation

In addition to the information required to perform a Tier 1 short-term analysis, a Tier 2 analysis requires the following information for stack sources:

1. the inside diameter of the stack at the exit point (m).

2. the stack gas exit velocity (m/s)

3. the stack gas exit temperature (K)

4. a determination of whether the area surrounding the modeled facility is urban or rural. This is usually assessed on the basis of land use in the vicinity of the facility. Refer to the "Guideline on Air Quality Models (Revised)"[6] for additional guidance on this determination.

5. downwash potential. Downwash effects must be included in dispersion estimates for point sources whenever the point of release is located on the roof of a building or structure, or within the lee of a nearby structure. The potential for downwash is determined in the following way. First, estimate the heights and maximum horizontal dimensions of the structures nearest the point of release. For each structure, determine which of these two dimensions is less, and call this length L. If the structure is less than 5L away from the source, then this structure may cause downwash. For every structure satisfying this criterion, calculate a height by multiplying L by 1.5, and adding this to the actual height of the structure. If any calculated height exceeds the height of the release, then downwash calculations must be made for that release.

Once these items are determined for each release being modeled, estimates of maximum concentrations from each release are obtained through individual SCREEN runs for each release. Recommendations for each SCREEN run are as follows:

1. Choose the default atmospheric temperature of 293K.

2. Area source emission rates reflect the total emission rate from divided by the area of the source.

3. For each release, exercise the automated distance array choosing as the minimum receptor distance the appropriate nearest fenceline distance for that release, and choosing 50 km as the maximum receptor distance. The maximum concentration for that release will then be chosen as the maximum at or beyond the nearest fenceline distance.

4. The option for flagpole receptors should not be used.

5. For each release, the maximum 1-hour concentration should be noted.

Using this approach with the Stack 1 example, the SCREEN model is exercised with the stack parameters specified in Section 3.2.1. The maximum short-term emission rate of 0.50 g/s (see Section 2.3.1), however, is used to estimate the maximum 1-hour source impact. The results of the SCREEN model indicate that the maximum 1-hour concentration is 38.8 $\mu g/m^3$, again occurring 165m downwind.

3.3.2   Acute Hazard Index Assessment

As in Tier 1, maximum 1-hour concentrations are divided by their acute threshold concentration values and summed to calculate the acute hazard index values. Again, this approach conservatively assumes that all worst-case impacts can occur simultaneously at the same location.

To illustrate this procedure, short-term impacts from the example plant are assessed using the hazard index approach. Again the acute threshold concentration values are taken as 200 and 100 $\mu g/m^3$, respectively. The results are:

Source	Compound	Max. 1-hr impact	Hazard Index
Stack 1	Pollutant A	34.8 $\mu g/m^3$	0.174
Stack 2	Pollutant A	70.5 $\mu g/m^3$	0.352
Stack 2	Pollutant B	29.9 $\mu g/m^3$	0.299
Stack 3	Pollutant B	50.0 $\mu g/m^3$	0.500
Stack 4	Pollutant B	60.4 $\mu g/m^3$	0.604
		Total Hazard Index	1.925

For this example, the acute hazard index estimated in Tier 2 is roughly 20% of that estimated for the same sources in Tier 1. However, since the total hazard index exceeds 1.0, further analysis at Tier 3 is indicated for health effects resulting from acute exposures.

4.0 TIER 3 ANALYSES

4.1 Introduction

Tier 3 analysis of a stationary source (or group of sources) of toxic pollutant(s) may be desired if the results of a Tier 2 analysis indicate an exceedance of a level of concern with respect to one or more of the following: (1) the maximum predicted cancer risk; (2) the maximum predicted chronic noncancer hazard index, or; (3) the maximum predicted acute hazard index. Tier 3 analysis of a stationary source (or group of sources) of toxic pollutant(s) is performed to provide the most scientifically-refined indication of the impact of that source. This Tier involves the utilization of site-specific source and plant layouts as well as meteorological information. In contrast to the previous Tiers, Tier 3 allows for a more realistic simulation of intermittent sources and combined source impacts. In addition, results from short-term analyses indicate not only if a risk level of concern can be exceeded, but how often that level of concern might be exceeded during an average year. Dispersion modeling for the Tier 3 analysis procedure is based on use of EPA's Industrial Source Complex (ISC) model[18] and as such utilizes many of the same techniques recommended in the "Guideline on Air Quality Models (Revised)"[6] approach to the dispersion modeling of criteria pollutants.

To facilitate the dispersion modeling of toxic air pollutants, the EPA has developed TOXLT (TOXic modeling system Long-Term)[5] for refined long-term analyses, and TOXST (TOXic modeling system Short-Term)[4] for refined short-term analyses. The TOXLT system incorporates the ISCLT (long-term) directly to calculate annual concentrations and the TOXST system incorporates the ISCST (short-term) model directly to calculate hourly concentrations. Codes and user's guides for both TOXLT and TOXST are available via electronic bulletin board (see Appendix A).

4.2 Long-Term Modeling

Long-term Tier 3 modeling using the TOXLT[5] modeling system to estimate maximum annual concentrations and maximum cancer risks. The TOXLT modeling system uses the ISCLT model to calculate these annual concentrations at receptor sites which are specified by the user. A post-processor called RISK subsequently calculates lifetime cancer risks and chronic noncancer hazard index values at each receptor.

4.2.1 Maximum Annual Concentration Estimation

In addition to the information required to perform a Tier 2 long-term analysis, the Tier 3 long-term analysis requires the following information:

1. five years of meteorological data from the nearest National Weather Service (NWS) station. These data are for the most recent, readily-available consecutive five year period. NWS data are available through the electronic bulletin board (see Appendix A). Alternatively, one or more years of meteorological data from on-site measurements may be substituted. These data should be obtained and quality-assured using procedures consistent with the "Guideline on Air Quality Modeling (Revised)."[6]

2. plant layout information, including all emission point and fenceline locations. This information should be sufficiently detailed to allow the modeler to specify emission point and fenceline receptor locations within 2 meters.

3. pollutant-specific data concerning deposition or decay half-life, if applicable.

Once these data have been obtained, an input file should be prepared for execution of the ISCLT model using the guidance available in the ISC User's Guide.[18] The ISCLT model should then be executed using the TOXLT system. Procedures utilized should also be consistent with the TOXLT User's Guide[5] (available via electronic bulletin board, see Appendix A). Specific recommendations concerning the development of these inputs include:

1. Annual emission rates should be converted to g/s for input. The TOXLT modeling system uses "base emission rates" and "emission rate multipliers" to specify the emission rate for each pollutant/source combination. Thus, for a given pollutant and source the emission rate equals the base emission rate (specified in the ISCLT input file) times the emission rate multiplier for that pollutant/source combination (specified in the RISK input file). In general, the input file to the ISCLT program should specify the same emission rates used in previous modeling tiers for each source, and emission rate multipliers of 1.0 should then be provided as inputs to the RISK post-processor. (This doesn't necessarily have to be the case, as long as the product of the emission rate provided as input to ISCLT and the emission rate multiplier provided as input to RISK equals the actual emission rate being modeled for each source.) In the case where more than one pollutant is being emitted from the same source, that source should only be included once in the ISCLT input file, and emission rate multipliers should be provided to the RISK post-processor for each pollutant being emitted from that source.

2. In general, each source should be modeled as a single ISCLT source group. However, all sources of a single pollutant may be grouped into a single ISCLT source group. Each source of more than one pollutant should be modeled as a single ISCLT source group by itself.

3. Input switches to the ISCLT model should be set to allow the creation of the master file inventory for post-processing. The regulatory default mode should be used. Choose the printed output option for tabulating the greatest impacts of each source.

4. STability ARray (STAR) summaries of the NWS meteorological data should be created using the STAR program (this program and a description of its use are available on the electronic bulletin board, see Appendix A). These should be included in the input file according to the ISCLT User's Guide.

5. A polar or rectangular receptor grid may be used, but with sufficient detail to accurately estimate the highest concentrations. The design of the receptor network should consider the long-term results of the earlier modeling tiers such that the highest resolution of receptors is in the vicinity of the highest predicted impacts. Additional receptors may need to be added in sufficient detail to accurately resolve the highest concentrations.

6. Where appropriate, direction-specific building downwash dimensions should be included for each radial direction.

The printed ISCLT output will indicate the top 10 impacts for each source group, while the master file inventory will contain all of the annual concentration predictions from each source group at each receptor.

Continuing with the examples from Tiers 1 and 2, TOXLT was utilized to perform site-specific ISCLT dispersion modeling for the 4 stacks in the example. Each of the stacks was modeled as an individual source group. A STAR summary of five years of meteorological data from the nearest NWS site was utilized along with specific source and plant boundary locations according to Figure 1 below. Stacks are represented in the Figure as open circles, with stacks 3 and 4 located at the same place. A rectangular receptor grid (indicated by the filled circles) with 50m spacing outside the plant boundary was used to obtain concentration predictions. Neither pollutant was presumed to decompose in the atmosphere.

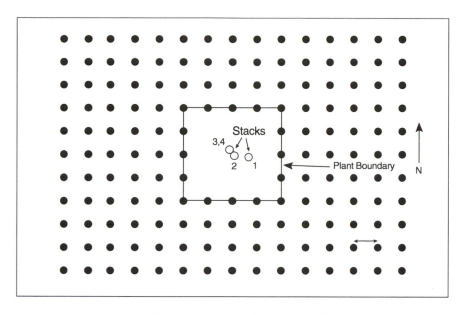

Figure 1. Schematic of Example Facility with Long-Term Impact Locations

The results of the dispersion modeling indicated the following maximum annual off-site concentrations for each of the source/pollutants combinations:

Source	Compound	Max. Impact	Location
Stack 1	Pollutant A	.788 µg/m³	X
Stack 2	Pollutant A	.305 µg/m³	Y
Stack 2	Pollutant B	.131 µg/m³	Y
Stack 3	Pollutant B	.172 µg/m³	Z
Stack 4	Pollutant B	.976 µg/m³	Z

It should be noted that the maximum concentrations from each source/receptor combination were not co-located. The positions of the maximum concentration from each source are indicated on Figure 1 corresponding to the letters X, Y, and Z in the table above. In general, the Tier 3 maximum concentration values are 25 to 30% as high as the Tier 2 values.

4.2.2 Cancer Risk Assessment

Concentrations from the ISCLT master file inventory are used by the RISK post-processor to calculate cancer risks at each receptor site in the ISCLT receptor array. RISK can then provide summaries of the calculated risks according to user specifications. Use of the RISK post-processor requires the following considerations:

1. As stated above, emission rate multipliers for each pollutant from each source should be provided as inputs to the RISK post-processor such that the product of the base emission rate input to ISCLT and the emission rate multiplier input to RISK equals the emission rate being modeled.

2. Unit cancer risk factors are provided to RISK either in the RISK post-processor input file or through an interactive process in TOXLT.

3. The RISK post-processor output options should be exercised to provide the total cancer risk at each receptor due to all pollutants, as well as individual pollutant or source contribution to these receptor-specific risks.

If the maximum predicted lifetime cancer risk in the receptor grid is less than the designated level of concern (e.g., 1×10^{-6}), placement of additional receptors in the ISCLT receptor array should be considered as a means of ensuring that the simulation is not underestimating maximum risk. If the maximum cancer risk in the receptor array is greater than the designated level of concern, additional runs of the RISK post-processor may be performed using reduced emission rate multipliers to assess the impacts of possible emission control scenarios. If the analysis shows no cancer risk greater than the designated level of concern and the receptor array is deemed adequate, the modeled source is considered to be in compliance with the specified criterion. In the case of noncompliance, it may be desirable on the part of the modeler to conduct a more refined analysis. See Section 5.0 if this document discusses some of the possibilities for further modeling refinements.

The output of the Risk post-processor for the example plant indicates that the maximum lifetime cancer risk outside the plant boundary is 4.2×10^{-7}, located at point W on Figure 1. Such a result would indicate that the facility would not cause a significant cancer risk to the public, according to the cancer risk level specified by the CAAA of 1990.

4.2.3   Chronic Noncancer Risk Assessment

In this assessment, concentrations from the ISCLT master file inventory are used by the RISK post-processor to calculate chronic noncancer hazard index values for a specific noncancer effect at each receptor site in the ISCLT receptor array. RISK can then provide summaries of the calculated index values according to user specifications. A separate risk simulation should be performed for each chronic noncancer effect being considered. Use of the RISK post-processor requires the following considerations:

1. As stated above, emission rate multipliers for each pollutant from each source should be provided as inputs to the RISK post-processor such that the product of the emission rate input to ISCLT and the emission rate multiplier input to RISK equals the actual emission rate being modeled.

2. Chronic threshold concentration values for the specific noncancer effect are provided to RISK either in the RISK post-processor input file or through an interactive process in TOXLT.

3. The RISK post-processor output options should be exercised to provide the total noncancer hazard index at each receptor due to all pollutants, as well as individual pollutant or source contribution to these receptor-specific hazard indices.

If the maximum hazard index value in the receptor grid exceeds 1.0, emission reduction scenarios can be performed (again, using reduced emission rate multipliers) to determine how this hazard index value can be reduced below 1.0. If the maximum hazard index value in the receptor grid does not exceed 1.0, the source(s) being modeled is considered to be in compliance with the specified criteria. In the case of non-compliance, it may be desirable on the part of the modeler to conduct a more refined analysis. See Section 5.0 if this document discusses some of the possibilities for further modeling refinements.

Using the chronic noncancer threshold concentration values for pollutants A and B of 20.0 and 5.0 $\mu g/m^3$, respectively, the RISK post-processor was exercised for the example facility to obtain a maximum hazard index value of 0.27 located at point Z on Figure 1. This result, which is approximately 30% of the Tier 2 result, would indicate that the facility does not present significant chronic noncancer risk in its current configuration.

4.3 Short-term Modeling

Short-term Tier 3 modeling uses the TOXST modeling system[4] to estimate maximum hourly concentrations and the receptor-specific expected annual number of exceedances of short-term concentration thresholds. For multiple pollutant scenarios, this amounts to the number of times the acute hazard index value exceeds 1.0. The model uses the ISCST model to calculate these hourly concentrations at receptor sites which are specified by the user. Acute hazard index values are subsequently calculated at each receptor by the TOXX post-processor, in which a Monte Carlo simulation is performed for intermittent sources to assess the average number of times per year the acute hazard index value exceeds 1.0 at each receptor.

4.3.1   Maximum Hourly Concentration Estimation

In addition to the information required to perform a Tier 2 analysis, the Tier 3 short-term analysis requires the following information:

1. five years of meteorological data from the nearest National Weather Service (NWS) station. These data are for the most recent, readily available consecutive five year period. NWS data are available through the electronic bulletin board (see Appendix A). Alternatively, one or more years of meteorological data from on-site measurements may be substituted. These data should be obtained and quality-assured using procedures consistent with the "Guideline on Air Quality Modeling (Revised)."[6]

2. plant layout information, including all emission point and fenceline locations. This information should be sufficiently detailed to allow the modeler to specify emission point and fenceline receptor locations within 2 meters of their actual locations.

3. pollutant-specific data concerning deposition or decay half-life, if applicable.

4. source-specific data concerning the annual average number of releases and their duration for all randomly-scheduled intermittent releases.

Once these data have been obtained, an input file should be prepared for execution of the ISCLT model using the guidance available in the ISC User's Guide.[18] The ISCST model should then be executed using the TOXST system. Procedures utilized should also be consistent with the TOXST User's Guide[5] (available through the electronic bulletin board, see Appendix A). Specific recommendations concerning the development of these inputs include:

1. Maximum hourly emissions rates are used for the analysis. The TOXST modeling system uses "base emission rates" and "emission rate multipliers" to specify the emission rate for each pollutant/source combination. Thus, for a given pollutant and source the emission rate equals the base emission rate (specified in the ISCST input file) times the emission rate multiplier for that pollutant/source combination (specified in the TOXX input file). The input file to the ISCST program should contain the same emission rates used in previous modeling tiers for each source, and the input file to the TOXX post-processor should be provided unit emission rate multipliers (1.0). If more than one pollutant is being emitted from the same source, that source may be included once in the ISCST input file with a unit emission rate (1.0) and the individual pollutant emission rates may be provided to the TOXX

post-processor. (It should be noted that this may complicate the interpretation of the printed ISCST output. Alternatively, multiple pollutants from the same source may be modeled as individual sources with actual emission rates in ISCST and unit emission rates in TOXX. This may require more computing time, but may allow direct interpretation of concentration predictions in the ISCST printed output. Regardless of which method is used, the modeler should take care that the product of the emission rate used in ISCST and the emission rate used in TOXX equals the emission rate of the pollutant and source being modeled.)

2. All continuous sources of the same pollutant should be modeled as one ISCST source group. Each intermittent source operating independently from one another should be modeled as a separate ISCST source group. All intermittent sources of the same pollutant emitting at the same tine may be modeled in the same ISCST source group. However, each source of more than one pollutant should be modelled as a source group by itself.

3. Input parameters in the ISCST input file should be set in accordance with the TOXST User's Guide.The regulatory default mode should be used. The ISCST output options should be chosen to provide summary results of the top 50 impacted receptors for each source group. (As noted earlier, if unit emission rates are being used in ISCST, interpretation of the concentration impacts as absolute may be inappropriate.)

4. Meteorological input files for ISCST may be created from NWS meteorological data using the RAMMET program (this program and a description of its use are available on the electronic bulletin board, see Appendix A).

5. A polar or rectangular receptor grid may be used, but with sufficient detail to accurately estimate the highest concentrations from each source. The design of the receptor network should consider the short-term results of the earlier modeling tiers such that the highest resolution of the receptors is in the vicinity of the highest predicted impacts. Additional receptors may need to be added in sufficient detail to accurately resolve the highest concentrations.

6. Where appropriate, direction-specific building downwash dimensions should be included for each radial direction.

7. The ISCST model option to create a TOXFILE output for post-processing should be chosen. The concentration threshold value (called "pcutoff") used to reduce the size of this binary concentration output file should be chosen appropriately to eliminate predicted concentration values below pos-

sible concern. Although it may be set higher, a good rule of thumb for setting this value is:

$$pcutoff = \frac{LACT}{\displaystyle\sum_{i=1}^{a}(Npol)_i}$$

where $LACT$ is the lowest acute concentration threshold value in the group of pollutants being modeled, and $Npol_i$ is the number of pollutants emitted from ISCST source group i.

The printed ISCST output will indicate the top 50 impacts for each ISCST source group, and the TOXFILE will contain all of the concentrations above the cutoff value from each ISCST source group at each receptor.

The ISCST model was exercised for the example facility. The maximum 1-hour concentrations for each source/pollutant combination were determined to be as follows:

Source	Compound	Max. Impact	Location
Stack 1	Pollutant A	34.5 µg/m³	Q
Stack 2	Pollutant A	67.9 µg/m³	R
Stack 2	Pollutant B	29.1 µg/m³	R
Stack 3	Pollutant B	39.2 µg/m³	S
Stack 4	Pollutant B	47.5 µg/m³	S

The locations of the predicted maximum 1-hour concentrations are shown in Figure 2. The maximum impacts from each source were only slightly lower than those from the Tier 2 analysis.

4.3.2   Acute Hazard Index Exceedance Assessment

Concentrations from the ISCST master file inventory are used by the TOXX post-processor to calculate acute hazard index values for each hour of a multiple-year simulation period at each receptor site in the ISCST receptor array. The program then counts the number of times a hazard index value exceeds 1.0 (an exceedance) and prints out a summary report which indicates the average number of times per year an exceedance occurs at each receptor. The use of the TOXX post-processor requires the following considerations:

1. As stated above, in most cases unit emission rate multipliers for each pollutant from each source are used as inputs to the TOXX post-processor.

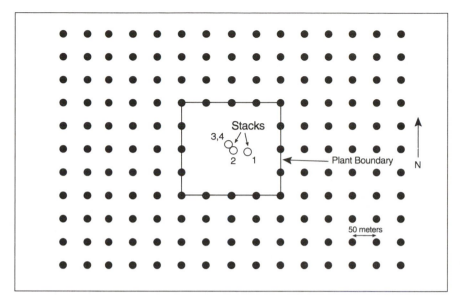

Figure 2. Schematic of Example Facility with Short-term Impact Locations

2. Acute threshold concentration values are provided to TOXX as the health effects thresholds in the TOXX post-processor input file.

3. The TOXX output option should be chosen to output the exceedances in polar grid format. Exceedance counts at discrete fenceline receptors will appear at the end of this table in the order in which discrete receptor locations were input to ISCST.

4. If only one pollutant is being modeled, the additive exceedance calculation option should not be chosen. If multiple pollutants are being modeled, the additive exceedance calculation option should be chosen. The TOXX post-processor should be set to perform 400 or more simulation years (maximum 1000). Unless otherwise specified by EPA guidance, background concentrations for toxic air pollutants should be set equal to 0.

5. The frequency of operation for each emission source is specified by providing values for the probability of the source switching on and the duration of the release. For each continuous emission, the probability of the source switching on is 1.0, and for each intermittent emission source, the probability of the source switching on is equal to the average number of releases per year divided by 8760 (the number of hours in a non-leap year).

The duration of release for each continuous source should be set equal to 1.0, and the duration of release for each intermittent release should be specified as the nearest integer hour which is not less than the release duration. (For example, if the average release duration is less than 1 hour, the duration of the release should be set equal to 1; if the average release duration is 3.2 hours, the duration of release should be set equal to 4.)

If the maximum number of acute hazard index exceedances in the receptor grid is less than some specified value (e.g., 0.1, equivalent to an average of 1 hourly exceedance every 10 years), the modeled source is considered to be in compliance with the acute threshold concentration criteria. However, resimulation with placement of additional receptors in the ISCST receptor array should be considered as a means of assuring that the simulation is not underestimating the maximum acute hazard index. If the maximum number of hazard index exceedances in the receptor array is greater than the specified value, additional runs of the TOXX post-processor with reduced emissions rate multipliers may be performed to assess the impacts of possible emission control scenarios. In the case of non-compliance, it may be desirable on the part of the modeler to conduct a more refined analysis. Section 5.0 of this document discusses such possibilities.

The TOXX post-processor was exercised for the example facility using the results form the ISCST simulation. The frequency of operation for each source ranged from 0.14 to 0.84, reflecting the actual yearly frequency of "on" time for each source. The output showed that none of the receptors experienced an impact resulting in a hazard index value of 1.0 or greater. Comparing this result with the Tier 2 result indicates that the hazard index never exceeds 1.0 because in a Tier 3 analysis the maximum impacts are seen not to occur at the same place and time. This indicates that the facility does not cause a significant health risk from acute exposure in its current configuration.

5.0 ADDITIONAL DETAILED ANALYSES

If any Tier 3 analyses indicate non-compliance with any of the user-specified criteria, it may be desireable to conduct an additional, more refined analysis. This may mean the use of on-site meteorological data or it may mean that a more appropriate modeling procedure is deemed applicable for the specific case. The determination of an appropriate alternative modeling procedure can only be made in a manner consistent with the approach outlined in the "Guideline on Air Quality Models (Revised)."[6]

In some cases, the EPA may allow exposure assessments to incorporate available information on actual locations of residences, potential residences, businesses, or population centers for the purpose of establishing the probability of human exposure to the predicted levels of toxic pollution near the source being modeled. In such cases, use of the Human Exposure Model (HEM II)[19] with the ISCLT dispersion model is preferred. Again, if the use of other modeling procedures is desired, the approval of a more appropriate alternative modeling procedure can only be made in a manner consistent with the approach outlined in Section 3.2 of the "Guideline on Air Quality Models (Revised)."[6]

6.0 SUMMARY OF DIFFERENCES BETWEEN MODELING TIERS

To summarize the major differences between the 3 modeling tiers described in this document, Table 3 below briefly lists the input requirements, output parameters, and assumptions associated with each tier. This Table may be used to quickly determine whether a given scenario may be modeled at any particular tier. Within each tier, cancer unit risk estimates, chronic noncancer concentration thresholds, and acute concentration thresholds are required to convert concentration predictions into cancer risks, chronic noncancer risks, and acute noncancer risks, respectively.

Modeling Tier	Input Requirements	Output Parameters	Major Assumptions
Tier 1	emission rate, stack height, minimum distance to fenceline	maximum off-site concentrations, worst-case cancer risk or worst-case noncancer hazard index (short- and long-term)	worst-case meteorology, worst-case downwash, worst-case stack parameters, short-term releases occur simultaneously, maximum impacts co-located, cancer and noncancer risks additive
Tier 2	emission rate, stack height, minimum distance to fenceline, stack velocity, stack temperature, stack diameter, rural/urban site classification, building dimensions for downwash calculations	maximum off-site concentrations, worst-case cancer risk and/or worst-case noncancer hazard index (short- and long-term)	worst-case meteorology, short-term releases occur simultaneously, maximum impacts co-located, cancer and non-cancer risks additive
Tier 3	emission rate, stack height, actual fenceline and release point locations, stack velocity, stack temperature, stack diameter, rural/urban site classification, local meteorological data, receptor locations for concentration predictions, frequency and duration of short-term (intermittent) releases	concentrations at each receptor point, long-term cancer risk estimates, chronic noncancer hazard index estimates at each receptor point, annual hazard index exceedance rate at each receptor.	cancer and noncancer risks additive

REFERENCES

1. Environmental Protection Agency, 1988. Glossary of Terms Related to Health, Exposure, and Risk Assessment. EPA-450/3-88-016. United States Environmental Protection Agency, Research Triangle Park, NC 27711.

2. Environmental Protection Agency, 1987. The Risk Assessment Guidelines of 1986. EPA-600/8-87-045. United States Environmental Protection Agency, Washington, DC 20460.

3. Brode, Roger W., 1988. Screening Procedures for Estimating the Air Quality Impact of Stationary Sources (Draft). EPA-450/4-88-010. United States Environmental Protection Agency, Research Triangle Park, NC 27711.

4. Environmental Protection Agency, 1992. Toxic Modeling System Short-Term (TOXST) User's Guide. EPA-450/4-92-002. United States Environmental Protection Agency, Research Triangle Park, NC 27711 (in preparation).

5. Environmental Protection Agency, 1992. Toxic Modeling System Long-Term (TOXLT) User's Guide. EPA-450/4-92-003. United States Environmental Protection Agency, Research Triangle Park, NC 27711 (in preparation).

6. Environmental Protection Agency, 1988. Guideline on Air Quality Models (Revised). EPA-450/2-78-027R. United States Environmental Protection Agency, Research Triangle Park, NC 27711.

7. Environmental Protection Agency, 1990. User's Guide to TSCREEN: A Model for Screening Toxic Air Pollutant Concentrations. EPA-450/4-90-013. United States Environmental Protection Agency, Research Triangle Park, NC 27711.

8. Environmental Protection Agency, 1991. Guidance on the Application of Refined Dispersion Models for Air Toxic Releases. EPA-450/4-91-007. United States Environmental Protection Agency, Research Triangle Park, NC 27711.

9. Catalano, J.A., D.B. Turner, and J.H. Novak, 1987. User's Guide for RAM - Second Edition. United States Environmental Protection Agency, Research Triangle Park, NC 27711.

10. Irwin, J.S., T. Chico, and J.A. Catalano. CDM 2.0 - Climatological Dispersion Model-User's Guide. United States Environmental Protection Agency, Research Triangle Park, NC 27711.

11. Environmental Protection Agency, 1991. Procedures for Establishing Emissions for Early Reduction Compliance Extensions. Draft. EPA-450/3-91-012a. United States Environmental Protection Agency, Research Triangle Park, NC 27711.

12. Environmental Protection Agency, 1978. Control of Volatile Organic Emissions from Manufacturers of Synthesized Pharmaceutical Products. EPA-450/2-78-029. United States Environmental Protection Agency, Research Triangle Park, NC 27711.

13. Environmental Protection Agency, 1980. Organic Chemical Manufacturing Volumes 1-10. EPA-450/3-80-023 through 028e. United States Environmental Protection Agency, Research Triangle Park, NC 27711.

14. Environmental Protection Agency, 1980. VOC Fugitive Emissions in Synthetic Organic Chem-

icals Manufacturing Industry - Background Information for Proposed Standards. EPA-450/3-80-033a. United States Environmental Protection Agency, Research Triangle Park, NC 27711.

15. Environmental Protection Agency, 1990. Protocol for the Field Validation of Emission Concentrations from Stationary Sources. EPA-450/4-980-015. United States Environmental Protection Agency, Research Triangle Park, NC 27711.

16. Pierce, T.E., Turner, D.B, Catalano, J.A., Hale, F.V., 1982. "PTPLU: A Single Source Gaussian Dispersion Algorithm." EPA-600/8-82-014. United States Environmental Protection Agency, Washington, DC 20460.

17. California Air Pollution Control Officers Association (CAPCOA), 1987. Toxic Air Pollutant Source Assessment Manual for California Air Pollution Control District and Applications for Air Pollution Control District Permits, Volumes 1 and 2. CAPCOA, Sacramento, CA.

18. Environmental Protection Agency, 1987. Industrial Source Complex (ISC) User's Guide- Second Edition (Revised), Volumes 1 and 2. EPA-450/4-88-002a and b. United States Environmental Protection Agency, Research Triangle Park, NC 27711.

19. Environmental Protection Agency, 1991. Human Exposure Model (HEM-II) User's Guide. EPA-450/4-91-010. United States Environmental Protection Agency, Research Triangle Park, NC 27711.

APPENDIX A
ELECTRONIC BULLETIN BOARD ACCESS INFORMATION

The Office of Air Quality Planning and Standards (OAQPS) of the EPA has developed an electronic bulletin board network to facilitate the exchange of information and technology associated with air pollution control. This network, entitled the OAQPS Technology Transfer Network (TTN), is comprised of individual bulletin boards that provide information on OAQPS organization, emission measurement methods, regulatory air quality models, emission estimation methods, Clean Air Act Amendments, training courses, and control technology methods. Additional bulletin boards will be implemented in the future.

The TTN service is free, except for the cost of the phone call, and may be accessed from any computer through the use of a modem and communications software. Anyone in the world wanting to exchange information about air pollution control can access the system, register as a system user, and obtain full access to all information areas on the network after a 1 day approval process. The system allows all users to peruse through information documents, download computer codes and user's guides, leave questions for others to answer, communicate with other users, leave requests for technical support from the OAQPS, or upload files for other users to access. The system is available 24 hours a day, 7 days a week, except for Monday, 8-12 a.m. EST, when the system is down for maintenance and backup.

The model codes and user's guides referred to in this document, in addition to the document itself, are all available on the TTN in the bulletin Board entitled SCRAM, short for Support Center for Regulatory Air Models. Procedures for downloading these codes and documents are also detailed in the SCRAM bulletin board.

Documentation on EPA-approved emission test methods is available on the TTN in the bulletin board entitled EMTIC, short for the Emission Measurement Testing Information Center. Procedures for reading or downloading these documents are also detailed in the EMTIC bulletin board.

The TTN may be accessed at the phone number (919)-541-5742, for users with 1200 or 2400 bps modems, or at the phone number (919)-541-1447, for users with a 9600 bps modem. The communications software should be configured with the following parameter settings: 8 data bits; 1 stop bit; and no (N) parity. Users will be asked to create their own case sensitive password, which they must remember to be able to access the network on future occasions. The entire network is menu-driven and extremely user-friendly, but any users requiring assistance may call the system operator at (919)-541-5384 during normal business hours EST.

APPENDIX B
REGIONAL METEOROLOGISTS/MODELING CONTACTS

Ian Cohen
EPA Region I (ATS-2311)
J.F.K. Federal Building
Boston, MA 02203-2211
FTS: 853-3229
Com: (617) 565-3225
E-mail: EPA9136
FAX: FTS 835-4939

James W. Yarbough
EPA Region VI (6T-AP)
1445 Ross Avenue
Dallas, TX 75202-2733
FTS: 255-7214
Com: (214) 255-7214
E-mail: EPA9663
FAX: FTS 255-2164

Rebecca Calby
EPA Region V (5AR-18J)
77 W. Jackson
Chicago, IL 60604
FTS: 886-6061
Com: (312) 886-6061
E-mail: EPA9553
FAX: FTS 886-5824

Robert Kelly
EPA Region II
26 Federal Plaza
New York, NY 10278
FTS: 264-2517
Com: (212)-264-2517
E-mail: EPA9261
FAX: FTS 264-7613

Richard L. Daye
EPA Region VII
726 Minnesota Avenue
Kansas City, KS 66101
FTS: 276-7619
Com: (913) 551-7619
E-mail: EPA9762
FAX: FTS 276-7065

Robert Wilson
EPA Region X (ES-098)
1200 Sixth Avenue
Seattle, WA 98101
FTS: 399-1530
Com: (206) 442-1530
E-mail: EPA9051
FAX: 399-0119

Alan J. Cimorelli
EPA Region III (3AM12)
841 Chestnut Building
Philadelphia, PA 19107
FTS: 597-6563
Com: (215) 597-6563
E-mail: EPA9358
FAX: FTS 597-7906

Larry Svoboda
EPA Region VIII (8AT-AP)
999 18th Street
Denver Place-Suite 500
Denver, CO 80202-2405
FTS: 776-5097
Com: (303) 293-0949
E-mail: EPA9853
FAX: FTS 330-7559

Lewis Nagler
EPA Region IV
345 Courtland Street, N.E.
Atlanta, GA 30365
FTS: 257-3864
Com: (404) 347-2864
E-mail: EPA9470
FAX: FTS 257-5207

Carol Bohnenkamp
EPA Region IX (A-2-1)
75 Hawthorne Street
San Francisco, CA 94105
FTS: 484-1238
Com: (415) 744-1238
E-mail: EPA9930
FAX: FTS 484-1076

TECHNICAL REPORT DATA

(Please read Instructions on reverse before completing)

1. REPORT NO. EPA-450/4-92-001	2.	3. RECIPIENT'S ACCESSION NO.
4. TITLE AND SUBTITLE A Tiered Modeling Approach for Assessing the Risks Due to Sources of Hazardous Air Pollutants		5. REPORT DATE
		6. PERFORMING ORGANIZATION CODE
7. AUTHOR(S) David E. Guinnup		8. PERFORMING ORGANIZATION REPORT NO.
9. PERFORMING ORGANIZATION NAME AND ADDRESS U.S. Environmental Protection Agency Office of Air Quality Planning and Standards Technical Support Division Research Triangle Park, NC 27711		10. PROGRAM ELEMENT NO.
		11. CONTRACT/GRANT NO.
12. SPONSORING AGENCY NAME AND ADDRESS		13. TYPE OF REPORT AND PERIOD COVERED
		14. SPONSORING AGENCY CODE
15. SUPPLEMENTARY NOTES		
16. ABSTRACT		

TECHNICAL REPORT DATA FORM

582

This document provides modeling guidance to support risk assessments as applied to stationary sources of hazardous air pollutants. The guidance focuses on procedures which may be used in support of the petition processes described in Title III of the Clean Air Act Amendments of 1990. The analysis approach described herein is a tiered one, in which each subsequent modeling tier requires additional site-specific information to produce a less conservative estimate of the risk associated with a given stationary source (or group of sources). The modeling approach begins with Tier 1 screening tables which require only source emission rates, stack heights, and nearest fenceline distances to estimate maximum cancer and/or noncancer risks. Tier 2 utilizes additional source parameters (including stack diameter, exit gas temperature and velocity, and nearby building dimensions) with the SCREEN computer program to develop more refined estimates of maximum risks. Tier 3 utilizes site-specific meteorological data, plant layout information, and release frequency data with the TOXST and TOXLT computer models to provide additional refinement to these assessments.

17. KEY WORDS AND DOCUMENT ANALYSIS

a. DESCRIPTORS	b. IDENTIFIERS/OPEN ENDED TERMS	c. COSATI Field/Group
Air Pollution Atmospheric Dispersion Modeling Risk Assessment		

18. DISTRIBUTION STATEMENT	19. SECURITY CLASS (Report) Unclassified	21. NO. OF PAGES
Release Unlimited	20. SECURITY CLASS (This Page) Unclassified	22. PRICE

EPA Form 2220-1 (Rev. 4-77) PREVIOUS EDITION IS OBSOLETE

TECHNICAL REPORT DATA FORM (*continued*)

APPENDIX

K

Science Advisory Board Memorandum on the Integrated Risk Information System and EPA Response

Honorable William K. Kelly
Administrator
U.S. Environmental Protection Agency
401 M Street, S.W.
Washington, D.C. 20460

Subject: Science Advisory Board's review of the Integrated Risk Information Systems

Dear Mr. Reilly:

The Environmental Health Committee of the Science Advisory Board (SAB) was given a presentation by EPA staff on the Integrated Risk Information System (IRIS) at its meeting on October 26, 1989. The presentation also included discussion of the activities of the Carcinogen Risk Assessment Verification Endeavor (CRAVE) and the RfD (Reference Dose) Review Group.

While it is our understanding that the IRIS was developed primarily for use within EPA, the Committee believes that the IRIS would be of great utility both within EPA and other organizations concerned with the potential health impacts of toxic chemicals in the environment. IRIS has the potential to provide a summary of toxicological data for a large number of chemicals in readily accessible form, either from an EPA on-line computer data bank, from access through existing routes such as the National Library of Medicine's TOXNET, or from regularly updated computer diskettes distributed to IRIS users. Many state and local regulatory agencies, as well as scientists working in the field of regulatory toxicology, would find IRIS to be a valuable reference source.

The IRIS files contain not only the toxicological data, but also EPA's summary of these data, which may be in the form of the weight-of-evidence characterization for carcinogenicity, unit risk numbers for substances judged to have sufficient evidence for carcinogenicity in animals or humans, and reference dose numbers. This type of information may be widely used both within EPA and by other environmental regulatory agencies as the basis for regulatory decisions. It is therefore very important that the information in IRIS be carefully reviewed for its accuracy, timeliness, and completeness, and that appropriate caveats regarding the data and EPA's evaluation of the data be included in the IRIS files.

We recommend that SAB reviews of Agency documents on specific substances be referenced in the IRIS files for these substances. A short summary of the SAB evaluation of EPA conclusions, especially as to the weight-of-evidence characterization, unit risk, or reference dose, should also be included in the IRIS file, and a short summary of any subsequent communication from the Administrator back to SAB in response to its evaluation.

We understand that <u>Federal Register</u> notices of proposed regulatory actions and final regulatory actions for chemicals in IRIS are now included in the regulatory summaries of IRIS files for those chemicals, a step forward which we commend. In the same vein, major EPA scientific reports such as health advisories, health assessment documents, criteria documents, and Risk Assessment Forum reports should also be cited in IRIS files, and we understand that this will occur in the future. Checks of the files for individual chemicals indicated that IRIS currently lacks citations to some key EPA reports on specific chemicals.

The current computer implementation of IRIS is somewhat cumbersome. For example, capabilities such as returning to earlier text in files or doing searches for specific words or phrases are not available in the current implementation. We understand that the computer implementation of IRIS will be upgraded, and we urge EPA to develop an implementation that is flexible, and "user friendly" for the spectrum of anticipated users both inside and outside of EPA. EPA should also consider the need for, and potential benefits from, developing more training materials and on-line help capabilities to assist users unfamiliar with IRIS to learn how to use the system. In any such efforts, EPA should remain cognizant that an increase in users should be expected, and the system designed accordingly.

The Agency needs an overall strategy on computerized lists of chemicals, one which takes into account the differing needs of various segments of the user community. While IRIS may be vary helpful for those wishing to know about the toxicological data, other users may simply wish to know what regulatory actions EPA has taken on a specific chemical, or how to deal with an emergency response in the event of chemical spills. EPA either has or is developing other computerized lists of chemicals, but the planning and coordination among these efforts could be improved. EPA should consider what computerized chemical lists are needed, and, more broadly, how modern computer and telecommunications technology can assist in the processes of risk assessment and risk management for the thousands of chemicals that are of interest to EPA. The Agency should then take steps to assure coordination, cross referencing, and standardization in access procedures for the various computerized lists of chemicals it is, and will be, developing.

The Environmental Health Committee is pleased to have had the opportunity to review IRIS and to offer its advice. We would appreciate your response to the major points we have raised:

1. Need for critical review of data for accuracy and completeness

2. Inclusion of SAB evaluation

3. Citation of major relevant EPA reports, including health advisories and other key documents

4. Implementation of improved electronic systems to allow more flexible handling of the data

5. Development of training materials and on-line help

6. Coordination, cross-referencing, and standardization of access to the various listings under development.

We will be pleased to assist the Agency further as it proceeds with the development of IRIS and other computerized chemical lists.

Dr. Raymond Loehr, Chairman
Science Advisory Board
Executive Committee

Dr. Arthur Upton, Chairman
Environmental Health Committee

Dr. Raymond Loehr
Chairman
Science Advisory Board
U.S. Environmental Protection Agency
401 M Street, S.W.
Washington, D.C. 20460

Dear Ray:

Thank you very much for your letter of March 14, 1990, and your comments on the Environmental Protection Agency's (EPA) Integrated Risk Information System (IRIS). I greatly appreciate, and share, the Science Advisory Board's interest in IRIS and its future.

As you correctly state, IRIS is an important risk information resource both for the Agency, and for other organizations concerned with the potential health impacts of toxic chemicals in the environment. Because the summary risk information on, to date, 397 chemicals represents authoritative EPA consensus positions on the adverse health effects of these chemicals, the Agency is aware of its obligation to the user community to provide system oversight and quality assurance. I share your concerns that the IRIS risk information be as accurate, timely, and complete as possible, and that appropriate discussion and/or caveats be included in the IRIS files.

In your letter you raise several interesting points which I have asked the Office of Health and Environmental Assessment in the Office of Research and Development, responsible for IRIS development and management, to reply. Please see the enclosure for a detailed response.

Again, thank you very much for your letter and your interest in the Agency's IRIS data base. We welcome your comments and appreciate your offer to work with the Agency as it proceeds with future development of IRIS.

Sincerely yours,

William K. Reilly

Enclosure

The Integrated Risk Information System (IRIS) is one of the Agency's major risk information resource tools containing summaries of health risks and EPA regulatory information on, to date, 397 chemicals. Updated monthly, it is used by EPA to provide high quality, timely scientific and technical information to Agency scientists, and promote Agency-wide coordination and consistency of risk assessments. Because IRIS, containing authoritative consensus EPA positions on chemical-specific potential adverse health effects, is used extensively both inside and outside the Agency, we recognize the need to maintain, and improve, the quality of the system, its access and delivery systems, and sufficient oversight. We welcome the Science Advisory Board's (SAB) interest and comments on IRIS and take this opportunity to respond to the major points raised in your March 14 letter.

1. **Data Review** - As you know, two Agency work groups develop the risk information summaries that appear in IRIS. Each work group is comprised of approximately 20 senior Agency scientists and statisticians from risk assessing program offices, laboratory facilities, and regional offices. During, and subsequent to, the work group deliberations, there are several levels of quality control and internal review built into the IRIS information development process. First, particularly in the case of the Carcinogenic Risk Assessment Verification Endeavor (CRAVE), an emphasis is placed on the use of external/ and or SAB peer-reviewed documents (e.g., Health Assessment Documents, Drinking Water Criteria Documents) to support these summaries and the quantitative risk values they contain. While the Reference Dose (RfD) Work Group process is different, actually developing the oral RfDs for each chemical, they use the same consensus procedures as the CRAVE Work Group. Also, the Oral RfD Work Group methodology has been peer-reviewed and receives SAB oversight.

Second, an extensive technical quality control process is part of each work group's operating procedures. Technical quality control includes internal work group draft Summary review, final Summary review, final check prior to IRIS loading, and a further check after the summary is online. This final consensus summary sheet development is the primary goal of the work groups and reflects the diligence and hard work of the group Chairs and members.

Third, an editorial quality check is conducted prior to loading on the System. This check, performed by a contractor, is being done on all chemical files currently on IRIS and on new files before they go online. It includes an edit for clarity, style, continuity, and typographical errors.

Finally, since 1986 when IRIS was made available to the Agency, and 1988 when IRIS was made available to the public, its use has grown far beyond earlier expectations. We acknowledge that additional oversight of the system is warranted. To that end, EPA's Risk Assessment Council, which is chaired by Deputy Administrator F. Henry Habicht II, has established a subcommittee for IRIS. This subcommittee, chaired by Dr. William H. Farland, Director, Office of Health and Environmental Assessment, Office of Research and Development, will address both generic and chemical-specific issues concerning IRIS and its associated work groups. Also, IRIS status will be an agenda item at each Council meeting.

These various levels of review and oversight help to assure that IRIS remains an important resource toll and that the quality and validity of the information continues to improve.

2. **SAB Evaluations** - Preliminary discussions with SAB Director Dr. Donald Barns regarding addition to IRIS of short summaries that could include SAB evaluation of and comments on the principal EPA documents that support the CRAVE and RfD findings, have taken place. The inclusion of the SAB information would underscore the argument that while individual IRIS summaries are not peer-reviewed, the reports and documents on which the summaries are based have received external review. The process and management details on how to accomplish this task will be worked out in conference with Dr. Barnes, Dr. Farland, and the IRIS staff.

3. **EPA Reports** - Only citations for EPA scientific reports and other references used in developing the RfD and/or CRAVE summaries are included in IRIS. Full bibliographies listing those references are currently being prepared and loaded on the System. Thus far, bibliographies for 251 chemicals are online, with 146 to go. Once the addition of all bibliographies are complete, a user will have citations for all reports, studies, and documents used by the two work groups. Also, summaries of Drinking Water Health Advisories are included on IRIS in Section III: HEALTH HAZARD ASSESSMENT FOR VARIED EXPOSURE DURATIONS. A backlog of the Drinking Water summaries currently exists. The IRIS staff is in the process of putting them on the system.

When IRIS was initially developed in 1986, EPA Regulatory Actions (Section IV) were part of the system. These regulatory action sections provide information, including applicable Federal Register citations, for the Clean Air Act, Safe Drinking Water Act, Clean Water Act, Federal Insecticide, Rodenticide and Fungicide Act, Toxic Substances Control Act, Resource Conservation and Recovery Act and the Superfund Reauthorization Act. Because this regulatory information is subject to change, we are aware that this section needs to be carefully reexamined to insure that it is up-to-date and complete. Working with the Risk Assessment Council's IRIS Subcommittee, the IRIS staff is in the process of developing a proposal to review and update the present regulatory action section. This work should commence in the near future.

4. **Delivery Systems** - Currently, Agency scientists and SAB members access IRIS using the EPA Electronic Mail system (EMAIL). IRIS on EMAIL is slow, cumbersome, and offers little or not reporting capabilities. In 1986, IRIS was a new Agency resource tool containing both qualitative and quantitative risk assessment information. The EMAIL delivery system, by design, obliged users to look at the whole chemical file, not just selected small sections, thus providing a wider chemical profile. At that time, there was concern that only the quantitative risk values would be accessed and not the qualitative discussion of the underlying studies, reports, assumptions, and limitations which is critical in evaluating and understanding the derivation of the risk values. As risk assessment methodologies have become more sophisticated, so too have IRIS users become more experienced and sophisticated in interpreting, evaluating, and using the IRIS risk information. Therefore, the time is right to provide them with a greatly enhanced delivery system that is fast, flexible, interactive, and user friendly.

On March 5, 1990, IRIS became available on the National Library of Medicine's (NLM) Toxicology Network (TOXNET). TOXNET is an online system that is highly regarded and easily accessed. IRIS on TOXNET provides many of the sophisticated functions requested by users. For more information on TOXNET, please refer to the enclosed NLM IRIS Fact Sheet.

Also, a Personal Computer (PC) based version of IRIS is being developed. The PC delivery system will provide the user with sophisticated user capabilities including easy movement within files, reliable keyword and string searches, reporting options, and a fast, accurate, and easily accessible system. We anticipate its availability in early 1991.

5. **Training Materials** - Your comments on the need for more and better developed IRIS training materials and online help are correct. The current user guide was inadequate for the users needs and did not provide clear, concise, and complete instructions. A revised user guide has been completed and the final version will be available both online and in paper copy by the end of May 1990. Also, development of new online help and other training materials are under consideration, including a revised case study, fact sheets, and interactive demonstration diskettes.

Training has been an important part of IRIS from its inception. A large training program both at Headquarters and in the Regions accompanied IRIS's availability in 1986. Presently, each Region has its own IRIS coordinator who conducts training as needed, and the IRIS staff conducts workshops and seminars both inside and outside the Agency on a regular basis. A joint symposium, sponsored by EPA and the Chemical Manufacturers Association, on IRIS and some of its underlying risk assessment methodologies is being considered as another opportunity to stress appropriate use of the system. Also, finalization and distribution of the PC version of IRIS will result in another round of intensive Agency-wide user training.

6. **List Coordination** - The Agency recently took a major step forward in coordinating and cross-referencing regulatory and regulation-like lists, by approving development of a pointer system that will contain references to all chemicals and other pollutants regulated by EPA and to all the lists on which each chemical or pollutant occurs. This system, tentatively called the Registry of Lists, is currently under development; a prototype should be built during this calendar year, and the system should be generally available in one to two years. It will be designed as a pointer system, telling users where other information is available, because each individual list has been compiled for different programmatic reasons, and there is generally not a uniform set of data elements across the lists. IRIS chemicals will be referred to explicitly in the Registry of Lists, and IRIS and the Registry will be compatible to ensure that IRIS users can get complete cross-reference information.

If you have further questions or comments regarding any of the responses included above, please contact Linda Tuxen, EPA IRIS Coordinator, at 202-382-5949 (FTS 382-5949).

L

Development of Data Used in Risk Assessment

This appendix provides additional information on the data needed to estimate different elements in the risk-characterization steps of emission characterization, transport and fate, exposure assessment, and assessment of toxicity.

EMISSION CHARACTERIZATION

The best approach to characterizing emissions is to measure the flux from each manufacturing, storage, use, or disposal facility. However, such flux measurements are generally not available, because sources are not uniform across geography or time, because they are so large (e.g., a several-square-block manufacturing site) that no point for measuring flux is apparent, or because flux measurements are so difficult and expensive, and require such detailed knowledge of local meteorology, as to be impractical. Therefore, most emission data are calculated or estimated from industry-wide averages applied to such things as "emission factors," process rates, quantities of chemical present at given locations, or numbers of individual components. Some information that might be needed to estimate and characterize emissions from a facility is provided in Table L-1. (Not all information is needed for all calculation methods.)

TRANSPORT AND FATE

Atmospheric-chemistry models are used to determine where emitted chemicals are transported and their characteristics when deposited. Several kinds of information are needed to estimate the transport and fate of pollutants:

TABLE L-1   Potential Data Needs for Calculation of Emissions

Process Vents
1. Volumetric flow rate of vent gas
2. Vent-gas discharge temperature
3. Concentration of individual or aggregate HAP
4. Operating hours per year of unit operation
5. Molecular weight of gas
6. Efficiency of control device
7. Production rate during measurement

Fugitive Emission
1. Numbers of pumps, valves, flanges, pressure-relief valves, open-ended lines, and compressors
2. Screening level
3. Weight % of HAPS in stream
4. Percent leaking equipment
5. Other HAPS characterization
6. Frequency of leak checking

Loading Emission
1. Type of cargo carrier
2. Mode of operation
3. Annual volume of liquid loaded
4. Temperature of liquid loaded
5. Weight in percent of HAP in loaded material
6. True vapor pressure of HAP loaded
7. Molecular weight of HAP
8. Efficiency of control device

Storage-Tank Emissions
1. Material stored
2. Diameter of tank
3. Rim seal type
4. Tank, roof, and shell color
5. Ambient temperature
6. Wind speed
7. Density and partial pressure of chemical
8. Molecular weight
9. Vapor pressure
10. Efficiency of control device
11. Type of storage tank
12. Annual throughput
13. Number and diameter of columns

Emission Factors
1. Magnitude of input into the process
2. Production level

Wastewater Sources
1. Volumetric flow rate of wastewater
2. Concentration
3. Production rate during flow determination
4. Production rate during concentration determination

Source: EPA, 1991c.

• Data on emissions of pollutants that result from production, storage, use, and disposal (discussed in previous section).

• Data on physical and chemical properties of pollutants (see Table L-2). For example, the vapor pressure of a chemical pollutant plays a major role in determining exchange of the chemical between the atmosphere and other environmental media. The vapor pressures of chemicals vary widely from those of gases (such as CO, CO_2, and SO_2), with vapor pressures of more than 1 atm, to those of aromatic compounds, organophosphates, dioxins, and other non-criteria pollutants, which are often in the range of 10^{-8}–10^{-3} atm. VOCs generally have vapor pressures of greater than 10^{-3} atm and semivolatile compounds vapor pressures of 10^{-8}-10^{-3} atm. Lead and other inorganic species are volatile as well. Water solubilityis important, because, with vapor pressure, it determines the distribution of a pollutant in the atmosphere. Water-soluble vapors, for example, might be efficiently scrubbed from air by rainfall or fog deposition—processes that can minimize human exposure, at least by inhalation. Suspended dust or aerosol particles can adsorb vapors of the pollutant and may also play a major role in determining the rate of exchange of chemicals between the atmosphere and other environmental media.

• Data on transformation, degradation, and sequestration of pollutants in the environment (Table L-2), including chemical, biologic, and physical data:

— Chemical data (e.g., for atmospheric oxidation and photochemical reactions). Chemical breakdown depends on molecular structure, and for some substances breakdown is rapid. If the chemical is susceptible to nucleophilic attack, oxidation, or hydroxylation, alterations can occur rapidly and change the potential exposure dramatically.

— Biologic data (e.g., on degradation by metabolic action of microorganisms). Alterations by biologically mediated reactions are enormously variable, and data are needed on products of alteration; for example, do emissions tend to become more toxic or less toxic?

— Physical data (e.g., on solubility and gravitational settlement). For particles, gravitational settlement or sedimentation increases with the aerodynamic diameter of the particle. Physical processes that occur in the atmosphere can affect particle-removal efficiency. Hydroscopic particles can increase in size because of the accumulation of water from the vapor phase in the atmosphere; this growth can help in their removal by sedimentation and washout.

• Data on rate of removal of pollutants by various routes. For example, the rate of catalytic oxidation of SO_2 decreases if the water concentration in the atmosphere falls below that necessary to maintain catalyst droplets. The critical point seems to be the percent relative humidity; above this, rates of catalytic oxidation increase dramatically. In clean air, SO_2 emissions are only very slowly oxidized via homogeneous reactions of the gas phase to SO_2 vapor. The development of the kind of information described here is important for the pre-

TABLE L-2 Physicochemical Properties of Chemical and Its Atmospheric Environment Important in Transport-Fate Calculations

Properties of Chemical	Properties of Environment
Physical properties:	Particulate load:
Molecular weight	For dust, other solid particulate matter
Density	For liquid aerosols
Vapor pressure (or boiling point)	Oxidant level
Water solubility	Temperature
Henry's constant (air-water distribution coefficient)	Relative humidity
Lipid solubility (or octanol-water distribution coefficient)	Amount and intensity of sunlight
	Amount and frequency of precipitation
Soil sorption constant	Meteorologic characteristics:
Chemical properties:	Ventilation
Rate constants for	Inversion
Oxidation	Surface cover:
Hydrolysis	Water
Photolysis	Vegetation
Microbial decomposition	Soil type
Other modes of decomposition	
Particle properties:	
Size	
Surface area	
Chemical composition	
Solubility	

diction of risk associated with environmental pollutants. Such data could be used to identify the most probable routes through the environment and provide clues to the rate of degradation (alteration) from source to receptor. Knowing the probable routes and sinks, one can identify populations that should have special attention in an evaluation of potential health effects. More refined approaches might include selecting or developing models to estimate transport and fate of pollutants.

• Data on types of models to predict the persistence, transport, and fate of pollutants, including their input requirements, degree of accuracy and precision, and method of validation. Several models of aerial dissipation have been reported.

EXPOSURE ASSESSMENT

To evaluate human exposure for risk-assessment purposes, information is needed on the following:

• Contaminants (e.g., types, in which media, at what concentrations, and for what durations).

• Exposed population (e.g., who is at risk, where, and under what circumstances; how long they are exposed and to what degree; and their intake of the contaminant from air, food, water, or through other relevant routes).

These are described in more depth below.

For the contaminant, the minimum data need include measured or estimated concentrations at the point of human contact for a specified duration. For air, concentration data are generated by sampling air and simultaneously or sequentially measuring the toxicant trapped at a given air flow rate and for a given period monitored. Beyond those generalities, analytical methods vary widely in specifics and in the key dimensions of accuracy (agreement with true value), precision (spread in data), and limit of detection. Errors can be large, particularly in trace analysis, so concerns are warranted about the quality of concentration data used in risk assessments. The following cautions are pertinent:

• All data should be collected with validated methods under strict quality-assurance and quality-control standards.
• A clear statement of uncertainty is fundamental to all analytic reports (Keith et al., 1983). Errors are likely to be greater with airborne trace-amount toxicants than with "criteria pollutants," which tend to occur at much higher concentrations. This is because the relative accuracy of instruments often decreases at low concentrations.
• A contaminant might be present but below the detection limit of the equipment. In this case, the concentration of the contaiminant should not be assumed to be zero. Rather, the detection limit (or some agreed-on fraction of it) should be used in the processing of data.
• Vapors must be discriminated from particle-bound residues in air monitoring, especially for toxicants of low to intermediate vapor pressure.
• Data on trace toxicants should be confirmed by mass spectrometry or other confirmatory method to increase confidence in the results.

For the *exposed population*, the nature of the harm must be defined. It is important to assess the various degrees of exposure and the numbers within each identifiable set of the population, such as sets defined by age or health status. In the absence of personal monitoring data, geographic, behavioral (e.g., activity-pattern), and demographic considerations will often allow estimation of the exposure, although the estimated exposure might not be directly related to an individual's exposure.

Because exposure to a specific chemical is rarely confined to a single route (although one route might dominate), the total exposure must be calculated by summing air (inhalation), dermal, and dietary (food and water) intakes. For example, pollutants that begin as "air pollutants" can generate substantial exposures through other media if they can move from air to water, soil, or vegetation.

A case in point is that of chlorinated hydrocarbons (polychlorinated biphenyls, toxaphene, DDT, etc.) in the Arctic; the mechanism was long-range transport in the air, but the exposure of indigenous peoples in the region is through the diet and results from the uptake of chemicals deposited in the food chain.

ASSESSMENT OF TOXICITY

A risk analysis must include an assessment of the toxicity of a chemical, i.e., of the potential hazard the public health. Such analysis can be based on a combination of experimental toxicity and human data. Clearly, information on the incidence of disease associated with known exposures to toxicants is the most useful for human risk assessment. It is also the least available, however, because it depends on the occurrence of some unplanned or unforeseen event (e.g., an accident or malfunction in a manufacturing facility) or it is collected for a narrowly defined population (e.g., a workforce) exposed at magnitudes and for durations well beyond what the general population experiences. For ethical (and also sometimes legal) reasons, controlled dose-response studies in humans are rare.

The human data that might be available for risk assessment are in three broad categories:

• *Clinical.* Outcome and disease data are reported for members of the general population, including, if known:
— A description of the outcome(s).
— The diagnostic criteria used.
— A description of individual characteristics that might affect outcomes (age, pre-existing illness, etc.).
— Exposure history, including dose and time frames.

The opinions of medical experts on the findings and the applicability of the results to the general population are also important in determining the usefulness of clinical evidence for risk assessment.

• *Toxicologic.* Outcome and disease data are reported for persons (usually volunteers, not members of the general population) after exposure under controlled experimental conditions, including:
— Description of the hypotheses tested.
— The criteria used to select the study groups.
— The relevance of the outcomes to the general population or specified subpopulations (e.g., potential high-risk groups).
— The diagnostic and detection methods.
— The experimental conditions.

— Personal characteristics that might affect exposure and outcome (e.g., age, sex, and pre-existing conditions).

In addition, the method of exposure (nature and composition of toxic agent, routes of exposure, media and means of exposure, time of exposure, and doses) and statistical evaluation (e.g., point and range estimates, measures of association and significance, and dose-response and time-response relations) should be described.

- *Epidemiologic.* Outcome and disease data are collected on groups of people in real-world settings. These data should be accompanied by:
    — A description of the hypotheses tested.
    — Criteria applied to select groups observed.
    — Study methods and target-group participation rates.
    — Diagnostic criteria for clearly defined outcomes.
    — Exposure history and characteristics, including period and doses relevant to outcome studied.
    — Evaluation of characteristics that might affect exposure and outcome (e.g., age, employment, activity patterns, and pre-existing health conditions).
    — Appropriate statistical analyses of comprehensive outcome measures (e.g., point and range estimates, dose-response data, time-response analysis, and measures of association and significance)
    — Interpretation of the findings, including analysis of generalizability, bias, and other confounding issues.

REFERENCES

EPA (U.S. Environmental Protection Agency). 1991. Procedures for Establishing Emissions for Early Reduction Compliance Extensions. Vol. 1. EPA-450/3-91-012a. U.S. Environmental Protection Agency, Washington, D.C.

Keith, L.H., G. Choudhary, and C. Rappe. 1983. Chlorinated Dioxins and Dibenzofurans in the Total Environment. Woburn, Mass.: Ann Arbor Science.

M

Charge to the Committee

The charge to the committee, as stated in Section 112(o) of the Clean Air Act Amendments of 1990 (CAAA-90), is as follows:

(1)   REQUEST OF THE ACADEMY.—Within 3 months of the date of enactment of the Clean Air Act Amendments of 1990, the Administrator shall enter into appropriate arrangements with the National Academy of Sciences to conduct a review of—

(A)   risk assessment methodology used by the Environmental Protection Agency to determine the carcinogenic risk associated with exposure to hazardous air pollutants from source categories and subcategories subject to the requirements of this section; and

(B)   improvements in such methodology.

(2)   ELEMENTS TO BE STUDIED.—In conducting such review, the National Academy of Sciences should consider, but not be limited to, the following—

(A)   the techniques used for estimating and describing the carcinogenic potency to humans of hazardous air pollutants; and

(B)   the techniques used for estimating exposure to hazardous air pollutants (for hypothetical and actual maximally exposed individuals as well as other exposed individuals).

(3)   OTHER HEALTH EFFECTS OF CONCERN.—To the extent practical, the Academy shall evaluate and report on the methodology for assessing the risk of adverse human health effects other than cancer for which safe thresh-

olds of exposure may not exist, including, but not limited to, inheritable genetic mutations, birth defects, and reproductive dysfunctions.

(4)   REPORT.—A report on the results of such review shall be submitted to the Senate Committee on Environmental and Public Works, the House Committee on Energy and Commerce, the Risk Assessment and Management Commission established by section 303 of the Clean Air Act Amendments of 1990 and the Administrator not later than 30 months after the date of enactment [May 15, 1993] of the Clean Air Act Amendments of 1990.

(5)   ASSISTANCE.—The Administrator shall assist the Academy in gathering any information the Academy deems necessary to carry out this subsection. The Administrator may use any authority under this Act to obtain information from any person and to require any person to conduct tests, keep and produce records, and make reports respecting research or other activities conducted by such person as necessary to carry out this subsection.

(6)   AUTHORIZATION.—Of the funds authorized to be appropriated to the Administrator by this Act, such amounts as are required shall be available to carry out this subsection.

(7)   GUIDELINES FOR CARCINOGENIC RISK ASSESSMENT.— The Administrator shall consider, but need not adopt, the recommendations contained in the report of the National Academy of Sciences prepared pursuant to this subsection and the views of the Science Advisory Board, with respect to such report. Prior to the promulgation of any standards under subsection (f), and after notice and opportunity for comment, the Administrator shall publish revised Guidelines for Carcinogenic Risk Assessment or a detailed explanation of the reasons that any recommendations contained in the report of the National Academy of Sciences will not be implemented. The publication of such revised Guidelines shall be a final Agency action for purposes of section 307.

N-1

The Case for "Plausible Conservatism" in Choosing and Altering Defaults

Adam M. Finkel

This Appendix was written by one member of our committee, who was asked to represent the viewpoint of those members of the committee who believe that EPA should choose and refine its default assumptions by continually evaluating them against two equally important standards: whether the assumption is scientifically plausible, and whether it is "conservative" and thus tends to safeguard public health in the face of scientific uncertainty. Indeed, these three themes of plausibility, uncertainty, and conservatism form most of the framework for the last six chapters of the CAPRA report, as reflected in the "crosscutting" chapters on model evaluation, uncertainty and variability, and on implementing an iterative risk assessment/management strategy. The particular way these themes should come together in the selection and modification of default assumptions is controversial; hence, the remainder of this appendix is organized into five parts: (1) a general discussion of what "conservatism" does and does not entail; (2) an enumeration of reasons why conservatism is appropriately part of the rationale for choosing and departing from defaults; (3) the specific plan proposed for EPA's consideration;[1] (4) a side-by-side analysis of this proposal against the competing principle of "maximum use of scientific information" (see Appendix N-2 following this paper); and (5) general conclusions.

---

[1] Although I will discuss and evaluate the general issue of conservatism in detail before I present our specific recommendations, I urge readers to consider whether the proposal detailed in this third section bears any resemblance to the kind of "conservatism for conservatism's sake" that critics decry.

WHAT IS "CONSERVATISM"?

The most controversial aspect of this proposal within the full committee was its emphasis on "conservatism" as one—not the only—organizing principle to judge—not to prejudge—the merits of defaults and their alternatives. Supporters of this proposal are well aware that there are strengths and weaknesses of the conservative orientation that make it one of the most hotly-contested topics in all of environmental policy analysis, but also believe that few topics have been surrounded by as much confusion and misinformation. Some observers of risk assessment appear to be convinced that EPA and other agencies have so overemphasized the principle of conservatism as to make most risk estimates alarmingly false and meaningless; others, including at least one member of this committee, have instead suggested that if anything, the claims of these critics tend to be more reflexive, undocumented by evidence, and exaggerated than are EPA's risk estimates themselves (Finkel, 1989). It is clear that partisans cannot agree on either the descriptive matter of whether risk assessment is too conservative or on the normative matter of how much conservatism (perhaps any at all) would constitute an excess thereof. However, at least some of the intensity marking this debate is due to a variety of misimpressions about what conservatism is and what its ramifications are. Before laying out the proposal, therefore, some of these definitional matters will be discussed.

First, a useful definition of conservatism should help clarify it in the face of the disparate charges leveled against it. *Conservatism is, foremost, one of several ways to generate risk estimates that allow risk management decisions to be made under conditions of uncertainty and variability.* Simply put, a risk assessment policy that ignored or rejected conservatism would strive to always represent risks by their "true values" irrespective of uncertainty (or variability), whereas any attempt to consider adding (or removing) some measure of conservatism would lead the assessor to confront the uncertainty. Incorporating "conservatism" merely means that from out of the uncertainty and/or variability, the assessor would deliberately choose an estimate that he believes is more likely to overestimate than to underestimate the risk.

Rationality in managing risks (as in any endeavor of private or social decision making) involves the attempt to maximize the benefit derived from choice under specific conditions in the world. If we do not know those conditions (uncertainty) or do not know to whom these conditions apply (human interindividual variability), we have to make the choice that *would* be optimal for a *particular* set of conditions and essentially hope for the best. If the true risk we are trying to manage is larger or smaller than we think it is (or if there are individuals for whom this is so) then our choice may be flawed, but we still have to choose. Unlike the search for scientific truth, where the "correct" action in the face of uncertainty is to reserve one's judgment, in managing risks decisions are inevitable, since reserving judgment is exactly equivalent to making the judgment that the *status quo* represents a desirable balance of economic costs expended (if any) and health risks remaining (if any). It is therefore vital that the risk assessment process handle uncertainties in a predictable way that is scientifically defensible, consistent with the Agency's statutory and public missions, and responsive to the needs of decision makers. Conservatism is a specific response to uncertainty that favors one type of error (overestimation) over its converse, but (especially if EPA follows the detailed prescriptions here) the fact that it admits that either type of error is possible is more important than the precise calculus it may use to balance those errors.

It is also crucial to understand what this asymmetry in favor of overestimation does and does not mean. Conservatism is *not* about valuing human lives above the money spent to comply with risk management decisions. Instead, it acknowledges that if there was no uncertainty in risk, society could "optimally" decide to spend a dollar or a billion dollars to save each life involved—conservatism is silent about this judgment. Assuming that society decides how it wishes to balance lives and dollars, conservatism only affects the decision at the margin, by deliberately preferring, from among the inevitable errors that uncertainty creates, to favor those errors which lead to *relatively* more dollars spent for the lives saved than those which lead to *relatively* fewer lives saved for the dollars spent.

Some would call this an orientation disposed to being "better safe than sorry" or a tendency towards "prudence," characterizations we do not dispute or shrink from. It is simply a matter of "good science" to admit that the true value of risk is surrounded by uncertainty, and that as a consequence, errors of overestimation or underestimation can still occur for whatever value of risk one chooses as the basis for risk management. Much detail about conservatism follows in this appendix of the report, but the essence of the disagreement between supporters of this proposal and supporters of the alternative position is simple; the former group believes that it is both prudent and scientifically justified to make reasonable attempts to favor errors of overestimation over those of underestimation. More importantly, it believes that not to do so would be both imprudent and scientifically questionable. This is no mere tautology, but encapsulates the disagreement with others who would argue that to eschew prudence is to advocate something "value-neutral" (and hence a morally superior position for scientists to espouse) and something more "scientific."

The controversies over conservatism are heightened by ambiguous definitions and uses of the term. The following section explains three dichotomies about the precise possible meanings of conservatism, in order to clarify some of the objections to it, and to foreshadow some of the features of this proposal for a principle of "plausible conservatism":

(1) The distinction between prudence and misestimation. When a particular estimate of risk is criticized as being "too conservative," that criticism can

mean one or both of two different things. The critic may actually mean that the assessor has chosen an estimate of risk which is designed to reduce the probability of errors of underestimation, but one which the critic deems overly zealous in that regard. In other words, one person's "prudence" may be another person's "overkill," although that distinction alone is purely one of differing personal values. On the other hand, the critic may instead mean that flaws in the estimation of a risk cause the estimate to be more skewed in the direction of prudence than the assessor himself intends, or than the risk manager comprehends. Such a criticism may not involve any personal value judgements. For example, the assessor may believe that a particular estimate falls at around the 95th percentile of the uncertainty distribution of the unknown risk; such an estimate would have a 5% probability of being an underestimate of the risk. If, in fact, the estimate given is so tilted towards minimizing underestimation that it falls at (say) the 99.9th percentile of the distribution, then the process would have built in more prudence than either party intended. It is possible that in many of the instances where EPA is under fire for allegedly being "too conservative," critics are espousing differing value judgments in addition to (or instead of) trying to point out disparities between the intended and actual level of conservatism. There is, as discussed below, little empirical evidence to suggest that EPA's potency, exposure, or risk estimates are markedly higher than estimates embodying a reasonable degree of prudence (i.e., the conventional benchmarks of the 95th or 99th percentiles that statisticians use). However, supporters of the proposal detailed in this appendix are clearly opposed to systematic misestimation, if and when it exists. We stress that our version of "plausible conservatism" in risk assessment does not allow EPA to adopt unreasonable assumptions or rely upon biased parameter values, and we believe that the entire committee's consensus recommendations in Chapters 9 and 10 will help combat this tendency, if it exists, and help shed light, rather than heat, on the question of whether EPA's risk estimates are more conservative than they are intended to be.

(2) *The distinction between conservatism as a response to uncertainty or as a response to variability.* This important distinction bears upon the legitimacy of criticisms of conservatism. The two issues of uncertainty and variability involve different motivations and produce different results, even though the same terms and mathematical procedures are used to deal with each and though they may at times be hard to separate operationally. This appendix of the report deals primarily with the former, and then generally with the subcategory of conservatism regarding model uncertainty. In this discussion of uncertainty issues, because we are dealing with lack of knowledge as to the true value of risk, the science-policy balancing of errors of underestimation and overestimation does suggest the common aphorism of "better safe than sorry." The science-policy response to variability, on the other hand, involves coping with differences among people in their exposures or susceptibilities to adverse effects—that is,

deciding with certainty (or with no additional uncertainty beyond that due to not knowing which models apply) who or what should be protected. In such cases, deciding how conservative to be in light of this variability is *not* about "being better safe than sorry," but involves a decision about who merits being safe and who may end up being sorry. Here, as elsewhere in the CAPRA report, the committee refrains from coming to policy judgments about how EPA should draw such lines in general or in particular. In this discussion, we simply stress that EPA should not let criticisms of its responses to uncertainty confuse it or necessarily cause it to rethink its responses to variability.[2]

(3) *The distinction between the "level of conservatism" and the "amount of conservatism".* Any estimate of an uncertain risk embodies conservatism, if any at all, in both relative and absolute senses. Here the new terms "level of conservatism" and "amount of conservatism" are coined to codify the difference, respectively, between the relative and absolute meanings of the term. The "level of conservatism" is a relative indicator of how unlikely the assessor deems it that the estimate will fall below the true value of risk; thus, a 99th percentile estimate embodies a higher "level of conservatism" (four percentile points higher) than does a 95th percentile estimate. The "amount of conservatism," in contrast, is an absolute measure of the mathematical difference between the estimate itself and the central tendency of the unknown quantity. Thus, it is quite possible to have a high "level" and a small "amount" simultaneously, or vice versa. For example, scientists might know the speed of light to a high degree of precision, and report a 99th percentile upper confidence limit of 186,301 miles/sec. and a "best estimate" of 186,300 miles/sec. (here the absolute amount of conservatism would be 1 mile/sec.). On the other hand, when uncertainty is large, even a modest "level" (say the 75th percentile) may introduce a large amount of conservatism in absolute terms. These two concepts are related in a straightforward manner with important policy implications. As scientific knowledge increases and uncertainty decreases, the absolute difference between the central tendency and any particular upper percentile will also decrease. Therefore, agencies could try and maintain a fixed level of conservatism over time and yet expect that the absolute amount of conservatism, and thus the practical impact of attempts to shift the balance in favor of overestimation, will become progressively less and less important. When uncertainty is reduced to minimal levels, the conservative estimate and the central tendencies will become so similar that the distinction be-

---

[2]For two reasons, we believe it is logically consistent to espouse a principle of "plausible conservatism" with regard to model uncertainty and not explicitly recommend the same response to variability: (1) as a pragmatic matter, we believe scientists have more that they alone can contribute to a discussion of how to choose among competing scientific theories than they have to contribute to a discussion of what kind of individuals EPA should try to protect; and (2) we believe the public has more clearly expressed a preference for "erring on the side of safety" when the truth is unknown than it has regarding how much protection to extend to the extremes of variability distributions.

comes practically irrelevant, although risk managers and public can remain assured that the *probability* of errors of underestimation remains constant and relatively small.

INHERENT ADVANTAGES OF "PLAUSIBLE CONSERVATISM"

It is perplexing to some members of this committee (and to many in the general populace) that the presumption that society should approach uncertain risks with a desire to be "better safe than sorry" has engendered so much skepticism. After all, perhaps it should instead be incumbent upon opponents of conservative defaults to defend their position that EPA ought to ignore or dilute plausible scientific theories that, if true, would mean that risks need to be addressed concertedly. That view, whatever its intellectual merits, seems at the outset not to give the public what it has consistently called for (explicitly in legislation and implicitly in the general conduct of professions ranging from structural engineering to medicine to diplomacy): namely, the attempt to guard against major errors that threaten health and safety. But the proposal for risk assessment based on "plausible conservatism" came about largely because of the wide variety of *other* factors supporting it, whether viewed through the lenses of logic, mathematics, procedure, or political economy. The following brief accounting of some of the virtues of a conservative orientation may seem somewhat superfluous, especially given the statements of earlier NRC committees on the topic.[3] However, this committee's decision not to endorse "plausible conservatism" by consensus has prompted this more thorough enumeration of some factors some members had thought were uncontroversial:

A. *"Plausible Conservatism" Reflects the Public's Preference Between Errors Resulting in Unnecessary Health Risks and those Resulting in Unnecessary Economic Expenditures.*

An examination of the two kinds of errors uncertainty in risk can cause supports the conclusion that society has not been indifferent between them. One type of error (caused by the overestimation of risk) leads to more resources invested than society would optimally invest if it knew the magnitude of the risk precisely. The other type (caused by underestimation of risk) leads to more lives lost (or more people subjected to unacceptably high individual risks) than society would tolerate if there was no uncertainty in risk. Whether the aversion to the latter type of error is due to the greater irreversibility of its consequences

---

[3]For example, consider this recent statement of the BEST Committee on Environmental Epidemiology (NRC, 1991): "public health policy requires that decisions be made despite incomplete evidence, with the aim of protecting public health in the future."

compared to the former,[4] the importance of regret (Bell, 1982) in most individual and social decision-making,[5] or other factors is beyond our capacity to answer. What matters is, do Congress and the public view risk management as a social endeavor that should strive both for scientific truth and for the prudent avoidance of unnecessary public health risks, and therefore do not view risk assessment as purely an exercise in coming as close to the "right answer" as possible? If this is so, then the competing proposal offered in Appendix N-2 espouses an unscientific value judgment, and one that also is unresponsive to social realities.

A counter-example may be illustrative here. In its recent indictment of conservatism in Superfund risk assessment, an industry coalition drew an extended analogy to link EPA's risk estimates with inflated predictions of the amount of time it would take someone to take a taxi ride to Dulles Airport (Hazardous Waste Cleanup Project, 1993). But this particular personal decision seems to be another prime example of where individuals and society would clearly prefer conservative estimates. As demonstrated below, *any* level of conservatism (positive, zero, or negative) corresponds to some underlying attitude towards errors of overestimation and underestimation. In this case, a conservative estimate of travel time simply means that the traveller regards each minute she arrives at the airport after the plane leaves as more costly to her than each minute of extra waiting time caused by arriving before the plane leaves. It is hardly surprising to conclude that a rational person would not be indifferent, but would rather be 10 minutes early than 10 minutes late to catch a plane. If, hypothetically, someone advising the traveller told her he wasn't sure whether the airline she chose would have a single ticket agent (and a 20-minute long line) or a dozen agents (and no line), it seems hard to believe that she would ask for a "best estimate" between zero and 20 minutes and allow only that much time (and even less likely she would assume that the long line simply couldn't happen). As long as the more "conservative" scenario was plausible, it would tend to dominate her thinking, simply because the decision problem is not about arriving at exactly the right moment, but about balancing the costs of a very early arrival against the *qualitatively* different costs of even a slightly late arrival. Again, reasonable people may differ widely about how large either asymmetry should be, but supporters of "plausible conservatism" are hard pressed to imagine not

---

[4]It is possible that profligacy in economic resources invested may also lead to adverse health consequences (MacRae, 1992). However, this "richer is safer" theory is based on controversial data (Graham et al., 1993), and at most offsets in an indirect way the more direct and irreversible consequences of underregulation in the eyes of the public.

[5]Anticipation of regret tends to make people choose courses of action that are less likely to leave them with the knowledge that they failed to take another available action that would have been much less damaging.

admitting that some adjustment to make catching the plane more likely, or reducing the risk more probable, aligns with the expressed desires of the public.

B.   *Conservative Defaults Help Increase the Chances that Risk Estimates Will Not be "Anti-Conservative."*

There are two different mathematical aspects of risk assessment under uncertainty that mitigate in favor of a conservative approach to selection of default options. Both factors tend to make risk estimates generated from conservative models less conservative than they might appear at first glance, and thus tip the balance further in favor of such models as minimally necessary to support prudent decisions.

Let us assume at the outset that the assessor and decision-maker both desire that at the very least, risk estimates should not be "anti-conservative," that is, not underestimate the mean (arithmetic average) of the true but unknown risk. The mean, after all, is the minimum estimator that a so called "risk-neutral" decision-maker (e.g., a person who is not actually trying to catch a plane, but who stands to win a wager if she arrives at the airport either just before *or* just after the plane leaves) would need in order to balance errors of overestimation and underestimation. In this regard, there exists a basic mathematical property of uncertain quantities that introduces an asymmetry. For non-negative quantities (such as exposures, potencies, or risks), the uncertainties are generally distributed in such a way that larger uncertainty increases the arithmetic mean, due to the disproportionate influence of the right-hand tail. For example, if the median (50th percentile) of such an uncertainty distribution was X, but the assessor believed that the standard error of that estimate was a factor of 10 in either direction, then the 90th percentile (19X) and the arithmetic mean (14X) would be nearly identical; if the uncertainty was a factor of 25 in either direction, the mean and the 95th percentile would be virtually identical (see Table 9-4). Some of the most familiar examples of the need to impose a moderate "level of conservatism" in order not to underestimate the mean come from empirical data that exhibit variability. For example, it is unlikely, even in a state that includes areas of high radon concentration, that a randomly selected home would have a radon concentration exceeding approximately 10 picocuries/liter. Yet the mean concentration for all homes in that state might equal or even exceed 10 because of the influence on the mean of the small number of homes with much higher levels.[6]

---

[6]This mathematical truism that the more uncertainty, the greater the level of conservatism required not to underestimate the mean, seriously undermines one of the major claims made by those who accuse EPA of "cascading conservatism." If each of a series of uncertain quantities is distributed in such a way that a reasonably conservative estimator (say, the 95th percentile) approximates or even falls below the mean of that quantity, then the more steps in the cascade the *less* conservative the output becomes with respect to the correct risk-neutral estimator.

The other basic mathematical advantage of introducing some conservatism into the scientific inferences that are made is the expectation that there may be other factors unknown to the assessor which would tend to increase uncertainty. This becomes a stronger argument for conservatism if one believes that more of these unknown influences would tend to increase than to decrease the true risk. Although it seems logical that factors science has not yet accounted for (such as unsuspected exposure pathways, additional mechanisms of toxicity, synergies among exposures, or variations in human susceptibility to carcinogenesis) would tend to add to the number or severity of pathways leading to exposure and/or greater risk, it is possible that "surprises" could also reveal humans to be more resistant to pollutants or less exposed than traditional analyses predict.

C. *"Plausible Conservatism" Fulfills the Statutory Mandate under which EPA Operates in the Air Toxics (and many other) Programs.*

The policy of preventive action in the face of scientific uncertainty has long been part of the Clean Air Act, as well as most of the other enabling legislation of EPA. Two key directives run through many of the sections of the Clean Air Act in this regard. First, various sections of the Act direct EPA to consider not merely substances that have been shown to cause harm, but those that "may reasonably be anticipated" to cause harm. As the D.C. Circuit court stated in its 1976 decision in *Ethyl Corp. v. EPA*, "commonly, reasonable medical concerns and theory long precede certainty. Yet the statutes and common sense demand regulatory action to prevent harm, even if the regulator is less than certain that harm is otherwise inevitable." Similarly, the Act has long required standards for air pollutants to provide "an ample margin of safety to protect public health." The leading case on the interpretation of Section 112, the 1987 case of *Natural Resources Defense Council v. EPA*, declared that

> In determining what is an "ample margin" the Administrator may, and perhaps must, take into account the inherent limitations of risk assessment and the limited scientific knowledge of the effects of exposure to carcinogens at various levels, and may therefore decide to set the level below that previously determined to be "safe."...[B]y its nature the finding of risk is uncertain and the Administrator must use his discretion to meet the statutory mandate.

Again, support for the idea that "plausible conservatism" is the most rational approach for EPA to take is not necessarily based on a reading of the various statutes. After all, it is possible that the statutes may be changed in the near or far future. However, it seems central to EPA's mission that the Agency consider whether it is necessary to prevent or minimize adverse events, even events of low probability. Therefore, the Agency inevitably will find it necessary to use risk assessment techniques that are sensitive enough to reflect the risks of those events. At a minimum, its techniques must explore the nature of possible extreme outcomes, as a prelude to science-policy choices as to whether to factor those extremes into its risk characterizations. *In essence, conservatism in the choice of default options is a way of making risk assessment a sensitive enough device to allow risk managers to decide to what extent they can fulfill the intent of the enabling legislation.* For this reason, members of the committee advanced the proposition, which proved eventually to be controversial within the committee, that "plausible conservatism" gives decisionmakers some of the information they need to make precautionary risk management decisions.

D.   *It Respects the Voice of Science, Not Only the Rights of Individual Scientists.*

By declaring that defaults would be chosen to be both scientifically supportable and health-protective, and that scientists would have to examine alternative models by these two criteria, EPA could help ensure that science will assume the leading role in defining evolving risk assessment methodology. Some have asserted that it shows disrespect for science to posit any standard for departure from defaults other than one that simply requires EPA to adopt "new and better science at the earliest possible time." But surely there is a generally inverse relationship between the amount of knowledgeable controversy over a new theory and the likely "staying power" and reliability of such "new science." At the extremes, EPA could either change its defaults over and over again with each new individual voice it hears complaining that a default is passé, or never change a default until absolute scientific unanimity had congealed and remained unshakable for some number of years. The "persuasive evidence" standard proposed here (see below) clearly falls between these two extremes. It reflects our belief that standards which rely more on scientific consensus than on the rights of individual scientists dissatisfied with the current situation are in fact more respectful of science as an institution.

The only cost to a standard that values scientific consensus over "heed the loudest voice you hear" is that advocates of "new science" need to persuade the mainstream of their colleagues that new is indeed better. This standard is in fact a bargain for scientists, because it buys credibility in the public arena and some degree of immunity against being undercut by the *next* new theory that comes along. And, in addition to this give-and-take principle that elevates respect for scientific decisions by valuing the concord of scientists, advocates of "new science" must appreciate that the twin standards of plausibility and conservatism in fact *remove* a major source of arbitrariness in EPA's science-policy apparatus. If the Agency merely held up its defaults as unconnected "rules we live by" and required scientists to prove them "wrong," then the charge of bureaucracy-over-science would have merit. But this recommendation for EPA to reaffirm or rethink the set of defaults as "the most conservative of the plausible spectrum" sends a clear signal to the scientific community that each default only has merit insofar as it embodies those twin concepts, and gives scientists two clear bases for challenging and improving the set of inference assumptions.

E.   *It Generates Routinely those Risk Estimates Essential to Various EPA Functions.*

The committee was also unable to reach agreement on the details of what roles "nonconservative" estimates should play in standard setting, priority setting, and risk communication, although the committee's recommendations in Chapter 9 reflect its belief that such estimates have utility in all of these arenas. However, no one has suggested that "nonconservative" estimates should drive out estimates produced via "plausible conservatism," but rather that they should supplement them. Indeed, the committee agrees that conservative estimates must be calculated for at least two important risk assessment purposes: (1) the foundation of the iterative system of risk assessment the committee has proposed is the screening-level analysis. Such analyses are solely intended to obviate the need for detailed assessment of risks that can to a high degree of confidence be deemed acceptable or *de minimis*. By definition, therefore, screening analyses must be conservative enough to eliminate the possibility that an exposure that indeed might pose some danger to health or welfare will fail to receive full scrutiny; and (2) even if EPA decided to use central-tendency risk estimates for standard-setting or other purposes, it would first have to explore the conservative end of the spectrum in order to have any clear idea where the expected value of the uncertain risk (as discussed above, the correct central-tendency estimate for a risk-neutral decision) actually falls. Because of the sensitivity of the expected value of a distribution to its right-hand tail, one cannot simply arrive at this midpoint in one step.[7]

For both reasons, risk assessment cannot proceed without the attempt to generate a conservative estimate, even if that estimate is only an input to a subsequent process. Therefore, the only argument among us is whether to modify or discard such estimates for some purposes other than screening or calculation of central tendencies, not whether they should be generated at all. Either way, a set of default assumptions embodying "plausible conservatism" must play some role.

F. *It Promotes an Orderly, Timely Process that Realistically Structures the Correct Incentives for Research.*

Many observers of risk assessment have pointed out that the scientific goal of "getting the right answer" for each risk assessment question conflicts directly with the regulatory and public policy goals of timeliness and striking a balance

---

[7]See Table 9-4 for various calculations showing how if the uncertainty is distributed continuously, the arithmetic mean can be very sensitive to the conservative percentiles. If instead, the uncertainty is dichotomous (say, the risk was either Y or zero depending on which of two models was correct), the expected value would depend *completely* on the value of Y and the subjective probability assigned to it. *In either case, the upper bound must be estimated before the mean can be.*

between limited resources available for research and those available for environmental protection itself. The committee agreed that too much emphasis on fine-tuning the science can lead to untoward delay; our real disagreement again comes down to the question of how to initiate and structure the process of modifying science-based inferences. As discussed in the preceding paragraph, one advantage of starting from a conservative stance and declaring the true central tendency as the ultimate goal is that it arguably is easier to move towards this desired midpoint (given the influence of the conservative possibility on it) than to start by trying to guess where that midpoint might be. There is also a procedural advantage to a conservative starting point, however, which stems from a frank assessment of the resources and natural motivations available to different scientific institutions. Some of us believe that an evaluation of the relative effort over the last decade or so devoted to positing and studying less conservative risk models (e.g., threshold and sublinear extrapolation models, cases where humans are less sensitive than test animals) *versus* the converse (e.g., synergies among exposures, cases where negative rodent tests might not spell safety for humans) reveals an asymmetry in research orientation, with the former type of research garnering much more resources and attention than the latter. This orientation is not necessarily either pernicious or unscientific, but EPA should make use of it rather than pretend it does not exist. The best way for the Agency to do so, we believe, is to begin with a stance of "plausible conservatism" and establish explicit procedures, based on peer review and full participation, that demonstrate convincingly that the Agency understands it must be receptive to new scientific information. This takes advantage of the tendency to preferentially test less conservative theories. Moreover, EPA must communicate to the public that a general tendency for risk estimates to become less conservative (in absolute terms) over time is not evidence of EPA bias, but of an open and mutual covenant between the Agency and the scientific community searching for better models.

G.   *It Reflects EPA's Fundamental Public Mission as a Scientific/Regulatory Agency.*

As discussed below, advocates of "best estimates" frequently fail to consider how difficult, error-prone, and value-laden the search for such desirable end points can be. Since CAPRA has been asked to suggest improvements in the methodology by which EPA assesses risks from exposures to hazardous air pollutants, it is also incumbent upon us at least to remark on the purpose of such risk estimates. Part of our disagreement on the entire set of defaults issues arises because there are two purposes for risk estimates: to accurately describe the true risks, if possible, and to identify situations where risks might be worth reducing. Other government agencies also have to serve the two masters of truth and decision, yet their use of analysis does not seem to arouse so much controversy. Military intelligence is an empirical craft that resembles risk assessment in its reliance on data and judgment, but there have been few exhortations that the

Department of Defense (DOD) should develop and rely on "best estimates" of the probability of aggression, rather than on accepted estimates of how high those probabilities might reasonably be. There is room for vigorous descriptive disagreement about the extent of conservatism in DOD predictions, and for normative argument about the propriety thereof, but these are questions of degree that do not imply DOD should abandon or downplay its public mission in favor of its "scientific" mission.[8]

SPECIFIC RECOMMENDATIONS TO IMPLEMENT THIS PRINCIPLE

Members of the committee who advocate that EPA should choose and modify its defaults with reference to the principle of "plausible conservatism" have in mind a very specific process to implement this principle, in order to accentuate its usefulness along the criteria discussed in the introduction to Part II of the report, and to minimize its potential drawbacks. In light of the controversy these four recommended procedures engendered within the committee, this section will emphasize what our vision of "plausible conservatism" does not involve or sanction, even though these features apparently were not sufficient to stanch the opposition to the proposal.

Step 1 In each instance within the emissions and exposure assessment or the toxicity assessment phase of risk assessment where two or more fundamentally different scientific (i.e., biological, physical, statistical, or mathematical) assumptions or models have been advanced to bridge a basic gap in our knowledge, EPA should first determine which of these models are deemed "plausible" by knowledgeable scientists. As an example, let us assume that scientists who believe benign rodent tumors can be surrogates for malignant tumors would admit that the opposite conclusion is also plausible, and *vice versa*. Then, from this "plausible set," *EPA should adopt (or should reaffirm) as a generic default that model or assumption which tends to yield risk estimates more conservative than the other plausible choices.* For example, EPA's existing statement (III.A.2 from the 1986 cancer guidelines) that chemicals may be radiomimetic at low doses, and thus that the linearized multistage model (LMS) is the appropriate default for exposure-response extrapolation, is not a statement of scientific fact, but is the preferred science-policy choice, for three reasons: (1) the scientific conclusion that the LMS model has substantial support in biologic theory and

---

[8]Note that these 7 advantages of conservatism are not an exhaustive list. Others that could have been discussed include: this proposal is close to what EPA already does; it jibes with the rest of the CAPRA report; it is also motivated by some pure management issues, notably the potential problem of a bias towards exaggeration in the cost figures that risk estimates are compared to.

observational data (so it cannot be rejected as "absolutely implausible"); (2) the scientific conclusion that no other extant model has so much *more* grounding in theory and observation so as to make the LMS fail a test of "relative plausibility"; and (3) the empirical observation that the LMS model gives more conservative results than other plausible models.[9]

Step 2   Armed with this set of scientifically supportable and health-protective models, EPA should then strive to amass and communicate information about the uncertainty and variability in the parameters that drive these models.[10] *The uncertainty distributions that result from such analyses will permit the risk manager to openly choose a level of conservatism concordant with the particular statutory, regulatory, and economic framework, confident that regardless of the level of conservatism chosen, the risk estimate will reflect an underlying scientific structure that is both plausible and designed to avoid the gross underestimation of risk.*   In Chapters 9 and 11, the committee supports this notion that the level of conservatism should be chosen quantitatively with reference to parameter uncertainty and variability, but qualitatively with reference to model uncertainty (i.e., under this proposal, models would be chosen to represent the "conservative end of the spectrum of plausible models"). Although the "plausible conservatism" proposal *per se* was not unanimously agreed to, the entire committee does share the concern that attempts to precisely fine-tune the level of conservatism implicit in the model structure may lead to implausible or illogical compromises that advance neither the values of prudence nor of scientific integrity.

Step 3   EPA should then undertake two related activities to ensure that its resulting risk estimates are not needlessly conservative, or misunderstood by

---

[9]EPA should be mindful of the distinction between "plausible as a general rule" and "plausible as an occasional exception" in choosing its generic defaults, and only consider the former at this stage (i.e., if a particular model is not plausible as a means of explaining the general case, it should be reserved for consideration in specific situations where a departure may be appropriate). For example, a more conservative model than the LMS model, a "superlinear" polynomial allowing for fractional powers of exposure (Bailar et al., 1988), may be plausible for certain individual chemicals but appears at present not to pass a consensus threshold of scientific plausibility as a generic rule to explain all exposure-response relationships. On the other hand, less conservative models such as the M-V-K model do cross this threshold as plausible-in-general but would not yet qualify as appropriate generic defaults under the "plausible conservatism" principle.

[10]As the committee discusses in its recommendations regarding "iteration," the level of effort devoted to supplanting point estimates of parameters with their corresponding uncertainty or variability distributions should be a function of the "tier" dictated by the type and importance of the risk management decision. For screening analyses, conservative point estimates within the rubric of the prevailing models will serve the needs of the decision, whereas for higher-tier analyses uncertainty distributions will be needed.

some or all of its audience. These steps are important even though by definition, risk estimates emerging from a framework of "plausible conservatism" cannot be ruled out as flatly impossible without some empirical basis (since they are based on a series of assumptions, each of which has some scientific support, the chain of assumptions must also be logically plausible, if perhaps unlikely). As some observers have pointed out, however, such estimates may be higher than some judge as necessary to support precautionary decisions (Nichols and Zeck-hauser, 1988; OMB, 1990). A quantitative treatment of uncertainty and an explicit choice of the level of conservatism with respect to parameter uncertainty, as recommended here and in Chapter 9, will help minimize this potential problem. EPA can mitigate these concerns still further by: (1) calibrating its risk estimates against available "reality checks," such as the upper confidence limit on human carcinogenic potency one can sometimes derive in the absence of positive epidemiologic data (Tollefson et al., 1990; Goodman and Wilson, 1991) or physical or observational constraints on the emissions estimates used or the ambient concentration estimates generated by the exposure models used; and (2) clearly communicating that its risk estimates are intended to be conservative (and are based on plausible but precautionary assumptions). In improving its risk communication, EPA should try to avoid either underestimating the level of conservatism (e.g., EPA's current tendency to imply that its estimates are "95th percentile upper bounds" when they really comprise several such inputs that, in combination with other nonconservative inputs, might still yield an output more conservative than the 95th percentile) or overstating the amount of conservatism (e.g., EPA's tendency to state that all its potency estimates "could be as low as zero" even in cases when there is little or no support for a threshold model or when the estimates are based on human data). In essence, the thrust of this step of our proposal is to further distinguish between the concepts of prudence and misestimation discussed above, and to discourage the latter practice so that critics of conservatism will have to come to grips with (or abandon) their opposition to the former.

Step 4 Finally, (a point to which the entire committee agreed) EPA should clarify its standard for how it decides it should replace an existing default assumption with an alternative (either as a general rule or for a specific substance or class of substances). Currently, EPA only uses language implying that each default shall remain in force "in the absence of evidence to the contrary," without any guidance as to what quality or quantity of evidence is sufficient to spur a departure or how to gauge these attributes (or, of course, any guidance if any principle other than one of evidentiary quality should govern the choice among alternatives). Here, a specific test for structuring departures from defaults is proposed. Specifically, EPA should go on record as supporting departures from defaults whenever *"there exists persuasive evidence, as reflected in a general consensus of knowledgeable scientists, that the alternative assumption (model) represents the conservative end of the spectrum of plausible assumptions (models)."* This language was carefully chosen, based on substantial debate within the committee, to achieve several objectives:

• to strike a balance between having defaults that are too rigid and ones that change too often (and that tend to change for unpredictable and perhaps even self-contradictory reasons). The requirement for "persuasive evidence," and the deference to scientific consensus as an indicator of this quality of evidence, yields an explicit standard that is neither as difficult to meet as "beyond a reasonable doubt" would be (a single scientific dissenter could thwart the process if EPA used this standard) nor as flexible and subject to backtracking as language such as "preponderance of the evidence" or "best available scientific opinion" would be. No other standard we considered seems to strike a better balance between elusive scientific unanimity and evanescent (and perhaps illusory) scientific plurality.

• to reaffirm the principle of "plausible conservatism" in the inferences made as time passes and scientific knowledge improves. If defaults changed solely on the basis of "correctness," there would be no continuity among the assumptions EPA uses in the way each attempts to cope with uncertainty (only the overconfident affirmation that each default is "correct" in spite of the uncertainty). Instead, this standard makes all assumptions/models comparable, whether they are holdovers from the 1986 guidelines or newly-adopted alternatives; they will all represent the choices deemed to be both supportable and health-protective. In other words, under this system the *level of conservatism* will remain constant (at least on a qualitative scale) while the *amount of conservatism* will generally decrease over time in lockstep with the progress of scientific knowledge.[11] Thus, control of pollutant sources can generally become less stringent over time without lessening the level of assurance that public health goals are being met.

• to encourage, under the iterative approach called for elsewhere in this report, the use of more data and more sophisticated models without cumbersome processes for approving their use. The "plausible conservatism" standard recommended here acknowledges that simplicity in risk assessment is useful for certain risk management purposes but is not an end in itself. Thus, the actual default model for certain atmospheric transport calculations might well be a more complex version of a simpler and more conservative screening model (e.g.,

---

[11]In special circumstances, a new scientific consensus may emerge that a model or assumption that is *more* conservative than the default is clearly plausible, either as a general rule or for specific chemicals or exposure scenarios. In such cases, the absolute amount of conservatism will increase. Although this asymmetry results in a *de facto* lower procedural threshold for adopting more conservative models than less conservative ones, the requirement implicit in the standard for a consensus about plausibility should limit the frequency with which the former type of departures will occur.

Lagrangian versus box models). Assessors would be free to use the simpler model for screening purposes without threatening the primacy of the more complex model for higher-tier risk assessments. An excellent example of this accommodation of multiple assumptions that each embody "plausible conservatism" might be the use of PBPK models in higher-tier assessments versus a generic scaling factor (such as body weight to the 0.67 or 0.75 power) in lower-tier assessments. Perhaps EPA should consider designating a particular PBPK model as the default option for interspecies scaling, while reiterating that the scaling factor (which is itself a simple pharmacokinetic model) would also be an appropriate default for less resource-intensive applications.[12]

• to encourage greater use of peer review and other mechanisms to increase the scientific community's role in the evolving selection of preferred models. Implicit in this standard is the intent that EPA should continue to use its Science Advisory Board and other expert bodies to determine when general scientific consensus exists. Workshops, public meetings, and other devices should increasingly be used to guarantee, as much as possible, that EPA's risk assessment decisions will be made with access to the best science available and the full participation of the entire expert community.

PITFALLS OF OUR PROPOSAL; COMPARISON WITH ALTERNATIVES

Some of the criticisms raised against conservatism in risk assessment have substantial merit, and are applicable to this proposal to include conservatism in the choice of default options. EPA can minimize some of these pitfalls by following other recommendations made in this appendix and elsewhere in the report. For example, the problem that conservatism can lead to incorrect risk comparisons and priority-setting decisions can be remedied in part by striving to make the "level of conservatism" explicit and roughly constant across assessments, and by generating additional estimates of central tendency (perhaps even derived via subjective weights applied to different basic biological theories) for use in ranking exercises only.[13] Similarly, there is a legitimate concern that the policy of conservatism can stifle research if EPA is perceived as uninterested in

---

[12]The only important caveat to this principle, which would apply to the transport model example as well as the PBPK example, is that with the addition of new model parameters (e.g., partition coefficients and rate constants in the PBPK case), the uncertainty and interindividual variability in those parameters must be estimated and incorporated into an explicit choice of a level of conservatism (see recommendation in Chapter 9).

[13]We note that risk ranking under uncertainty is a complicated and error-prone process, regardless of whether conservative, average, or other point estimates are used to summarize each risk. The medians or means of two risk distributions can be in one rank order while the upper bounds could well be in the opposite order; no single ranking alone is correct.

any new information that might show the risk has been overstated; the emphasis here on scientific consensus does tend to slow the adoption of less conservative models at their early stages of development, but this should neither discourage thorough research nor discourage researchers from submitting quality data which EPA could readily incorporate into its existing model structure regardless of what effect it would have on the risk estimate.

The fundamental concern about conservatism is that it has led to systematic exaggeration of all environmental health problems and has encouraged wasting of scarce resources on trivial risks. The latter part of this charge is a subjective matter of economic and social policy that falls outside this committee's purview. And while the former concern is an empirical one, it has sparked a vigorous debate that is far from resolved. On one side, those convinced that EPA's procedures yield estimates far above the true values of risk can cite numerous examples where individual assumptions seem to each contribute more and more conservatism (Nichols and Zeckhauser, 1988; OMB, 1990; Hazardous Waste Cleanup Project, 1993). Others believe the evidence shows that current procedures embody a mix of conservative, neutral, and anti-conservative assumptions, and that the limited observational "reality checks" available suggest that existing exposure, potency, and risk estimates are in fact not markedly conservative (Allen et al., 1988; Bailar et al., 1988; Goodman and Wilson, 1991; Finley and Paustenbach, in press; Cullen, in press).

The practical and constructive question EPA must grapple with, however, is not whether "plausible conservatism" is ideal, but whether it is preferable to the alternative(s). The primary alternative to this proposal (Appendix N-2) directs EPA risk assessors to use defaults on the basis of the "best available scientific information," with the apparent goal of generating central-tendency estimates (CTEs) of risk. According to proponents of this approach, there is a clear boundary line between the "objective" activity of risk assessment and the value-laden activity of risk management, and the imposition of conservatism (if any) should occur in the latter phase, with managers adding "margins of safety" to make precautionary decisions out of the CTEs. In comparing this proposal with the alternative, it is important to consider the two fundations of the latter approach, the CTE (or "most scientific estimate") and the margin of safety, and ask whether either concept is really as appealing as it may sound.

The margin of safety idea is problematic, for one obvious reason: *it is only through exploring the conservative models and parameter values that analysts or managers can have any idea what they are trying to be "safe" from.* Perhaps it would be ideal for the manager rather than the assessor always to tailor the level of conservatism, but in reality, only the assessor can initially determine for the manager what a "conservative decision" would entail, because the assessor has the access to information on the spectrum of plausible values of risk. Applying any kind of generic safety factor to CTEs of risk would certainly result in a haphazard series of decisions, some (much) more conservative than a reasonable degree of prudence would call for, others (much) less so. Besides, taken as a whole the committee's report returns some discretion and responsibility to the risk manager that assessors have admittedly usurped in the past by presenting point estimates alone. The committee's emphasis on quantitative uncertainty and variability analysis gives risk managers the ability to tailor decisions so that the degree of protection (and the confidence with which it can be ensured) are only as stringent as they desire. But in the narrow area of model uncertainty, this proposal deems it unwise to encourage risk managers to *guess at* what a protective decision would be, by censoring information about models which, although conservative, are still deemed by experts to be plausibly true.

The CTE also has potentially fatal problems associated with it. Even if the models used to construct CTEs are based on "good science," we have argued (above) that these estimates are not designed to predict the expected value of potency, exposure, or risk (for which the conservative end of the spectrum must be explored and folded in), but instead are surrogates for other central-tendency estimators such as the median or mode (maximum likelihood). These latter estimators generally do not even give neutral weight to errors of underestimation and overestimation, and hence must be regarded as "anti-conservative." But advocates of CTEs have also failed to consider the problems of the models from whence they come. The following are four examples, illustrating four archetypes of central-tendency estimation, which suggest that on a case-by-case basis, "good science" may not be all its proponents advertise it to be:

Case 1: *"More science" merely means more data.* Some of the alternative CTE estimates advocated by critics of conservatism are alleged to be more scientific because they make use of "all the data at hand." This distinction is hardly a cut-and-dried one, however. For example, consider the current EPA default of using the bioassay result from the most sensitive of the (usually no more than four) sex-species combination of rodent tested. Call this potency estimate "A," and the alternative that could be derived by pooling all (four) data sets as "ABCD." Assuming that we know very little about the relative susceptibilities of different varieties of rodents *versus* the average human (in general or for the particular substance at issue), we must logically admit that it is possible the true risk to the average human may be greater than that implied by A, less than that implied by ABCD, or somewhere in between. One could prefer ABCD to A on the basis of a different value judgment about the costs of overestimation and underestimation, but the only "scientific" difference is that ABCD makes use of more data. But "purchasing" an array of data is akin to buying cards in a blackjack game: "more is better" only holds true as long as all the individual elements are valuable rather than otherwise. Assuming rodent varieties A through D differ significantly (or we wouldn't be quarreling over the two estimators), then humans must either be most like variety A or most like one of the other three. If the former, then data points B, C, and D *dilute* and ruin what is already in fact the "best estimate"; if the latter, then more is indeed better (in the sense of moving us closer to the truth). Therefore, EPA's true dilemma is whether the additional data are more likely to hurt or to help, and *this too is a policy judgment about balancing estimation errors,* not a simple matter of "good science." However, as a matter of process and of implementation, there is a clear difference between a policy of choosing A and a policy of choosing ABCD. The former policy sets up incentives to actually advance the scientific foundation and get to the truth of which sex/species is the best predictor in specific or in general; when such information becomes available, "good science" will justifiably carry the day. On the other hand, the latter policy only encourages additional rote application of current bioassay designs to generate more data that assessors can pool.

A related example in the exposure assessment arena is discussed in Chapter 10. A CTE of approximately 7 years of exposure to a typical stationary source of toxic air pollutants is indeed based on much more data (in this case, data on the variation in the number of years a person stays at one residence before moving) than is the standard 70-year assumption EPA has used. But as noted in Chapter 10, those data, although valid at face value, may speak to a different question than the one EPA must address. To ensure that individual lifetime risk is correctly calculated in a nation containing thousands of such sources, EPA would need to consider data not only on years at one residence, but also on the likelihood (that we consider substantial) that when someone moved away from proximity to a source, he or she would move to an area where there is still exposure to the same or similar carcinogens. In both examples, "a great deal more data" (on interspecies susceptibility or on autocorrelation of exposure rates as people move, respectively) would certainly be preferable to EPA's *status quo* assumption, but questions arise as to whether "a little more data" help or hurt the realism of the calculations.

Case 2: *"More science" means constructing chimeras out of incompatible theories.* One brand of CTE that has gained some currency in recent years allegedly provides a means of incorporating all of the plausible scientific models, much as meta-analysis incorporates all of the available epidemiologic studies or bioassays on a particular compound. Unfortunately, there may be a world of difference between pooling related data sets and averaging incompatible theories. In Chapter 9, we discuss the obvious pitfalls of such hybrid CTEs, which arguably confuse rather than enrich the information base from which the risk manager can choose a course of action. For example, when faced with two conflicting theories about the potency of TCDD, EPA arguably should not have tried to change its potency estimate to "split the difference" between the two theories and make it appear that new science had motivated this change (Finkel, 1988). Rather, EPA could have achieved the same risk management objective by loosening regulatory standards on TCDD if it felt it could justify this on the grounds that there was a significant probability that the existing risk estimate was excessively conservative. The committee could not agree on what sort of advice to give decisionmakers when some risk is either zero (or nearly zero) or is at some unacceptably high level X, depending on which of two fundamentally incompatible biologic theories is in fact the correct one. The committee did agree, however, that analysis should certainly not report only a point estimate of risk equal to $(1-p)$X, where p is the subjective probability assigned to the chance that the risk is (near) zero. In the specific context of default options, this proposal remains that EPA should retain its "plausible conservative" default until scientific consensus emerges that the alternative model supplants the default at the conservative end of the plausible set of model choices.

Case 3: *"More science" means introducing more data-intensive models without considering uncertainty or variability in the parameters that drive them.* This particular problem should be easy to rectify by incorporating the committee's recommendations in Chapters 9 and 10, but it is mentioned here because to date EPA has considered several departures from defaults (e.g., the case of methylene chloride, at least as interpreted by Portier and Kaplan, 1989) in which the level of conservatism may have changed abruptly because the parameters of the default model were assessed conservatively, but the parameters in the new model were either CTEs or point estimates of unknown conservatism. All of the burden should not fall upon purveyors of new models, however; EPA needs to level the playing field itself by systematically exploring the conservatism inherent in the parameters of its default models (for example, as we discuss in Chapter 11, is the surface area or 3/4 power correction a conservative estimate of interspecies scaling, or something else?).

Case 4: *"More science" is clearly an improvement but not airtight.* It is noteworthy that the most detailed case-specific reassessment of a default assumption, the CIGA case discussed in Chapter 6, has recently been called into question on the grounds that the new science casts serious doubt upon EPA's default as applied to existing animal data, but does not itself provide unimpeachable support for an alternative risk estimate (Melnick, 1993). We do not presume to reach any conclusion about this dispute, or about its implications for the general process of departing from defaults. As a matter of process, the CIGA case would probably meet the "persuasive evidence" test recommended here, and therefore one should not necessarily characterize EPA's acceptance of this new science as a mistake in policy. However, for purposes of risk communication, EPA should understand and emphasize that scientific consensus in issues such as these does not necessarily imply scientific truth.

CONCLUSIONS

In summary, EPA's choice between competing principles for choosing and departing from defaults has important and provocative implications for four areas of environmental science and EPA programs.

• *Values.* The choice between "plausible conservatism" and "best science" is inescapably one of science and of values. As this Appendix shows, both principles rely in part on science and decision theory, and both embody specific sets of value judgments. Unfortunately, some of the critics of "plausible conservatism" have shone a spotlight on the values inherent in that position while ignoring the value judgments inherent in the alternatives. We have argued that in many cases (and especially in the most practically important cases, where a dichotomy exists about whether a model predicting unacceptably high risk or a model predicting zero risk is correct), CTEs are difficult to derive and may not be meaningful. But even if CTEs were free of these pitfalls, one must recognize that choosing them over conservative estimates is a value-laden choice, and indeed that choosing *among* the three or more different brands of CTE, which may differ from each other by orders of magnitude in some cases, also requires value choices. The mode (maximum likelihood estimator) is a CTE with a particular purpose in decision theory; it maximizes the probability that the decision flowing from this estimate will be "right," without regard either to the direction or magnitude of deviations from this ideal. The median CTE seeks to balance the probability of the two types of error, again without regard to the magnitude of either; this, too, represents a particular value orientation. Finally, the mean attempts to balance the (unweighted) product of the probability and magnitude of errors of either type. The conservative estimate rounds out this set of possible choices, as it simply seeks to balance a weighted product of the probability and magnitude of error (see Figure N1-1, which gives examples of what purpose each estimator serves). Thus, the choice of risk assessment estimates can certainly be explicit, and can and should be unbiased in the sense of deriving from an open and honest process, but it *cannot* be wholly objective or value-neutral. Because leaving EPA with no principle for choosing among these estimators would itself be a value-laden decision, some members of the committee have advocated "plausible conservatism," cognizant that this is an *alternative* judgment.

• *Science.* The tendency of critics of our position to hold up terms such as "best science" or "credible science" as the alternative should not confuse readers into inferring that any alternatives must espouse "bad science" or "incredible science." Defaults which are not credible have no place in either of the proposals advocated in these appendices. Supporters of "plausible conservatism" believe defaults based on this principle have additional merits beyond their credibility, and that "best science" ought to be more than data for data's sake or anti-conservatism for its own sake. Looking to the future, none of the members of the committee wishes to "freeze" risk assessment science in its current incarnation, or to suppress information about new scientific ideas that challenge existing ones. The intelligent question is not *whether* to improve the science, but *how* and *when* to include it in risk characterization for risk management. Again, in the vexing paradigm case where an alternative model incompatible with the default predicts vastly smaller or zero risk, the operational decision involves whether to leap to the new risk characterization, to develop a hybrid of the two theories, or to proceed cautiously until general scientific consensus supports the alternative. This section of the chapter has explored reasons why the last alternative is preferable, and if thoughtfully implemented, will fuel rather than freeze scientific research.

- *EPA Practice.* As the previous chapter indicates, EPA has always had to walk the fine line between too much flexibility and too much rigidity. Even though the committee is concerned that EPA has not had an underlying rationale for its decisions to this point, on balance some of us feel that its individual decisions to depart from defaults have managed this tension admirably well. We are unaware of serious charges that EPA has been unreceptive to new scientific information; indeed, according to some of the references cited above, even in the methylene chloride and CIGA cases, EPA has arguably been too quick to adopt "new science."

- *Risk Management.* This chapter has explored in some detail the intertwining, rather than the boundary, of risk assessment methodology and risk management practice. None of the suggestions made here violate the principle that risk management concerns should not unduly influence the conduct of risk assessment; they only reinforce the point that risk assessment exists to provide useful answers to the questions risk managers choose to ask. And whatever one thinks about the advisability of a clear attempt to separate risk assessment from risk management, this discussion has shown that a "plausible conservatism" orientation is no more violative of that boundary than a central tendency orientation would or could be. Furthermore, both "plausible conservatism" and the "best science" alternative leave vast room for risk managers to exercise their rightful discretion, particularly in the selection of decision alternatives and the integration of information external to the risk assessment (e.g., cost and efficiency estimates, public concerns) on which real decisions often hinge. Finally, we hope to have dispelled the *false choice* that others have posited between valuing science and the values held by scientists. Surely as scientists or otherwise, our values include respect for public health precaution, for predictably and order, and for striking a thoughtful and appropriate balance between the inevitable errors that uncertainty causes. We could decide that other values outweigh these, but we cannot rationalize such a choice by costuming it in the garb of "good science." "Plausible conservatism" embraces the idea that assessors and managers need not abandon either their valuing of science or their values as scientists. Supporters of this principle hope that EPA will follow this path, even though it is presented here as a recommendation that the full committee could not agree to.

AFTERTHOUGHTS

The alternative view which follows (Appendix N-2) was written after this Appendix was completed. Together, these two statements reflect reasoned disagreement which I hope will provide EPA with "grist for the mill" to help it resolve important questions about risk assessment principles and model uncertainty. However, there are a number of inconsistencies and misinterpretations in Appendix N-2 that I believe cloud this debate. Some of the ambiguity stems from the lack of responsiveness to important issues raised in this Appendix. For example, Appendix N-2 asserts that "risk managers should not be restricted by value judgments made during risk assessment," but nowhere does it explain how this vision could be realized, in light of the assertions herein that a vague call for "full use of scientific information" must either impose a set of value judgments of its own or else restrict risk assessors to presenting every conceivable interpretation of every model, data set, and observation.[14] Similarly, the statement that "risk characterizations must be as accurate as possible," and the implicit equating of accuracy with the amount of data amassed, responds neither to the assertion that accuracy may not be the most appropriate response to uncertainty nor to the four examples in Appendix N-1 showing that "more science" may lead to less accuracy as well as substitute risk-neutral or risk-prone value judgments for risk-averse ones.

There are legitimate reasons for concern about a principle of "plausible conservatism," concerns that, if anything, might have been strengthened by more specificity in Appendix N-2 about the putative merits of an alternative. But in at least three respects, the material in N-2 misinterprets the stated intent of the "plausible conservatism" proposal, thus making a fair comparison impossible.

(1) Proponents of the "plausible conservatism" approach assuredly do not believe that "the fundamental output of a risk assessment is [or should be] a single estimate of risk: one number." This "red herring" permeates Appendix N-2 despite the clear statements in Appendix N-1 that default options only provide a scaffolding upon which all of the uncertainties and variabilities contingent on the models selected must be assessed and communicated. In fact, Chapter 9 of the report states quite clearly the committee's view that risk assessors must *abandon* their reliance on single point estimates and instead routinely provide quantitative descriptions of uncertainty (preferably via probability distributions). Indeed, three of the four specific recommendations in Appendix N-1 for implementing the "plausible conservatism" proposal reinforce the purpose of Chapter

---

[14]In fact, the Appendix contradicts itself a few pages later when it states that "weighing the plausibility of alternatives is a highly judgmental evaluation that must be carried out by scientists." This is a clear call for scientists to play a role in science policy, which Appendix N-1 clearly *endorses*, but then the authors of N-2 return to the "hands off" view and re-contradict themselves with the admonition that "scientists should not attempt to resolve risk management disputes by influencing the choice of default options."

9 by emphasizing "the uncertainty distributions that result" from proper risk assessments conducted according to guidelines containing default options.[15] The only uncertainty not accounted for by estimating risk using default models unless there are specific reasons to replace them is the subjective probability that the other model(s) is/are correct. Even though this additional uncertainty may be substantial, the committee agreed in Chapter 9 that at present, both the methodology for coming up with the subjective weights and the theory for how to meaningfully "average" irreconcilable models are sufficiently rudimentary that a single risk characterization covering all plausible models would be a precarious basis for risk management and communication. Thus, however the proposal in Appendix N-1 and Appendix N-2 differ, they do not advocate different "fundamental outputs of risk assessment."

(2) The authors of Appendix N-2 characterize their recommendation that "risk managers can and should override conservative default value judgments in the risk assessment process whenever they believe it is appropriate public policy to do so" as one that "contrasts sharply with the approach advocated by Dr. Finkel." For the record, nothing in Appendix N-1 advocates constraining the activities of risk managers in any such way. Indeed, the last paragraph of Appendix N-1 speaks to the "rightful discretion" risk managers should have to supplement, subordinate, or discard quantitative risk information when they deem this necessary to make sound decisions. If a risk characterization (again, a distribution, not just a number) emerging from the "plausibly conservative" models chosen suggests a significant risk, the manager can still let other concerns (economics, feasibility, equity, or even lack of confidence in the scientific underpinning of the risk assessment) justify not reducing the risk. What proponents of "plausible conservatism" object to, and what Appendix N-2 either leaves open or endorses (it is unclear), is for someone (a scientist? a manager?) to declare as a matter of science that a risk *already is acceptable*, simply because other models may exist that give more sanguine risk predictions than do the conservative defaults.

(3) Despite all their criticism of how even examining "conservatism" intrudes into policy, the authors of Appendix N-2 admit they "do not object to ["plausible conservatism"] for selecting the default options," only to its use in deciding when to displace an option. What justification could make the same principle appropriate at the outset but objectionable from then on? Their stated objection is that it will "freeze risk characterizations at the level determined by

---

[15]A substantial amount of uncertainty may be contributed by the parameters that drive risk models, even before interindividual variability is taken into account. For example, even if one specifies that the linearized multistage model must be used, the uncertainty in cancer potency due only to random sampling error in the typical bioassay can span five orders of magnitude at a 90 percent confidence level (Guess et al., 1977).

the conservative default options." Appendix N-1, however, argues that consensus processes in science neither intend to nor result in the "freezing" of science, only in "freezing out" unpersuasive or poor-quality science until it improves. It seems, then, that the authors of Appendix N-2 really do object to reliance on prudence and conservatism for the initial selection of the defaults, and only tolerate the existing defaults because of their expectation that they could be abandoned speedily.

In contrast to some of the issues raised above, where there really is less disagreement that Appendix N-2 indicates, here there is more controversy than Appendix N-2 admits to. Our lack of consensus on this most fundamental issue—how to choose and how to modify default options—is what caused the committee to decide not to recommend any principles for meeting these challenges.

REFERENCES

Allen, B.C., K.S. Crump, and A.M. Shipp. 1988. Correlation between carcinogenic potency of chemicals in animals and humans. Risk Anal. 8:531-544.
Bailar, J.C., III, E.A. Crouch, R. Shaikh, and D. Spiegelman. 1988. One-hit models of carcinogenesis: Conservative or not? Risk Anal. 8:485-497.
Bell, D. 1982. Regret in decision-making under uncertainty. Operations Res. 30:961-981.
Cullen, A. In press. Measures of compounding conservatism is probablistic risk assessment. Risk Anal.
EPA (U.S. Environmental Protection Agency). 1986. Guidelines for carcinogen risk assessment. Fed. Regist. 51:33992-34003.
Finkel, A. 1988. Dioxin: Are we safer now than before? Risk Anal. 8:161-165.
Finkel, A. 1989. Is risk assessment really too "conservative?": Revising the revisionists. Columbia J. Environ. Law 14:427-467.
Finley, B., and D. Paustenbach. In press. The benefits of probabilistic techniques in health risk assessment: Three case studies involving contaminated air, water, and soil. Paper presented at the National Academy of Sciences Symposium on Improving Exposure Assessment, Feb. 14-16, 1992, Washington, D.C. Risk Anal.
Goodman, G., and R. Wilson. 1991. Quantitative prediction of human cancer risk from rodent carcinogenic potencies: A closer look at the epidemiological evidence for some chemicals not definitively carcinogenic in humans. Regul. Toxicol. Pharmacol. 14:118-146.
Graham, J.D., B.-H. Chang, and J.S. Evans. 1992. Poorer is riskier. Risk Anal. 12:333-337.
Hazardous Waste Cleanup Project. 1993. Exaggerating Risk: How EPA's Risk Assessments Distort the Facts at Superfund Sites Throughout the United States. Hazardous Waste Cleanup Project, Washington, D.C.
MacRae, J.B., Jr. 1992. Statement of James B. MacRae, Jr., Acting Administrator, Office of Information and Regulatory Affairs, U.S. Office of Management and Budget. Hearing before the Committee on Government Affairs, U.S. Senate, March 19, Washington, D.C.
Melnick, R.L. 1992. An alternative hypothesis on the role of chemically induced protein droplet ($\alpha2\mu$-globulin) nephropathy in renal carcinogenesis. Regul. Toxicol. Pharmacol. 16:111-125.
Nichols, A., and R. Zeckhauser. 1988. The perils of prudence: How conservative risk assessments distort regulation. Regul. Toxicol. Pharmacol. 8:61-75.

NRC (National Research Council). 1991. Environmental Epidemiology. Public Health and Hazardous Wastes, Vol. 1. Washington, D.C.: National Academy Press.

OMB (U.S. Office of Management and Budget). 1990. Regulatory Program of the U.S. Government, April 1, 1990-March 31, 1991. Washington, D.C.: U.S. Government Printing Office.

Portier, C.J., and N.L. Kaplan. 1989. The variability of safe dose estimates when using complicated models of the carcinogenic process. A case study: Methylene chloride. Fundam. Appl. Toxicol. 13:533-544.

Tollefson, L., R.J. Lorentzen, R.N. Brown, and J.A. Springer. 1990. Comparison of the cancer risk of methylene chloride predicted from animal bioassay data with the epidemiologic evidence. Risk Anal. 10:429-435.

APPENDIX

N-2

Making Full Use of Scientific Information in Risk Assessment

Roger O. McClellan and D. Warner North

INTRODUCTION

This appendix is written in response to Appendix N-1 written by Adam Finkel, which is included in the CAPRA report at the request of the committee. That appendix advocates a principle of "plausible conservatism" for choosing and altering default assumptions and in making cancer risk estimates. It describes this principle as an alternative to the use of best available science and calculation of central tendency risk estimates. This appendix proposes an alternative view to Appendix N-1. We present a different framing of the issue of making full use of science in risk assessment, as opposed to increasing the use of conservative value judgments as described in Appendix N-1.

EPA already practices what we interpret as plausible conservatism in the selection of default options. As set forth in the 1986 *Guidelines for Carcinogen Risk Assessment*, EPA has selected its default options to be scientifically plausible and protective of human health. EPA's cancer potency estimates are intended to be plausible upper bounds on risk. Neither we nor others on the CAPRA Committee have asserted that these EPA risk assessment procedures are inappropriate. Rather CAPRA has sought to strengthen EPA's risk assessment process through further refinements. One of the potential refinements is an explicit standard for departure from defaults. We have concerns that using plausible conservatism as the standard for departure from defaults, as advocated in Appendix N-1, may not be useful and appropriate.

A major theme of the CAPRA report is that of an iterative approach to risk assessment. EPA should carry out risk assessments at multiple levels, with more detail and more use of site and substance-specific data in the upper tiers of an iterative process. While simple procedures and single-number estimates are appropriate for screening purposes in lower tiers of risk assessment, explicit disclosure of uncertainty and results from multiple scientifically plausible models are encouraged as part of upper tier risk assessment.

It is assumed in Appendix N-1 that the fundamental output of a risk assessment is a single estimate of risk: one number. We take a very different view, that risk assessment is a process for summarizing the available scientific information in both qualitative and quantitative form, for risk managers and for interested members of the public. Thus, regulatory decisions on managing risks should **not** be driven solely by single number risk estimates, but rather by a more comprehensive characterization of available scientific information, including uncertainties. We believe the CAPRA report strongly supports this latter interpretation.

An important aspect of risk management is the management of research directed at improving risk assessments by reducing uncertainty, permitting conservative assumptions to be superseded by more accurate models and observational data. The tiered approach to risk assessment and explicit consideration of both model and parameter uncertainties will facilitate identification of the opportunities for research that are most important for achieving the nation's health protection, environmental, and economic goals. We view debate over which conservative assumption to use in risk assessment as a poor substitute for an effective process to identify and pursue research that will improve regulatory decisions by reducing both the uncertainties and the need for the conservative assumptions.

ORGANIZATION OF THIS APPENDIX

In this appendix, we discuss: 1) the role of risk assessment in supporting societal decisions on managing risk; 2) the use of "plausible conservatism" in selecting default options and alternatives to default options; 3) the use of an iterative approach in which specific science displaces default options; 4) the need for risk characterizations to be matched to their intended uses, and why a single quantitative estimate of risk may not be adequate; 5) why the process for conducting science-based risk assessments should be integrated and comprehensive; and 6) how risk assessments can serve an important role in guiding research to improve future risk assessments.

The Role of Risk Assessment in Supporting
Societal Decisions on Managing Risk

The development of risk assessments is one part of a larger process by which societal decisions and actions concerning risks are made. Risk assessments are that phase of the overall process in which all of the available information concerning exposure to the agent(s), the agent's(s') ability to cause adverse responses, and exposure-dose-response relationships, is synthesized into a risk characterization whose degree of comprehensiveness is matched to the intended use of the risk characterization. When specific data are not available, default options based on general scientific knowledge and risk assessment policy are used. The risk characterization product of the risk assessment is then used as input along with a diverse array of other information to make a wide range of risk-based decisions, as for example, whether to limit exposure to the agent(s) (and if so, to what extent). These risk-based decisions may on occasion involve a comparison of risks between agents causing similar adverse responses or, more broadly, disease. In other cases, the risk characterization may be used as input to decisions as to how to allocate economic or other societal resources. Clearly, risk characterizations must be as accurate as possible because of the potential importance of the decisions concerning health (and disease) and allocation of scarce societal resources.

This appendix proposes that risk characterizations should be developed by a well-documented process that makes full use of the available scientific data. When specific data are not available, the process should use default options and other assumptions that are clearly identified. The end-product risk characterization should be reported with a degree of comprehensiveness matched to its intended use and in a form that can be readily understood by decision makers and interested members of the public. One intent of the process is to avoid the introduction of *unidentified* bias that would either under-estimate or over-estimate the risk being characterized. The approach we advocate emphasizes scientific plausibility with regard to the use of alternative models and appropriate disclosure of uncertainties.

The approach we are advocating contrasts sharply with the approach advocated by Dr. Finkel, which introduces into the risk assessment process an additional standard: whether the alternative based on the scientific information yields a plausible, conservative estimate of risk. A default option would be displaced only if it is found to be no longer plausible, or if a plausible alternative gives a higher estimate of risk. Thus, judgments on the extent of conservatism would largely determine the result from the risk assessment process. It is our opinion that value judgments as to the degree of conservatism should not have such a large influence on the output of the risk assessment process. We believe that EPA should make these value judgments consistently according to established guidelines where such judgments are necessary (e.g., choice of default options), and should disclose the use of such judgments fully to risk managers and to the public.

The value judgments are most appropriately dealt with as part of the risk management or risk decision-making phase of the overall process. In particular, risk managers should not be restricted by value judgments made during risk assessment. Risk managers can and should override conservative default value judgments in the risk assessment process whenever they believe it is appropriate public policy to do so. Such departures should be clearly identified as policy and not as science. Risk managers must assume full responsibility for making such overrides and for explaining their reasoning to the interested and affected members of the public.

Use of "Plausible Conservatism" in Selecting Default Options and Alternatives to Default Options

It has been noted that inference guidelines, or default options as they are typically called in this report, are generic guidelines used when the necessary scientific information is not available. These guidelines are based on general scientific knowledge and applied to assure consistency in the development of multiple risk assessments. It is our understanding that EPA has selected default options that are scientifically plausible and conservative in the sense that they are intended to avoid underestimating health risks. Hence, these generic guidelines generally follow the principle of "plausible conservatism" as we believe it is described in Appendix N-1. We do not object to this approach for selecting the default options.

We do object to the use of "plausible conservatism" as a criterion in deciding when specific science can be used to replace a default option. The use of "plausible conservatism" as the test for displacing default options places an excessively high hurdle for the new science. The use of "plausible conservatism" will therefore discourage the conduct of research to generate the scientific information that might displace the use of the default option. The result will be to freeze risk characterizations at the level determined by the conservative default options.

As specific science is developed and used to replace default options, the result will typically be a reduction both in the estimates of risk and the extent of uncertainty in the risk estimates. The replacement of default options with specific science was illustrated in Chapter 6 using formaldehyde as an example. In this case the initial risk estimate, which was based on a default option for relating exposure to response, i.e., the cancer risk, had a plausible upper bound estimate of 0.016 (1.6×10^{-2}) at 1 ppm. The lower limit may be zero. Thus, there was a wide range of uncertainty, from 0 to 0.016. In successive iterations as new scientific information was incorporated on delivered dose to target tissue using data on DNA-protein cross-links, first from rats and then from monkeys, the upper bound on risk at 1 ppm was reduced to 2.8×10^{-3} and then to 3.3×10^{-4}. For neither of these iterations can a lower bound estimate of zero be excluded. Thus, at the last iteration the range of uncertainty has been reduced to 0 to 3.3×10^{-4}. This is a substantial reduction from the 0 to 1.6×10^{-2} calculated based on the default options.

In this example the departure from the default options was far more plausible than the original default options. The DNA-protein cross-links provide a direct measurement of a biomarker for the extent to which the formaldehyde is penetrating into tissues where cancers might be induced.

In many other situations, the difference in plausibility between the default and the alternative using specific scientific information may be less apparent. It is our judgment that weighing the plausibility of alternatives is a highly judgmental evaluation that must be carried out by scientists. We believe it would be a mistake to try to define a sharp threshold for plausibility. Such a sharp threshold will stifle research and impede communication about uncertainties. When an alternative approach is judged plausible, but the default option also plausible, it will be appropriate for the risk estimates from both approaches to be conveyed to the risk manager, as CAPRA has recommended.

Better criteria for departures from defaults are needed. However, we believe that scientific judgment will remain at the heart of the process for determining that a default option should be displaced, either for a specific substance, or for a class of substances. Chapter 6 provides several examples of instances in which departures from defaults have been accepted, or considered and rejected as not yet adequately supported by scientific information, based on outside scientific peer review through the EPA Science Advisory Board. In our opinion EPA's process for making such judgments works reasonably well—although there is clearly room for improvement. More research directed at the important uncertainties should permit more departures from defaults, based upon adequate support from the scientific information obtained through the research.

We view the extent of conservatism in risk assessment guidelines as a policy issue to be determined by EPA, most appropriately through notice and comment rulemaking in the same manner as when EPA risk assessment guidelines were adopted in 1986. The proposal in Appendix N-1 does not give precise guidance for establishing default options or for departing from these defaults. Scientists may disagree as to whether a model is plausible or not plausible, and lack of plausibility will be very difficult to establish outside the range of observed data. The usual choice will be between simple models whose structure is assumed, (e.g., low dose linearity) vs. more complex models based on knowledge of biological and pathobiological processes. Both alternatives may be judged plausible. However, the biologically based models may be more valuable because they incorporate more information and provide a better basis for discriminating on the extent of the risk posed by different chemicals at relevant levels of human exposure.

We are also concerned that recommendations from CAPRA on policy issues could be inappropriate and subject to misinterpretation. Therefore, we believe it is inappropriate for the National Research Council to recommend default options to EPA. NRC recommendations might be perceived as being based on solely on science, but such would not be the case; such recommendations would reflect value judgments that scientists are no more qualified to make than other citizens. However, it is appropriate for NRC to point out where default options are needed, so that these policy questions can be addressed by the regulatory agency. For example, should the same cancer potency be used for all chemicals in a class (discussed at the end of Chapter 6)? Should the same cancer potency be applied to all people, or should sensitive subgroups be treated separately (discussed in Chapter 10)? It is our position that judgments on what are the appropriate defaults should be made by the regulatory agency, and not by the members of an NRC committee.

There are broader questions of risk assessment and risk management policy that CAPRA has declined to address. There is much dispute and inconsistency on the appropriate basis for regulating toxic chemicals, especially carcinogens. Some within the scientific community believe that Congress and the regulatory agencies have gone much too far in regulating some chemicals (e.g., synthetic pesticide residues in processed food) and not far enough in regulating other chemicals (indoor radon and other indoor air toxicants). We believe that such disputes and inconsistencies should be addressed using risk assessment for communication, to inform those with decision responsibility what science can and cannot say about the magnitude of the risks posed by chemicals to health and the environment. Scientists should not attempt to resolve risk management disputes by influencing the choice of default options or the criteria for departure from default options.

The Use of an Iterative Approach, in which Specific Science Displaces Default Options and Provides a Means to Improve Risk Assessments and Reduce Uncertainty in Risks

The CAPRA report advocates the conduct of iterative risk assessments matched to decision-making needs. This approach recognizes that EPA must deal with at least 189 hazardous air pollutants, many with limited data and, perhaps, posing low risks. EPA needs an approach for carrying out iterative risk assessment on hazardous air pollutants, and Chapter 12 builds upon EPA's planned methodology to describe such an approach. As a part of this approach, EPA must develop a system for prioritizing these chemicals so that the limited funds available may be used most effectively to protect human health. Because of differences in the available data and the differences in the magnitude of the risk posed by different chemicals, EPA should not deal with each chemical the same way. The highly quantitative formal techniques described in CAPRA Chapters 9, 10, 11 are not intended for every chemical, but only for supporting the most important and difficult regulatory decisions, for which advanced analytical concepts and procedures may be needed. The sophistication and complexity of these methods add to the difficulty of communicating to regulatory decision makers and to the public. EPA needs a risk assessment process that can deal effectively, cheaply, and quickly with most of the chemicals, while permitting more sophisticated and data-intensive risk assessment in situations where the additional time, expense, analytical sophistication, and risk communication difficulties are warranted by the importance of the regulatory decisions.

Risk Characterizations Must Be as Clear and Comprehensive as Practical, Given Their Intended Uses; and A Single Quantitative Estimate of Risk May Not Be Adequate

The risk assessment process and the resulting risk characterization should be matched to the intended use of the risk characterization. (Recall the discussion in the preceding section of the need for an iterative approach.) Obviously, the degree of comprehensiveness that can be achieved for a given risk characterization will be dependent on the extent of the scientific information available.

For the chemicals with the least amount of data the risk characterization may be a qualitative, narrative summary of the limited available information. For chemicals with more extensive data, such as several bioassays, the risk characterization may include a plausible upper bound risk estimate, using the 95% upper confidence limit computed from the bioassay data set that yields the highest risk estimate (e.g., most sensitive strain, sex, species, and tumor end point) and a conservative and relatively crude exposure estimate.

For the most extensive data sets, it may be possible to provide multiple risk calculations corresponding to alternative models and data sets corresponding to individuals and populations. These data may be organized in the form of one or more probability distributions, from which a probability distribution on risk is computed. The probability distribution on risk may be summarized by using expected values or other summary statistics computed through Monte Carlo analysis or other probabilistic analysis techniques. Such central tendency estimates will be helpful supplements to upper and lower bound calculations (more generally, statistical confidence limits) to assist decision makers and the public in understanding the implications of the probability distributions. Such analysis based on the most extensive available data for cancer potency and exposure has not, to our knowledge, been carried out in support of a major regulatory decision, but the procedures involved are illustrated in Appendices (Texaco and ENSR articles) and in the scientific literature (Wallsten and Whitfield, 1989; Howard et al., 1972).

The proposal in Appendix N-1 for plausible conservatism seems to assume that the output of risk assessment is a single risk number that can be used for regulatory decision making. We oppose this aspect of his proposal, especially for the upper tiers of risk assessment. The goal for risk assessment should be to inform decision makers and the public, not to give them a number.[1] To the

---

[1] In Appendix N-1, Dr. Finkel uses an example of when to leave for the airport to illustrate his advocacy of conservative estimates, and we use the same example to make the point that single-number estimates may be inadequate as a summary of information for purposes of decision making. The decision on when to leave for the airport depends on the information about how long it will take to get to the airport, an uncertain quantity. It is our judgment that most decision makers would not wish to have this uncertainty summarized as a single estimated travel time, as he has asserted. Rather, we believe that decision makers prefer to have a description of the possibilities and their

extent that risk assessment provides only one number, based on conservative assumptions, then the group that determines which conservative assumptions shall be used will determine regulatory policy. Thus, the discretion of the risk manager will be preempted by the risk assessment process.

The EPA Science Advisory Board Report on Dioxins (EPA, 1989) stressed the importance of replacing linear extrapolation with a biologically based model, and that the default of linearity might cause risk to be overestimated or underestimated. The SAB encouraged EPA to consider revisions in the regulatory standard based on policy and on the scientific uncertainties. SAB did not support changing the single number risk estimate on the basis of the scientific information then available.

It can be argued theoretically that for decision making, the best single number will be the expected value—the average over the probability distribution. However, we believe that the distribution is better than the any single number. If an average value is to be used, misinterpretation should be minimized, and for more than a decade EPA's risk estimates have generally been upper bounds. (Only a few risk estimates based on human epidemiology have represented conceptual departures—for example, lung cancer from indoor radon, where the health risk estimate comes from extrapolation of observed lung cancer incidence in uranium miners.)

In Appendix N-1, the example of a substance that may pose an unacceptably high risk of X, or zero, depending on which of two incompatible biologic theories is true. Such a situation is clearly one in which risk managers will wish to learn about this critically important uncertainty as to which theory is correct. Within the risk management context if not within risk assessment, it may be useful to characterize the judgment of knowledgeable scientists in terms of a subjective probability. Suppose there is a consensus among scientists that the probability is p that the risk is at or near zero. In our judgement, the decision maker will wish to understand this characterization of the risk: a probability p that the risk is at or near zero and a probability 1-p that the risk is at the high level X. We believe it inappropriate to summarize this situation by presenting only the expected value of (1-p)X as the estimate of risk for the decision maker. The probability distribution should be used for the risk characterization, not one

---

likelihood. For example, an estimate of the travel time under normal conditions might be supplemented by a description of possible delays and the probabilities that such delays might occur. Such an analysis might be quite simple, with only a few sources of delay considered, or quite complex, requiring a computer to calculate the probability distribution on the time from departure to boarding the airplane. In presenting the analysis, an assessor might highlight the most important uncertainties (e.g., "The normal driving time is approximately 30 minutes, with a probability of 20% that traffic delays might add between 10 and 30 minutes. The probability that travel time by taxi to the airport would exceed one hour is judged to be less than 5%.").

risk estimate. Decision makers and the public should have little difficulty understanding this simple characterization.

The Process for Conducting Science-Based Risk Assessments Should Be Integrated and Comprehensive

The process being advocated for the conduct of science-based risk assessments builds on the general principles outlined in the 1983 NRC Committee Report (the Red Book). We reaffirm these general principles and build on them in proposing a process for the conduct of risk assessments. The general principles we believe to be appropriate include:

• A paradigm linking exposure to dose to response can be used as a structure for integrating data to characterize the risk of a specific pollutant. For characterizing the risk associated with a specific source the paradigm is readily expanded to include a source to exposure linkage.
• Scientific information, to the extent it is available, should be used as much as feasible in the risk assessment process.
• When differences of scientific opinion exist on the use or interpretation of scientific information or hypotheses, these should be clearly documented in the risk assessment process and the impact on risk characterization identified.
• Guidelines are necessary to structure the interpretation and use of scientific information, including consideration of specific scientific information and to guide actions when information is incomplete or absent in particular assessments.
• The guidelines should include clearly identified default options (e.g., the preferred inference option chosen on the basis of risk assessment policy that appears to be the best choice in the absence of data to the contrary).
• Guidelines should promote the use of specific information and departures from the use of default options. Departures from defaults should be based on the scientific validity of the data and models, as judged by scientists.
• All scientific data, scientific assumptions, scientific hypotheses, default options, and the specific risk assessment methodology used should be clearly documented in each risk assessment. Where differences of scientific opinion exist, these differences should be clearly described.
• The resulting risk characterization, including quantitative estimates of risk and probabilistic descriptions of risk, should be communicated to the risk manager in as clear and comprehensive a manner as possible, as appropriate for the intended use of the risk characterization.

Risk Assessments Can Serve an Important Role in Guiding Research to Improve Future Risk Assessments

It is our opinion that risk assessments can have a major role in guiding research to improve the scientific basis for future risk assessments. This will require a new attitude recognizing that the risk assessment process should yield not only a risk characterization but also identify the unanswered questions which, if addressed with research, could have the potential for reducing the uncertainty in the estimates of risk as schematically related in Figure N-1. This process of identifying research needs (opportunities) may be informal or formalized as in the use of sensitivity analyses. Having identified the major sources of uncertainty, the question may be asked as to whether the issue can be addressed with current research technologies and, if so, the potential cost and time required to carry out the research. These costs and time estimates can then be balanced against the potential value of the information in making decisions on proceeding with the targeted research effort.

A recent OTA report, *Researching Health Risks* (OTA, 1993), addressed the issue of conducting targeted research of this kind both as related to specific chemicals but also as a means of improving risk assessment methodology. Obviously, the two go hand-in-hand with research on specific chemicals (in which they serve as useful probes) addressing generic toxicological/risk assessment issues while also providing highly relevant information applicable to the specific chemical.

The most important risk management decisions will involve large potential impacts on public health and large economic consequences from control actions. Such decisions should involve a careful review of the underlying science. Risk managers may wish to consider whether to act with present information, which may involve large uncertainties in the public health consequences, or to delay the decision for a period of time which research is carried out to reduce these uncertainties and therefore provides a better basis for decision. It is our belief that Congress could do much more to encourage EPA, other federal agencies such as NIEHS, and private sector organizations to plan and carry out research to reduce important uncertainties on the health consequences of toxic air contaminants. Such research might take a decade or more to complete, but research started now might provide significant new information supporting departures from defaults that could save billions of dollars in control costs while providing even better protection of public health.

Scientific knowledge of the mechanism by which toxic substances cause cancer and other chronic health impacts is evolving rapidly. However, much of this research is aimed at understanding and treating the health impacts, rather than understanding the relationship of the health impacts to the relatively low levels of exposure to toxic substances in the ambient air. The most important uncertainties are those for which the value of information is high, because reso-

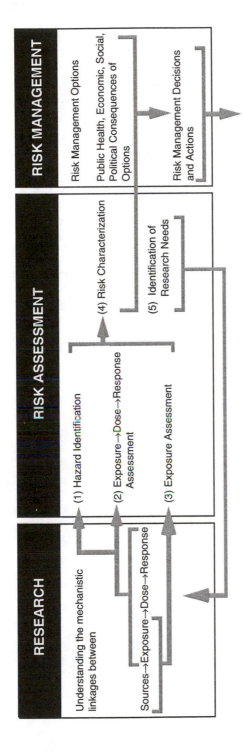

FIGURE N-1 NAS/NRC risk assessment/management paradigm. Source: Adapted from NRC, 1983a.

lution of the uncertainties are likely to change decisions, leading to substantial benefits in improved public health and reduced control costs (OTA, 1993). More targeted research designed to avoid costly regulations based on conservative default options in risk assessment should pay very large economic dividends, while at the same time allowing better management of the substances that do present substantial risks to public health.

REFERENCES

EPA (U.S. Environmental Protection Agency). 1986. Guidelines for carcinogen risk assessment. Fed. Regist. 51:33992-34003.

EPA (U.S. Environmental Protection Agency). 1989. Letter report to EPA administrator, William Reilly, from the Science Advisory Board, Nov. 28. SAB-EC-90-003. U.S. Environmental Protection Agency, Washington, D.C.

Howard, R.A., J.E. Matheson, and D.W. North. 1972. The decision to seed hurricanes. Science 176:1191-1202.

NRC (National Research Council). 1983. Risk Assessment in the Federal Government: Managing the Process. Washington, D.C.: National Academy Press.

OTA (U.S. Office of Technology Assessment). 1993. Researching Health Risks. U.S. Office of Technology Assessment, Washington, D.C.

Whitfield, R.G., and T.S. Wallsten. 1989. A risk assessment for selected lead-induced health effects: An example of a general methodology. Risk Anal. 9:197-207.

Index